T0141815

Continuous Time Modeling in the Behavioral and Related Sciences

Kees van Montfort • Johan H. L. Oud •
Manuel C. Voelkle
Editors

Continuous Time Modeling in the Behavioral and Related Sciences

Editors
Kees van Montfort
Marketing and Supply Chain Management
Nyenrode Business University
Breukelen, The Netherlands

Johan H. L. Oud
Behavioural Science Institute
University of Nijmegen
Nijmegen, The Netherlands

Manuel C. Voelkle
Department of Psychology
Humboldt-Universität zu Berlin
Berlin, Germany

ISBN 978-3-030-08401-1 ISBN 978-3-319-77219-6 (eBook)
https://doi.org/10.1007/978-3-319-77219-6

Mathematics Subject Classification (2010): 37N40, 62M10, 62P15, 62P25, 65F60, 91B99, 91D30, 93E99, 97M70

This Springer imprint is published by the registered company Springer Nature Switzerland AG
The registered company address is: Gewerbestrasse 11, 6330 Cham, Switzerland

Preface

Over the past decades behavioral scientists have increasingly realized the potential of longitudinal data to address specific research questions, in particular those of a cause-effect nature. As a consequence, the methodology of longitudinal research and analysis has made much progress and an increasing number of large-scale longitudinal (time series and panel) data sets have become available for analysis. However, in accordance with the way longitudinal data are collected, at a restricted number of discrete time points, the statistical analysis is typically based on discrete time models. As argued by the authors in the present book, a series of problems is connected to this type of models, which make their results highly questionable. One main issue is the dependence of discrete time parameter estimates on the chosen time interval in dynamic modeling, which leads to incomparability of results across different observation intervals and, if unaccounted for, may even lead to contradictory conclusions.

Continuous time modeling, in particular by means of differential equations, offers a powerful solution to these problems, yet the use of continuous time models in the behavioral and related sciences such as psychology, sociology, economics, and medicine is still rare. Fortunately, recent initiatives to introduce and adapt continuous time models in a behavioral science context are gaining momentum. The purpose of the book is to assess the state of the art and to bring together the different initiatives. Furthermore, we emphasize the applicability of continuous time methods in applied research and practice.

Bergstrom, one of the pioneers of continuous time modeling in econometrics, credits Bartlett for the first significant contribution to the problem of estimating continuous time stochastic models from discrete data. As Bartlett succinctly stated in 1946: "The discrete nature of our observations in many economic and other time series does not reflect any lack of continuity in the underlying series. ... An unemployment index does not cease to exist between readings, nor does Yule's pendulum cease to swing" (p. 31).

At the time of Bartlett's contribution, intensive discussions took place in econometrics on the interpretation, identification, and estimation of nonrecursive simultaneous equations models. Simultaneity was considered by Bergstrom and others as an artifact from short-term causal effects between variables within relatively long discrete time intervals (often years in economics), being in the case of nonrecursive simultaneity reciprocal effects within such intervals. In his 1966 publication "Nonrecursive models as discrete approximations to systems of stochastic differential equations," Bergstrom further concretized this interpretation of nonrecursiveness by describing the underlying causal effects by a system of stochastic differential equations. In fact, he connected this system to the discrete time data in two ways: the exact discrete model (EDM) and the approximate discrete model (ADM). Whereas the EDM connects the system to the discrete time data by highly nonlinear restrictions, the ADM uses the approximate linear restrictions of a (recursive or nonrecursive) simultaneous equations model. This book is greatly indebted to Bergstrom, as most of the chapters more or less explicitly continue and elaborate on the kind of exact or approximate methods, started by Bergstrom.

To implement the exact approach, traditionally maximum likelihood by means of filtering techniques, in particular Kalman filtering, is used and several of the chapters apply these techniques. In behavioral science, the application of maximum likelihood by means of structural equation modeling (SEM) became very popular and several other chapters make use of this alternative procedure.

In total, the book contains 16 chapters which cover a vast range of continuous time modeling approaches, going from one closely mimicking traditional linear discrete time modeling to highly nonlinear state space modeling techniques. Each chapter describes the type of research questions and data that the approach is most suitable for, provides detailed statistical explanations of the models, and includes one or more applied examples. To allow readers to implement the various techniques directly, accompanying computer code is made available online.

The book addresses the great majority of researchers in the behavioral and related sciences, who are interested in longitudinal data analysis. This includes readers who are involved in research in psychology, sociology, economics, education, management, and medical sciences. It is meant as a reference work for scientists and students working with longitudinal data and wanting to apply continuous time methods. The book also provides an overview of various recent developments for methodologists and statisticians. Especially for PhD students it offers the means to carry out analyses of longitudinal data in continuous time. Readers are supposed to have knowledge of statistics as taught at Master's or early PhD level.

We thank the authors for their willingness to contribute to the book, Eva Hiripi of Springer for her help in realizing the project, and Yoram Clapper, Leonie Richter, and Vincent Schmeits for their excellent typesetting work.

Breukelen, The Netherlands — Kees van Montfort
Nijmegen, The Netherlands — Johan H. L. Oud
Berlin, Germany — Manuel C. Voelkle

Contents

Contributors

Frits D. Bijleveld Vrije Universiteit Amsterdam School of Business and Economics, Department of Econometrics, Amsterdam, The Netherlands

SWOV Institute for Road Safety Research, The Hague, The Netherlands

Steven M. Boker University of Virginia, Department of Psychology, Charlottesville, VA, USA

Andreas M. Brandmaier Max Planck Institute for Human Development, Center for Lifespan Psychology, Berlin, Germany

Max Planck UCL Centre for Computational Psychiatry and Ageing Research, Berlin, Germany

Marcus J. Chambers Department of Economics, University of Essex, Colchester, UK

Meng Chen The Pennsylvania State University, Department of Human Development and Family Studies, State College, PA, USA

Sy-Miin Chow The Pennsylvania State University, Department of Human Development and Family Studies, State College, PA, USA

David A. Cole Vanderbilt University, Department of Psychology & Human Development, Peabody College of Education and Human Development, Nashville, TN, USA

Jacques J. F. Commandeur Vrije Universiteit Amsterdam School of Business and Economics, Department of Econometrics, Amsterdam, The Netherlands

SWOV Institute for Road Safety Research, The Hague, The Netherlands

Pascal R. Deboeck University of Utah, Department of Psychology, College of Social and Behavioral Sciences, Salt Lake City, UT, USA

Charles C. Driver Max Planck Institute for Human Development, Center for Lifespan Psychology, Berlin, Germany

Emilio Ferrer University of California, Davis, Department of Psychology, Davis, CA, USA

Joseph E. Gonzales University of Massachusetts, Lowell, Department of Psychology, Lowell, MA, USA

Ellen L. Hamaker Department of Methodology and Statistics, Faculty of Social Sciences, Utrecht University, Utrecht, The Netherlands

Michael D. Hunter School of Psychology, Georgia Institute of Technology, Atlanta, GA, USA

Siem Jan Koopman Vrije Universiteit Amsterdam School of Business and Economics, Department of Econometrics, Amsterdam, The Netherlands

CREATES, Aarhus University, Aarhus, Denmark

Tinbergen Institute, Amsterdam, The Netherlands

Rebecca M. Kuiper Department of Methodology and Statistics, Faculty of Social Sciences, Utrecht University, Utrecht, The Netherlands

Omar Licandro University of Nottingham, School of Economics, Nottingham, UK

IAE-CSIC and Barcelona GSE, Barcelona, Spain

J. Roderick McCrorie School of Economics and Finance, University of St Andrews, St Andrews, UK

Heiner Meulemann University of Cologne, Institute for Sociology and Social Psychology, Cologne, Germany

Robert G. Moulder University of Virginia, Department of Psychology, Charlottesville, VA, USA

Nynke M. D. Niezink Carnegie Mellon University, Department of Statistics & Data Science, Pittsburgh, PA, USA

Zita Oravecz Pennsylvania State University, Department of Human Development and Family Studies, State College, PA, USA

Johan H. L. Oud Behavioural Science Institute, University of Nijmegen, Nijmegen, The Netherlands

Kristopher J. Preacher Vanderbilt University, Department of Psychology & Human Development, Peabody College of Education and Human Development, Nashville, TN, USA

Luis A. Puch Universidad Complutense de Madrid, Department of Economic Analysis and ICAE, Madrid, Spain

Nilam Ram Pennsylvania State University, Department of Human Development and Family Studies, State College, PA, USA

Jesús Ruiz Department of Economics and ICAE, Universidad Complutense de Madrid, Madrid, Spain

Oisín Ryan Department of Methodology and Statistics, Faculty of Social Sciences, Utrecht University, Utrecht, The Netherlands

Hermann Singer FernUniversität in Hagen, Department of Economics, Hagen, Germany

Tom A. B. Snijders University of Groningen, Department of Sociology, Groningen, The Netherlands

University of Oxford, Nuffield College, Oxford, UK

Joel S. Steele Portland State University, Department of Psychology, Portland, OR, USA

Michael A. Thornton Department of Economics and Related Studies, University of York, York, UK

Stacey S. Tiberio Oregon Social Learning Center, Eugene, OR, USA

Helgi Tómasson University of Iceland, Faculty of Economics, Reykjavík, Iceland

Kees van Montfort Nyenrode Business Universiteit, Department of Marketing and Supply Chain Management, Breukelen, The Netherlands

Erasmus University Rotterdam, Department of Biostatistics, Rotterdam, The Netherlands

Manuel C. Voelkle Humboldt-Universität zu Berlin, Department of Psychology, Berlin, Germany

Max Planck Institute for Human Development, Center for Lifespan Psychology, Berlin, Germany

Sunčica Vujić University of Antwerp, Department of Economics, Antwerp, Belgium

University of Bath, Department of Economics, Bath, UK

Julie Wood Pennsylvania State University, Department of Human Development and Family Studies, State College, PA, USA

Chapter 1
First- and Higher-Order Continuous Time Models for Arbitrary N Using SEM

Johan H. L. Oud, Manuel C. Voelkle, and Charles C. Driver

1.1 Introduction

This chapter is closely based on an earlier article in *Multivariate Behavioral Research* (Oud et al. 2018). The chapter extends the previous work by the analysis of the well-known Wolfer sunspot data along with more detailed information on the relation between the discrete time and continuous time model in the so-called exact discrete model (EDM). The previously published MBR article provides additional information on the performance of CARMA(p, q) models by means of several simulations, an empirical example about the relationships between mood at work and mood at home along with a subject-group-reproducibility test. However, to improve readability, we repeat the basic introduction to CARMA(p, q) modeling in the present chapter.

Electronic Supplementary Material The online version of this article (https://doi.org/10.1007/978-3-319-77219-6_1) contains supplementary material, which is available to authorized users.

J. H. L. Oud (✉)
Behavioural Science Institute, University of Nijmegen, Nijmegen, The Netherlands
e-mail: j.oud@pwo.ru.nl

M. C. Voelkle
Department of Psychology, Humboldt-Universität zu Berlin, Berlin, Germany
e-mail: manuel.voelkle@hu-berlin.de

C. C. Driver
Max Planck Institute for Human Development, Berlin, Germany
e-mail: driver@mpib-berlin.mpg.de

K. van Montfort et al. (eds.), *Continuous Time Modeling in the Behavioral and Related Sciences*, https://doi.org/10.1007/978-3-319-77219-6_1

Time series analysis has a long history. Although the roots can be traced back to the earlier work of statisticians such as Yule, Kendall, and Durbin (see the historical overview in Mills 2012), it was the landmark publication "Time Series Analysis, Forecasting and Control" by Box and Jenkins (1970) that made time series analysis popular in many fields of science. Examples are Harvey (1989) and Lütkepohl (1991) in economics or Gottman (1981) and McCleary and Hay (1980) in the social sciences.

Time series analysis has greatly benefited from the introduction of the state space approach. The state space approach stems from control engineering (Kalman 1960; Zadeh and Desoer 1963) and sharply distinguishes the state of a system, which is a vector of latent variables driven by the system dynamics in the state transition equation, from the observations. The measurement part of the state space model specifies the relation between the observed and the underlying latent variables. It turns out that any Box-Jenkins autoregressive and moving average (ARMA) model as well as any extended ARMAX model, in which exogenous variables are added to the model, can be represented as a state space model (Caines 1988; Deistler 1985; Harvey 1981; Ljung 1985). However, the state space representation is much more flexible, allows one to formulate many time series models that cannot easily be handled by the Box-Jenkins ARMA approach, and makes important state space modeling techniques such as the Kalman filter and smoother accessible for time series analysis (Durbin and Koopman 2001). Therefore, the ARMA model in continuous time presented in this chapter, called a CARMA model (Brockwell 2004; Tómasson 2011; Tsai and Chan 2000), will in this chapter be formulated as a state space model.

Structural equation modeling (SEM) was introduced by Jöreskog (1973, 1977) along with the first SEM software: LISREL (Jöreskog and Sörbom 1976). The strong relationships between the state space approach and SEM were highlighted by Oud (1978) and Oud et al. (1990). Both consist of a measurement part and an explanatory part, and in both the explanatory part specifies the relationships between latent variables. Whereas in the state space approach the latent explanatory part is a recursive dynamic model, in SEM it is a latent structural equation model. As explained in detail by Oud et al. (1990), SEM is sufficiently general to allow specification of the state space model as a special case and estimation of its parameters by maximum likelihood and other estimation procedures offered by SEM. By allowing arbitrary measurement error structures, spanning the entire time range of the model, SEM further enhances the flexibility of the state space approach (Voelkle et al. 2012b).

An important drawback of almost all longitudinal models in SEM and time series analysis, however, is their specification in discrete time. Oud and Jansen (2000), Oud and Delsing (2010), and Voelkle et al. (2012a) discussed a series of problems connected with discrete time models, which make their application in practice highly questionable. One main problem is the dependence of discrete time results on the chosen time interval. This leads, first, to incomparability of results over different observation intervals within and between studies. If unaccounted for, it can easily lead to contradictory conclusions. In a multivariate model, one researcher could find a positive effect between two variables x and y, while another researcher finds a negative effect between the same variables, just because of a different observation interval length in discrete time research. Second, because results depend on the specific length of the chosen observation interval, the use of equal intervals in discrete time studies does not solve the problem (Oud and Delsing 2010). Another interval might have given different results to both researchers. Continuous time analysis is needed to make the different and possibly contradictory effects in discrete time independent of the interval for equal as well as unequal intervals.

The CARMA time series analysis procedure to be presented in this chapter is based on SEM continuous-time state-space modeling, developed for panel data by Oud and Jansen (2000) and Voelkle and Oud (2013). The first of these publications used the nonlinear SEM software package Mx (Neale 1997) to estimate the continuous time parameters, the second the ctsem program (Driver et al. 2017). ctsem interfaces to OpenMx (Boker et al. 2011; Neale et al. 2016), which is a significantly improved and extended version of Mx. In both publications the kernel of the model is a multivariate stochastic differential equation, and in both maximum likelihood estimation is performed via the so-called exact discrete model EDM (Bergstrom 1984). The EDM uses the exact solution of the stochastic differential equation to link the underlying continuous time parameters exactly to the parameters of the discrete time model describing the data. The exact solution turns out to impose nonlinear constraints on the discrete time parameters. Several authors tried to avoid the implementation of nonlinear constraints by using approximate procedures, although acknowledging the advantages of the exact procedure (e.g., Gasimova et al. 2014; Steele and Ferrer 2011a,b). In an extensive simulation study, Oud (2007) compared, as an example, the multivariate latent differential equation procedure (MLDE; Boker et al. 2004), which is an extension of the local linear approximation (LLA; Boker 2001), to the exact procedure. The exact procedure was found to give considerably lower biases and root-mean-square error values for the continuous time parameter estimates.

Instead of SEM, most time series procedures, particularly those based on the state space approach (Durbin and Koopman 2001; Hannan and Deistler 1988; Harvey 1989), use filtering techniques for maximum likelihood estimation of the model. Singer (1990, 1991, 1998) adapted these techniques for continuous time modeling of panel data. In a series of simulations, Oud (2007) and Oud and Singer (2008) compared the results of SEM and filtering in maximum likelihood estimation of various continuous time models. It turned out that in case of identical models, being appropriate for both procedures, the parameter estimates as well as the associated standard errors are equal. Because both procedures maximize the likelihood, this ought to be the case. Considerable technical differences between the two procedures, however, make it nevertheless worthwhile to find that results indeed coincide. In view of its greater generality and its popularity outside of control theory, especially in social science, SEM can thus be considered a useful alternative to filtering. Making use of the extended SEM framework underlying OpenMx (Neale et al. 2016), both SEM and the Kalman filter procedure, originally developed in control theory, are implemented in ctsem.

An argument against SEM as an estimation procedure for time series analysis could be that it is known and implemented as a large N procedure, while time series analysis is defined for $N = 1$ or at least includes $N = 1$ as the prototypical case. In the past, different minimum numbers of sample units have been proposed for SEM, such as $N \geq 200$, at least 5 units per estimated parameter, at least 10 per variable, and so on, with all recommendations being considerably larger than $N = 1$. Recently, Wolf et al. (2013), on the basis of simulations, settled on a range of minimum values from 30 to 460, depending on key model properties. For small N, in particular $N = 1$, leading to a nonpositive definite sample covariance matrix, some SEM programs simply refuse to analyze the data or they change the input data before the analysis. For example, LISREL gives a warning ("Matrix to be analyzed is not positive definite") and then increases the values on the diagonal of the data covariance matrix S to get this matrix positive definite ("Ridge option taken with ridge constant = ..."), before starting the analysis.

As made clear by Hamaker et al. (2003), Singer (2010), and Voelkle et al. (2012b), however, the long-standing suggestion that a SEM analysis cannot be performed because of a nonpositive definite covariance matrix S is wrong. Nothing in the likelihood function requires the sample covariance matrix S to be positive definite. It is the model implied covariance matrix Σ that should be positive definite, but this depends only on the chosen model and in no way on the sample size N. For the same reason, minimum requirements with regard to N do not make sense, and $N = 1$ is an option as well as any other N. The quality of the estimates as measured, for example, by their standard errors, depends indeed on the amount of data. The amount of data, however, can be increased both by the number of columns in the data matrix (i.e., the number of time points T in longitudinal research) and by the number of rows (i.e., the number of subjects N in the study). The incorrect requirement of a positive definite S for maximum likelihood estimation (Jöreskog and Sörbom 1996) may be explained by the modified likelihood function employed in LISREL and other SEM programs. This contains the quantity $\log |S|$ (see Equation (3) in Voelkle et al. 2012b), and thus these programs cannot handle $N = 1$ and N smaller than the number of columns in S. By employing the basic "raw data" or "full-information" maximum likelihood function (RML or FIML), OpenMx avoids this problem and (a) allows any N, including $N = 1$, (b) permits the nonlinear constraints of the EDM, (c) allows any arbitrary missing data pattern under the missing at random (MAR) assumption, and (d) allows individually varying observation intervals in continuous time modeling by means of so-called definition variables (Voelkle and Oud 2013).

Until now, however, the SEM continuous time procedure has not yet been applied on empirical $N = 1$ data, although the possibility is discussed and proven to be statistically sound by Singer (2010), and the software to do so is now readily available (Driver et al. 2017). Thus, the aim of the present chapter is, first, to discuss the state space specification of CARMA models in a SEM context. Second, to show that the SEM continuous time procedure is appropriate for $N = 1$ as well as for $N > 1$. Third, the proposed analysis procedure using ctsem will be applied on the well-known Wolfer sunspot data. This $N = 1$ data set has been analyzed by several continuous time analysts before. The results from ctsem will be compared to those found previously.

1.2 Continuous Time Model

1.2.1 Basic Model

In discrete time, the multivariate autoregressive moving-average model ARMA (p, q) with p the maximum lag of the dependent variables vector $\mathbf{y}_t$ and q the maximum lag of the error components vector $\mathbf{e}_t$ reads for a model with, for example, $p = 2$ and $q = 1$

$$\mathbf{y}_t = \mathbf{F}_{t,t-1}\mathbf{y}_{t-1} + \mathbf{F}_{t,t-2}\mathbf{y}_{t-2} + \mathbf{G}_t\mathbf{e}_t + \mathbf{G}_{t,t-1}\mathbf{e}_{t-1}. \tag{1.1}$$

The autoregressive part with **F**-matrices specifies the lagged effects of the dependent variables, while the moving-average part with **G**-matrices handles the incoming errors and lagged errors. Assuming the errors in the vectors $\mathbf{e}_t$, $\mathbf{e}_{t-1}$ to be independently standard normally distributed (having covariance matrices **I**) and $\mathbf{G}_t$, $\mathbf{G}_{t,t-1}$ lower-triangular, the moving-average effects $\mathbf{G}_t\mathbf{e}_t$, $\mathbf{G}_{t,t-1}\mathbf{e}_{t-1}$ get covariance matrices $\mathbf{Q}_t = \mathbf{G}_t\mathbf{G}_t'$, $\mathbf{Q}_{t,t-1} = \mathbf{G}_{t,t-1}\mathbf{G}_{t,t-1}'$, which may be nondiagonal and with arbitrary variances on the diagonal. Specifying moving-average effects is no less general than the covariance matrices. Any covariance matrix **Q** can be written as $\mathbf{Q} = \mathbf{G}\mathbf{G}'$ in terms of a lower-triangular matrix (Cholesky factor) **G**. In addition, estimating **G** instead of directly **Q** has the advantage of avoiding possible negative variance estimates showing up in the direct estimate of **Q**.

The moving-average part $\mathbf{G}_t\mathbf{e}_t + \mathbf{G}_{t,t-1}\mathbf{e}_{t-1}$ in Eq. (1.1) may also be written as $\mathbf{G}_{t,t-1}\mathbf{e}_{t-1} + \mathbf{G}_{t,t-2}\mathbf{e}_{t-2}$ with the time indices shifted backward in time from t and $t-1$ to $t-1$ and $t-2$. Replacing the instantaneous error component $\mathbf{G}_t\mathbf{e}_t$ by the lagged one $\mathbf{G}_{t,t-1}\mathbf{e}_{t-1}$ (and $\mathbf{G}_{t,t-1}\mathbf{e}_{t-1}$ by $\mathbf{G}_{t,t-2}\mathbf{e}_{t-2}$) could be considered more appropriate, if the errors are taken to stand for the unknown causal influences on the system, which need some time to operate and to affect the system. The fact that the two unobserved consecutive error components get other names but retain their previous values will result in an observationally equivalent (equally fitting) system. Although equivalent, the existence of different representations in discrete time (forward or instantaneous representation in terms of t and $t-1$ and backward or lagging representation in terms of $t-1$ and $t-2$) is nevertheless unsatisfactory. The forward representation puts everything that happens in between t and $t-1$ forward in time at t, the backward representation puts the same information backward in time at $t-1$. From a causal standpoint, though, the backward representation is no less problematic than the forward representation, since it is anticipating effects that in true time will happen only later.

The ambiguous representation in discrete time of the behavior between t and $t-1$ in an ARMA(p, q) model disappears in the analogous continuous time CARMA(p, q) model, which reads for $p = 2$ and $q = 1$

$$\frac{\mathrm{d}^2\mathbf{y}(t)}{\mathrm{d}t^2} = \mathbf{F}_0\mathbf{y}(t) + \mathbf{F}_1\frac{\mathrm{d}\mathbf{y}(t)}{\mathrm{d}t} + \mathbf{G}_0\frac{\mathrm{d}\mathbf{W}(t)}{\mathrm{d}t} + \mathbf{G}_1\frac{\mathrm{d}^2\mathbf{W}(t)}{\mathrm{d}t^2}. \tag{1.2}$$

Writing $\mathbf{y}(t)$ instead of $\mathbf{y}_t$ emphasizes the development of $\mathbf{y}$ across continuous time. The role of successive lags in discrete time is taken over by successive derivatives in continuous time. The causally unsatisfactory instantaneous and lagging representations meet, so to speak, in the derivatives, which instead of using a discrete time interval $\Delta t = t - (t-1) = 1$, let the time interval go to zero: $\Delta t \to 0$. Equation (1.2) is sometimes written as

$$\mathbf{F}_0\mathbf{y}(t) + \mathbf{F}_1\frac{\mathrm{d}\mathbf{y}(t)}{\mathrm{d}t} + \mathbf{F}_2\frac{\mathrm{d}^2\mathbf{y}(t)}{\mathrm{d}t^2} = \mathbf{G}_0\frac{\mathrm{d}\mathbf{W}(t)}{\mathrm{d}t} + \mathbf{G}_1\frac{\mathrm{d}^2\mathbf{W}(t)}{\mathrm{d}t^2}, \tag{1.3}$$

with $\mathbf{F}_2 = -\mathbf{I}$ and opposite signs for $\mathbf{G}_0, \mathbf{G}_1$, making it clear that the CARMA(2,1) model has $\mathbf{F}_2$ as the highest degree matrix in the autoregressive part and $\mathbf{G}_1$ as the highest degree in the moving-average part.

The next subsection will show in more detail how discrete time and continuous time equations such as (1.1) and (1.2) become connected as $\Delta t \to 0$. The error process in continuous time is the famous Wiener process[1] $\mathbf{W}(t)$ or random walk through continuous time. Its main defining properties are the conditions of independently and normally distributed increments, $\Delta\mathbf{W}(t) = \mathbf{W}(t) - \mathbf{W}(t - \Delta t)$, having mean $\mathbf{0}$ and covariance matrix $\Delta t\mathbf{I}$. This means that the increments with arbitrary Δt are standard normally distributed for $\Delta t = 1$ as assumed for $\mathbf{e}_t, \mathbf{e}_{t-1}$ in discrete time. Likewise, the role of lower-triangular $\mathbf{G}_0, \mathbf{G}_1$ in (1.2) is analogous to the role of $\mathbf{G}_t, \mathbf{G}_{t,t-1}$ in discrete time. Derivative $\mathrm{d}\mathbf{W}(t)/\mathrm{d}t$ (white noise) does not exist in the classical sense but can be defined in the generalized function sense, and also integral $\int_{t_0}^{t} \mathbf{G}\mathrm{d}\mathbf{W}(t)$ can be defined rigorously (Kuo 2006, pp. 104–105 and pp. 260–261). Integrals are needed to go back again from the continuous time specification in Eq. (1.2) to the observed values in discrete time.

We will now show how the CARMA(2,1) model in Eq. (1.2) and the general CARMA(p, q) model can be formulated as special cases of the continuous time state space model. The continuous time state space model consists of two equations: a latent dynamic equation (1.4) with so-called drift matrix $\mathbf{A}$ and diffusion matrix $\mathbf{G}$ and a measurement equation (1.5) with loading matrix $\mathbf{C}$ and measurement error

[1] In this chapter we follow the common practice to write the Wiener process by capital letter $\mathbf{W}$, although it is here a vector whose size should be inferred from the context.

vector $\mathbf{v}(t)$:

$$\frac{\mathrm{d}\mathbf{x}(t)}{\mathrm{d}t} = \mathbf{A}\mathbf{x}(t) + \mathbf{G}\frac{\mathrm{d}\mathbf{W}(t)}{\mathrm{d}t}, \tag{1.4}$$

$$\mathbf{y}(t) = \mathbf{C}\mathbf{x}(t) + \mathbf{v}(t). \tag{1.5}$$

In general, the variables in state vector $\mathbf{x}(t)$ are assumed to be latent and only indirectly measured by the observed variables in $\mathbf{y}(t)$ with measurement errors in $\mathbf{v}(t)$. The measurement error vector $\mathbf{v}(t)$ is assumed independent of $\mathbf{x}(t)$ and normally distributed: $\mathbf{v}(t) \sim N(\mathbf{0}, \mathbf{R})$. For the initial state $\mathbf{x}(t_0)$, we assume $\mathbf{x}(t_0) \sim N(\boldsymbol{\mu}_{\mathbf{x}(t_0)}, \boldsymbol{\Phi}_{\mathbf{x}(t_0)})$. Often, but not necessarily, it is assumed $E[\mathbf{x}(t_0)] = \boldsymbol{\mu}_{\mathbf{x}(t_0)} = \mathbf{0}$. The latter would imply that the model has an equilibrium state: $E[\mathbf{x}(t)] = E[\mathbf{x}(t_0)] = \mathbf{0}$, and in case all eigenvalues of $\mathbf{A}$ have negative real part, $\mathbf{0}$ is the stable equilibrium state in the model.

In state space form, the observed second-order model CARMA(2,1) in Eq. (1.2) gets a state vector $\mathbf{x}(t) = [\mathbf{x}_1(t)'\ \mathbf{x}_2(t)']'$, which is two times the size of the observed vector $\mathbf{y}(t)$. The first part $\mathbf{x}_1(t)$ is not directly related to the observed variables and thus belongs to the latent part of the state space model. This special case of the state space model equates the second part to the observed vector: $\mathbf{y}(t) = \mathbf{x}_2(t)$. Equation (1.2) then follows from state space model (1.4)–(1.5) by specification

$$\mathbf{A} = \begin{bmatrix} \mathbf{0} & \mathbf{F}_0 \\ \mathbf{I} & \mathbf{F}_1 \end{bmatrix}, \ \mathbf{G} = \begin{bmatrix} \mathbf{G}_0 & \mathbf{0} \\ \mathbf{G}_1 & \mathbf{0} \end{bmatrix}, \ \mathbf{C} = \begin{bmatrix} \mathbf{0} & \mathbf{I} \end{bmatrix}, \ \mathbf{v}(t) = \mathbf{0}. \tag{1.6}$$

Applying (1.6) we find first

$$\frac{\mathrm{d}\mathbf{x}_1(t)}{\mathrm{d}t} = \mathbf{F}_0\mathbf{x}_2(t) + \mathbf{G}_0\frac{\mathrm{d}\mathbf{W}(t)}{\mathrm{d}t}, \tag{1.7}$$

$$\frac{\mathrm{d}\mathbf{x}_2(t)}{\mathrm{d}t} = \mathbf{x}_1(t) + \mathbf{F}_1\mathbf{x}_2(t) + \mathbf{G}_1\frac{\mathrm{d}\mathbf{W}(t)}{\mathrm{d}t} \Rightarrow$$

$$\frac{\mathrm{d}^2\mathbf{x}_2(t)}{\mathrm{d}t^2} = \frac{\mathrm{d}\mathbf{x}_1(t)}{\mathrm{d}t} + \mathbf{F}_1\frac{\mathrm{d}\mathbf{x}_2(t)}{\mathrm{d}t} + \mathbf{G}_1\frac{\mathrm{d}^2\mathbf{W}(t)}{\mathrm{d}t^2}. \tag{1.8}$$

Substituting (1.7) into the implication in (1.8) gives

$$\frac{\mathrm{d}^2\mathbf{x}_2(t)}{\mathrm{d}t^2} = \mathbf{F}_0\mathbf{x}_2(t) + \mathbf{F}_1\frac{\mathrm{d}\mathbf{x}_2(t)}{\mathrm{d}t} + \mathbf{G}_0\frac{\mathrm{d}\mathbf{W}(t)}{\mathrm{d}t} + \mathbf{G}_1\frac{\mathrm{d}^2\mathbf{W}(t)}{\mathrm{d}t^2}, \tag{1.9}$$

which for $\mathbf{y}(t) = \mathbf{x}_2(t)$ leads to the CARMA(2,1) model in (1.2).

The specification in the previous paragraph can be generalized to find the CARMA(p, q) model in state space form in (1.10), where $r = \max(p, q+1)$. $\mathbf{G}$ leads to diffusion covariance matrix $\mathbf{Q} = \mathbf{G}\mathbf{G}'$, which has Cholesky factor-based covariance matrices $\mathbf{G}_i\mathbf{G}_i{}'$ on the diagonal and in case $q > 0$ off-diagonal matrices

$\mathbf{G}_i\mathbf{G}_j{}'$ $(i, j = 0, 1, \ldots, r-1)$. In the literature one often finds the alternative state space form (1.11) (see e.g., Tómasson 2011; Tsai and Chan 2000). Here the moving average matrices $\mathbf{G}_1, \mathbf{G}_2, \ldots, \mathbf{G}_{r-1}$ are rewritten as $\mathbf{G}_i = \mathbf{H}_i\mathbf{G}_0$ in terms of corresponding matrices $\mathbf{H}_1, \mathbf{H}_2, \ldots, \mathbf{H}_{r-1}$, specified in the measurement part of the state space

$$\mathbf{A} = \begin{bmatrix} \mathbf{0} & \mathbf{0} & \cdots & \mathbf{0} & \mathbf{F}_0 \\ \mathbf{I} & \mathbf{0} & \ddots & \mathbf{0} & \mathbf{F}_1 \\ \mathbf{0} & \mathbf{I} & \ddots & \mathbf{0} & \mathbf{F}_2 \\ \vdots & \vdots & \ddots & \vdots & \vdots \\ \mathbf{0} & \mathbf{0} & \ldots & \mathbf{I} & \mathbf{F}_{r-1} \end{bmatrix}, \quad \mathbf{G} = \begin{bmatrix} \mathbf{G}_0 & \mathbf{0} & \cdots & \mathbf{0} & \mathbf{0} \\ \mathbf{G}_1 & \mathbf{0} & \ddots & \mathbf{0} & \mathbf{0} \\ \mathbf{G}_2 & \mathbf{0} & \ddots & \mathbf{0} & \mathbf{0} \\ \vdots & \vdots & \ddots & \vdots & \vdots \\ \mathbf{G}_{r-1} & \mathbf{0} & \ldots & \mathbf{0} & \mathbf{0} \end{bmatrix},$$

$$\mathbf{C} = \begin{bmatrix} \mathbf{0} & \mathbf{0} & \mathbf{0} & \cdots & \mathbf{I} \end{bmatrix}, \ \mathbf{v}(t) = \mathbf{0}. \tag{1.10}$$

model and $\mathbf{G}_0$. For two reasons we prefer (1.10) in the case of CARMA(p, q) models with $q > 0$. The derivation of (1.11) requires the matrices $\mathbf{H}_i$ and $\mathbf{F}_j$ to commute, which in practice restricts the applicability to the univariate case. In addition, using the measurement part for the moving average specification would make it difficult to specify at the same time measurement parameters. For CARMA(p,0) models (all $\mathbf{H}_i = \mathbf{0}$), we prefer (1.11), because each state variable in state vector $\mathbf{x}(t)$ is easily interpretable as the derivative of the previous one. The interpretation of the state variables in (1.7)–(1.8) is less simple.

$$\mathbf{A} = \begin{bmatrix} \mathbf{0} & \mathbf{I} & \mathbf{0} & \cdots & \mathbf{0} \\ \mathbf{0} & \mathbf{0} & \mathbf{I} & \ddots & \mathbf{0} \\ \vdots & \vdots & \vdots & \ddots & \vdots \\ \mathbf{0} & \mathbf{0} & \mathbf{0} & \ldots & \mathbf{I} \\ \mathbf{F}_0 & \mathbf{F}_1 & \mathbf{F}_2 & \cdots & \mathbf{F}_{r-1} \end{bmatrix}, \quad \mathbf{G} = \begin{bmatrix} \mathbf{0} & \mathbf{0} & \mathbf{0} & \cdots & \mathbf{0} \\ \mathbf{0} & \mathbf{0} & \mathbf{0} & \ddots & \mathbf{0} \\ \vdots & \vdots & \vdots & \ddots & \vdots \\ \mathbf{0} & \mathbf{0} & \mathbf{0} & \ldots & \mathbf{0} \\ \mathbf{0} & \mathbf{0} & \mathbf{0} & \cdots & \mathbf{G}_0 \end{bmatrix},$$

$$\mathbf{C} = \begin{bmatrix} \mathbf{I} & \mathbf{H}_1 & \mathbf{H}_2 \cdots \mathbf{H}_{r-1} \end{bmatrix}, \ \mathbf{v}(t) = \mathbf{0}. \tag{1.11}$$

The fact that the general CARMA(p, q) model fits seamlessly into the state space model means that all continuous time time series problems in modeling and estimation can be handled by state space form (1.4)–(1.5). The state space approach in fact reformulates higher-order models as a first-order model, and this will be applied in the sequel.

1.2.2 Connecting Discrete and Continuous Time Model in the EDM

The EDM combines the discrete time and continuous time model and does so in an exact way. It is by the EDM that the exact procedure in this chapter differentiates from many approximate procedures found in the literature. We show how the exact connection looks like between the general continuous time state space model and its discrete time counterpart, derived from it. The first-order models ARMA(1,0) and CARMA(1,0) in state space form differ from the general discrete and continuous time state space model only in a simpler measurement equation. So, having made the exact connections between the general state space models and thus between ARMA(1,0) and CARMA(1,0) and knowing that each CARMA(p, q) model can be written as a special case of the general state space model, the exact connections between CARMA(p, q) and ARMA(p, q), where the latter is derived from the former, follow. Next we consider the question of making an exact connection between an arbitrary ARMA(p^*, q^*) model and a CARMA(p, q) model, where the degrees p^* and p as well as q^* and q need not be equal.

Comparing discrete time equation (1.1) to the general discrete time state space model in (1.12)–(1.13), one observes that the latter becomes immediately the ARMA(1,0) model for $\mathbf{y}_t = \mathbf{x}_t$ but is more flexible in time handling.

$$\mathbf{x}_t = \mathbf{A}_{\Delta t}\mathbf{x}_{t-\Delta t} + \mathbf{G}_{\Delta t}\mathbf{e}_{t-\Delta t} \tag{1.12}$$

$$\mathbf{y}_t = \mathbf{C}\mathbf{x}_t + \mathbf{v}_t. \tag{1.13}$$

Inserting arbitrary lag Δt instead of fixed lag $\Delta t = 1$ enables us to put discrete time models with different intervals (e.g., years and months) on the same time scale and to connect them to the common underlying continuous time model for $\Delta t \to 0$.

State equation (1.12) can be put in the equivalent difference quotient form

$$\frac{\Delta \mathbf{x}_t}{\Delta t} = \mathbf{A}_{*\Delta t}\mathbf{x}_{t-\Delta t} + \mathbf{G}_{\Delta t}\frac{\mathbf{e}_{t-\Delta t}}{\Delta t}$$
$$\text{for } \mathbf{A}_{*\Delta t} = (\mathbf{A}_{\Delta t} - \mathbf{I})/\Delta t \text{ implying } \mathbf{A}_{\Delta t} = \mathbf{I} + \mathbf{A}_{*\Delta t}\Delta t. \tag{1.14}$$

So we have the discrete time state space model in two forms: difference quotient form (1.14) and solution form (1.12). Equation (1.12) is called the solution of (1.14), because it describes the actual state transition across time in accordance with (1.14) and is so said to satisfy the difference quotient equation. Note that analogously the general continuous time state space model (1.4)–(1.5) immediately accommodates the special CARMA(1,0) model for $\mathbf{y}(t) = \mathbf{x}(t)$ and can be put in two forms: stochastic differential equation (1.4) and its solution (1.15) (Arnold 1974; Singer 1990):

$$\mathbf{x}(t) = e^{\mathbf{A}\Delta t}\mathbf{x}(t - \Delta t) + \int_{t-\Delta t}^{t} e^{\mathbf{A}(t-s)}\mathbf{G}d\mathbf{W}(s). \tag{1.15}$$

In the exact discrete model EDM the connection between discrete and continuous time is made by means of the solutions, which in both cases describe the actual transition from the previous state at $t - \Delta t$ to the next state at t. The EDM thus combines both models and connects them exactly by the equalities:

$$\mathbf{A}_{\Delta t} = \mathrm{e}^{\mathbf{A}\Delta t} \quad \text{and} \quad \mathbf{Q}_{\Delta t} = \int_{t-\Delta t}^{t} \mathrm{e}^{\mathbf{A}(t-s)}\mathbf{Q}\mathrm{e}^{\mathbf{A}'(t-s)}\mathrm{d}(s)\,. \tag{1.16}$$

While discrete time autoregression matrix $\mathbf{A}_{\Delta t}$ and continuous time drift matrix $\mathbf{A}$ are connected via the highly nonlinear matrix exponential, the errors are indirectly connected by their covariance matrices $\mathbf{Q}_{\Delta t} = \mathbf{G}_{\Delta t}\mathbf{G}'_{\Delta t}$ and $\mathbf{Q} = \mathbf{G}\mathbf{G}'$. In estimating, after finding the drift matrix $\mathbf{A}$ on the basis of $\mathbf{A}_{\Delta t}$, next on the basis of $\mathbf{G}_{\Delta t}$ the diffusion matrix $\mathbf{G}$ is found.

The connection between discrete and continuous time becomes further clarified by two definitions of the matrix exponential $\mathrm{e}^{\mathbf{A}\Delta t}$. The standard definition (1.17)

$$\begin{aligned}\mathrm{e}^{\mathbf{A}\Delta t} &= \sum\nolimits_{k=0}^{\infty} (\mathbf{A}\Delta t)^k/k! \\ &= \mathbf{I} + \mathbf{A}\Delta t + \left(\frac{1}{2}\mathbf{A}^2\Delta t^2 + \frac{1}{6}\mathbf{A}^3\Delta t^3 + \frac{1}{24}\mathbf{A}^4\Delta t^4 + \cdots\right)\end{aligned} \tag{1.17}$$

shows the rather complicated presence of $\mathbf{A}$ in $\mathbf{A}_{\Delta t}$ and the formidable task to extract continuous time $\mathbf{A}$ from discrete time $\mathbf{A}_{\Delta t}$ in an exact fashion. However, it also shows that for $\Delta t \to 0$, the quantity between parentheses becomes arbitrarily small, and the linear part $\mathbf{I} + \mathbf{A}\Delta t$ could be taken as an approximation of $\mathrm{e}^{\mathbf{A}\Delta t}$, leading precisely to $\mathbf{A}_{*\Delta t}$ in the difference quotient equation (1.14) as approximation of $\mathbf{A}$ in the differential equation (1.4). Depending on the length of the interval Δt, however, the quality of this approximation can be unduly bad in practice. An alternative definition is based on oversampling (Singer 2012), meaning that the total time interval between measurements is divided into arbitrary small subintervals: $\delta = \frac{\Delta t}{D}$ for $D \to \infty$. The definition relies on the multiplication property of autoregression: $\mathbf{x}_t = \mathbf{A}_{0.5\Delta t}\mathbf{x}_{t-0.5\Delta t}$ and $\mathbf{x}_{t-0.5\Delta t} = \mathbf{A}_{0.5\Delta t}\mathbf{x}_{t-\Delta t} \Rightarrow \mathbf{x}_t = \mathbf{A}_{0.5\Delta t}\mathbf{A}_{0.5\Delta t}\mathbf{x}_{t-\Delta t} = \mathbf{A}_{\Delta t}\mathbf{x}_{t-\Delta t}$ which is equally valid in the continuous time case: $\mathbf{x}_t = \mathrm{e}^{\mathbf{A}(0.5\Delta t)}\mathrm{e}^{\mathbf{A}(0.5\Delta t)}\mathbf{x}_{t-\Delta t} = \mathrm{e}^{\mathbf{A}\Delta t}\mathbf{x}_{t-\Delta t}$, leading to multiplicative definition:

$$\mathrm{e}^{\mathbf{A}\Delta t} = \lim_{D\to\infty}\prod_{d=0}^{D-1}(\mathbf{I} + \mathbf{A}\delta_d) \text{ for } \delta_d = \frac{\Delta t}{D}. \tag{1.18}$$

$\mathbf{I} + \mathbf{A}\Delta t$ in (1.17), becoming $\mathbf{I} + \mathbf{A}\delta$ in (1.18), is no longer an approximation of $\mathrm{e}^{\mathbf{A}\delta}$ but becomes equal to $\mathrm{e}^{\mathbf{A}\delta}$ for $D \to \infty$ and $\delta \to 0$, while then $\mathbf{A}_{*\delta} = (\mathrm{e}^{\mathbf{A}\delta} - \mathbf{I})/\delta$ in difference equation (1.14) becomes equal to drift matrix $\mathbf{A}$ in differential equation (1.4). So, the formidable task to extract $\mathbf{A}$ from $\mathbf{A}_{\Delta t}$ is performed here in the same simple way as done by just taking approximate $\mathbf{A}_{*\Delta t}$ from $\mathbf{A}_{\Delta t}$ in (1.14), but instead of once over the whole interval Δt, it is repeated over many small subintervals, becoming for sufficiently large D equal to the exact procedure in terms of (1.17). We write $\mathbf{A}_{*\Delta t} \approx \mathbf{A}$, but one should keep in mind that by taking a sufficiently small interval in terms of the right-hand side of (1.18), one can get $\mathbf{A}_{*\Delta t}$ as close to $\mathbf{A}$ as wanted. An adapted version[2] of the oversampling procedure is described in Voelkle and Oud (2013).

Let us illustrate the connection between CARMA(1,0) and ARMA(1,0) by an example. If $\mathbf{A} = \begin{bmatrix} -1.0 & 0.2 \\ 0.3 & -1.5 \end{bmatrix}$, the exact connection in the EDM is for $\Delta t = 1$ made by $\mathbf{A}_{\Delta t=1} = \mathrm{e}^{\mathbf{A}\Delta t} = \mathrm{e}^{\mathbf{A}} = \begin{bmatrix} 0.377 & 0.058 \\ 0.088 & 0.231 \end{bmatrix}$. $\mathbf{A}$ in the differential equation may be compared to $\mathbf{A}_{*\Delta t=1} = (\mathrm{e}^{\mathbf{A}\Delta t} - \mathbf{I})/\Delta t = \mathrm{e}^{\mathbf{A}} - \mathbf{I} = \begin{bmatrix} -0.623 & 0.058 \\ 0.088 & -0.769 \end{bmatrix}$ in the difference equation for $\Delta t = 1$. For $\Delta t = 0.1$ we get $\mathbf{A}_{*\Delta t=0.1} = (\mathrm{e}^{\mathbf{A}\times 0.1} - \mathbf{I})/0.1 = \begin{bmatrix} -0.949 & 0.177 \\ 0.264 & -1.390 \end{bmatrix}$ which is much closer to $\mathbf{A}$ and for $\Delta t = 0.001$, $\mathbf{A}_{*\Delta t=0.001} = (\mathrm{e}^{\mathbf{A}\times 0.001} - \mathbf{I})/0.001 = \begin{bmatrix} -0.999 & 0.200 \\ 0.300 & -1.499 \end{bmatrix}$ becomes virtually equal to $\mathbf{A}$.

Making an exact connection between an ARMA(p^*, q^*) model and a model CARMA(p, q) such that the ARMA process $\{\mathbf{y}_t; t = 0, t = \Delta t, t = 2\Delta t, \ldots\}$ generated by ARMA(p^*, q^*) is a subset of the CARMA process $\{\mathbf{y}(t); t \geqslant 0\}$ generated by CARMA(p, q) is called "embedding" in the literature. The degrees p^* and q^* of the embedded model and p and q of the embedding model need not be equal. Embeddability is a much debated issue. Embedding is not always possible and need not be unique. Embedding is clearly possible for the case of ARMA(1,0) model $\mathbf{y}_t = \mathbf{A}_{\Delta t}\mathbf{y}_{t-\Delta t} + \mathbf{G}_{\Delta t}\mathbf{e}_{t-\Delta t}$ derived from CARMA(1,0) model $\frac{\mathrm{d}\mathbf{y}(t)}{\mathrm{d}t} = \mathbf{A}\mathbf{y}(t) + \mathbf{G}\frac{\mathrm{d}\mathbf{W}(t)}{\mathrm{d}t}$ with $\mathbf{A}_{\Delta t} = \mathrm{e}^{\mathbf{A}\Delta t}$, $\mathbf{Q}_{\Delta t} = \int_{t-\Delta t}^{t} \mathrm{e}^{\mathbf{A}(t-s)}\mathbf{Q}\mathrm{e}^{\mathbf{A}'(t-s)}\mathrm{d}s$, $\mathbf{Q}_{\Delta t} = \mathbf{G}_{\Delta t}\mathbf{G}'_{\Delta t}$, $\mathbf{Q} = \mathbf{G}\mathbf{G}'$ as shown above. The same is true for the higher-order ARMA(p, q) model, derived from CARMA(p, q). However, in general it is nontrivial to prove embeddability and to find the parameters of the CARMA(p, q) model embedding an ARMA(p^*, q^*) process. For example, not all ARMA(1,0) processes have a CARMA(1,0) process in which it can be embedded. A well-known example is the simple univariate

[2]The adapted version uses (1.18A)

$$\mathrm{e}^{\mathbf{A}\Delta t} = \lim_{D\to\infty} \prod_{d=0}^{D-1} \left[\left(\mathbf{I} - \frac{1}{2}\mathbf{A}\delta_d\right)^{-1} \left(\mathbf{I} + \frac{1}{2}\mathbf{A}\delta_d\right) \right] \text{ for } \delta_d = \frac{\Delta t}{D}, \tag{1.18A}$$

which converges much more rapidly than (1.18). It is based on the approximate discrete model (ADM), described by Oud and Delsing (2010), which just as the EDM goes back to Bergstrom (1984). Computation of the matrix exponential by (1.18) or (1.18A) has the advantage over the diagonalization method (Oud and Jansen 2000) that no assumptions with regard to the eigenvalues need to be made. Currently by most authors the Padé-approximation (Higham 2009) is considered the best computation method, which therefore is implemented in the most recent version of ctsem.

process $y_t = a_{\Delta t} y_{t-\Delta t} + g_{\Delta t} e_{t-\Delta t}$ with $-1 < a_{\Delta t} < 0$, because there does not exist any a for which $a_{\Delta t} = e^{a\Delta t}$ can be negative. However, Chan and Tong (1987) showed that for this ARMA(1,0) process with $-1 < a_{\Delta t} < 0$, a higher order CARMA(2,1) process can be found, in which it can be embedded.

Also embeddability need not be unique. Different CARMA models may embed one and the same ARMA model. A classic example is "aliasing" in the case of matrices $\mathbf{A}$ with complex conjugate eigenvalue pairs $\lambda_{1,2} = \alpha \pm \beta i$ with i the imaginary unit (Hamerle et al. 1991; Phillips 1973). Such complex eigenvalue pairs imply processes with oscillatory movements. Adding $\pm k2\pi/\Delta t$ to β leads for arbitrary integer k to a different $\mathbf{A}$ with a different oscillation frequency but does not change $\mathbf{A}_{\Delta t} = e^{\mathbf{A}\Delta t}$ and so may lead to the same ARMA model. The consequence is that the CARMA model cannot uniquely be determined (identified) by the ARMA model and the process generated by it. Fortunately, the number of aliases in general is limited in the sense that there exists only a finite number of aliases that lead for the same ARMA model to a real $\mathbf{G}$ and so to a positive definite $\mathbf{Q}$ in the CARMA model (Hansen and Sargent 1983). The size of the finite set additionally depends on the observation interval Δt, a smaller Δt leading to less aliases. The number of aliases may also be limited by sampling the observations in the discrete time process at unequal intervals (Oud and Jansen 2000; Tómasson 2015; Voelkle and Oud 2013).

An important point with regard to the state space modeling technique of time series is the latent character of the state. Even in the case of an observed ARMA(p, q) or CARMA(p, q) model of such low dimension as $p = 2$, we have seen that part of the state is not directly connected to the data. This has especially consequences for the initial time point. Suppose for the ARMA(2,1) model in state space form (1.19)–(1.20),

$$\begin{bmatrix} \mathbf{x}_{1,t-\Delta t} \\ \mathbf{x}_{1,t} \end{bmatrix} = \begin{bmatrix} \mathbf{0} & \mathbf{I} \\ \mathbf{F}_{t,t-2\Delta t} & \mathbf{F}_{t,t-\Delta t} \end{bmatrix} \begin{bmatrix} \mathbf{x}_{1,t-2\Delta t} \\ \mathbf{x}_{1,t-\Delta t} \end{bmatrix} + \begin{bmatrix} \mathbf{0} & \mathbf{0} \\ \mathbf{0} & \mathbf{G}_{t,t-\Delta t} \end{bmatrix} \mathbf{e}_{t-\Delta t},$$

$$\mathbf{x}_t = \mathbf{A}_{\Delta t} \quad \mathbf{x}_{t-\Delta t} + \mathbf{G}_{\Delta t} \quad \mathbf{e}_{t-\Delta t}, \tag{1.19}$$

$$\mathbf{y}_t = \begin{bmatrix} \mathbf{0} & \mathbf{I} \end{bmatrix} \mathbf{x}_t, \tag{1.20}$$

the initial time point, where the initial data are located, is $t_0 = t - \Delta t$. It means that there are no data directly or indirectly connected to (the lagged) part of $\mathbf{x}_{t-\Delta t}$. The initial parameters related to this part can nevertheless be estimated but become highly dependent on the model structure, and the uncertainty will be reflected in high standard errors. It does not help to start the model at later time point $t_0 + \Delta t$. That would result in data loss, since the $\mathbf{0}$ in (1.20) simply eliminates the lagged part of $\mathbf{x}_{t-\Delta t}$ without any connection to the data. Similar remarks apply to the initial derivative $\mathrm{d}\mathbf{x}_1(t)/\mathrm{d}t$

$$\begin{bmatrix} \frac{\mathrm{d}\mathbf{x}_1(t)}{\mathrm{d}t} \\ \frac{\mathrm{d}^2\mathbf{x}_1(t)}{\mathrm{d}t^2} \end{bmatrix} = \begin{bmatrix} \mathbf{0} & \mathbf{I} \\ \mathbf{F}_0 & \mathbf{F}_1 \end{bmatrix} \begin{bmatrix} \mathbf{x}_1(t) \\ \frac{\mathrm{d}\mathbf{x}_1(t)}{\mathrm{d}t} \end{bmatrix} + \begin{bmatrix} \mathbf{0} & \mathbf{0} \\ \mathbf{0} & \mathbf{G}_0 \end{bmatrix} \frac{\mathrm{d}\mathbf{W}(t)}{\mathrm{d}t},$$

$$\frac{\mathrm{d}\mathbf{x}(t)}{\mathrm{d}t} = \mathbf{A} \quad \mathbf{x}(t) + \mathbf{G} \frac{\mathrm{d}\mathbf{W}(t)}{\mathrm{d}t}, \tag{1.21}$$

$$\mathbf{y}(t) = \begin{bmatrix} \mathbf{I} & \mathbf{0} \end{bmatrix} \mathbf{x}(t), \tag{1.22}$$

in (1.21)–(1.22), which is not directly connected to the data and cannot be computed at the initial time point. Again, the related initial parameters can be estimated in principle. One should realize, however, that dependent on the model structure, the number of time points analyzed and the length of the observation intervals, these initial parameter estimates can become extremely unreliable. In a simulation study of a CARMA(2,0) model with oscillating movements, Oud and Singer (2008) found in the case of long interval lengths extremely large standard errors for the estimates related to the badly measured initial $\mathrm{d}\mathbf{x}_1(t)/\mathrm{d}t$. This lack of data and relative unreliability of estimates are the price one has to pay for choosing higher-order ARMA(p, q) and CARMA(p, q) models.

1.2.3 *Extended Continuous Time Model*

The extended continuous time state space model reads

$$\frac{\mathrm{d}\mathbf{x}(t)}{\mathrm{d}t} = \mathbf{A}\mathbf{x}(t) + \mathbf{B}\mathbf{u}(t) + \boldsymbol{\gamma} + \mathbf{G}\frac{\mathrm{d}\mathbf{W}(t)}{\mathrm{d}t}, \tag{1.23}$$

$$\mathbf{y}_{t_i} = \mathbf{C}\mathbf{x}(t_i) + \mathbf{D}\mathbf{u}(t_i) + \boldsymbol{\kappa} + \mathbf{v}_{t_i}. \tag{1.24}$$

In comparison to the basic model in (1.4)–(1.5), the extended model exhibits one minor notational change and two major additions. The minor change is in the measurement equation and is only meant to emphasize the discrete time character of the data at the discrete time points t_i ($i = 0, \ldots, T - 1$) with $\mathbf{x}(t_i)$ and $\mathbf{u}(t_i)$ sampling the continuous time vectors $\mathbf{x}(t)$ and $\mathbf{u}(t)$ at the observation time points. One major addition are the effects $\mathbf{B}\mathbf{u}(t)$ and $\mathbf{D}\mathbf{u}(t)$ of fixed exogenous variables in vector $\mathbf{u}(t)$. The other is the addition of random subject effect vectors $\boldsymbol{\gamma}$ and $\boldsymbol{\kappa}$ to the equations. While the (statistically) fixed variables in $\mathbf{u}(t)$ may change across time (time-varying exogenous variables), the subject-specific effects $\boldsymbol{\gamma}$ and $\boldsymbol{\kappa}$ with possibly a different value for each subject in the sample are assumed to be constant across time but normally distributed random variables: $\boldsymbol{\gamma} \sim N(\mathbf{0}, \boldsymbol{\Phi}_{\boldsymbol{\gamma}})$, $\boldsymbol{\kappa} \sim N(\mathbf{0}, \boldsymbol{\Phi}_{\boldsymbol{\kappa}})$. To distinguish them from the changing states, the constant random effects in $\boldsymbol{\gamma}$ are called traits. Because trait vector $\boldsymbol{\gamma}$ is modeled to influence $\mathbf{x}(t)$ continuously, before as well as after t_0, $\boldsymbol{\Phi}_{\mathbf{x}(t_0),\boldsymbol{\gamma}}$, the covariance matrix between initial state and traits cannot in general be assumed zero. The additions in state equation (1.23) lead to the following extended solution:

$$\mathbf{x}(t) = \mathrm{e}^{\mathbf{A}(t-t_0)}\mathbf{x}(t_0) + \int_{t_0}^{t} \mathrm{e}^{\mathbf{A}(t-s)}\mathbf{B}\mathbf{u}(s)\mathrm{d}(s) + \mathbf{A}^{-1}\big[\mathrm{e}^{\mathbf{A}(t-t_0)} - \mathbf{I}\big]\boldsymbol{\gamma} + \int_{t_0}^{t} \mathrm{e}^{\mathbf{A}(t-s)}\mathbf{G}\mathrm{d}\mathbf{W}(s). \tag{1.25}$$

1.2.4 Exogenous Variables

We have seen that the basic model in the case of stability (all eigenvalues of $\mathbf{A}$ having negative real part) has $\mathbf{0}$ as stable equilibrium state. Exogenous effects $\mathbf{Bu}(t)$ and $\mathbf{Du}(t)$ accommodate nonzero constant as well as nonconstant mean trajectories $E[\mathbf{x}(t)]$ and $E[\mathbf{y}(t)]$ even in the case of stability. By far the most popular exogenous input function is the unit function, $e(s) = 1$ for all s over the interval, with the effect $\mathbf{b}_e$ called intercept and integrating over interval $[t_0, t)$ into $\int_{t_0}^{t} \mathrm{e}^{\mathbf{A}(t-s)}\mathbf{b}_e e(s)\mathrm{d}(s) = \mathbf{A}^{-1}[\mathrm{e}^{\mathbf{A}(t-t_0)} - \mathbf{I}]\mathbf{b}_e$. In the measurement equation, the effect of the unit variable in $\mathbf{D}$ is called measurement intercept or origin and allows measurement instruments to have scales with different starting points in addition to the different units specified in $\mathbf{C}$.

Useful in describing sudden changes in the environment is the intervention function, a step function that takes on a certain value a until a specific time point t' and changes to value b at that time point until the end of the interval: $i(s) = a$ for all $s < t'$, $i(s) = b$ for all $s \geqslant t'$. An effective way of handling the step or piecewise constant function is a two-step procedure, in which Eq. (1.26) is applied twice: first with $\mathbf{u}(t_0)$ containing the step function value of the first step before t' and next with $\mathbf{u}(t_0)$ containing the step function value of the second step.

$$\mathbf{x}(t) = \mathrm{e}^{\mathbf{A}(t-t_0)}\mathbf{x}(t_0) + \mathbf{A}^{-1}\left[\mathrm{e}^{\mathbf{A}(t-t_0)} - \mathbf{I}\right]\mathbf{B}\mathbf{u}(t_0) + \mathbf{A}^{-1}\left[\mathrm{e}^{\mathbf{A}(t-t_0)} - \mathbf{I}\right]\boldsymbol{\gamma} + \int_{t_0}^{t} \mathrm{e}^{\mathbf{A}(t-s)}\mathbf{G}\mathrm{d}\mathbf{W}(s). \tag{1.26}$$

In the second step, the result $\mathbf{x}(t)$ of the first step is inserted as $\mathbf{x}(t_0)$. The relatively simple solution equation (1.26) has much more general applicability, though, than just for step functions. It can be used to approximate any exogenous behavior function in steps and approximate its effect arbitrarily closely by oversampling (dividing the observation interval in smaller intervals) and choosing the oversampling intervals sufficiently small.[3]

The handling of exogenous variables takes another twist, when it is decided to endogenize them. The problem with oversampling is that it is often not known, how the exogenous behavior function looks like in between observations. By endogenizing the exogenous variables, they are handled as random variables, added to the state vector, and in the same way as the other state variables related to their past values in an autoregressive fashion. Advantages of endogenizing are its nonapproximate nature and the fact that the new state variables may not only be modeled to influence the other state variables but also to be reciprocally influenced by them.

In addition to differentiating time points within subjects, exogenous variables also enable to differentiate subjects in case of an $N > 1$ sample. Suppose the first element of $\mathbf{u}(t)$ is the unit variable, corresponding in $\mathbf{B}$ with first column $\mathbf{b}_e$, and the second element is a dummy variable differentiating boys and girls (boys 0 at all-time points and girls 1 at all-time points) and corresponding to second column $\mathbf{b}_d$. Supposing all remaining variables have equal values, the mean or expectation $E[\mathbf{x}(t)]$ of girls over the interval $[t_0, t)$ will then differ by the amount of $\mathbf{A}^{-1}[\mathrm{e}^{\mathbf{A}(t-t_0)} - \mathbf{I}]\mathbf{b}_d$ from the one of boys. This amount will be zero for $t - t_0 = 0$, but the regression-like analysis procedure applied at initial time point t_0 allows to distinguish different initial means $E[\mathbf{x}(t_0)]$ for boys and girls. Thus, the same dummy variable at t_0 may impact both the state variables at t_0 and according to the state space model over the interval $t - t_0$ the state variables at the next observation time point.

1.2.5 Traits

Although, as we have just seen, there is some flexibility in the mean or expected trajectory, because subjects in different groups can have different mean trajectories, it would nevertheless be a strange implication of the model, if a subject's expected

[3]A better approximation than a step function is given by a piecewise linear or polygonal approximation (Oud and Jansen 2000; Singer 1992). Then we write $\mathbf{u}(t)$ in (1.23) as $\mathbf{u}(t) = \mathbf{u}(t_0) + (t - t_0)\mathbf{b}_{(t_0,t]}$ and (1.26) becomes:

$$\mathbf{x}(t) = \mathrm{e}^{\mathbf{A}(t-t_0)}\mathbf{x}(t_0) + \mathbf{A}^{-1}\left[\mathrm{e}^{\mathbf{A}(t-t_0)} - \mathbf{I}\right]\mathbf{B}\mathbf{u}(t_0) + \left\{\mathbf{A}^{-2}\left[\mathrm{e}^{\mathbf{A}(t-t_0)} - \mathbf{I}\right] - \mathbf{A}^{-1}(t - t_0)\right\}\mathbf{B}\mathbf{b}_{(t_0,t]}$$
$$+ \mathbf{A}^{-1}\left[\mathrm{e}^{\mathbf{A}(t-t_0)} - \mathbf{I}\right]\boldsymbol{\gamma} + \int_{t_0}^{t} \mathrm{e}^{\mathbf{A}(t-s)}\mathbf{G}\mathrm{d}\mathbf{W}(s). \tag{1.26A}$$

current and future behavior is totally dependent on the group of which he or she is modeled to be a member. It should be noted that the expected trajectories are not only interesting per se, but they also play a crucial role in the estimated latent sample trajectory of a subject, defined as the conditional mean $E[\mathbf{x}(t)|\mathbf{y}]$, where $\mathbf{y}$ is the total data vector of the subject (Kalman smoother), or $E[\mathbf{x}(t)|\mathbf{y}[t_0, t]]$, where $\mathbf{y}[t_0, t]$ is all data up to and including t (Kalman filter). In a model without traits, the subject regresses toward (in the case of a stable model) or egresses from (in an unstable model) the mean trajectory of its group. The consequences are particularly dramatic for predictions, because then after enough time is elapsed, the subject's trajectory in a stable model will be coinciding with its group trajectory.

From solution equation (1.25), it becomes clear, however, that in the state-trait model, each subject gets its own mean trajectory that differs from the group's mean. After moving the initial time point of a stable model sufficiently far into the past, $t_0 \rightarrow -\infty$, the subject's expected trajectory is

$$E[\mathbf{x}(t)|\boldsymbol{\gamma}] = \int_{-\infty}^{t} \mathrm{e}^{\mathbf{A}(t-s)}\mathbf{B}\mathbf{u}(s)\mathrm{d}(s) - \mathbf{A}^{-1}\boldsymbol{\gamma}, \tag{1.27}$$

which keeps a subject-specific distance $-\mathbf{A}^{-1}\boldsymbol{\gamma}$ from the subject's group mean trajectory $E[\mathbf{x}(t)] = \int_{-\infty}^{t} \mathrm{e}^{\mathbf{A}(t-s)}\mathbf{B}\mathbf{u}(s)\mathrm{d}(s)$. As a result the subject's sample trajectory regresses toward its own mean instead of its group mean. A related advantage of the state-trait model is that it clearly distinguishes trait variance (diagonals of $\boldsymbol{\Phi}_{\boldsymbol{\gamma}}$), also called unobserved heterogeneity between subjects, from stability. Because in a pure state model ($\boldsymbol{\gamma} = \mathbf{0}$) all subject-specific mean trajectories coincide with the group mean trajectory, trait variance and stability are confounded in the sense that an actually nonzero trait variance leads to a less stable model (eigenvalues of $\mathbf{A}$ having less negative real part) as a surrogate for keeping the subject-specific mean trajectories apart. In a state-trait model, however, stability is not hampered by hidden heterogeneity.

It should be noted that the impact of the fixed and random effects $\mathbf{B}\mathbf{u}(t)$ and $\boldsymbol{\gamma}$ in the state equation (1.23) is quite different from that of $\mathbf{D}\mathbf{u}(t_i)$ and $\boldsymbol{\kappa}$ in the measurement equation (1.24). The latter is a one-time snapshot event with no consequences for the future. It just reads out in a specific way the current contents of the system's state. However, the state equation is a dynamic equation where influences may have long-lasting and cumulative future effects that are spelled out by Eq. (1.25) or (1.26). In particular, the traits $\boldsymbol{\gamma}$ differ fundamentally from the nondynamic or "random measurement bias" $\boldsymbol{\kappa}$, earlier proposed for panel data by Goodrich and Caines (1979), Jones (1993), and Shumway and Stoffer (2000).

1.3 Model Estimation by SEM

As emphasized above, if the data are collected in discrete time, we need the EDM to connect the continuous time parameter matrices to the discrete time parameter matrices describing the data. The continuous time model in state space form contains eight parameter matrices that are connected to the corresponding discrete time matrices as shown in (1.28). In (1.28) $\otimes$ is the Kronecker product and the row operator puts the elements of the $\mathbf{Q}$ matrix row-wise in a column vector, whereas irow stands for the inverse operation.

While there are only eight continuous time parameter matrices, there may be many more discrete time parameter matrices. This is typically the case for the dynamic matrices $\mathbf{A}_{\Delta t_{i,j}}, \mathbf{B}_{\Delta t_{i,j}}, \mathbf{Q}_{\Delta t_{i,j}}$. The observation time points t_i $(i = 0, \ldots, T-1)$ may differ for different subjects j $(j = 1, \ldots, N)$ but also the observation intervals $\Delta t_{i,j} = t_{i,j} - t_{i-1,j}$ $(i = 1, \ldots, T-1)$ between the observation time points. Different observation intervals can lead to many different discrete time matrices $\mathbf{A}_{\Delta t_{i,j}}, \mathbf{B}_{\Delta t_{i,j}}, \mathbf{Q}_{\Delta t_{i,j}}$ but all based on the same underlying continuous time matrices $\mathbf{A}$, $\mathbf{B}$, $\mathbf{Q}$.

$$
\begin{aligned}
&\mathbf{A} && \mathbf{A}_{\Delta t_{i,j}} = \mathrm{e}^{\mathbf{A}\Delta t_{i,j}} \\
&\mathbf{B} && \mathbf{B}_{\Delta t_{i,j}} = \mathbf{A}^{\circ}{}_{\Delta t_{i,j}}\mathbf{B} \quad \text{for } \mathbf{A}^{\circ}{}_{\Delta t_{i,j}} = \mathbf{A}^{-1}(\mathrm{e}^{\mathbf{A}\Delta t_{i,j}} - \mathbf{I}) \\
&\mathbf{Q} = \mathbf{G}\mathbf{G}' && \mathbf{Q}_{\Delta t_{i,j}} = \int_{t-\Delta t_{i,j}}^{t} \mathrm{e}^{\mathbf{A}(t-s)}\mathbf{Q}\mathrm{e}^{\mathbf{A}'(t-s)}\mathrm{d}(s) \\
& && \qquad = \mathrm{irow}[\mathbf{A}_{\#}^{-1}(\mathrm{e}^{\mathbf{A}_{\#}\Delta t_{i,j}} - \mathbf{I})\mathrm{row}\,\mathbf{Q}] \\
& && \qquad \text{for } \mathbf{A}_{\#} = \mathbf{A} \otimes \mathbf{I} + \mathbf{I} \otimes \mathbf{A} \\
&\mathbf{C} && \mathbf{C}_{t_{i,j}} = \mathbf{C} \\
&\mathbf{D} && \mathbf{D}_{t_{i,j}} = \mathbf{D} \\
&\mathbf{R} && \mathbf{R}_{t_{i,j}} = \mathbf{R} \\
&\boldsymbol{\mu}_{\mathbf{x}(t_0)} = E[\mathbf{x}(t_0)] = \mathbf{B}_{t_0}\mathbf{u}(t_0) \\
&\boldsymbol{\Phi}_{\mathbf{x}(t_0)} = E[(\mathbf{x}(t_0) - \boldsymbol{\mu}_{\mathbf{x}(t_0)})(\mathbf{x}(t_0) - \boldsymbol{\mu}_{\mathbf{x}(t_0)})']
\end{aligned}
\tag{1.28}
$$

The most extreme case is that none of the intervals is equal to any other interval, a situation a traditional discrete time analysis would be unable to cope with but is unproblematic in continuous time analysis (Oud and Voelkle 2014).

The initial parameter matrices $\boldsymbol{\mu}_{\mathbf{x}(t_0)}$ and $\boldsymbol{\Phi}_{\mathbf{x}(t_0)}$ deserve special attention. In a model with exogenous variables, the initial state mean may take different values in different groups defined by the exogenous variables. Since the mean trajectories $E[\mathbf{x}(t)]$ may be deviating from each other because of exogenous influences after t_0, it is natural to let them already differ at t_0 as a result of past influences. These differences are defined regression-wise by $E[\mathbf{x}(t_0)] = \mathbf{B}_{t_0}\mathbf{u}(t_0)$ with $\mathbf{B}_{t_0}\mathbf{u}(t_0)$

absorbing all unknown past influences. For example, if $\mathbf{u}(t_0)$ consists of two variables, the unit variable (1 for all subjects) and a dummy variable defining gender (0 for boys and 1 for girls), there will be two means $E[\mathbf{x}(t_0)]$, one for the boys and one for the girls. If $\mathbf{u}(t_0)$ contains only the unit variable, the single remaining vector $\mathbf{b}_{t_0}$ in $\mathbf{B}_{t_0}$ will become equal to the initial mean: $E[\mathbf{x}(t_0)] = \mathbf{b}_{t_0}$. Because the instantaneous regression matrix $\mathbf{B}_{t_0}$ just describes means and differences as a result of unknown effects from before t_0, it should not be confused with the dynamic $\mathbf{B}$ and as a so-called predetermined quantity in estimation not undergo any constraint from $\mathbf{B}$. Similarly, $\mathbf{\Phi}_{\mathbf{x}(t_0)}$ should not undergo any constraint from the continuous time diffusion covariance matrix $\mathbf{Q}$.

A totally new situation for the initial parameters arises, however, if we assume the system to be in equilibrium. Equilibrium means first $\boldsymbol{\mu}_{\mathbf{x}(t)} = \boldsymbol{\mu}_{\mathbf{x}(t_0)}$ as well as all exogenous variables $\mathbf{u}(t) = \mathbf{u}(t_0)$ being constant. Evidently, the latter is the case, if the only exogenous variable is the unit variable, reducing $\mathbf{B}$ to a vector of intercepts, but also if it contains additional gender or any other additional exogenous variables, differentiating subjects from each other but constant in time. The assumption of equilibrium, $\boldsymbol{\mu}_{\mathbf{x}(t)} = \boldsymbol{\mu}_{\mathbf{x}(t_0)}$,—possibly but not necessarily a stable equilibrium—leads to equilibrium value

$$\boldsymbol{\mu}_{\mathbf{x}(t)} = \boldsymbol{\mu}_{\mathbf{x}(t_0)} = -\mathbf{A}^{-1}\mathbf{B}\mathbf{u}_c, \tag{1.29}$$

with $\mathbf{u}_c$ the value of the constant exogenous variables $\mathbf{u}(t) = \mathbf{u}(t_0) = \mathbf{u}_c$. If we assume the system to be stationary, additionally $\mathbf{\Phi}_{\mathbf{x}(t)} = \mathbf{\Phi}_{\mathbf{x}(t_0)}$ is assumed to be in equilibrium, leading to equilibrium value

$$\mathbf{\Phi}_{\mathbf{x}(t)} = \mathbf{\Phi}_{\mathbf{x}(t_0)} = \text{irow}[-\mathbf{A}_{\#}^{-1}\text{row}\,\mathbf{Q}]\,. \tag{1.30}$$

The novelty of the stationarity assumption is that the initial parameters are totally defined in terms of the dynamic parameters as is clearly seen from (1.29) and (1.30). It means, in fact, that the initial parameters disappear and the total number of parameters to be estimated is considerably reduced. Although attractive and present as an option in ctsem (Driver et al. 2017), the stationarity assumption is quite restrictive and can be unrealistic in practice.

To estimate the EDM as specified in (1.28) by SEM, we put all variables and matrices of the EDM into SEM model

$$\boldsymbol{\eta} = \boldsymbol{B}\boldsymbol{\eta} + \boldsymbol{\zeta} \quad \text{with} \quad \boldsymbol{\Psi} = E(\boldsymbol{\zeta}\boldsymbol{\zeta}'), \tag{1.31}$$

$$\mathbf{y} = \mathbf{\Lambda}\boldsymbol{\eta} + \boldsymbol{\epsilon} \quad \text{with} \quad \mathbf{\Theta} = E(\boldsymbol{\epsilon}\boldsymbol{\epsilon}'). \tag{1.32}$$

The SEM model consists of two equations, structural equation (1.31) and measurement equation (1.32), in terms of four vectors, $\boldsymbol{\eta}, \boldsymbol{\zeta}, \mathbf{y}, \boldsymbol{\epsilon}$, and four matrices, $\boldsymbol{B}, \boldsymbol{\Psi}, \mathbf{\Lambda}, \mathbf{\Theta}$. From Eqs. (1.31)–(1.32), one easily derives the model implied mean $\boldsymbol{\mu}$

and covariance matrix $\mathbf{\Sigma}$ and next the raw maximum likelihood equation (1.33) (see e.g., Bollen 1989)

$$\mathrm{RML} = \sum_{j=1}^{N} \Big[m_j \log(2\pi) + \log(|\mathbf{\Sigma}_j|) + (\mathbf{y}_j - \boldsymbol{\mu}_j)' \mathbf{\Sigma}_j^{-1} (\mathbf{y}_j - \boldsymbol{\mu}_j) \Big]. \tag{1.33}$$

The subscript j makes the SEM procedure extremely flexible by allowing any number of subjects, including $N = 1$, and any missing value pattern for each of the subjects j, as the number of variables m_j ($m = pT$), the data vector $\mathbf{y}_j$, the mean vector $\boldsymbol{\mu}_j$, and the covariance matrix $\mathbf{\Sigma}_j$ may all be subject specific. In case of missing values, the corresponding rows and columns of the missing elements for that subject j are simply deleted.

For obtaining the maximum likelihood estimates of the EDM, it suffices to show how the SEM vectors $\boldsymbol{\eta}, \mathbf{y}, \boldsymbol{\zeta}, \boldsymbol{\epsilon}$, and matrices, $\boldsymbol{B}, \mathbf{\Lambda}, \mathbf{\Psi}, \mathbf{\Theta}$ include the variables and matrices of the EDM. This is done in (1.34). In the vector of exogenous variables $\mathbf{u}(t) = [\mathbf{u}_c\ \mathbf{u}_v(t)]$, we distinguish two parts: the part $\mathbf{u}_c$, consisting of the unit variable and, for example, gender and other variables that differ between subjects but are constant across time, and the part $\mathbf{u}_v(t)$ that at least for one subject in the sample is varying across time. Exogenous variables like weight and income, for example, are to be put into $\mathbf{u}_v(t)$. We abbreviate the dynamic errors $\int_{t_{i,j}-\Delta t_{i,j}}^{t_{i,j}} \mathrm{e}^{\mathbf{A}(t_{i,j}-s)}\mathbf{G}\mathrm{d}\mathbf{W}(s)$ to $\mathbf{w}(t_{i,j} - \Delta t_{i,j})$. The mean sum of squares and cross-products matrix of $\mathbf{u}_j$ over sample units is called $\mathbf{\Phi}_{\mathbf{u}}$, and the mean sum of cross-products between $\mathbf{x}(t_{0,j}) - \boldsymbol{\mu}_{\mathbf{x}(t_{0,j})}$ and $\mathbf{u}_j$ is called $\mathbf{\Phi}_{\mathbf{x}(t_0),\mathbf{u}}$. The latter must be estimated, though, if the state is latent.

The traits $\boldsymbol{\gamma}$ and $\boldsymbol{\kappa}$ are not explicitly displayed but can be viewed as a special kind of constant zero-mean exogenous variables $\mathbf{u}_{c,j}$ in $\mathbf{u}_j$, whose covariance matrices $\mathbf{\Phi}_{\boldsymbol{\gamma}}$ and $\mathbf{\Phi}_{\boldsymbol{\kappa}}$ in $\mathbf{\Phi}_{\mathbf{u}}$ as well as $\mathbf{\Phi}_{\mathbf{x}(t_0),\boldsymbol{\gamma}}$ and $\mathbf{\Phi}_{\mathbf{x}(t_0),\boldsymbol{\kappa}}$ in $\mathbf{\Phi}_{\mathbf{x}(t_0),\mathbf{u}}$ are not fixed quantities but have to be estimated. These latent variables have no loadings in $\mathbf{\Lambda}$ and have $\mathbf{B}_{c,t_0} = \mathbf{0}$ in $\boldsymbol{B}$. For $\boldsymbol{\gamma}$ the $\mathbf{B}_{c,\Delta t_{i,j}}$ in $\boldsymbol{B}$ are replaced by $\mathbf{A}^{\circ}{}_{\Delta t_{i,j}}$ (see (1.28)) and for $\boldsymbol{\kappa}$ the $\mathbf{D}_{c,t_{i,j}}$ in $\mathbf{\Lambda}$ by $\mathbf{I}$.

$$\begin{aligned}
\boldsymbol{\eta} &= [\mathbf{x}'_j\ \mathbf{u}'_j]' \ \text{ with } \mathbf{x}_j = [\mathbf{x}'(t_{0,j})\ \mathbf{x}'(t_{1,j})\ \cdots\ \mathbf{x}'(t_{T-1,j})]' \\
&\qquad\qquad\qquad \text{and } \mathbf{u}_j = [\mathbf{u}'_{c,j}\ \mathbf{u}'_v(t_{0,j})\ \mathbf{u}'_v(t_{1,j})\ \cdots \mathbf{u}'_v(t_{T-1,j})]', \\
\boldsymbol{\zeta} &= [\mathbf{w}'_j\ \mathbf{u}'_j]' \ \text{ with } \mathbf{w}_j = [\mathbf{x}'(t_{0,j}) - \boldsymbol{\mu}'_{\mathbf{x}(t_{0,j})}\ \mathbf{w}'(t_{1,j} - \Delta t_{1,j}) \\
&\qquad\qquad\qquad \cdots\ \mathbf{w}'(t_{T-1,j} - \Delta_{T-1,j})]', \\
\mathbf{y} &= [\mathbf{y}'_j\ \mathbf{u}'_j]' \ \text{ with } \mathbf{y}_j = [\mathbf{y}'(t_{0,j})\ \mathbf{y}'(t_{1,j})\ \cdots\ \mathbf{y}'(t_{T-1,j})]', \\
\boldsymbol{\epsilon} &= [\mathbf{v}'_j\ \mathbf{0}']' \ \text{ with } \mathbf{v}_j = [\mathbf{v}'(t_{0,j})\ \mathbf{v}'(t_{1,j})\ \cdots\ \mathbf{v}'(t_{T-1,j})]',
\end{aligned}$$

$$
\boldsymbol{B} = \begin{bmatrix}
\mathbf{0} & \mathbf{0} & \cdots & \mathbf{0} & \mathbf{B}_{c,t_0} & \mathbf{B}_{v,t_0} & \mathbf{0} & \cdots & \mathbf{0} \\
\mathbf{A}_{\Delta t_{1,j}} & \mathbf{0} & \cdots & \mathbf{0} & \mathbf{B}_{c,\Delta t_{1,j}} & \mathbf{B}_{v,\Delta t_{1,j}} & \mathbf{0} & \cdots & \mathbf{0} \\
\mathbf{0} & \mathbf{A}_{\Delta t_{2,j}} & \cdots & \mathbf{0} & \mathbf{B}_{c,\Delta t_{2,j}} & \mathbf{0} & \mathbf{B}_{v,\Delta t_{2,j}} & \cdots & \mathbf{0} \\
\vdots & \vdots & \ddots & \vdots & \vdots & \vdots & \vdots & \ddots & \vdots \\
\mathbf{0} & \mathbf{0} & \cdots & \mathbf{A}_{\Delta t_{T-1,j}} & \mathbf{B}_{c,\Delta t_{T-1,j}} & \mathbf{0} & \mathbf{0} & \cdots & \mathbf{B}_{v,\Delta t_{T-1,j}} \\
\mathbf{0} & \mathbf{0} & \cdots & \mathbf{0} & \mathbf{0} & \mathbf{0} & \mathbf{0} & \cdots & \mathbf{0} \\
\mathbf{0} & \mathbf{0} & \cdots & \mathbf{0} & \mathbf{0} & \mathbf{0} & \mathbf{0} & \cdots & \mathbf{0} \\
\mathbf{0} & \mathbf{0} & \cdots & \mathbf{0} & \mathbf{0} & \mathbf{0} & \mathbf{0} & \cdots & \mathbf{0} \\
\mathbf{0} & \mathbf{0} & \cdots & \mathbf{0} & \mathbf{0} & \mathbf{0} & \mathbf{0} & \cdots & \mathbf{0}
\end{bmatrix},
$$

$$
\boldsymbol{\Psi} = \begin{bmatrix}
\boldsymbol{\Phi}_{\mathbf{x}(t_0)} & \mathbf{0} & \cdots & \mathbf{0} & \boldsymbol{\Phi}_{\mathbf{x}(t_0),\mathbf{u}} \\
\mathbf{0} & \mathbf{Q}_{\Delta t_{1,j}} & \cdots & \mathbf{0} & \mathbf{0} \\
\vdots & \vdots & \ddots & \vdots & \vdots \\
\mathbf{0} & \mathbf{0} & \cdots & \mathbf{Q}_{\Delta t_{T-1,j}} & \mathbf{0} \\
\boldsymbol{\Phi}'_{\mathbf{x}(t_0),\mathbf{u}} & \mathbf{0} & \cdots & \mathbf{0} & \boldsymbol{\Phi}_{\mathbf{u}}
\end{bmatrix},
$$

$$
\boldsymbol{\Lambda} = \begin{bmatrix}
\mathbf{C}_{t_{0,j}} & \mathbf{0} & \cdots & \mathbf{0} & \mathbf{D}_{c,t_{0,j}} & \mathbf{D}_{v,t_{0,j}} & \mathbf{0} & \cdots & \mathbf{0} \\
\mathbf{0} & \mathbf{C}_{t_{1,j}} & \cdots & \mathbf{0} & \mathbf{D}_{c,t_{1,j}} & \mathbf{0} & \mathbf{D}_{v,t_{1,j}} & \cdots & \mathbf{0} \\
\vdots & \vdots & \ddots & \vdots & \vdots & \vdots & \vdots & \ddots & \vdots \\
\mathbf{0} & \mathbf{0} & \cdots & \mathbf{C}_{t_{T-1,j}} & \mathbf{D}_{c,t_{T-1,j}} & \mathbf{0} & \mathbf{0} & \cdots & \mathbf{D}_{v,t_{T-1,j}} \\
\mathbf{0} & \mathbf{0} & \cdots & \mathbf{0} & \mathbf{I} & \mathbf{0} & \mathbf{0} & \cdots & \mathbf{0} \\
\mathbf{0} & \mathbf{0} & \cdots & \mathbf{0} & \mathbf{0} & \mathbf{I} & \mathbf{0} & \cdots & \mathbf{0} \\
\mathbf{0} & \mathbf{0} & \cdots & \mathbf{0} & \mathbf{0} & \mathbf{0} & \mathbf{I} & \cdots & \mathbf{0} \\
\mathbf{0} & \mathbf{0} & \cdots & \mathbf{0} & \mathbf{0} & \mathbf{0} & \mathbf{0} & \ddots & \mathbf{0} \\
\mathbf{0} & \mathbf{0} & \cdots & \mathbf{0} & \mathbf{0} & \mathbf{0} & \mathbf{0} & \cdots & \mathbf{I}
\end{bmatrix},
$$

$$
\boldsymbol{\Theta} = \begin{bmatrix}
\mathbf{R}_{t_{0,j}} & & \boldsymbol{\Theta}_{\mathrm{cov}} & & \mathbf{0} \\
& \mathbf{R}_{t_{1,j}} & & & \mathbf{0} \\
\boldsymbol{\Theta}'_{\mathrm{cov}} & & \ddots & & \vdots \\
& & & \mathbf{R}_{t_{T-1,j}} & \mathbf{0} \\
\mathbf{0} & \mathbf{0} & \cdots & \mathbf{0} & \mathbf{0}
\end{bmatrix}. \tag{1.34}
$$

For the crucial property of measurement invariance, we need to specify

$$
\begin{aligned}
\mathbf{C}_{t_{0,j}} &= \mathbf{C}_{t_{1,j}} = \cdots = \mathbf{C}_{t_{T-1,j}} = \mathbf{C}, \\
\mathbf{D}_{c,t_{0,j}} &= \mathbf{D}_{c,t_{1,j}} = \cdots = \mathbf{D}_{c,t_{T-1,j}} = \mathbf{D}_c, \\
\mathbf{D}_{v,t_{0,j}} &= \mathbf{D}_{v,t_{1,j}} = \cdots = \mathbf{D}_{v,t_{T-1,j}} = \mathbf{D}_v, \\
\mathbf{R}_{t_{0,j}} &= \mathbf{R}_{t_{1,j}} = \cdots = \mathbf{R}_{t_{T-1,j}} = \mathbf{R}.
\end{aligned}
\tag{1.35}
$$

Although measurement invariance is important for substantive reasons, statistically speaking, the assumption of strict measurement invariance may be relaxed if necessary. An additional advantage of the SEM approach is the possibility of specifying measurement error covariances across time in $\mathbf{\Theta}_{\mathrm{cov}}$. This can be done in ctsem by the MANIFESTTRAIT option. Measurement instruments often measure specific aspects, which they do not have in common with other instruments and can be taken care of by freeing corresponding elements in $\mathbf{\Theta}_{\mathrm{cov}}$. In view of identification, however, one should be very cautious in choosing elements of $\mathbf{\Theta}_{\mathrm{cov}}$ to be freed.

1.4 Analysis of Sunspot Data: CARMA(2,1) on $N = 1$, $T = 167$

The Wolfer sunspot data from 1749 to 1924 is a famous time series of the number of sunspots that has been analyzed by many authors. Several of them applied a CARMA(2,1) model to the series, the results of which are summarized in Table 1.1 together with the results of ctsem. The sunspot data, which are directly available in R (R Core Team 2015) by "sunspot.year," are monthly data averaged over the years 1700–1988. The data from 1749 to 1924 are analyzed by the authors in Table 1.1 as a stationary series. Stationarity means that the initial means and (co)variances are in the stable equilibrium position of the model, which is defined in terms of the other model parameters. An option of ctsem lets the initial means and (co)variances be constrained in terms of these parameters. ctsem version 1.1.6 and R version 3.3.2 were used for our analyses.[4]

As shown in Table 1.1, differences in parameter estimates between ctsem and previously reported analyses are small and likely caused by numerical imprecision. The most recent estimates by Tómasson (2011) and our results are closer to each other than to the older ones reported by Phadke and Wu (1974) and Singer (1991). The results in Table 1.1 imply an oscillatory movement of the sunspot numbers. Complex eigenvalues of the matrix **A** lead to oscillatory movements. The

[4]The programming code of the analysis is available as supplementary material at the book website http://www.springer.com/us/book/9783319772189.

Table 1.1 Parameter estimates of CARMA(2,1) model on yearly sunspot data 1749–1924 by several authors and ctsem; $N = 1, T = 176$

	Parameter				
	f_0	f_1	g_0	g_1	μ
Phadke and Wu (1974)	−0.359	−0.327	15.70	9.94	44.8
Singer (1991)	−0.360	−0.330	15.72	9.94	44.6
Tómasson (2011)	−0.357	−0.327	15.52	10.01	44.8
ctsem (Driver et al. 2017)	−0.357	−0.327	15.52	10.02	44.9

eigenvalues of $\mathbf{A} = \begin{bmatrix} 0 & 1 \\ f_0 & f_1 \end{bmatrix}$ are

$$\lambda_{1,2} = f_1/2 \pm j\sqrt{-f_1^2/4 - f_0}, \tag{1.36}$$

where j is the imaginary number $\sqrt{-1}$. The eigenvalues are complex, because $-{f_1}^2/4 > f_0$ for all 4 sets of parameter estimates in Table 1.1. The period of the oscillation Tp is computed by

$$Tp = 2\pi/\sqrt{-{f_1}^2/4 - f_0} \tag{1.37}$$

resulting in a period of 10.9 years for all 4 sets of parameter estimates, despite the small differences in parameter estimates.

1.5 Conclusion

As noted by Prado and West (2010), “In many statistical models the assumption that the observations are realizations of independent random variables is key. In contrast, time series analysis is concerned with describing the dependence among the elements of a sequence of random variables” (p. 1). Without doubt, SEM for a long period took position in the first group of models, which hampered the development of $N = 1$ modeling and time series analysis in an SEM context. The present chapter attempts to reconcile both perspectives by putting time series of independently drawn subjects in one and the same overall SEM model, while using continuous time state space modeling to simultaneously account for the dependence between observations in each time series over time. The present article explained in detail how this may be achieved for first- and higher-order CARMA(p, q) models in an extended SEM framework.

Attempts to combine time series of different subjects in a common model are rare in traditional time series analysis and state space modeling. A first, rather isolated proposal was done by Goodrich and Caines (1979). They call a data set consisting of

$N > 1$ time series "cross-sectional," thereby using this term in a somewhat different meaning from what is customary in social science. They give a consistency proof for state space model parameter estimates in this kind of data in which "the number T of observations on the transient behavior is fixed but the number N of independent cross-sectional samples tends to infinity" (p. 403). As a matter of fact, an important advantage of $N > 1$ models is to not be forced to $T \rightarrow \infty$ asymptotics, which at least in the social sciences is often unrealistic. Arguably, there are not many processes with, for example, exactly the same parameter values over the whole time range until infinity. As argued by Yu (2014, p. 738), an extra advantage offered by continuous time modeling in this respect is that asymptotics can be applied on the time dimension, even if T is taken as fixed. Supposing the discretely observed data to be recorded at $0, \Delta t, 2\Delta t, n\Delta t (= T)$, this so-called "in-fill" asymptotics takes T as fixed but lets $n \rightarrow \infty$ in continuous time. By letting N as well as n go to infinity, a kind of double asymptotics results, which may be particularly useful for typical applications in the social sciences, where it is often hard to argue that T will approach infinity.

References

Arnold, L. (1974). *Stochastic differential equations*. New York: Wiley.

Bergstrom, A. R. (1984). Continuous time stochastic models and issues of aggregation over time. In Z. Griliches & M. D. Intriligator (Eds.), *Handbook of econometrics* (Vol. 2, pp. 1145–1212). Amsterdam: North-Holland. https://doi.org/10.1016/S1573-4412(84)02012-2

Boker, S. M. (2001). Differential structural equation modeling of intraindividual variability. In L. M. Collins & A. G. Sayer (Eds.), *New methods for the analysis of change* (pp. 5–27). Washington, DC: American Psychological Association. https://doi.org/10.1007/s11336-010-9200-6

Boker, S. M., Neale, M., Maes, H., Wilde, M., Spiegel, M., Brick, T., ... Fox, J. (2011). OpenMx: An open source extended structural equation modeling framework. *Psychometrika, 76*(2), 306–317. https://doi.org/10.1007/s11336-010-9200-6

Boker, S. M., Neale, M., & Rausch, J. (2004). Latent differential equation modeling with multivariate multi-occasion indicators. In K. van Montfort, J. H. L. Oud, & A. Satorra (Eds.), *Recent developments on structural equation models* (pp. 151–174). Amsterdam: Kluwer.

Bollen, K. A. (1989). *Structural equations with latent variables*. New York: Wiley. https://doi.org/10.1002/9781118619179

Box, G. E. P., & Jenkins, G. M. (1970). *Time serie analysis: Forecasting and control*. Oakland, CA: Holden-Day.

Brockwell, P. J. (2004). Representations of continuous-time ARMA processes. *Journal of Applied Probability, 41*(a), 375–382. https://doi.org/10.1017/s0021900200112422

Caines, P. E. (1988). *Linear stochastic systems*. New York: Wiley.

Chan, K., & Tong, H. (1987). A note on embedding a discrete parameter ARMA model in a continuous parameter ARMA model. *Journal of Time Series Analysis, 8*, 277–281. https://doi.org/10.1111/j.1467-9892.1987.tb00439.x

Deistler, M. (1985). General structure and parametrization of arma and state-space systems and its relation to statistical problems. In E. J. Hannan, P. R. Krishnaiah, & M. M. Rao (Eds.), *Handbook of statistics: Volume 5. Time series in the time domain* (pp. 257–277). Amsterdam: North Holland. https://doi.org/10.1016/s0169-7161(85)05011-8

Driver, C. C., Oud, J. H., & Voelkle, M. C. (2017). Continuous time structural equation modeling with R package ctsem. *Journal of Statistical Software, 77*, 1–35. https://doi.org/10.18637/jss.v077.i05

Durbin, J., & Koopman, S. J. (2001). *Time series analysis by state space methods*. Oxford: Oxford University Press. https://doi.org/10.1093/acprof:oSo/9780199641178.001.0001

Gasimova, F., Robitzsch, A., Wilhelm, O., Boker, S. M., Hu, Y., & Hülür, G. (2014). Dynamical systems analysis applied to working memory data. *Frontiers in Psychology, 5*, 687 (Advance online publication). https://doi.org/10.3389/fpsyg.2014.00687

Goodrich, R. L., & Caines, P. (1979). Linear system identification from nonstationary cross-sectional data. *IEEE Transactions on Automatic Control, 24*, 403–411. https://doi.org/10.1109/TAC.1979.1102037

Gottman, J. M. (1981). *Time-series analysis: A comprehensive introduction for social scientists*. Cambridge: Cambridge University Press.

Hamaker, E. L., Dolan, C. V., & Molenaar, C. M. (2003). ARMA-based SEM when the number of time points T exceeds the number of cases N: Raw data maximum likelihood. *Structural Equation Modeling, 10*, 352–379. https://doi.org/10.1207/s15328007sem1003-2

Hamerle, A., Nagl, W., & Singer, H. (1991). Problems with the estimation of stochastic differential equations using structural equations models. *Journal of Mathematical Sociology, 16*(3), 201–220. https://doi.org/10.1080/0022250X.1991.9990088

Hannan, E. J., & Deistler, M. (1988). *The statistical theory of linear systems*. New York: Wiley.

Hansen, L. P., & Sargent, T. J. (1983). The dimensionality of the aliasing problem. *Econometrica, 51*, 377–388. https://doi.org/10.2307/1911996

Harvey, A. C. (1981). *Time series models*. Oxford: Philip Allen.

Harvey, A. C. (1989). *Forecasting, structural time series models and the Kalman filter*. Cambridge: Cambridge University Press.

Higham, N. J. (2009). The scaling and squaring method for the matrix exponential revisited. *SIAM Review, 51*, 747–764. https://doi.org/10.1137/090768539

Jones, R. H. (1993). *Longitudinal data with serial correlation; a state space approach*. London: Chapman & Hall. https://doi.org/10.1007/978-1-4899-4489-4

Jöreskog, K. G. (1973). A general method for estimating a linear structural equation system. In A. S. Goldberger & O. D. Duncan (Eds.), *Structural equation models in the social sciences* (pp. 85–112). New York: Seminar Press.

Jöreskog, K. G. (1977). Structural equation models in the social sciences: Specification, estimation and testing. In P. R. Krishnaiah (Ed.), *Applications of statistics* (pp. 265–287). Amsterdam: North Holland.

Jöreskog, K. G., & Sörbom, D. (1976). *LISREL III: Estimation of linear structural equation systems by maximum likelihood methods: A FORTRAN IV program*. Chicago: National Educational Resources.

Jöreskog, K. G., & Sörbom, D. (1996). *LISREL 8: User's reference guide*. Chicago: Scientific Software International.

Kalman, R. E. (1960). A new approach to linear filtering and prediction problems. *Journal of Basic Engineering, 82*, 35–45 (Trans. ASME, ser. D).

Kuo, H. H. (2006). *Introduction to stochastic integration*. New York: Springer.

Ljung, L. (1985). Estimation of parameters in dynamical systems. In E. J. Hannan, P. R. Krishnaiah, & M. M. Rao (Eds.), *Handbook of statistics: Volume 5. Time series in the time domain* (pp. 189–211). Amsterdam: North Holland. https://doi.org/10.1016/s0169-7161(85)05009-x

Lütkepohl, H. (1991). *Introduction to multiple time series analysis*. Berlin: Springer.

McCleary, R., & Hay, R. (1980). *Applied time series analysis for the social sciences*. Beverly Hills: Sage.

Mills, T. C. (2012). *A very British affair: Six Britons and the development of time series analysis during the 20th century*. Basingstoke: Palgrave Macmillan.

Neale, M. C. (1997). *Mx: Statistical modeling* (4th ed.). Richmond, VA: Department of Psychiatry.

Neale, M. C., Hunter, M. D., Pritikin, J. N., Zahery, M., Brick, T. R., Kirkpatrick, R. M., ... Boker, S. M. (2016). OpenMx 2.0: Extended structural equation and statistical modeling. *Psychometrika, 81*, 535–549. https://doi.org/10.1007/s11336-014-9435-8

Oud, J. H. L. (1978). *Systeem-methodologie in sociaal-wetenschappelijk onderzoek [Systems methodology in social science research]*. Unpublished doctoral dissertation, Radboud University Nijmegen, Nijmegen.

Oud, J. H. L. (2007). Comparison of four procedures to estimate the damped linear differential oscillator for panel data. In K. van Montfort, J. H. L. Oud, & A. Satorra (Eds.), *Longitudinal models in the behavioral and related sciences* (pp. 19–39). Mahwah, NJ: Erlbaum.

Oud, J. H. L., & Delsing, M. J. M. H. (2010). Continuous time modeling of panel data by means of SEM. In K. van Montfort, J. H. L. Oud, & A. Satorra (Eds.), *Longitudinal research with latent variables* (pp. 201–244). New York: Springer. https://doi.org/10.1007/978-3-642-11760-2-7

Oud, J. H. L., & Jansen, R. A. R. G. (2000). Continuous time state space modeling of panel data by means of SEM. *Psychometrika, 65*, 199–215. https://doi.org/10.1007/BF02294374

Oud, J. H. L., & Singer, H. (2008). Continuous time modeling of panel data: SEM versus filter techniques. *Statistica Neerlandica, 62*, 4–28. https://doi.org/10.1111/j.1467-9574.2007.00376.x

Oud, J. H. L., van den Bercken J. H., & Essers, R. J. (1990). Longitudinal factor score estimation using the Kalman filter. *Applied Psychological Measurement, 14*, 395–418. https://doi.org/10.1177/014662169001400406

Oud, J. H. L., & Voelkle, M. C. (2014). Do missing values exist? Incomplete data handling in cross-national longitudinal studies by means of continuous time modeling. *Quality & Quantity, 48*, 3271–3288. https://doi.org/10.1007/s11135-013-9955-9

Oud, J. H. L., Voelkle, M. C., & Driver, C. C. (2018). SEM based CARMA time series modeling for arbitrary *N*. *Multivariate Behavioral Research, 53*(1), 36–56. https://doi.org/10.1080/00273171.1383224

Phadke, M., & Wu, S. (1974). Modeling of continuous stochastic processes from discrete observations with application to sunspots data. *Journal of the American Statistical Association, 69*, 325–329. https://doi.org/10.1111/j.1467-9574.2007.00376.x

Phillips, P. C. B. (1973). The problem of identification in finite parameter continuous time models. In A. R. Bergstrom (Ed.), *Statistical inference in continuous time models* (pp. 135–173). Amsterdam: North-Holland.

Prado, R., & West, M. (2010). *Time series: Modeling, computation, and inference*. Boca Raton: Chapman & Hall.

R Core Team. (2015). *R: A language and environment for statistical computing*. Vienna, Austria: R Foundation for Statistical Computing. Retrieved from http://www.R-project.org/

Shumway, R. H., & Stoffer, D. (2000). *Time series analysis and its applications*. New York: Springer. https://doi.org/10.1007/978-1-4419-7865-3

Singer, H. (1990). *Parameterschätzung in zeitkontinuierlichen dynamischen Systemen [Parameter estimation in continuous time dynamic systems]*. Konstanz: Hartung-Gorre.

Singer, H. (1991). *LSDE-A program package for the simulation, graphical display, optimal filtering and maximum likelihood estimation of linear stochastic differential equations: User's guide*. Meersburg: Author.

Singer, H. (1992). The aliasing-phenomenon in visual terms. *Journal of Mathematical Sociology, 17*, 39–49. https://doi.org/10.1080/0022250X.1992.9990097

Singer, H. (1998). Continuous panel models with time dependent parameters. *Journal of Mathematical Sociology, 23*, 77–98.

Singer, H. (2010). SEM modeling with singular moment part I: ML estimation of time series. *Journal of Mathematical Sociology, 34*, 301–320. https://doi.org/10.1080/0022250X.2010.509524

Singer, H. (2012). SEM modeling with singular moment part II: ML-estimation of sampled stochastic differential equations. *Journal of Mathematical Sociology, 36*, 22–43. https://doi.org/10.1080/0022250X.2010.532259

Steele, J. S., & Ferrer, E. (2011a). Latent differential equation modeling of self-regulatory and coregulatory affective processes. *Multivariate Behavioral Research, 46*, 956–984. https://doi.org/10.1080/00273171.2011.625305

Steele, J. S., & Ferrer, E. (2011b). Response to Oud & Folmer: Randomness and residuals. *Multivariate Behavioral Research, 46*, 994–1003. https://doi.org/10.1080/00273171.625308

Tómasson, H. (2011). *Some computational aspects of Gaussian CARMA modelling*. Vienna: Institute for Advanced Studies.

Tómasson, H. (2015). Some computational aspects of Gaussian CARMA modelling. *Statistical Computation, 25*, 375–387. https://doi.org/10.1007/s11222-013-9438-9

Tsai, H., & Chan, K.-S. (2000). A note on the covariance structure of a continuous-time ARMA process. *Statistica Sinica, 10*, 989–998.

Voelkle, M. C., & Oud, J. H. L. (2013). Continuous time modelling with individually varying time intervals for oscillating and non-oscillating processes. *British Journal of Mathematical and Statistical Psychology, 66*, 103–126. https://doi.org/10.1111/j.2044-8317.2012.02043.x

Voelkle, M. C., Oud, J. H. L., Davidov, E., & Schmidt, P. (2012a). An SEM approach to continuous time modeling of panel data: Relating authoritarianism and anomia. *Psychological Methods, 17*, 176–192. https://doi.org/10.1037/a0027543

Voelkle, M. C., Oud, J. H. L., Oertzen, T. von, & Lindenberger, U. (2012b). Maximum likelihood dynamic factor modeling for arbitrary *N* and *T* using SEM. *Structural Equation Modeling: A Multidisciplinary Journal, 19*, 329–350.https://doi.org/10.1080/10705511.2012.687656

Wolf, E. J., Harrington, K. M., Clark, S. L., & Miller, M. W. (2013). Sample size requirements for structural equation models: An evaluation of power, bias, and solution propriety. *Educational and Psychological Measurement, 73*, 913–934. https://doi.org/10.1177/0013164413495237

Yu, J. (2014). Econometric analysis of continuous time models: A survey of Peter Philips's work and some new results. *Econometric Theory, 30*, 737–774. https://doi.org/10.1017/S0266466613000467

Zadeh, L. A., & Desoer, C. A. (1963). *Linear system theory: The state space approach*. New York: McGraw-Hill.

Chapter 2
A Continuous-Time Approach to Intensive Longitudinal Data: What, Why, and How?

Oisín Ryan, Rebecca M. Kuiper, and Ellen L. Hamaker

2.1 Introduction

The increased availability of intensive longitudinal data—such as obtained with ambulatory assessments, experience sampling, ecological momentary assessments, and electronic diaries—has opened up new opportunities for researchers to investigate the *dynamics* of psychological *processes*, that is, the way psychological variables evolve, vary, and relate to one another over time (cf. Bolger and Laurenceau 2013; Chow et al. 2011; Hamaker et al. 2005). A useful concept in this respect is that of people being dynamic systems whose current state depends on their preceding states. For instance, we may be interested in the relationship between momentary stress and anxiety. We can think of stress and anxiety as each defining an axis in a two-dimensional space, and let the values of stress and anxiety at each moment in time define a position in this space. Over time, the point that represents a person's momentary stress and anxiety moves through this two-dimensional space, and our goal is to understand the lawfulness that underlies these movements.

Electronic Supplementary Material The online version of this article (https://doi.org/10.1007/978-3-319-77219-6_2) contains supplementary material, which is available to authorized users.

O. Ryan (✉) · R. M. Kuiper · E. L. Hamaker
Department of Methodology and Statistics, Faculty of Social and Behavioural Sciences, Utrecht University, Utrecht, The Netherlands
e-mail: o.ryan@uu.nl; r.m.kuiper@uu.nl; e.l.hamaker@uu.nl

K. van Montfort et al. (eds.), *Continuous Time Modeling in the Behavioral and Related Sciences*, https://doi.org/10.1007/978-3-319-77219-6_2

There are two frameworks that can be used to describe such movements: (1) the *discrete-time* (DT) framework, in which the passage of time is treated in discrete steps, and (2) the *continuous-time* (CT) framework, in which time is viewed as a continuous variable. Most psychological researchers are at least somewhat familiar with the DT approach, as it is the basis of the vast majority of longitudinal models used in the social sciences. In contrast, CT models have gained relatively little attention in fields such as psychology: This is despite the fact that many psychological researchers have been advocating their use for a long time, claiming that the CT approach overcomes practical and conceptual problems associated with the DT approach (e.g., Boker 2002; Chow et al. 2005; Oud and Delsing 2010; Voelkle et al. 2012). We believe there are two major hurdles that hamper the adoption of the CT approach in psychological research. First, the estimation of CT models typically requires the use of specialized software (cf. Chow et al. 2007; Driver et al. 2017; Oravecz et al. 2016) or unconventional use of more common software (cf. Boker et al. 2010a, 2004; Steele and Ferrer 2011). Second, the results from CT models are not easily understood, and researchers may not know how to interpret and represent their findings.

Our goal in this chapter is twofold. First, we introduce readers to the perspective of psychological processes as CT processes; we focus on the *conceptual reasons* for which the CT perspective is extremely valuable in moving our understanding of processes in the right direction. Second, we provide a didactical description of how to interpret the results of a CT model, based on our analysis of an empirical dataset. We examine the direct interpretation of model parameters, examine different ways in which the dynamics described by the parameters can be understood and visualized, and explain how these are related to one another throughout. We will restrict our primary focus to the simplest DT and CT models, that is, first-order (vector) autoregressive models and first-order differential equations.

The organization of this chapter is as follows. First, we provide an overview of the DT and CT models under consideration. Second, we discuss the practical and conceptual reasons researchers should adopt a CT modeling approach. Third, we illustrate the use and interpretation of the CT model using a bivariate model estimated from empirical data. Fourth, we conclude with a brief discussion of more complex models which may be of interest to substantive researchers.

2.2 Two Frameworks

The relationship between the DT and CT frameworks has been discussed extensively by a variety of authors. Here, we briefly reiterate the main issues, as this is vital to the subsequent discussion. For a more thorough treatment of this topic, the reader is referred to Voelkle et al. (2012). We begin by presenting the first-order vector autoregressive model in DT, followed by the presentation of the first-order differential equation in CT. Subsequently, we show how these models are connected and discuss certain properties which can be inferred from the parameters of the model. For simplicity, and without loss of generalization, we describe single-subject DT and CT models, in terms of observed variables. Extensions for multiple-subject data, and extensions for latent variables, in which the researchers can account for measurement error by additionally specifying a measurement model, are readily available (in the case of CT models, see, e.g., Boker et al. 2004; Driver et al. 2017; Oravecz and Tuerlinckx 2011).

2.2.1 *The Discrete-Time Framework*

DT models are those models for longitudinal data in which the passage of time is accounted for only with regard to the order of observations. If the true data-generating model for a process is a DT model, then the process only takes on values at discrete moments in time (e.g., hours of sleep per day or monthly salary). Such models are typically applied to data that consist of some set of variables measured repeatedly over time. These measurements typically show autocorrelation, that is, serial dependencies between the observed values of these variables at consecutive measurement occasions. We can model these serial dependencies using (discrete-time) autoregressive equations, which describe the relationship between the values of variables observed at consecutive measurement occasions.

The specific type of DT model that we will focus on in this chapter is the first-order vector autoregressive (VAR(1)) model (cf. Hamilton 1994). Given a set of V variables of interest measured at N different occasions, the VAR(1) describes the relationship between $\boldsymbol{y}_\tau$, a $V \times 1$ column vector of variables measured at occasion τ (for $\tau = 2, \ldots, N$) and the values those same variables took on at the preceding measurement occasion, the vector $\boldsymbol{y}_{\tau-1}$. This model can be expressed as

$$\boldsymbol{y}_\tau = \boldsymbol{c} + \boldsymbol{\Phi}\, \boldsymbol{y}_{\tau-1} + \boldsymbol{\epsilon}_\tau, \tag{2.1}$$

where $\boldsymbol{\Phi}$ represents a $V \times V$ matrix with autoregressive and cross-lagged coefficients that regress $\boldsymbol{y}_\tau$ on $\boldsymbol{y}_{\tau-1}$. The $V \times 1$ column vector $\boldsymbol{\epsilon}_\tau$ represents the variable-specific random shocks or innovations at that occasion, which are normally distributed with mean zero and a $V \times V$ variance-covariance matrix $\boldsymbol{\Psi}$. Finally, $\boldsymbol{c}$ represents a $V \times 1$ column vector of intercepts.

In the case of a stationary process, the mean $\boldsymbol{\mu}$ and the variance-covariance matrix of the variables $\boldsymbol{y}_\tau$ (generally denoted $\boldsymbol{\Sigma}$) do not change over time.[1] Then, the vector $\boldsymbol{\mu}$ represents the long-run expected values of the random variables, $E(\boldsymbol{y}_\tau)$, and is a function of the vector of intercepts and the matrix with lagged regression

[1]The variance-covariance matrix of the variables $\boldsymbol{\Sigma}$ is a function of both the lagged parameters and the variance-covariance matrix of the innovations, $vec(\boldsymbol{\Sigma}) = (\boldsymbol{I} - \boldsymbol{\Phi} \otimes \boldsymbol{\Phi})^{-1} vec(\boldsymbol{\Psi})$, where $vec(.)$ denotes the operation of putting the elements of an $N \times N$ matrix into an $NN \times 1$ column matrix (Kim and Nelson 1999, p. 27).

coefficients, that is, $\boldsymbol{\mu} = (\boldsymbol{I} - \boldsymbol{\Phi})^{-1}\boldsymbol{c}$, where $\boldsymbol{I}$ is a $V \times V$ identity matrix (cf. Hamilton 1994). In terms of a V-dimensional dynamical system of interest, $\boldsymbol{\mu}$ represents the *equilibrium position* of the system. By definition, τ is limited to positive integers; that is, there is no 0.1th or 1.5th measurement occasion.

Both the single-subject and multilevel versions of the VAR(1) model have frequently been used to analyze intensive longitudinal data of psychological variables, including symptoms of psychopathology, such as mood- and affect-based measures (Bringmann et al. 2015, 2016; Browne and Nesselroade 2005; Moberly and Watkins 2008; Rovine and Walls 2006). In these cases, the autoregressive parameters ϕ_{ii} are often interpreted as reflecting the *stability*, *inertia*, or *carry-over* of a particular affect or behavior (Koval et al. 2012; Kuppens et al. 2010, 2012). The cross-lagged effects (i.e., the off-diagonal elements ϕ_{ij} for $i \neq j$) quantify the lagged relationships, sometimes referred to as the *spillover*, between different variables in the model. These parameters are often interpreted in substantive terms, either as predictive or Granger-causal relationships between different aspects of affect or behavior (Bringmann et al. 2013; Gault-Sherman 2012; Granger 1969; Ichii 1991; Watkins et al. 2007). For example, if the standardized cross-lagged effect of $y_{1,\tau-1}$ on $y_{2,\tau}$ is larger than the cross-lagged effect of $y_{2,\tau-1}$ on $y_{1,\tau}$, researchers may draw the conclusion that y_1 is the driving force or dominant variable of that pair (Schuurman et al. 2016). As such, substantive researchers are typically interested in the (relative) magnitudes and signs of these parameters.

2.2.2 The Continuous-Time Framework

In contrast to the DT framework, which treats values of processes indexed by observation τ, the CT framework treats processes as *functions* of the continuous variable time t: The processes being modeled are assumed to vary continuously with respect to time, meaning that these variables may take on values if observed at any imaginable moment. CT processes can be modeled using a broad class of differential equations, allowing for a wide degree of diversity in the types of dynamics that are being modeled. It is important to note that many DT models have a differential equation counterpart. For the VAR(1) model, the CT equivalent is the first-order stochastic differential equation (SDE), where stochastic refers to the presence of random innovations or shocks.

The first-order SDE describes how the position of the V-dimensional system at a certain point in time, $\boldsymbol{y}(t)$, relative to the equilibrium position $\boldsymbol{\mu}$, is related to the *rate of change* of the process with respect to time (i.e., $\frac{d\boldsymbol{y}(t)}{dt}$) in that same instant. The latter can also be thought of as a vector of *velocities*, describing in what direction and with what magnitude the system will move an instant later in time (i.e., the ratio of the change in position over some time interval, to the length of that time interval, as the length of the time interval approaches zero). The first-order SDE can be expressed as

$$\frac{d\boldsymbol{y}(t)}{dt} = \boldsymbol{A}(\boldsymbol{y}(t) - \boldsymbol{\mu}) + \boldsymbol{G}\frac{d\boldsymbol{W}(t)}{dt} \tag{2.2}$$

where $\boldsymbol{y}(t), \frac{d\boldsymbol{y}(t)}{dt}$ and $\boldsymbol{\mu}$ are $V \times 1$ column vectors described above, $\boldsymbol{y}(t) - \boldsymbol{\mu}$ represents the position as a deviation from the equilibrium, and the $V \times V$ matrix $\boldsymbol{A}$ represents the *drift* matrix relating $\frac{d\boldsymbol{y}(t)}{dt}$ to $(\boldsymbol{y}(t) - \boldsymbol{\mu})$. The diagonal elements of $\boldsymbol{A}$, relating the position in a certain dimension to the velocity in that same dimension, are referred to as auto-effects, while the off-diagonal elements are referred to as cross-effects. The second part on the right-hand side of Eq. (2.2) represents the stochastic part of the model: $\boldsymbol{W}(t)$ denotes the so-called Wiener process, broadly speaking a continuous-time analogue of a random walk. This stochastic element has a variance-covariance matrix $\boldsymbol{G}\boldsymbol{G}' = \boldsymbol{Q}$, which is often referred to as the *diffusion* matrix (for details see Voelkle et al. 2012).

The model representation in Eq. (2.2) is referred to as the differential form as it includes the derivative $\frac{d\boldsymbol{y}(t)}{dt}$. The same model can be represented in the integral form, in which the derivatives are integrated out, sometimes referred to as the solution of the derivative model. The integral form of this particular first-order differential equation is known as the CT-VAR(1) or Ornstein-Uhlenbeck model (Oravecz et al. 2011). In this form, we can describe the same system but now in terms of the positions of the system (i.e., the values the variables take on) at different points in time. For notational simplicity, we can represent $\boldsymbol{y}(t) - \boldsymbol{\mu}$ as $\boldsymbol{y^c}(t)$, denoting the position of the process as a deviation from its equilibrium.

The CT-VAR(1) model can be written as

$$\boldsymbol{y^c}(t) = \boldsymbol{e}^{\boldsymbol{A}\Delta t}\boldsymbol{y^c}(t - \Delta t) + \boldsymbol{w}(\Delta t) \tag{2.3}$$

where $\boldsymbol{A}$ has the same meaning as above, the $V \times 1$ vector $\boldsymbol{y^c}(t - \Delta t)$ represents the position as a deviation from equilibrium some time interval Δt earlier, $\boldsymbol{e}$ represents the matrix exponential function, and the $V \times 1$ column vector $\boldsymbol{w}(\Delta t)$ represents the stochastic innovations, the integral form of the Wiener process in Eq. (2.2). These innovations are normally distributed with a variance-covariance matrix that is a function of the time interval between measurements Δt, the drift matrix $\boldsymbol{A}$, and the diffusion matrix $\boldsymbol{Q}$ (cf. Voelkle et al. 2012).[2] As the variables in the model have been centered around their equilibrium, we omit any intercept term. The relationship between lagged variables, that is, the relationships between the positions of the centered variables in the multivariate space, separated by some time interval Δt, is an (exponential) function of the drift matrix $\boldsymbol{A}$ *and* the length of that time interval.

[2]Readers should note that there are multiple different possible ways to parameterize the CT stochastic process in integral form, and also multiple different notations used (e.g., Oravecz et al. 2011; Voelkle et al. 2012).

2.2.3 Relating DT and CT Models

It is clear from the integral form of the first-order SDE given in Eq. (2.3) that the relationship between lagged values of variables is dependent on the length of the time interval between these lagged values. As such, if the DT-VAR(1) model in Eq. (2.1) is fitted to data generated by the CT model considered here, then the autoregressive and cross-lagged effects matrix $\boldsymbol{\Phi}$ will be a function of the time interval Δt between the measurements. We denote this dependency by writing $\boldsymbol{\Phi}(\Delta t)$. This characteristic of the DT model has been referred to as the lag problem (Gollob and Reichardt 1987; Reichardt 2011).

The precise relationship between the CT-VAR(1) and DT-VAR(1) effect matrices is given by the well-known equality

$$\boldsymbol{\Phi}(\Delta t) = e^{\boldsymbol{A}\Delta t}. \tag{2.4}$$

Despite this relatively simple relationship, it should be noted that taking the exponential of a matrix is not equivalent to taking the exponential of each of the elements of the matrix. That is, any lagged effect parameter $\phi_{ij}(\Delta t)$, relating variable i and variable j across time points, is not only dependent on the corresponding CT cross-effect a_{ij} but is a nonlinear function of the interval and *every other element* of the matrix $\boldsymbol{A}$. For example, in the bivariate case, the DT cross-lagged effect of $y_1(t-\Delta t)$ on $y_2(t)$, denoted $\phi_{21}(\Delta t)$, is given by

$$\frac{a_{21}\left(e^{\frac{1}{2}\left(a_{11}+a_{22}+\sqrt{a_{11}^2+4a_{12}a_{21}-2a_{11}a_{22}+a_{22}^2}\right)\Delta t}-e^{\frac{1}{2}\left(a_{11}+a_{22}-\sqrt{a_{11}^2+4a_{12}a_{21}-2a_{11}a_{22}+a_{22}^2}\right)\Delta t}\right)}{\sqrt{a_{11}^2+4a_{12}a_{21}-2a_{11}a_{22}+a_{22}^2}} \tag{2.5}$$

where e represents the scalar exponential. In higher-dimensional models, these relationships quickly become intractable. For a derivation of Eq. (2.5), we refer readers to Appendix.

This complicated nonlinear relationship between the elements of $\boldsymbol{\Phi}$ and the time interval has major implications for applied researchers who wish to interpret the parameters of a DT-VAR(1) model in the substantive terms outlined above. In the general multivariate case, the size, sign, and relative strengths of both autoregressive and cross-lagged effects may differ depending on the value of the time interval used in data collection (Deboeck and Preacher 2016; Dormann and Griffin 2015; Oud 2007; Reichardt 2011). As such, conclusions that researchers draw regarding the stability of processes and the nature of how different processes relate to one another may differ greatly depending on the time interval used.

While the relationship in Eq. (2.4) describes the DT-VAR(1) effects matrix we would find given the data generated by a CT-VAR(1) model, the reader should note that not all DT-VAR(1) processes have a straightforward equivalent representation as a CT-VAR(1). For example, a univariate discrete-time AR(1) process with a negative autoregressive parameter cannot be represented as a CT-AR(1) process; as the exponential function is always positive, there is no A that satisfies Eq. (2.4) for $\Phi < 0$. As such, we can refer to DT-VAR(1) models with a CT-VAR(1) equivalent as those which exhibit "positive autoregression." We will focus throughout on the CT-VAR(1) as the data-generating model.[3]

2.2.4 Types of Dynamics: Eigenvalues, Stability, and Equilibrium

Both the DT-VAR(1) model and the CT-VAR(1) model can be used to describe a variety of different types of dynamic behavior. As the dynamic behavior of a system is always understood with regard to how the variables in the system move in relation to the equilibrium position, often dynamic behaviors are described by differentiating the type of equilibrium position or *fixed point* in the system (Strogatz 2014). In the general multivariate case, we can understand these different types of dynamic behavior or fixed points with respect to the *eigenvalues* of the effects matrices $\boldsymbol{A}$ or $\boldsymbol{\Phi}$ (see Appendix for a more detailed explanation of the relationship between these two matrices and eigenvalues). In this chapter we will focus on *stable* processes, in which, given a perturbation, the system of interest will inevitably return to the equilibrium position. We limit our treatment to these types of processes, because we believe these are most common and most relevant for applied researchers. A brief description of other types of fixed points and how they relate to the eigenvalues of the effects matrix $\boldsymbol{A}$ is given in the discussion section—for a more complete taxonomy, we refer readers to Strogatz (2014, p. 136).

In DT settings, stable processes are those for which the absolute values of the eigenvalues of $\boldsymbol{\Phi}$ are smaller than one. In DT applications researchers also typically discuss the need for *stationarity*, that is, time-invariant mean and variance, as introduced above. Stability of a process ensures that stationarity in relation to the mean and variance hold. For CT-VAR(1) processes, stability is ensured if the real parts of the eigenvalues of $\boldsymbol{A}$ are negative. It is interesting to note that the equilibrium position of stable processes can be related to our observed data in various ways: In some applications $\boldsymbol{\mu}$ is constrained to be equal to the mean of the observed values (e.g., Hamaker et al. 2005; Hamaker and Grasman 2015), while in others the equilibrium can be specified a priori or estimated to be equal to some (asymptotic) value (e.g., Bisconti et al. 2004).

[3]In general, there is no straightforward CT-VAR(1) representation of DT-VAR(1) models with real, negative eigenvalues. However it may be possible to specify more complex continuous-time models which do not exhibit positive autoregression. Notably, Fisher (2001) demonstrates how a DT-AR(1) model with negative autoregressive parameter can be modeled with the use of two continuous-time (so-called) Itô processes.

We can further distinguish between dynamic processes that have real eigenvalues, complex eigenvalues, or, in the case of systems with more than two variables, a mix of both. In the section "Making Sense of CT Models," we will focus on the interpretation of a CT-VAR(1) model with real, negative, non-equal eigenvalues. We can describe the equilibrium position of this system as a *stable node*. In the discussion section, we examine another type of system which has been the focus of psychological research, sometimes described as a damped linear oscillator (e.g., Boker et al. 2010b), in which the eigenvalues of $\boldsymbol{A}$ are complex, with a negative real part. The fixed point of such a system is described as a *stable spiral*. Further detail on the interpretation of these two types of systems is given in the corresponding sections.

2.3 Why Researchers Should Adopt a CT Process Perspective

There are both practical and theoretical benefits to CT model estimation over DT model estimation. Here we will discuss three of these practical advantages which have received notable attention in the literature. We then discuss the fundamental conceptual benefits of treating psychological processes as continuous-time systems.

The first practical benefit to CT model estimation is that the CT model deals well with observations taken at unequal intervals, often the case in experience sampling and ecological momentary assessment datasets (Oud and Jansen 2000; Voelkle and Oud 2013; Voelkle et al. 2012). Many studies use random intervals between measurements, for example, to avoid participant anticipation of measurement occasions, potentially resulting in unequal time intervals both within and between participants. The DT model, however, is based on the assumption of equally spaced measurements, and as such estimating the DT model from unequally spaced data will result in an estimated $\boldsymbol{\Phi}$ matrix that is a blend of different $\boldsymbol{\Phi}(\Delta t)$ matrices for a range of values of Δt.

The second practical benefit of CT modeling over DT modeling is that, when measurements are equally spaced, the lagged effects estimated by the DT models are not generalizable beyond the time interval used in data collection. Several different researchers have demonstrated that utilizing different time intervals of measurement can lead researchers to reach very different conclusions regarding the values of parameters in $\boldsymbol{\Phi}$ (Oud and Jansen 2000; Reichardt 2011; Voelkle et al. 2012). The CT model has thus been promoted as facilitating better comparisons of results between studies, as the CT effects matrix $\boldsymbol{A}$ is independent of time interval (assuming a sufficient frequency of measurement to capture the relevant dynamics).

Third, the application of CT models allows us to explore how cross-lagged effects are expected to change depending on the time interval between measurements, using the relationship expressed in Eq. (2.4). Some authors have used this relationship to identify the time interval at which cross-lagged effects are expected to reach a maximum (Deboeck and Preacher 2016; Dormann and Griffin 2015). Such information could be used to decide upon the "optimal" time interval that should be used in gathering data in future research.

While these practical concerns regarding the use of DT models for CT processes are legitimate, there may be instances in which alternative practical solutions can be used, without necessitating the estimation of a CT model. For instance, the problem of unequally spaced measurements in DT modeling can be addressed by defining a time grid and adding missing data to your observations, to make the occasions approximately equally spaced in time. Some simulation studies indicate that this largely reduces the bias that results from using DT estimation of unequally spaced data (De Haan-Rietdijk et al. 2017).

Furthermore, the issue of comparability between studies that use different time intervals can be solved, in certain circumstances, by a simple transformation of the estimated $\boldsymbol{\Phi}$ matrix, described in more detail by Kuiper and Ryan (2018). Given an estimate of $\boldsymbol{\Phi}(\Delta t)$, we can solve for the underlying $\boldsymbol{A}$ using Eq. (2.4). This is known as the "indirect method" of CT model estimation (Oud et al. 1993). However this approach cannot be applied in all circumstances, as it involves using the matrix logarithm, the inverse of the matrix exponential function. As the matrix logarithm function in the general case does not give a unique solution, this method is only appropriate if both the estimated $\boldsymbol{\Phi}(\Delta t)$ and true underlying $\boldsymbol{A}$ matrices have real eigenvalues only (for further discussion of this issue, see Hamerle et al. 1991).

However, the CT perspective has added value above and beyond the potential practical benefits discussed above. Multiple authors have argued that psychological phenomena, such as stress, affect, and anxiety, do not vary in discrete steps over time but likely vary and evolve in a continuous and smooth manner (Boker 2002; Gollob and Reichardt 1987). Viewing psychological processes as CT dynamic systems has important implications for the way we conceptualize the influence of psychological variables on each other. Gollob and Reichardt (1987) give the example of a researcher who is interested in the effect of taking aspirin on headaches: This effect may be zero shortly after taking the painkiller, substantial an hour or so later, and near zero again after 24 h. All of these results may be considered as accurately portraying the effect of painkillers on headaches *for a specific time interval*, although each of these intervals considered separately represents only a snapshot of the process of interest.

It is only through examining the underlying dynamic trajectories, and exploring how the cross-lagged relationships evolve and vary as a function of the time interval, that we can come to a more complete picture of the dynamic system of study. We believe that—while the practical benefits of CT modeling are substantial—the conceptual framework of viewing psychological variables as CT processes has the potential to transform longitudinal research in this field.

2.4 Making Sense of CT Models

In this section, we illustrate how researchers can evaluate psychological variables as dynamic CT processes by describing the interpretation of the drift matrix parameters $\boldsymbol{A}$. We describe multiple ways in which the dynamic behavior of the model in general, as well as specific model parameters, can be understood. In order to aid researchers who are unfamiliar with this type of analysis, we take a broad approach in which we incorporate the different ways in which researchers interested in dynamical systems and similar models interpret their results. For instance, Boker and colleagues (e.g., Boker et al. 2010b) typically interpret the differential form of the model directly; in the econometrics literature, it is typical to plot specific trajectories using impulse response functions (Johnston and DiNardo 1997); in the physics tradition, the dynamics of the system are inspected using vector fields (e.g., Boker and McArdle 1995); the work of Voelkle, Oud, and others (e.g., Deboeck and Preacher 2016; Voelkle et al. 2012) typically focuses on the integral form of the equation and visually inspecting the time interval dependency of lagged effects.

We will approach the interpretation of a single CT model from these four angles and show how they each represent complimentary ways to understand the same system. For ease of interpretation, we focus here on a bivariate system; the analysis of larger systems is addressed in the discussion section.

2.4.1 Substantive Example from Empirical Data

To illustrate the diverse ways in which the dynamics described by the CT-VAR(1) model can be understood, we make use of a substantive example. This example is based on our analysis of a publicly available single-subject ESM dataset (Kossakowski et al. 2017). The subject in question is a 57-year-old male with a history of major depression. The data consists of momentary, daily, and weekly items relating to affective states. The assessment period includes a double-blind phase in which the dosage of the participant's antidepression medication was reduced. We select only those measurements made in the initial phases of the study, before medication reduction; it is only during this period that we would expect the system of interest to be stable. The selected measurements consist of 286 momentary assessments over a period of 42 consecutive days. The modal time interval between momentary assessments was 1.766 h (interquartile range of 1.250–3.323).

For our analysis we selected two momentary assessment items, "I feel down" and "I am tired," which we will name Down (Do) and Tired (Ti), respectively. Feeling down is broadly related to assessments of negative affect (Meier and Robinson 2004), and numerous cross-sectional analyses have suggested a relationship between negative affect and feelings of physical tiredness or fatigue (e.g., Denollet and De Vries 2006). This dataset afforded us the opportunity to investigate the links between these two processes from a dynamic perspective. Each variable was

Table 2.1 Parameter estimates from the substantive example

Parameter	Value	Std. error
a_{11}	−0.995	0.250
a_{21}	0.375	0.441
a_{12}	0.573	0.595
a_{22}	−2.416	1.132
q_{11}	1.734	0.612
q_{21}	−0.016	0.650
q_{22}	4.606	1.374

standardized before the analysis to facilitate ease of interpretation of the parameter estimates. Positive values of Do indicate that the participant felt down more than average, negative values indicate below-average feelings of being down and likewise for positive and negative values of Ti.

The analysis was conducted using the *ctsem* package in R (Driver et al. 2017). Full details of the analysis, including R code, can be found in the online supplementary materials. Parameter estimates and standard errors are given in Table 2.1, including estimates of the stochastic part of the CT model, represented by the diffusion matrix $\boldsymbol{Q}$. The negative value of γ_{21} indicates that there is a negative covariance between the stochastic input and the rates of change of Do and Ti; in terms of the CT-VAR(1) representation, there is a negative covariance between the residuals of Do and Ti in the same measurement occasion. Further interpretation of the diffusion matrix falls beyond the scope of the current chapter. As the analysis is meant as an illustrative example only, we will throughout interpret the estimated drift matrix parameter as though they are true population parameters.

2.4.2 *Interpreting the Drift Parameters*

The drift matrix relating the processes Down ($Do(t)$) and Tired ($Ti(t)$) is given by

$$\boldsymbol{A} = \begin{bmatrix} -0.995 & 0.573 \\ 0.375 & -2.416 \end{bmatrix}. \tag{2.6}$$

As the variables are standardized, the equilibrium position is $\boldsymbol{\mu} = [0, 0]$ (i.e., $E[Do(t)] = E[Ti(t)] = 0$). The drift matrix $\boldsymbol{A}$ describes how the position of the system at any particular time t (i.e., $Do(t)$ and $Ti(t)$) relates to the *velocity* or *rate of change* of the process, that is, how the position of the process is changing. The system of equations which describe the dynamic system made up of Down and Tired is given by

$$\begin{bmatrix} E[\frac{dDo(t)}{dt}] \\ E[\frac{dTi(t)}{dt}] \end{bmatrix} = \begin{bmatrix} -0.995 & 0.573 \\ 0.375 & -2.416 \end{bmatrix} \begin{bmatrix} Do(t) \\ Ti(t) \end{bmatrix} \tag{2.7}$$

such that

$$E\left[\frac{dDo(t)}{dt}\right] = -0.995Do(t) + 0.573Ti(t) \tag{2.8}$$

$$E\left[\frac{dTi(t)}{dt}\right] = 0.375Do(t) - 2.416Ti(t) \tag{2.9}$$

where the rates of change of Down and Tired at any point in time are both dependent on the positions of both Down and Tired at that time.

Before interpreting any particular parameter in the drift matrix, we can determine the type of dynamic process under consideration by inspecting the eigenvalues of $\boldsymbol{A}$. The eigenvalues of $\boldsymbol{A}$ are $\lambda_1 = -2.554$ and $\lambda_2 = -0.857$; since both eigenvalues are negative, the process under consideration is stable. This means that if the system takes on a position away from equilibrium (e.g., due to a random shock from the stochastic part of the model on either Down or Tired), the system will inevitably return to its equilibrium position over time. It is for this reason that the equilibrium position or fixed point in stable systems is also described as the *attractor point*, and stable systems are described as *equilibrium-reverting*. As the eigenvalues of the system are real-valued as well as negative, the system returns to equilibrium with an *exponential decay*; when the process is far away from the equilibrium, it takes on a greater velocity, that is, moves faster toward equilibrium. We can refer to the type of fixed point in this system as a *stable node* (Strogatz 2014).

Typical of such an equilibrium-reverting process, we see negative CT auto-effects $a_{11} = -0.995$ and $a_{22} = -2.416$. This reflects that, if either variable in the system takes on a position away from the equilibrium, they will take on a velocity of opposite sign to this deviation, that is, a velocity which returns the process to equilibrium. For higher values of $Do(t)$, the rate of change of $Do(t)$ is of greater (negative) magnitude, that is, the velocity toward the equilibrium is higher. In addition, the auto-effect of $Ti(t)$ is more than twice as strong (in an absolute sense) as the auto-effect of $Do(t)$. If there were no cross-effects present, this would imply that $Ti(t)$ returns to equilibrium faster than $Do(t)$; however, as there are cross-effects present, such statements cannot be made in the general case from inspecting the auto-effects alone.

In this case the cross-effects of $Do(t)$ and $Ti(t)$ on each other's rates of change are positive rather than negative. Moreover, the cross-effect of $Ti(t)$ on the rate of change of $Do(t)$ ($a_{12} = 0.573$) is slightly stronger than the corresponding cross-effect of $Do(t)$ on the rate of change of $Ti(t)$ ($a_{21} = 0.375$). These cross-effects quantify the force that each component of the system exerts on the other. However, depending on what values each variable takes on at a particular point in time t, that is, the position of the system in each of the $Do(t)$ and $Ti(t)$ dimensions, this may translate to $Do(t)$ pushing $Ti(t)$ to return faster to its equilibrium or to deviate away from its equilibrium position and vice versa. To better understand both the cross-effects and auto-effects described by $\boldsymbol{A}$, it is helpful to visualize the possible trajectories of our two-dimensional system.

2.4.3 *Visualizing Trajectories*

We will now describe and apply two related tools which allow us to visualize the trajectories of the variables in our model over time: impulse response functions and vector fields. These tools can help us to understand the dynamic system we are studying, by exploring the dynamic behavior which results from the drift matrix parameters.

2.4.3.1 Impulse Response Functions

Impulse response functions (IRFs) are typically used in the econometrics literature to aid in making forecasts based on a DT-VAR model. The idea behind this is to allow us to explore how an impulse to one variable in the model at occasion τ will affect the values of both itself and the other variables in the model at occasions $\tau+1$, $\tau+2$, $\tau+3$, and so on. In the stochastic systems we focus on in this chapter, we can conceptualize these impulses as random perturbations or innovations or alternatively as external interventions in the system.[4] IRFs thus represent the trajectories of the variables in the model over time, following a particular impulse, assuming no further stochastic innovations (see Johnston and DiNardo 1997, Chapter 9).

To specify impulses in an IRF, we generally assign a value to a single variable in the system at some initial occasion, $y_{i,\tau}$. The corresponding values of the other variables at the initial occasion $y_{j,\tau}$, $j \neq i$ are usually calculated based on, for instance, the covariance in the stochastic innovations, $\boldsymbol{\Psi}$, or the stable covariance between the processes $\boldsymbol{\Sigma}$. Such an approach is beneficial in at least two ways: first, it allows researchers to specify impulses which are more likely to occur in an observed dataset; second, it aids researchers in making more accurate future predictions or forecasts. For a further discussion of this issue in relation to DT-VAR models, we refer the reader to Johnston and DiNardo (1997, pp. 298–300). Below, we will take a simplified approach and specify bivariate impulses at substantively interesting values.

The IRF can easily be extended for use with the CT-VAR(1) model. We can calculate the impulse response of our system by taking the integral form of the CT-VAR(1) model in Eq. (2.3) and (a) plugging in the $\boldsymbol{A}$ matrix for our system, (b) choosing some substantively interesting set of impulses $\boldsymbol{y}(t = 0)$, and (c) calculating $\boldsymbol{y}(t)$ for increasing values of $t > 0$. To illustrate this procedure, we will specify four substantively interesting sets of impulses. The four sets of impulses shown here include $\boldsymbol{y}(0) = [1, 0]$, reflecting what happens when $Do(0)$ takes on a positive value 1 standard deviation above the persons mean, while $Ti(0)$ is at equilibrium; $\boldsymbol{y}(0) = [0, 1]$ reflecting when $Ti(0)$ takes on a positive value of corresponding size while $Do(0)$ is at equilibrium; $\boldsymbol{y}(0) = [1, 1]$ reflecting what

[4] Similar functions can be used for deterministic systems (those without a random innovation part); however in these cases the term *initial value* is more typically used.

happens when $Do(0)$ and $Ti(0)$ both take on values 1 standard deviation above the mean; and $\mathbf{y}(0) = [1, -1]$ reflecting what happens when $Do(0)$ and $Ti(0)$ take on values of equal magnitude but opposite valence (1SD more and 1SD less than the mean, respectively). Figure 2.1a–d contains the IRFs for both processes in each of these four scenarios.

Examining the IRFs shows us the equilibrium-reverting behavior of the system: Given any set of starting values, the process eventually returns, in an exponential fashion, to the bivariate equilibrium position where both processes take on a value of zero.

In Figure 2.1a, we can see that when $Ti(t)$ is at equilibrium and $Do(0)$ takes on a value of plus one, then $Ti(t)$ is pushed away from equilibrium in the same (i.e., positive) direction. In substantive terms, when our participant is feeling down at a

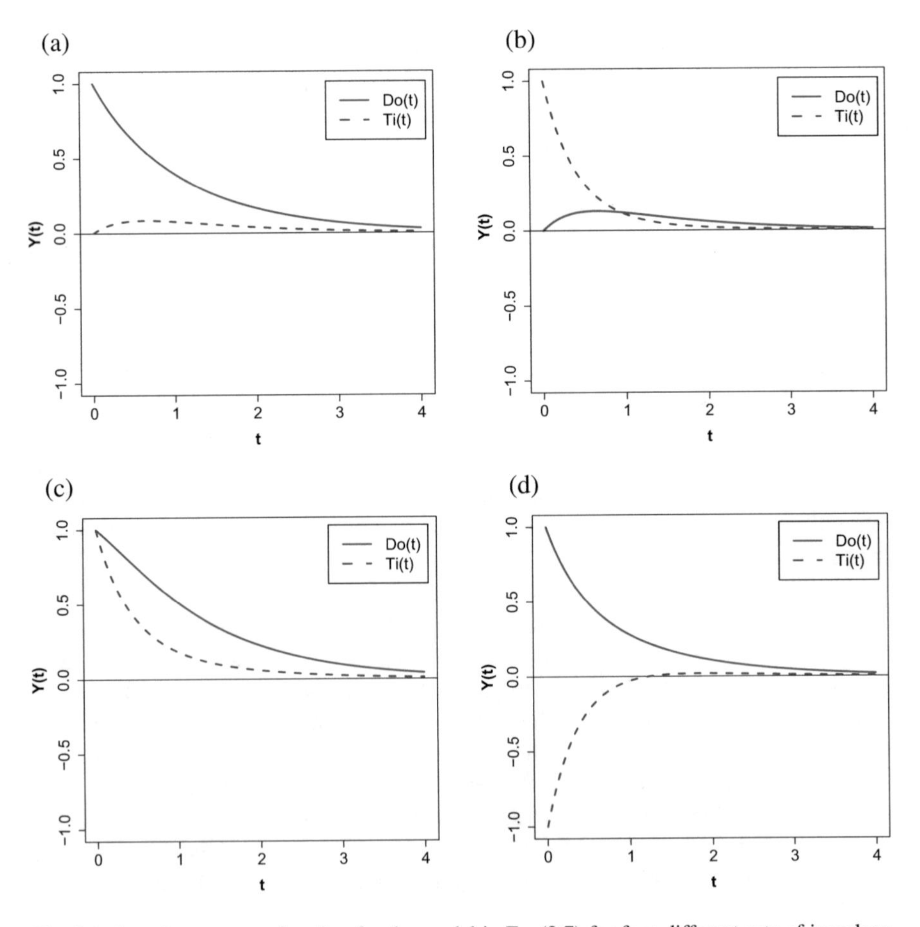

Fig. 2.1 Impulse response function for the model in Eq. (2.7) for four different sets of impulses; red solid line = $Do(t)$ and blue dashed line = $Ti(t)$. (**a**) $Do(0) = 1, Ti(0) = 0$, (**b**) $Do(0) = 0, Ti(0) = 1$, (**c**) $Do(0) = 1, Ti(0) = 1$, (**d**) $Do(0) = 1, Ti(0) = -1$

particular moment, he begins to feel a little tired. Eventually, both $Do(t)$ and $Ti(t)$ return to equilibrium due to their negative auto-effects. The feelings of being down and tired have returned to normal around $t = 4$, that is, 4 h after the initial impulse; stronger impulses ($|Do(0)| > 1$) will result in the system taking longer to return to equilibrium, and weaker impulses ($|Do(0)| < 1$) would dissipate quicker.

Figure 2.1b shows the corresponding reaction of $Do(t)$ at equilibrium to a positive value of $Ti(0)$. We can further see that the deviation of $Do(t)$ in Fig. 2.1b is greater than the deviation of $Ti(t)$ in Fig. 2.1a: a positive value of $Ti(t)$ exerts a greater push on $Do(t)$ than vice versa, because of the greater cross-effect of $Ti(t)$ on $Do(t)$. In this case this strong cross-effect, combined with the relatively weaker auto-effect of $Do(t)$, results in $Do(t)$ taking on a higher value than $Ti(t)$ at around $t = 1$, 1 h after the initial impulse. Substantively, when our participant is feeling physically tired at a particular moment (Fig. 2.1b), he begins to feel down over the next couple of hours, before eventually these feelings return to normal (again in this case, around 4 h later).

Figure 2.1c further demonstrates the role of the negative auto-effects and positive cross-effects in different scenarios. In Fig. 2.1c, both processes take on positive values at $t = 0$; the positive cross-effects result in both processes returning to equilibrium at a slower rate than in Fig. 2.1a, b. In substantive terms this means that, when the participant is feeling very down, and very tired, it takes longer for the participant to return to feeling normal. Here also the stronger auto-effect of $Ti(t)$ than $Do(t)$ is evident: although both processes start at the same value, an hour later $Ti(1)$ is much closer to zero than $Do(1)$, that is, $Ti(t)$ decays faster to equilibrium than $Do(t)$. In substantive terms, this tells us that when the participant is feeling down and physically tired, he recovers much quicker from the physical tiredness than he does from feeling down.

In Fig. 2.1d, we see that $Do(0)$ and $Ti(0)$ taking on values of opposite signs result in a speeding up of the rate at which each variable decays to equilibrium. The auto-effect of $Do(t)$ is negative, which is added to by the positive cross-effect of $Ti(t)$ multiplied by the negative value of $Ti(0)$. This means that $Do(0)$ in Fig. 2.1d takes on a stronger negative velocity, in comparison to Fig. 2.1a or c. A positive value for $Do(0)$ has a corresponding effect of making $Ti(0)$ take on an even stronger positive velocity. Substantively, this means that when the participant feels down, but feels less tired (i.e., more energetic) than usual, both of these feelings wear off and return to normal quicker than in the other scenarios we examined. The stronger auto-effect of $Ti(t)$, in combination with the positive cross-effect of $Do(t)$ on $Ti(t)$, actually results in $Ti(t)$ shooting past the equilibrium position in the $Ti(t)$ dimension ($Ti(t) = 0$) and taking on positive values around $t = 1.5$, before the system as a whole returns to equilibrium. Substantively, when the participant is feeling initially down but quite energetic, we expect that he feels a little bit more tired than usual about an hour and half later, before both feelings return to normal.

2.4.3.2 Vector Fields

Vector fields are another technique which can be used to visualize the dynamic behavior of the system by showing potential trajectories through a bivariate space. In our case the two axes of this space are $Do(t)$ and $Ti(t)$. The advantage of vector fields over IRFs in this context is that in one plot it shows how, for a range of possible starting positions, the process is expected to move in the (bivariate) space a moment later. For this reason, the vector field is particularly useful in bivariate models with complex dynamics, in which it may be difficult to obtain the full picture of the dynamic system from a few IRFs alone. Furthermore, by showing the dynamics for a grid of values, we can identify areas in which the movement of the process is similar or differs.

To create a vector field, $E[\frac{dy(t)}{dt}]$ is calculated for a grid of possible values for $y_1(t)$ and $y_2(t)$ covering the full range of the values both variables can take on. The vector field for $Do(t)$ and $Ti(t)$ is shown in Fig. 2.2. The base of each arrow represents a potential position of the process $\boldsymbol{y}(t)$. The head of the arrow represents where the process will be if we take one small step in time forward, that is the value of $\boldsymbol{y}(t + \Delta t)$ as Δt approaches zero. In other words, the arrows in this vector field represent the information of two derivatives, $dDo(t)/dt$ and $dTi(t)/dt$. Specifically, the direction the arrow is pointing is a function of the sign (positive or

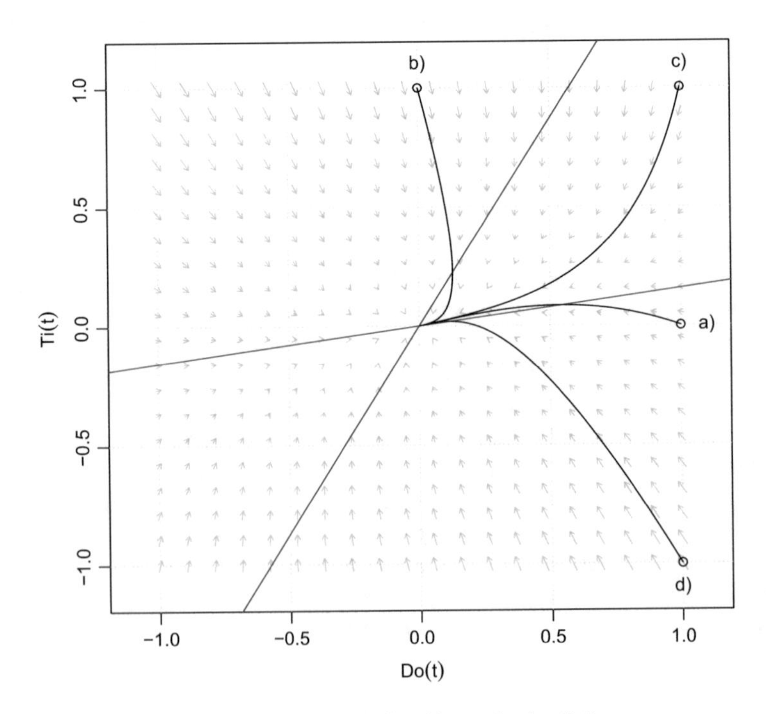

Fig. 2.2 Vector field for $Do(t)$ and $Ti(t)$, including blue and red nullclines

negative) of the derivatives, while the length of the arrow represents the magnitude of this movement and is a function of the absolute values of the derivative(s).

If an arrow in the vector field is completely vertical, this means that, for that position, taking one small step forward in time would result in a change in the system's position along the $Ti(t)$ axis (i.e., a change in the value of Tired), but not along the $Do(t)$ axis (i.e., $dDo(t)/dt = 0$ and $dTi(t)/dt \neq 0$). The converse is true for a horizontal arrow (i.e., $dDo(t)/dt \neq 0$ and $dTi(t)/dt = 0$). The two lines in Fig. 2.2, blue and red, identify at which positions $dDo(t)/dt = 0$ and $dTi(t)/dt = 0$, respectively; these are often referred to as *nullclines*. If the nullclines are not perfectly perpendicular to one another, this is due to the presence of at least one cross-effect. The point at which these nullclines cross represents the equilibrium position in this two-dimensional space, here located at $Do(t) = 0, Ti(t) = 0$. The crossing of these nullclines splits the vector field in four quadrants, each of which is characterized by a different combination of negative and positive values for $dDo(t)/dt$ and $dTi(t)/dt$. The top left and bottom right quadrants represent areas in which the derivatives are of opposite sign, $dDo(t)/dt > 0$ & $dTi(t)/dt < 0$ and $dDo(t)/dt < 0$ & $dTi(t)/dt > 0$, respectively. The top right and bottom left quadrants represent areas where the derivatives are of the same sign, $dDo(t)/dt < 0$ & $dTi(t) < 0$ and $dDo(t)/dt > 0$ & $dTi(t) > 0$, respectively.

By tracing a path through the arrows, we can see the trajectory of the system of interest from any point in the possible space of values. In Fig. 2.2, we include the same four bivariate trajectories as we examined with the IRFs. Instead of the IRF representation of two variables whose values are changing, the vector field represents this as the movement of one process in a two-dimensional space. For instance, the trajectory starting at $Do(t) = 0$ and $Ti(t) = 1$ begins in the top left quadrant, where $dDo(t)/dt$ is positive and $dTi(t)/dt$ is negative; this implies that the value of Down will increase, and the value of Tired will decrease (as can be seen in Fig. 2.1b). Instead of moving directly to the equilibrium along the $Ti(t)$ dimension, the system moves away from equilibrium along the $Do(t)$ dimension, due to the cross-effect of $Ti(t)$ on $Do(t)$, until it moves into the top right quadrant. In this quadrant, $dDo(t)/dt$ and $dTi(t)/dt$ are both negative; once in this quadrant, the process moves toward equilibrium, tangent to the $dDo(t)/dt$ nullcline. The other trajectories in Fig. 2.2 analogously describe the same trajectories as in Fig. 2.1a, c, d.

In general, the trajectories in this vector field first decay the quickest along the $Ti(t)$ dimension and the slowest along the $Do(t)$ dimension. This can be clearly seen in trajectories (b), (c), and (d). Each of these trajectories first changes steeply in the $Ti(t)$ dimension, before moving to equilibrium at a tangent to the red ($\frac{dDo(t)}{dt}$) nullcline. This general property of the bidimensional system is again related to the much stronger auto-effect of $Ti(t)$ and the relatively small cross-effects. In a technical sense, we can say that $Do(t)$ represents the "slowest eigendirection" (Strogatz 2014, Chapter 5).

2.4.4 Inspecting the Lagged Parameters

Another way to gain insight into the processes of interest is by determining the relationships between lagged positions of the system, according to our drift matrix. To this end, we can use Eq. (2.4) to determine $\boldsymbol{\Phi}(\Delta t)$ for some Δt. For instance, we can see that the autoregressive and cross-lagged relationships between values of competence and exhaustion given $\Delta t = 1$ are

$$\boldsymbol{\Phi}(\Delta t = 1) = \begin{bmatrix} 0.396 & 0.117 \\ 0.077 & 0.106 \end{bmatrix}. \quad (2.10)$$

For this given time interval, the cross-lagged effect of Down on Tired ($\phi_{21}(\Delta t = 1) = 0.077$) is smaller than the cross-lagged effect of Tired on Down ($\phi_{12}(\Delta t = 1) = 0.117$). However, as shown in Eq. (2.5), the value of each of these lagged effects changes in a nonlinear way depending on the time interval chosen. To visualize this, we can calculate $\boldsymbol{\Phi}(\Delta t)$ for a range of Δt and represent this information graphically in a lagged parameter plot, as in Fig. 2.3. From Fig. 2.3, we can see that both cross-lagged effects reach their maximum (and have their

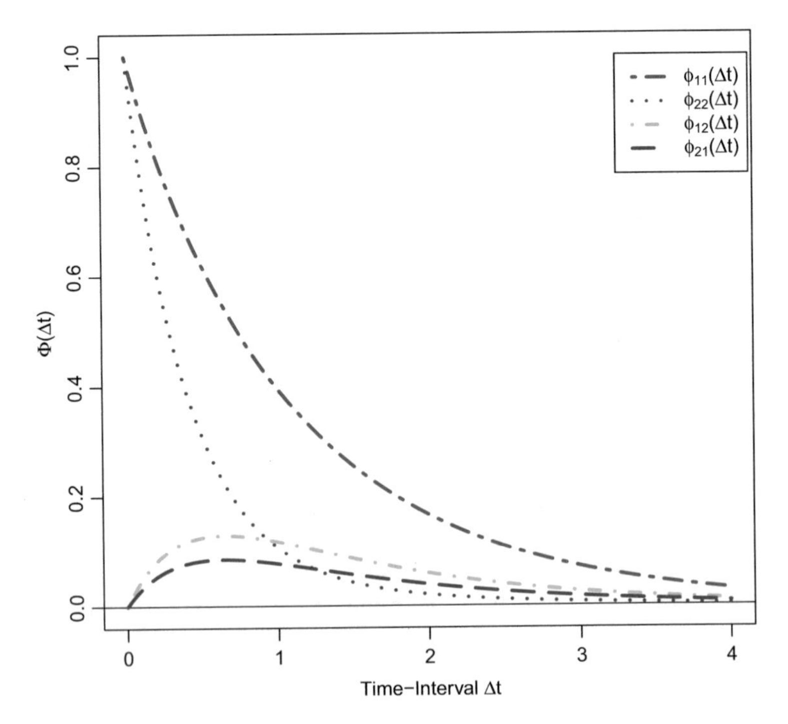

Fig. 2.3 The elements of $\boldsymbol{\Phi}(\Delta t)$ for the bivariate example (i.e., $\phi_{11}(\Delta t)$, $\phi_{12}(\Delta t)$, $\phi_{21}(\Delta t)$, $\phi_{22}(\Delta t)$) plotted for a range of values for Δt

maximum difference) at a time interval of $\Delta t = 0.65$; furthermore, we can see that the greater cross-effect (a_{12}) results in a stronger cross-lagged effect $\phi_{12}(\Delta t)$ for a range of Δt. Moreover, we can visually inspect how the size of each of the effects of interest, as well as the difference between these effects, varies according to the time interval. From a substantive viewpoint, we could say that the effect of feeling physically tired has the strongest effect on feelings of being down around 40 min later.

While the shape of the lagged parameters may appear similar to the shapes of the trajectories plotted in the IRFs, lagged parameter plots and IRFs represent substantively different information. IRFs plot the positions of each variable in the system as they change over time, given some impulse ($\boldsymbol{y}(t)$ vs t given some $\boldsymbol{y}(0)$). In contrast, lagged parameter plots show how the lagged relationships change depending on the length of the time interval between them, independent of impulse values ($\boldsymbol{e}^{\boldsymbol{A}\Delta t}$ vs Δt). The lagged relationships can be thought of as the components which go into determining any specific trajectories.

2.4.5 *Caution with Interpreting Estimated Parameters*

It is important to note that in the above interpretation of CT models, we have treated the matrix $\boldsymbol{A}$ as known. In practice of course, researchers should take account of the uncertainty in parameter estimates. For example, the *ctsem* package also provides lagged parameter plots with credible intervals to account for this uncertainty.

Furthermore, researchers should be cautious about extrapolating beyond the data. For instance, when we consider a vector field, we should be careful about interpreting regions in which there is little or no observed data (cf. Boker and McArdle 1995). The same logic applies for the interpretation of IRFs for impulses that do not match observed values. Moreover, we should also be aware that interpreting lagged parameter plots for time intervals much shorter than those we observe data at is a form of extrapolation: It relies on strong model-based assumptions, such as ruling out the possibility of a high-frequency higher-order process (Voelkle and Oud 2013; Voelkle et al. 2012).

2.5 Discussion

In this chapter we have set out to clarify the connection between DT- and CT-VAR(1) models and how we can interpret and represent the results from these models. So far we have focused on single-subject, two-dimensional, first-order systems with a stable node equilibrium. However, there are many ways in which these models can be extended, to match more complicated data and/or dynamic behavior. Below we consider three such extensions: (a) systems with more than two dimensions (i.e., variables), (b) different types of fixed points resulting from non-real eigenvalues of the drift matrix, and (c) moving from single-subject to multilevel datasets.

2.5.1 Beyond Two-Dimensional Systems

In the empirical illustration, we examined the interpretation of a drift matrix in the context of a bivariate CT-VAR(1) model. Notably, the current trend in applications of DT-VAR(1) models in psychology has been to focus more and more on the analysis of large systems of variables, as typified, for example, by the dynamic network approach of Bringmann et al. (2013, 2016). The complexity of these models grows rapidly as the number of variables is added: To estimate a full drift matrix for a system of three variables, we must estimate nine unique parameters, in contrast to four drift matrix parameters for a bivariate system. In addition, we must estimate a three-by-three covariance matrix for the residuals, rather than a two-by-two matrix.

The relationship between the elements of $\boldsymbol{A}$ and $\boldsymbol{\Phi}(\Delta t)$ becomes even less intuitive once the interest is in a system of three variables, because the lagged parameter values are dependent on the drift matrix as a whole, as explained earlier. This means that both the relative sizes and the signs of the cross-lagged effects may differ depending on the interval: The same lagged parameter may be negative for some time intervals and positive for others, and zero elements of $\boldsymbol{A}$ can result in corresponding non-zero elements of $\boldsymbol{\Phi}$ (cf. Aalen et al. 2017, 2016, 2012; Deboeck and Preacher 2016). Therefore, although we saw in our bivariate example that, for instance, negative CT cross-effects resulted in negative DT cross-lagged effects, this does not necessarily hold in the general case (Kuiper and Ryan 2018).

Additionally, substantive interpretation of the lagged parameters in systems with more than two variables also becomes less straightforward. For example, Deboeck and Preacher (2016) and Aalen et al. (2012, 2016, 2017) argue that the interpretation of $\boldsymbol{\Phi}(\Delta t)$ parameters in mediation models (with three variables and a triangular $\boldsymbol{A}$ matrix) as direct effects may be misleading: Deboeck and Preacher argue that instead they should be interpreted as *total* effects. This has major consequences for the practice of DT analyses and the interpretation of its results.

2.5.2 Complex and Positive Eigenvalues

The empirical illustration is characterized by a system with negative, real, non-equal eigenvalues, which implies that the fixed point in the system is a *stable node*. In theory, however, there is no reason that psychological processes must adhere to this type of dynamic behavior. We can apply the tools we have defined already to understand the types of behavior that might be described by other types of drift matrices. Notably, some systems may have drift matrices with complex eigenvalues, that is, eigenvalues of the form $\alpha \pm \omega i$, where $i = \sqrt{-1}$ is the imaginary number,

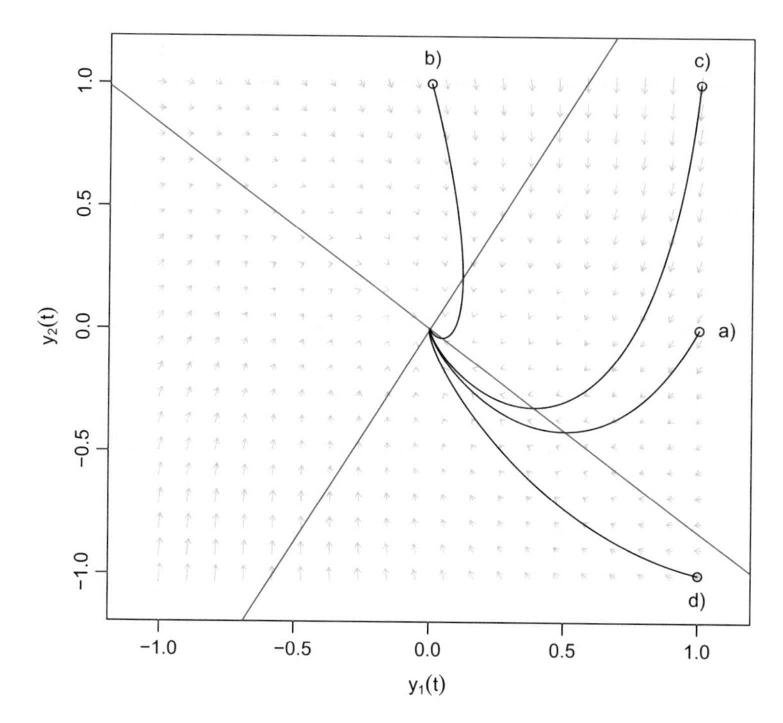

Fig. 2.4 Vector field for a stable spiral corresponding to a drift matrix with negative real part complex eigenvalues

$\omega \neq 0$, α is referred to as the real part, and ωi as the imaginary part of the eigenvalue. If the real component of these eigenvalues is negative ($\alpha < 0$), then the system is still stable, and given a deviation it will return eventually to a resting state at equilibrium. However, unlike the systems we have described before, these types of systems spiral or oscillate around the equilibrium point, before eventually coming to rest. Such systems have been described as *stable spirals*, or alternatively as *damped* (linear or harmonic) *oscillators* (Boker et al. 2010b; Voelkle and Oud 2013).

A vector field for a process which exhibits this type of stable spiral behavior is shown in Fig. 2.4, with accompanying trajectories. The drift matrix corresponding to this vector field is

$$\boldsymbol{A} = \begin{bmatrix} -0.995 & 0.573 \\ -2.000 & -2.416 \end{bmatrix} \tag{2.11}$$

which is equivalent to our empirical example above but with the value of a_{21} altered from 0.375 to -2.000. The eigenvalues of this matrix are $\lambda_1 = -1.706 + 0.800i$ and $\lambda_2 = -1.706 - 0.800i$. In contrast to our empirical example, we can see that the trajectories follow a spiral pattern; the trajectory which starts at $y_1(t) = 1$, $y_2(t) = 1$ actually overshoots the equilibrium in the $Ti(t)$ dimension before spiraling back once in the bottom quadrant. There are numerous examples of psychological systems that are modeled as damped linear oscillators using second-order differential equations, which include the first- and second-order derivatives (cf., Bisconti et al. 2004; Boker et al. 2010b; Boker and Nesselroade 2002; Horn et al. 2015). However, as shown here, such behavior may also result from a first-order model.

Stable nodes and spirals can be considered the two major types of stable fixed points, as they occur whenever the real part of the eigenvalues of $\boldsymbol{A}$ is negative, that is, $\alpha < 0$. Many other types of stable fixed points can be considered as special cases: when we have real, negative eigenvalues that are exactly equal, the fixed point is called a *stable star node* (if the eigenvectors are distinct) or a *stable degenerate node* (if the eigenvectors are not distinct). In contrast, if the real part of the eigenvalues of $\boldsymbol{A}$ is positive, then the system is unstable, also referred to as non-stationary or a unit root in the time series literature (Hamilton 1994). This implies that, given a deviation, the system will not return to equilibrium; in contrast to stable systems, in which trajectories are *attracted* to the fixed point, the trajectories of unstable systems are *repelled* by the fixed point. As such we can also encounter unstable nodes, spirals, star nodes, and degenerate nodes. The estimation and interpretation of unstable systems in psychology may be a fruitful ground for further research.

Two further types of fixed points may be of interest to researchers; in the special case where the eigenvalues of $\boldsymbol{A}$ have an imaginary part and no real part ($\alpha = 0$), the fixed point is called a *center*. In a system with a center fixed point, trajectories spiral around the fixed point without ever reaching it. Such systems exhibit oscillating behavior, but without any damping of oscillations; certain biological systems, such as the circadian rhythm, can be modeled as a dynamic system with a center fixed point. Such systems are on the borderline between stable and unstable systems, sometimes referred to as *neutrally stable*; trajectories are neither attracted to or repelled by the fixed point. Finally, a *saddle point* occurs when the eigenvalues of $\boldsymbol{A}$ are real but of opposite sign (one negative, one positive). Saddle points have one stable and one unstable component; only trajectories which start exactly on the stable axis return to equilibrium, and all others do not. Together spirals, nodes, and saddle points cover the majority of the space of possible eigenvalues for $\boldsymbol{A}$. Strogatz (2014) describes the different dynamic behavior generated by different combinations of eigenvalues of $\boldsymbol{A}$ in greater detail.

2.5.3 Multilevel Extensions

The time series literature (such as from the field of econometrics) as well as the dynamic systems literature (such as from the field of physics) tends to be concerned with a single dynamic system, either because there is only one case ($N = 1$) or because all cases are exact replicates (e.g., molecules). In psychology however, we typically have data from more than one person, and we also know that people tend to be highly different. Hence, when we are interested in modeling their longitudinal data, we should take their differences into account somehow. The degree to which this can be done depends on the number of time points we have per person. In traditional panel data, we typically have between two and six waves of data. In this case, we should allow for individual differences in means or intercepts, in order to separate the between-person, stable differences from the within-person dynamic process, while assuming the lagged relationships are the same across individuals (cf. Hamaker et al. 2015).

In contrast, experience sampling data and other forms of intensive longitudinal data consist of many repeated measurement per person, such that we can allow for individual differences in the lagged coefficients. This can be done by either analyzing the data of each person separately or by using a dynamic multilevel model in which the individuals are allowed to have different parameters (cf. Boker et al. 2016; Driver and Voelkle 2018). Many recent studies have shown that there are substantial individual differences in the dynamics of psychological phenomena and that these differences can be meaningfully related to other person characteristics, such as personality traits, gender, age, and depressive symptomatology, but also to later health outcomes and psychological well-being (e.g., Bringmann et al. 2013; Kuppens et al. 2010, 2012).

While the current chapter has focused on elucidating the interpretation of a single-subject CT-VAR(1) model, the substantive interpretations and visualization tools we describe here can be applied in a straightforward manner to, for example, the fixed effects estimated in a multilevel CT-VAR(1) model or to individual-specific parameters estimated in a multilevel framework. The latter would however lead to an overwhelming amount of visual information. The development of new ways of summarizing the individual differences in dynamics, based on the current tools, is a promising area.

2.5.4 *Conclusion*

There is no doubt that the development of dynamical systems modeling in the field of psychology has been hampered by the difficulty in obtaining suitable data to model such systems. However this is a barrier that recent advances in technology will shatter in the coming years. Along with this new source of psychological data, new psychological theories are beginning to emerge, based on the notion of psychological processes as dynamic systems. Although the statistical models needed to investigate these theories may seem exotic or difficult to interpret at first, they reflect the simple intuitive and empirical notions we have about psychological processes: Human behavior, emotion, and cognition fluctuate continuously over time, and the models we use should reflect that. We hope that our treatment of CT-VAR(1) models and their interpretation will help researchers to overcome the knowledge barrier to this approach and can serve as a stepping stone toward a broader adaptation of the CT dynamical system approach to psychology.

Acknowledgment We thank an editor and anonymous reviewer for helpful comments that led to improvements in this chapter. The work of the authors was supported by grants from the Netherlands Organization for Scientific Research (NWO Onderzoekstalent 406-15-128) to Oisín Ryan and Ellen Hamaker, and (NWO VENI 451-16-019) to Rebecca Kuiper.

Appendix: Matrix Exponential

Similar to the scalar exponential, the matrix exponential can be defined as an infinite sum

$$\boldsymbol{e}^{\boldsymbol{A}} = \sum_{k=0}^{\infty} \frac{1}{k!} \boldsymbol{A}^k$$

The exponential of a matrix is not equivalent to taking the scalar exponential of each element of the matrix, unless that matrix is diagonal. The exponential of a matrix can be found using an eigenvalue decomposition

$$\boldsymbol{A} = \boldsymbol{V}\boldsymbol{D}\boldsymbol{V}^{-1}$$

where $\boldsymbol{V}$ is a matrix of eigenvectors of $\boldsymbol{A}$ and $\boldsymbol{D}$ is a diagonal matrix of the eigenvalues of $\boldsymbol{A}$ (cf. Moler and Van Loan 2003). The matrix exponential of $\boldsymbol{A}$ is given by

$$\boldsymbol{e}^{\boldsymbol{A}} = \boldsymbol{V}\boldsymbol{e}^{\boldsymbol{D}}\boldsymbol{V}^{-1}$$

where $\boldsymbol{e}^{\boldsymbol{D}}$ is the diagonal matrix whose entries are the scalar exponential of the eigenvalues of $\boldsymbol{A}$. When we want to solve for the matrix exponential of a matrix multiplied by some constant Δt, we get

$$\boldsymbol{e}^{\boldsymbol{A}\Delta t} = \boldsymbol{V}\boldsymbol{e}^{\boldsymbol{D}\Delta t}\boldsymbol{V}^{-1} \tag{2.12}$$

Take it that we have a 2×2 square matrix given by

$$\boldsymbol{A} = \begin{bmatrix} a & b \\ c & d \end{bmatrix}$$

and we wish to solve for $\boldsymbol{e}^{\boldsymbol{A}\Delta t}$. The eigenvalues of $\boldsymbol{A}$ are given by

$$\lambda_1 = \frac{1}{2}\left(a + d - \sqrt{a^2 + 4bc - 2ad + d^2}\right)$$
$$\lambda_2 = \frac{1}{2}\left(a + d + \sqrt{a^2 + 4bc - 2ad + d^2}\right)$$

where we will from here on denote

$$R = \sqrt{a^2 + 4bc - 2ad + d^2}$$

for notational simplicity. The exponential of the diagonal matrix made up of eigenvalues multiplied by the constant Δt is given by

$$\boldsymbol{e}^{\boldsymbol{D}\Delta t} = \begin{bmatrix} e^{\frac{1}{2}(a+d-R)\Delta t} & 0 \\ 0 & e^{\frac{1}{2}(a+d+R)\Delta t} \end{bmatrix}$$

The matrix of eigenvectors of $\boldsymbol{A}$ is given by

$$\boldsymbol{V} = \begin{bmatrix} \frac{a-d-R}{2c} & \frac{a-d+R}{2c} \\ 1 & 1 \end{bmatrix}$$

assuming $c \neq 0$, with inverse

$$\boldsymbol{V}^{-1} = \begin{bmatrix} \frac{-c}{R} & \frac{a-d+R}{2R} \\ \frac{c}{R} & \frac{-a+d+R}{2R} \end{bmatrix}.$$

Multiplying $\boldsymbol{V}\boldsymbol{e}^{\boldsymbol{D}}\boldsymbol{V}^{-1}$ gives us

$$\boldsymbol{e}^{\boldsymbol{A}\Delta t} = \begin{bmatrix} \frac{R-a+d}{2R}e^{\lambda_1\Delta t} + \frac{R+a-d}{2R}e^{\lambda_2\Delta t} & \frac{b(-e^{\lambda_1\Delta t}+e^{\lambda_2\Delta t})}{R} \\ \frac{c(-e^{\lambda_1\Delta t}+e^{\lambda_2\Delta t})}{R} & \frac{R+a-d}{2R}e^{\lambda_1\Delta t} + \frac{R-a+d}{2R}e^{\lambda_2\Delta t} \end{bmatrix} \quad (2.13)$$

Note that we present here only a worked out example for a 2×2 square matrix. For larger square matrices (representing models with more variables), the eigenvalue decomposition remains the same although the terms for the eigenvalues, eigenvectors, and determinants become much less feasible to present.

References

Aalen, O., Gran, J., Røysland, K., Stensrud, M., & Strohmaier, S. (2017). Feedback and mediation in causal inference illustrated by stochastic process models. *Scandinavian Journal of Statistics, 45*, 62–86. https://doi.org/10.1111/sjos.12286

Aalen, O., Røysland, K., Gran, J., Kouyos, R., & Lange, T. (2016). Can we believe the DAGs? A comment on the relationship between causal DAGs and mechanisms. *Statistical Methods in Medical Research, 25*(5), 2294–2314. https://doi.org/10.1177/0962280213520436

Aalen, O., Røysland, K., Gran, J., & Ledergerber, B. (2012). Causality, mediation and time: A dynamic viewpoint. *Journal of the Royal Statistical Society: Series A (Statistics in Society), 175*(4), 831–861.

Bisconti, T., Bergeman, C. S., & Boker, S. M. (2004). Emotional well-being in recently bereaved widows: A dynamical system approach. *Journal of Gerontology, Series B: Psychological Sciences and Social Sciences, 59*, 158–167. https://doi.org/10.1093/geronb/59.4.P158

Boker, S. M. (2002). Consequences of continuity: The hunt for intrinsic properties within parameters of dynamics in psychological processes. *Multivariate Behavioral Research, 37*(3), 405–422. https://doi.org/10.1207/S15327906MBR3703-5

Boker, S. M., Deboeck, P., Edler, C., & Keel, P. (2010a). Generalized local linear approximation of derivatives from time series. In S. Chow & E. Ferrar (Eds.), *Statistical methods for modeling human dynamics: An interdisciplinary dialogue* (pp. 179–212). Boca Raton, FL: Taylor & Francis.

Boker, S. M., & McArdle, J. J. (1995). Statistical vector field analysis applied to mixed cross-sectional and longitudinal data. *Experimental Aging Research, 21*, 77–93. https://doi.org/10.1080/03610739508254269

Boker, S. M., Montpetit, M. A., Hunter, M. D., & Bergeman, C. S. (2010b). Modeling resilience with differential equations. In P. Molenaar & K. Newell (Eds.), *Learning and development: Individual pathways of change* (pp. 183–206). Washington, DC: American Psychological Association. https://doi.org/10.1037/12140-011

Boker, S. M., Neale, M., & Rausch, J. (2004). Latent differential equation modeling with multivariate multi-occasion indicators. In K. van Montfort, J. H. L. Oud, & A. Satorra (Eds.), *Recent developments on structural equation models* (pp. 151–174). Dordrecht: Kluwer.

Boker, S. M., & Nesselroade, J. R. (2002). A method for modeling the intrinsic dynamics of intraindividual variability: Recovering parameters of simulated oscillators in multi-wave panel data. *Multivariate Behavioral Research, 37*, 127–160.

Boker, S. M., Staples, A. D., & Hu, Y. (2016). Dynamics of change and change in dynamics. *Journal for Person-Oriented Research, 2*(1–2), 34. https://doi.org/10.17505/jpor.2016.05

Bolger, N., & Laurenceau, J.-P. (2013). *Intensive longitudinal methods: An introduction to diary and experience sampling research*. New York, NY: The Guilford Press.

Bringmann, L., Lemmens, L., Huibers, M., Borsboom, D., & Tuerlinckx, F. (2015). Revealing the dynamic network structure of the beck depression inventory-ii. *Psychological Medicine, 45*(4), 747–757. https://doi.org/10.1017/S0033291714001809

Bringmann, L., Pe, M., Vissers, N., Ceulemans, E., Borsboom, D., Vanpaemel, W., . . . Kuppens, P. (2016). Assessing temporal emotion dynamics using networks. *Assessment, 23*(4), 425–435. https://doi.org/10.1177/1073191116645909

Bringmann, L., Vissers, N., Wichers, M., Geschwind, N., Kuppens, P., Peeters, . . . Tuerlinckx, F. (2013). A network approach to psychopathology: New insights into clinical longitudinal data. *PLoS ONE, 8*, e60188. https://doi.org/10.1371/journal.pone.0060188

Browne, M. W., & Nesselroade, J. R. (2005). Representing psychological processes with dynamic factor models: Some promising uses and extensions of ARMA time series models. In A. Maydue-Olivares & J. J. McArdle (Eds.), *Psychometrics: A festschrift to Roderick P. McDonald* (pp. 415–452). Mahwah, NJ: Lawrence Erlbaum Associates.

Chow, S., Ferrer, E., & Hsieh, F. (2011). *Statistical methods for modeling human dynamics: An interdisciplinary dialogue*. New York, NY: Routledge.

Chow, S., Ferrer, E., & Nesselroade, J. R. (2007). An unscented Kalman filter approach to the estimation of nonlinear dynamical systems models. *Multivariate Behavioral Research, 42*(2), 283–321. https://doi.org/10.1080/00273170701360423

Chow, S., Ram, N., Boker, S., Fujita, F., Clore, G., & Nesselroade, J. (2005). Capturing weekly fluctuation in emotion using a latent differential structural approach. *Emotion, 5*(2), 208–225.

De Haan-Rietdijk, S., Voelkle, M. C., Keijsers, L., & Hamaker, E. (2017). Discrete- versus continuous-time modeling of unequally spaced ESM data. *Frontiers in Psychology, 8*, 1849. https://doi.org/10.3389/fpsyg.2017.01849

Deboeck, P. R., & Preacher, K. J. (2016). No need to be discrete: A method for continuous time mediation analysis. *Structural Equation Modeling: A Multidisciplinary Journal, 23*(1), 61–75.

Denollet, J., & De Vries, J. (2006). Positive and negative affect within the realm of depression, stress and fatigue: The two-factor distress model of the Global Mood Scale (GMS). *Journal of Affective Disorders, 91*(2), 171–180. https://doi.org/10.1016/j.jad.2005.12.044

Dormann, C., & Griffin, M. A. (2015). Optimal time lags in panel studies. *Psychological Methods, 20*(4), 489. https://doi.org/10.1037/met0000041

Driver, C., Oud, J. H. L., & Voelkle, M. (2017). Continuous time structural equation modelling with r package ctsem. *Journal of Statistical Software, 77*, 1–35. https://doi.org/10.18637/jss.v077.i05

Driver, C. C., & Voelkle, M. C. (2018). Hierarchical Bayesian continuous time dynamic modeling. *Psychological Methods*. Advance online publication. http://dx.doi.org/10.1037/met0000168

Fisher, M. (2001). *Modeling negative autoregression in continuous time*. http://www.markfisher.net/mefisher/papers/continuous_ar.pdf

Gault-Sherman, M. (2012). It's a two-way street: The bidirectional relationship between parenting and delinquency. *Journal of Youth and Adolescence, 41*, 121–145. https://doi.org/10.1007/s10964-011-9656-4

Gollob, H. F., & Reichardt, C. S. (1987). Taking account of time lags in causal models. *Child Development, 58*, 80–92. https://doi.org/10.2307/1130293

Granger, C. W. J. (1969). Investigating causal relations by econometric models and cross-spectral methods. *Econometrica, 37*, 424–438. https://doi.org/10.2307/1912791

Hamaker, E. L., Dolan, C. V., & Molenaar, P. C. M. (2005). Statistical modeling of the individual: Rationale and application of multivariate time series analysis. *Multivariate Behavioral Research, 40*(2), 207–233. https://doi.org/10.1207/s15327906mbr4002_3

Hamaker, E. L., & Grasman, R. P. P. P. (2015). To center or not to center? Investigating inertia with a multilevel autoregressive model. *Frontiers in Psychology, 5*, 1492. https://doi.org/10.3389/fpsyg.2014.01492

Hamaker, E. L., Kuiper, R., & Grasman, R. P. P. P. (2015). A critique of the cross-lagged panel model. *Psychological Methods, 20*(1), 102–116. https://doi.org/10.1037/a0038889

Hamerle, A., Nagl, W., & Singer, H. (1991). Problems with the estimation of stochastic differential equations using structural equations models. *Journal of Mathematical Sociology, 16*(3), 201–220. https://doi.org/10:1080=0022250X:1991:9990088

Hamilton, J. D. (1994). *Time series analysis*. Princeton, NJ: Princeton University Press.

Horn, E. E., Strachan, E., & Turkheimer, E. (2015). Psychological distress and recurrent herpetic disease: A dynamic study of lesion recurrence and viral shedding episodes in adults. *Multivariate Behavioral Research, 50*(1), 134–135. https://doi.org/10.1080/00273171.2014.988994

Ichii, K. (1991). Measuring mutual causation: Effects of suicide news on suicides in Japan. *Social Science Research, 20*, 188–195. https://doi.org/10.1016/0049-089X(91)90016-V

Johnston, J., & DiNardo, J. (1997). *Econometric methods* (4th ed.). New York, NY: McGraw-Hill.

Kim, C.-J., & Nelson, C. R. (1999). *State-space models with regime switching: Classical and Gibbs-sampling approaches with applications*. Cambridge, MA: The MIT Press. https://doi.org/10.2307/2669796

Kossakowski, J., Groot, P., Haslbeck, J., Borsboom, D., & Wichers, M. (2017). Data from critical slowing down as a personalized early warning signal for depression. *Journal of Open Psychology Data, 5*(1), 1.

Koval, P., Kuppens, P., Allen, N. B., & Sheeber, L. (2012). Getting stuck in depression: The roles of rumination and emotional inertia. *Cognition and Emotion, 26*, 1412–1427.

Kuiper, R. M., & Ryan, O. (2018). Drawing conclusions from cross-lagged relationships: Reconsidering the role of the time-interval. *Structural Equation Modeling: A Multidisciplinary Journal*. https://doi.org/10.1080/10705511.2018.1431046

Kuppens, P., Allen, N. B., & Sheeber, L. B. (2010). Emotional inertia and psychological maladjustment. *Psychological Science, 21*(7), 984–991. https://doi.org/10.1177/0956797610372634

Kuppens, P., Sheeber, L. B., Yap, M. B. H., Whittle, S., Simmons, J., & Allen, N. B. (2012). Emotional inertia prospectively predicts the onset of depression in adolescence. *Emotion, 12*, 283–289. https://doi.org/10.1037/a0025046

Meier, B. P., & Robinson, M. D. (2004). Why the sunny side is up: Associations between affect and vertical position. *Psychological Science, 15*(4), 243–247. https://doi.org/10.1111/j.0956-7976.2004.00659.x

Moberly, N. J., & Watkins, E. R. (2008). Ruminative self-focus and negative affect: An experience sampling study. *Journal of Abnormal Psychology, 117*, 314–323. https://doi.org/10.1037/0021-843X.117.2.314

Moler, C., & Van Loan, C. (2003). Nineteen dubious ways to compute the exponential of a matrix, twenty-five years later. *SIAM Review, 45*(1), 3–49.

Oravecz, Z., & Tuerlinckx, F. (2011). The linear mixed model and the hierarchical Ornstein–Uhlenbeck model: Some equivalences and differences. *British Journal of Mathematical and Statistical Psychology, 64*(1), 134–160. https://doi.org/10.1348/000711010X498621

Oravecz, Z., Tuerlinckx, F., & Vandekerckhove, J. (2011). A hierarchical latent stochastic difference equation model for affective dynamics. *Psychological Methods, 16*, 468–490. https://doi.org/10.1037/a0024375

Oravecz, Z., Tuerlinckx, F., & Vandekerckhove, J. (2016). Bayesian data analysis with the bivariate hierarchical Ornstein-Uhlenbeck process model. *Multivariate Behavioral Research, 51*(1), 106–119. https://doi.org/10.1080/00273171.2015.1110512

Oud, J. H. L. (2007). Continuous time modeling of reciprocal relationships in the cross-lagged panel design. In S. M. Boker & M. J. Wenger (Eds.), *Data analytic techniques for dynamic systems in the social and behavioral sciences* (pp. 87–129). Mahwah, NJ: Lawrence Erlbaum Associates.

Oud, J. H. L., & Delsing, M. J. M. H. (2010). Continuous time modeling of panel data by means of SEM. In K. van Montfort, J. H. L. Oud, & A. Satorra (Eds.), *Longitudinal research with latent variables* (pp. 201–244). New York, NY: Springer. https://doi.org/10.1007/978-3-642-11760-2-7

Oud, J. H. L., & Jansen, R. A. (2000). Continuous time state space modeling of panel data by means of SEM. *Psychometrika, 65*(2), 199–215. https://doi.org/10.1007/BF02294374

Oud, J. H. L., van Leeuwe, J., & Jansen, R. (1993). Kalman filtering in discrete and continuous time based on longitudinal lisrel models. In *Advances in longitudinal and multivariate analysis in the behavioral sciences* (pp. 3–26). Nijmegen: ITS.

Reichardt, C. S. (2011). Commentary: Are three waves of data sufficient for assessing mediation? *Multivariate Behavioral Research, 46*(5), 842–851.

Rovine, M. J., & Walls, T. A. (2006). Multilevel autoregressive modeling of interindividual differences in the stability of a process. In T. A. Walls & J. L. Schafer (Eds.), *Models for intensive longitudinal data* (pp. 124–147). New York, NY: Oxford University Press. https://doi.org/10.1093/acprof:oso/9780195173444.003.0006

Schuurman, N. K., Ferrer, E., de Boer-Sonnenschein, M., & Hamaker, E. L. (2016). How to compare cross-lagged associations in a multilevel autoregressive model. *Psychological methods, 21*(2), 206–221. https://doi.org/10.1037/met0000062

Steele, J. S., & Ferrer, E. (2011). Latent differential equation modeling of selfregulatory and coregulatory affective processes. *Multivariate Behavioral Research, 46*(6), 956–984. https://doi.org/10.1080/00273171.2011.625305

Strogatz, S. H. (2014). *Nonlinear dynamics and chaos: With applications to physics, biology, chemistry, and engineering*. Boulder, CO: Westview press.

Voelkle, M., & Oud, J. H. L. (2013). Continuous time modelling with individually varying time intervals for oscillating and non-oscillating processes. *British Journal of Mathematical and Statistical Psychology, 66*(1), 103–126. https://doi.org/10.1111/j.2044-8317.2012.02043.x

Voelkle, M., Oud, J. H. L., Davidov, E., & Schmidt, P. (2012). An SEM approach to continuous time modeling of panel data: Relating authoritarianism and anomia. *Psychological Methods, 17*, 176–192. https://doi.org/10.1037/a0027543

Watkins, M. W., Lei, P.-W., & Canivez, G. L. (2007). Psychometric intelligence and achievement: A cross-lagged panel analysis. *Intelligence, 35*, 59–68. https://doi.org/10.1016/j.intell.2006.04.005

Chapter 3
On Fitting a Continuous-Time Stochastic Process Model in the Bayesian Framework

Zita Oravecz, Julie Wood, and Nilam Ram

3.1 Introduction

Process modeling offers a robust framework for developing, testing, and refining substantive theories through mathematical specification of the mechanisms and/or latent processes that produce observed data. Across many fields of application (e.g., chemistry, biology, engineering), process modeling uses detailed mathematical models to obtain accurate description and explanation of equipment and phenomena, and to support prediction and optimization of both intermediate and final outcomes. In the social and behavioral sciences, the process modeling approach is being used to obtain insight into the complex processes underlying human functioning. In particular, the approach offers a way to describe substantively meaningful components of behavior using mathematical functions that map directly to theoretical concepts. In cognitive science, for example, drift diffusion models (see, e.g., Ratcliff and Rouder 1998) have been used to derive rate of information accumulation, non-decision time, bias, and decision boundaries from observed data on reaction time and correctness of response. Applied to data about individuals' decisions in gambling tasks, process models are used to describe people's tendency to take risks, their response consistency, and their memory for payoffs (Wetzels

Electronic Supplementary Material The online version of this article (https://doi.org/10.1007/978-3-319-77219-6_3) contains supplementary material, which is available to authorized users.

Z. Oravecz (✉) · J. Wood · N. Ram
The Pennsylvania State University, State College, PA, USA
e-mail: zzo1@psu.edu; jfw5255@psu.edu; nur5@psu.edu

K. van Montfort et al. (eds.), *Continuous Time Modeling in the Behavioral and Related Sciences*, https://doi.org/10.1007/978-3-319-77219-6_3

et al. 2010). In sociology, multinomial tree-based process models have been used to infer cultural consensus, latent ability, willingness to guess, and guessing bias from individuals' judgments on a shared knowledge domain (Oravecz et al. 2014). In psychology, process models are proving especially useful for study of regulatory and interpersonal processes. Process models based on principles governing thermostats and reservoirs are being used to describe regulation of negative affect and stress (Chow et al. 2005; Deboeck and Bergeman 2013), and process models based on physical principles of pendulums are being used to describe interaction and coordination between partners of a dyad (e.g., wife-husband, mother-child; see in Chow et al. 2010; Ram et al. 2014; Thomas and Martin 1976). In this paper we present and illustrate how a specific process model, a multilevel Ornstein-Uhlenbeck model, can be used to describe and study moment-to-moment continuous-time dynamics of affect captured in ecological momentary assessment studies.

Process modeling is, of course, sometimes challenging. Mathematically precise descriptions of humans' behavior are often complex, with many parameters, non-linear relations, and multiple layers of between-person differences that require consideration within a multilevel framework (e.g., repeated measures nested within persons nested within dyads or groups). The complexity of the models often means that implementation within classical statistical frameworks is impractical. Thankfully, the Bayesian statistical inference framework (see, e.g., Gelman et al. 2013) has the necessary tools. The algorithms underlying the Bayesian estimation framework are designed for highly dimensional problems with non-linear dependencies.

Consider, for example, the multilevel extension of continuous-time models where all dynamic parameters are allowed to vary across people, thus describing individual differences in intraindividual variation or velocity of changes. Driver et al. (2017) offer an efficient and user-friendly R package for estimation of multilevel continuous-time models, cast in the structural equation modeling framework. However, at the time of writing, person differences are only allowed in the intercepts and not in the variance or velocity parameters. A flexible Bayesian extension of the package that allows for all dynamics parameters to be person-specific is, though, in progress (Driver and Voelkle 2018). Importantly, flexible Markov chain Monte Carlo methods (Robert and Casella 2004) at the core of Bayesian estimation also provide for simultaneous estimation of model parameters and regression coefficients within a multilevel framework that supports identification and examination of interindividual differences in intraindividual dynamics (i.e., both time-invariant and time-varying covariates).

In the sections that follow, we review how process models are used to analyze longitudinal data obtained from multiple persons (e.g., ecological momentary assessment data), describe the mathematical details of the multilevel Ornstein-Uhlenbeck model, and illustrate through empirical example how a Bayesian implementation of this model provides for detailed and new knowledge about individuals' affect regulation.

3.1.1 The Need for Continuous-Time Process Models to Analyze Intensive Longitudinal Data

Psychological processes continuously organize behavior in and responses to an always-changing environment. In attempting to capture these processes as they occur in situ, many researchers make use of a variety of experience sampling, daily diary, ecological momentary assessment (EMA), and ambulatory assessment study designs. In EMA studies, for example, self-reports and/or physiological measurements are collected multiple times per day over an extended period of time (i.e., weeks) from many participants as they go about their daily lives—thus providing ecological validity and reducing potential for reporting bias in the observations (Shiffman et al. 2008; Stone and Shiffman 1994). The data obtained are considered *intensive longitudinal data* (ILD, Bolger and Laurenceau 2013; Walls and Schafer 2006), in that they contain many replicates across both persons and time, and support modeling of interindividual differences in intraindividual dynamics. ILD, however, also present unique analytical challenges. First, the data are often unbalanced. The number of measurements is almost never equal across participants because the intensive nature of the reporting means that study participants are likely to miss at least some of the prompts and/or (particularly in event contingent designs) provide a different number of reports because of natural variation in exposure to the phenomena of interest. Second, the data are often time-unstructured. Many EMA studies purposively use semi-random time sampling to reduce expectation biases in reporting and obtain what might be considered a representative sampling of individuals' context. Although the in situ and intensive nature of the data obtained in these studies provides for detailed description of the processes governing the moment-to-moment continuous-time dynamics of multiple constructs (e.g., affect valence, affect arousal), the between-participant differences in data collection schedule (length and timing) make it difficult to use traditional statistical modeling tools that assume equally spaced measurements or equal number of measurements. Modeling moment-to-moment dynamics in the unbalanced and time-unstructured data being obtained in EMA studies requires continuous-time process models (see more discussion in Oud and Voelkle 2014b).

3.1.2 The Need for Continuous-Time Process Models to Capture Temporal Changes in Core Affective States

To illustrate the benefits of a process modeling approach and the utility of continuous-time process models, we will analyze data from an EMA study in which participants reported on their core affect (Russell 2003) in the course of living their everyday lives. In brief, core affect is a neurophysiological state defined by an integral blend of valence (level of pleasantness of feeling) and arousal (level of physiological activation) levels. Core affect, according to the theory of constructed emotion (see, e.g., Barrett 2017), underlies all emotional experience and changes continuously over time. People can consciously access their core affect by indicating the level of valence and arousal of their current experience. Our empirical example makes use of data collected as part of an EMA study where $N = 52$ individuals reported on their core affect (valence and arousal) for 4 weeks, six times a day at semi-random time intervals. Specifically, participants' awake time was divided into six equal-length time intervals within which a randomly timed text prompt arrived asking participants about their current levels of core affect (along with other questions related to general well-being). The intensive longitudinal data, obtained from four individuals, are shown in Fig. 3.1. Interval-to-interval changes in arousal and valence are shown in gray and blue, respectively. Some individual differences are immediately apparent: the four people differ in terms of the center of the region in which their core affect fluctuates, the extent of fluctuation, and the degree of overlap in arousal and valence. The process modeling goal is to develop a mathematical specification of latent processes that underlie the moment-to-moment dynamics of core affect and how those dynamics differ across people.

Based on reviews of empirical studies of temporal changes in emotions and affect, Kuppens et al. (2010) proposed the *DynAffect* framework, wherein intraindividual affective dynamics are described in terms of three key elements: affective baseline, homeostatic regulatory force, and variation around the baseline. Using a process modeling approach, these features are translated into a mathematical description of core affect dynamics—a continuous-time process model. Important

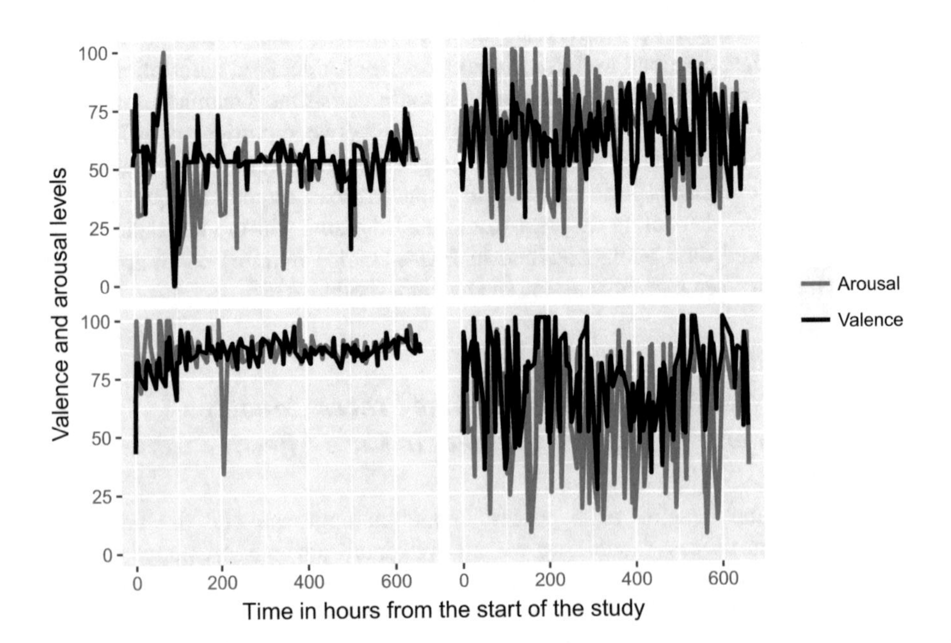

Fig. 3.1 Changes in core affect over time from four participants

general features of the mathematical parameterization include the following: (1) the latent person-level temporal dynamics of core affect are explicitly mapped to the substantive theory, (2) measurement noise in observed data is accommodated through addition of a measurement model, and (3) simultaneous modeling of person and population-level characteristics (e.g., organization of interindividual differences) is accommodated within a multilevel framework. Previous studies using similar process models have confirmed the utility of the approach for studying interindividual differences in intraindividual dynamics. In particular, it has been shown that people who score high on the neuroticism scale of Big Five personality model show lower baseline pleasantness and increased fluctuation (Oravecz et al. 2011), people who tend to apply reappraisal as emotion regulation strategy show higher levels of moment-to-moment arousal regulation (Oravecz et al. 2016), and older people tend to have higher arousal baseline with less fluctuation (Wood et al. 2017). In this chapter we add new information about how interindividual differences in the temporal dynamics of core affect are related to interindividual differences in trait-level emotional well-being (i.e., relatively stable characteristics of individuals' emotional patterns).

3.2 The Ornstein-Uhlenbeck Process to Describe Within-Person Latent Temporal Dynamics

3.2.1 The Stochastic Differential Equation Definition of the Ornstein-Uhlenbeck Process

As noted earlier, process modeling requires specification of a mathematical model that describes the mechanisms and/or latent processes that produce observed data. Here, three key features of the temporal dynamics of core affect are described by using an Ornstein-Uhlenbeck process model (OU; Uhlenbeck and Ornstein 1930), the parameters of which will be estimated using EMA data that are considered noisy measurements of this latent process. Let us denote the position of the latent process at time t with $\theta(t)$. The OU process can be defined as a solution of the following first-order stochastic differential equation (Langevin equation):

$$d\theta(t) = \beta(\mu - \theta(t))dt + \sigma dW(t). \tag{3.1}$$

Parameter $\theta(t)$ represents the latent variable evolving in time, and Eq. (3.1) describes the change in this latent variable, $d\theta(t)$, with respect to time t. As can be seen, changes in the latent state are a function of μ, which represents the baseline level of the process, β, which represents attractor or regulatory strength, $W(t)$, which is the position of a standard Wiener process (also known as Brownian motion; Wiener 1923) at time t, and σ, which scales the added increments ($dW(t)$) from the Wiener process (together, also called the innovation process). The Wiener process evolves in continuous time, following a random trajectory, uninfluenced by its previous positions. If you consider Eq. (3.1) as a model for temporal changes in a person's latent core affect dynamics, μ corresponds to the baseline or homeostatic "goal," and β quantifies the strength of the regulation toward this goal.

The first part of the right-hand side of Eq. (3.1), $\beta(\mu - \theta(t))dt$, describes the deterministic part of the process dynamics. In this part of the model, the degree and direction of change in $\theta(t)$ is determined by the difference between the state of the process at time t, $\theta(t)$, and the baseline, μ, which is scaled by an attractor strength coefficient β (also called drift). We only consider stable processes here, meaning that the adjustment created by the deterministic part of the dynamics is always toward the baseline and never away from it. This is achieved in the current parameterization by restricting the range of β to be positive.[1] More specifically, when the process is above baseline ($\theta(t) > \mu$), the time differential dt is scaled by a negative number ($\mu - \theta(t) < 0$); therefore the value of $\theta(t)$ will decrease toward baseline as a function of the magnitude of the difference, scaled by β. Similarly, when the process is below baseline ($\mu > \theta(t)$), time differential dt part is positive, and the value of $\theta(t)$ will increase toward baseline. As such, the OU process is a *mean-reverting process*: the current value of $\theta(t)$ is always adjusted toward the baseline, μ, which is therefore characterized as the attractor point of the system. The magnitude of the increase or decrease in $\theta(t)$ is scaled proportional to distance from the baseline by β, which defines the attractor strength. When β goes to zero, the OU process approaches a Wiener process, that is, a continuous-time random walk process. When β becomes very large (goes to infinity), the OU process fluctuates around the baseline μ with a certain variance (stationary variance, see next paragraph).

The second part of the right-hand side of Eq. (3.1), $\sigma dW(t)$, describes the stochastic part of the process dynamics. This part adds random input to the system, the magnitude of which is scaled by β and σ. Parameter σ can be transformed into a substantively more interesting parameter γ, by scaling it with the regulation strength β, that is $\gamma = \frac{\sigma^2}{2\beta}$. The γ parameter expresses the within-person fluctuation around baseline due to inputs to the system-defined as all affect-provoking biopsychosocial (BPS) influences internal and external to the individual. As such, γ can be viewed substantively as the degree of BPS-related reactivity in the core affect system or the extent of input that constantly alters the system. Parameter γ is the stationary variance of the OU process: if we let the process evolve over a long time ($t \rightarrow \infty$), the OU process converges to a stationary distribution, a normal distribution with mean μ and variance γ, given that we have a stable process (i.e., $\beta > 0$; see above).

Together, the deterministic and stochastic parts of the model describe how the latent process of interest (e.g., core affect) changes over time: it is continuously drawn toward the baseline while also being disrupted by the stochastic inputs. Psychological processes for which this type of perturbation and mean reversion can be an appropriate model include emotion dynamics and affect regulation (Gross

[1]More intricate dynamics with unrestricted range β include exploding processes with repellers.

2002), semantic foraging (i.e., search in semantic memory, see Hills et al. 2012), and so on. In our current example on core affect, the deterministic part of the OU process is used to describe person-specific characteristics of a self-regulatory system, and the stochastic part of the process is used to describe characteristics of the biopsychosocial inputs that perturb that system—and invoke the need for regulation.

To illustrate the three main parameters of the Ornstein-Uhlenbeck process, we simulated data with different μ, γ, and β values, shown in Fig. 3.2. Baseline (μ) levels are indicated with dotted gray lines. In the first row of the plot matrix, only the baseline, μ, differs between the two plots (set to 50 on the left and 75 on the right), and γ and β are kept the same (100 and 1, respectively). In the second row, only the level of the within-person variance, γ, differs between the two plots (set to 100 on the right and 200 on the left), while μ and β are kept the same (50 and 1, respectively). Finally, in the last row, we manipulate only the level of mean-reverting regulatory force, with a low level of regulation set on the left plot ($\beta = 0.1$) and high regulation on the right ($\beta = 5$). The baseline and the BPS-reactivity kept the same ($\mu = 50$ and $\gamma = 100$). As can be seen, the process on the left wanders away from the baseline and tends to stay longer in one latent position. Descriptively, low β (i.e., weak regulation) on the left produces θ with high autocorrelation (e.g., high

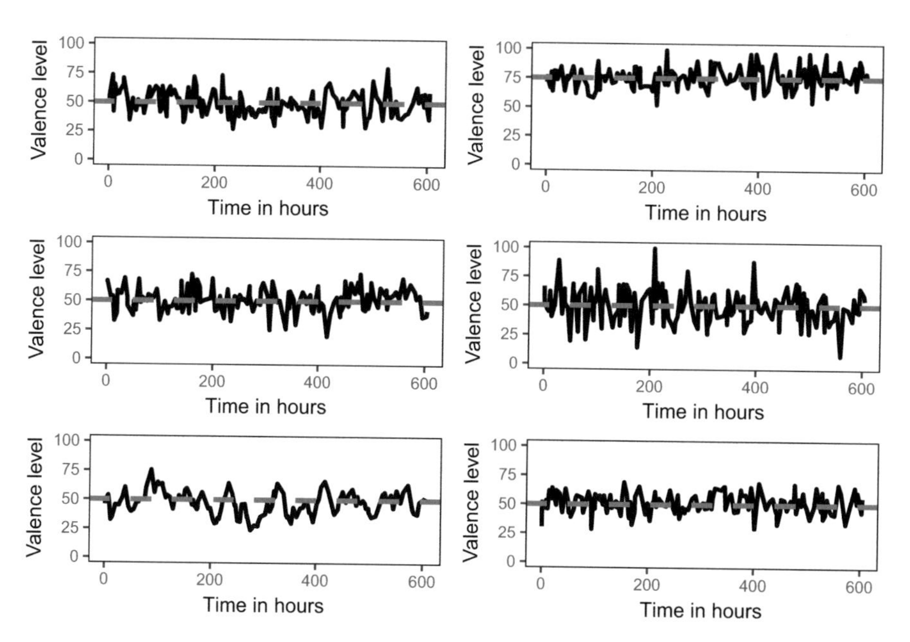

Fig. 3.2 Six Ornstein-Uhlenbeck process trajectories. Each trajectory in the plot matrix consists of 150 irregularly spaced data points. In the first row, the μ was set to 50 for the left plot and 75 for the right plot, with γ kept at 100 and β at 1 for both plots. In the second row, μ and β were kept the same (50 and 1), and γ was set to 100 for the right and 200 for the left. For the last row, μ and γ were kept the same (50 and 100), and β-s were set to 0.1 for the left plot and 5 for the right plot

"emotion inertia"), while high β (i.e., strong regulation) on the right produces θ with low autocorrelation (e.g., low "emotion inertia").

3.2.2 *The Position Equation of the Ornstein-Uhlenbeck Process*

Once the temporal dynamics have been articulated in mathematical form, parameters of the model can estimated from empirical data. With the OU model, however, the estimation algorithm for obtaining parameters in Eq. (3.1) would require approximation of the derivatives from the data shown in Fig. 3.1, potentially introducing some approximation errors. Instead, we take a different approach, solving the stochastic integral in Eq. (3.1) and then estimating the parameters in the integrated model directly from the actual observed data.

We can integrate over Eq. (3.1), to get the value of the latent state θ (i.e., the position of the process) at time t, after some time difference Δ:

$$\theta(t) = \mu + e^{-\beta\Delta}(\theta(t-\Delta) - \mu) + \sigma e^{-\beta\Delta} \int_{t-\Delta}^{t} e^{\beta u} dW(u). \tag{3.2}$$

The integral in Eq. (3.2) is a stochastic integral, taken over the Wiener process. For the OU process, the above stochastic integral was solved based on Itô calculus (see, e.g., Dunn and Gipson 1977), resulting in the conditional distribution of OU process positions, more specifically:

$$\theta(t) \mid \theta(t-\Delta) \sim N(\mu + e^{-\beta\Delta}(\theta(t-\Delta) - \mu),\ \gamma - \gamma e^{-2\beta\Delta}). \tag{3.3}$$

Equation (3.3) is the conditional distribution of the position of the OU process, $\theta(t)$, based on the previous position of the process at $\theta(t - \Delta)$ after elapsed time Δ and based on its three process parameters, μ, γ, and β, described earlier. Equation (3.3) is a particularly useful representation of the OU process as it can be used to formulate the likelihood function for the OU model without the need of approximating derivatives in Eq. (3.1).

The mean of the distribution presented in Eq. (3.3) is a compromise between the baseline μ and the distance of the process from its baseline $(\theta(t) - \mu)$, scaled by $e^{-\beta\Delta}$. The larger β and/or Δ is, the closer this exponential expression gets to 0, and the mean of Eq. (3.3) will get closer to μ. When β and/or Δ are small, the exponential part approaches 1, and the mean will be closer to $\theta(t)$. In fact, β controls the continuous-time exponential autocorrelation function of the process; larger β-s correspond to lower autocorrelation and more centralizing behavior. Naturally, autocorrelation also decreases with the passage of time (higher values of Δ). Figure 3.3 shows a graphical illustration of the continuous-time autocorrelation function of the OU process. Larger values of β correspond to faster regulation to

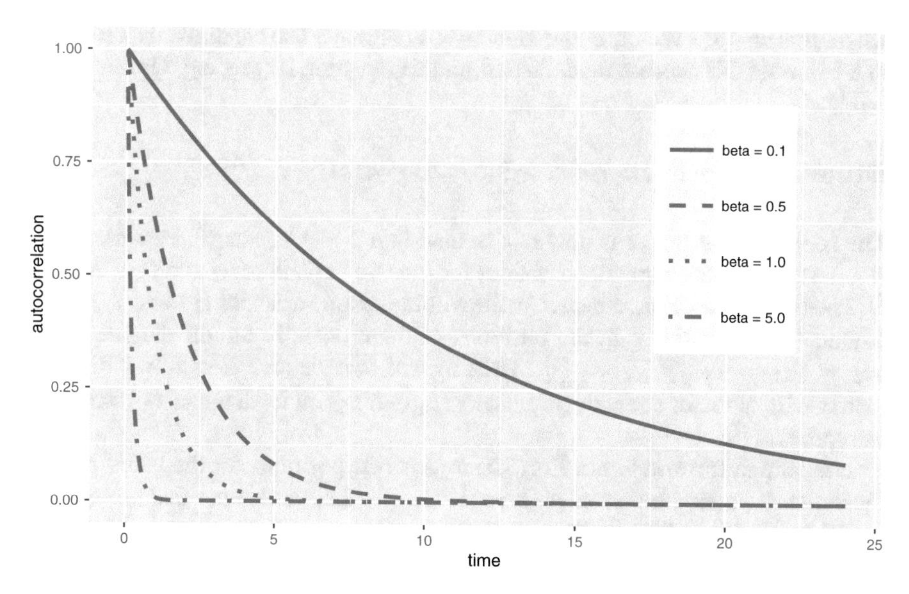

Fig. 3.3 Continuous-time autocorrelation function of the OU process

baseline μ, therefore less autocorrelation in the positions of θ over time. Smaller β values correspond to more autocorrelation over time.

The variance of the process presented in Eq. (3.3) is $\gamma - \gamma e^{-2\beta\Delta}$. We can rearrange this expression to the form of $\gamma(1 - e^{-2\beta\Delta})$. Now if we consider a large Δ value (long elapsed time), the exponential part of this expression goes to 0. Therefore γ represents all the variation in the process—it is the stationary (long run) variance, as described above. The moment-to-moment variation is governed by γ but scaled by the elapsed time and the attractor strength.

3.2.3 Extending the Ornstein-Uhlenbeck Process to Two Dimensions

Thus far, we have presented the model with respect to a univariate $\theta(t)$. Many psychological processes, however, involve multiple variables. Core affect, for example, is defined by levels of both valence and arousal. Process models of core affect, thus, might also consider how the two component variables covary. We can straightforwardly extend into multivariate space by extending $\theta(t)$ into a multivariate vector, $\boldsymbol{\Theta}$, and the corresponding multivariate (n-dimensional) extension of Eq. (3.1) is:

$$d\boldsymbol{\Theta}(t) = B(\boldsymbol{\mu} - \boldsymbol{\Theta}(t))dt + \boldsymbol{\Sigma} d\mathbf{W}(t), \quad (3.4)$$

where $\boldsymbol{\mu}$ is an $n \times 1$ vector, B and $\boldsymbol{\Sigma}$ are $n \times n$ matrices. The conditional distribution of a bivariate OU process model based on this equation is (see, e.g, Oravecz et al. 2011):

$$\boldsymbol{\Theta}(t)|\boldsymbol{\Theta}(t-\Delta) \sim N_2\left(\boldsymbol{\mu} + e^{-B\Delta}(\boldsymbol{\Theta}(t-\Delta) - \boldsymbol{\mu}), \boldsymbol{\Gamma} - e^{-B\Delta}\boldsymbol{\Gamma}e^{-B^{\mathrm{T}}\Delta}\right). \quad (3.5)$$

The latent state at time t is now represented in a 2×1 vector $\boldsymbol{\Theta}$. Parameter $\boldsymbol{\mu}$ is 2×1 vector representing the baselines for the two variables (μ_1 and μ_2); $\boldsymbol{\Gamma}$ is a 2×2 stationary covariance matrix, with within-person variances (γ_1 and γ_2) in the diagonal and covariance in the positions of the process in the off-diagonal, with $\gamma_{12} = \gamma_{21} = \rho_\gamma\sqrt{\gamma_1\gamma_2} = \frac{\sigma_{12}}{\beta_1\beta_2}$. Drift matrix B is defined as a 2×2 diagonal matrix with attractor strength parameters (β_1 and β_2) on its diagonals (^{T} stands for transpose).

It is straightforward to use Eq. (3.5) to describe positions for the latent process dynamics for a single person p ($p = 1, \ldots, P$). Let us assume that we want to model n_p positions for person p, at times $t_{p,1}, t_{p,2}, \ldots, t_{p,s}, \ldots, t_{p,n_p}$. The index s denotes the sth measurement occasion of person p. Now elapsed time Δ can be written specifically as elapsed time between two of these time points: $t_{p,s} - t_{p,s-1}$. We let all OU parameters be person-specific, and then Eq. (3.5) becomes:

$$\begin{aligned}\boldsymbol{\Theta}(t_{p,s})|\boldsymbol{\Theta}(t_{p,s-1}) \sim N_2\big(&\boldsymbol{\mu}_{p,s} + e^{-B_p(t_{p,s}-t_{p,s-1})}(\boldsymbol{\Theta}(t_{p,s-1}) - \boldsymbol{\mu}_{p,s}), \quad (3.6)\\ &\boldsymbol{\Gamma}_p - e^{-B_p(t_{p,s}-t_{p,s-1})}\boldsymbol{\Gamma}_p e^{-B_p^{\mathrm{T}}(t_{p,s}-t_{p,s-1})}\big),\end{aligned}$$

where $\boldsymbol{\Theta}(t_{p,s}) = (\Theta_1(t_{p,s}), \Theta_2(t_{p,s}))^{\mathrm{T}}$ and the rest of the parameters are defined as before in Eq. (3.5).

We note here that our choice of notation of the OU parameters was inspired by literature on modeling animal movements with OU processes (see, e.g., Blackwell 1997; Dunn and Gipson 1977). However, in time-series literature in many of the works referenced so far, the following formulation of the multivariate stochastic differential equation is common:

$$d\boldsymbol{\eta}(t) = (A\boldsymbol{\eta}(t) + \mathbf{b})dt + Gd\mathbf{W}(t),$$

A simple algebraic rearrangement of our Eq. (3.4) gives:

$$d\boldsymbol{\Theta}(t) = (-B\boldsymbol{\Theta}(t) + B\boldsymbol{\mu})dt + \boldsymbol{\Sigma}d\mathbf{W}(t).$$

If we work out the correspondence between the terms, we find that differences for the latent states ($\boldsymbol{\Theta}(t) = \boldsymbol{\eta}(t)$) and the scaler for the effect of the stochastic fluctuations ($\boldsymbol{\Sigma} = G$) are only on the level of the notation. With respect to the drift matrix across the two formulations, $A = -B$, the correspondence is straightforward (the sign only matters when stationarity constraints are to be implemented). The only real difference between the two formulations concerns the $\mathbf{b}$ and $\boldsymbol{\mu}$ parameters:

in the formulation introduced in this chapter, $\boldsymbol{\mu}$ has a substantively interesting interpretation, since it represents a homeostatic baseline to which the process most often returns. In contrast, the **b** parameter, in the more typical SDE formulation, only denotes the intercept of the stochastic differential equation. Our process baseline parameter derives from the typical SDE formulation as $\boldsymbol{\mu} = -A^{-1}\mathbf{b}$.

3.2.4 *Accounting for Measurement Error*

In many social and behavioral science applications, it is reasonable to assume that observed data are actually noisy measurements of the latent underlying process. Therefore, we add a state-space model extension to link the latent OU process variables to the observed data. Equation (3.6) is considered a transition equation, describing changes on the latent level. The observed data is denoted as $\mathbf{Y}(t_{p,s}) = (Y_1(t_{p,s}), Y_2(t_{p,s}))^{\mathrm{T}}$ at time point $t_{p,s}$ for person p at observation s. The measurement equation is then specified as:

$$\mathbf{Y}(t_{p,s}) = \boldsymbol{\Theta}(t_{p,s}) + \boldsymbol{e}_{p,s} \tag{3.7}$$

with the error in measurement distributed as $\boldsymbol{e}_{p,s} \sim N_2(\mathbf{0}, E_p)$, with the off-diagonals of E_p fixed to 0, and person-specific measurement error variances, $\epsilon_{1,p}$ and $\epsilon_{2,p}$, on the diagonals.

3.3 A Multilevel/Hierarchical Extension to the Ornstein-Uhlenbeck Process

The above sections have outlined how the OU model can be used to describe intraindividual dynamics for a given person, p. Also of interest is how individuals differ from one another—interindividual differences in the intraindividual dynamics, and how those differences are related to other between-person differences. The multilevel framework allows for inferences on the hierarchical (or population, or group) level while accommodating individual differences in a statistically coherent manner (Gelman and Hill 2007; Raudenbush and Bryk 2002). The multilevel structure of the model parameters assumes that parameters of the same type share a certain commonality expressed by their superordinate population distributions. In brief, when estimating OU parameters for multiple persons $p = 1, \ldots, P$ in the multilevel framework, we pool information across participants by placing the parameters into a distribution with specific shape (e.g., Gaussian). Treating the person-specific parameters as random variables drawn from a particular interindividual difference distribution thus improves the recovery of the person-level parameters. Further, the multilevel modeling framework allows for a straightforward way to include covariates in the model, without needing to resort to a two-stage analysis (i.e., first estimating person-level model parameters, then exploring their association with the covariates through correlation or regression analysis). We will include both time-varying covariates (TVCs) and time-invariant covariates (TICs) in the model. The modeling of process model parameters such as the regulatory (attractor) force as a function of covariates has not yet been a focus in continuous-time models, although in many cases it can be done in a straightforward manner. In the next paragraphs, we outline how reasonable population distributions are chosen for each person-specific process model parameter.

3.3.1 Specifying the Population Distribution for the Baseline

The two-dimensional baseline, $\boldsymbol{\mu}_{p,s}$, shown in Eq. (3.6), can be made a function of both time-varying and time-invariant covariates. For the TICs, let us assume that K covariates are measured and x_{jp} denotes the score of person p on covariate j ($j = 1, \ldots, k$). All person-specific covariate scores are collected into a vector of length $K + 1$, denoted as $\boldsymbol{x}_p = (x_{p,0}, x_{p,1}, x_{p,2}, \ldots, x_{p,K})^{\mathrm{T}}$, with $x_{p,0} = 1$, to allow for an intercept term. Regarding the TVCs, suppose that we repeatedly measure person p on D time-varying covariates which are collected in a vector $\boldsymbol{z}_{p,s} = (z_{p,s,1}, \ldots, z_{p,s,D})^{\mathrm{T}}$, where index s stands for sth measurement occasion for person p. In order to avoid collinearity problems, no intercept is introduced in the vector $\boldsymbol{z}_{p,s}$.

For the applied example, we use time of the self-report (e.g., time of day) as an indicator of timing within a regular diurnal cycle as TVC ($\boldsymbol{z}_{p,s}$). We expected that some people will show low levels of valence and arousal in the morning, with the baseline increasing and decreasing in a quadratic manner over the course of the day. We also consider interindividual differences in gender and self-reported emotional health as TICs ($\boldsymbol{x}_p$). We expected that person-specific baselines, regulatory force, and extent of BPS input would all differ systematically across gender and linearly with general indicators of individuals' emotional health.

The level 2 (population-level) distribution of $\boldsymbol{\mu}_{p,s}$ with regression on the time-invariant and time-varying covariates person-specific random variation can be written as follows:

$$\boldsymbol{\mu}_{p,s} \sim N_2\left(\boldsymbol{\Delta}_{p\mu}\boldsymbol{z}_{p,s} + \mathbf{A}_{\mu}\boldsymbol{x}_p, \boldsymbol{\Sigma}_{\mu}\right). \tag{3.8}$$

The covariance matrix $\boldsymbol{\Sigma}_{\mu}$ is defined as:

$$\boldsymbol{\Sigma}_{\mu} = \begin{bmatrix} \sigma^2_{\mu_1} & \sigma_{\mu_1\mu_2} \\ \sigma_{\mu_1\mu_2} & \sigma^2_{\mu_2} \end{bmatrix}, \tag{3.9}$$

where $\sigma^2_{\mu_1}$ and $\sigma^2_{\mu_2}$ quantify the unexplained (by covariates) individual differences in baseline values in the two dimensions (in our case valence and arousal), and the covariance parameter $\sigma_{\mu_1\mu_2}$ describes how person-specific levels of valence and arousal covary on the population level, that is, it provides a *between-person measure of covariation* in the core affect dimensions. The TVC regression coefficient matrix $\boldsymbol{\Delta}_{p\mu}$ has dimensions $2D \times P$, allowing between-person differences in the time-varying associations. The TIC regression coefficient matrix $\mathbf{A}_{\mu}$ is of dimensions $2 \times (K+1)$, containing the regression weights for the time-invariant covariates $\boldsymbol{x}_p$.

3.3.2 Specifying the Population Distribution for the Regulatory Force

The regulatory or attractor force is parameterized as a diagonal matrix B_p, with diagonal elements β_{1p} and β_{2p} representing the levels of regulation for the two dimensions. By definition this matrix needs to be positive definite to ensure that there is always an adjustment toward the baseline and never away from it, implying that the process is stable and stationary. This constraint will be implemented by constraining both β_{1p} and β_{2p} to be positive. The population distributions for these two will be set up with a lognormal population distribution. For β_{1p} this is,

$$\beta_{1p} \sim LN\left(\boldsymbol{x}_p^{\mathrm{T}}\boldsymbol{\alpha}_{\beta_1}, \sigma^2_{\beta_1}\right).$$

The mean of this distribution is written as the product of time-invariant covariates and their corresponding regression weights, with the vector $\boldsymbol{\alpha}_{\beta_1}$ containing the (fixed) regression coefficients and parameter $\sigma^2_{\beta_1}$ representing the unexplained interindividual variation in the regulatory force, in the first dimensions. The specification and interpretation of the parameters for the second dimension follow the same logic.

3.3.3 Specifying the Population Distribution for the BPS Input

The 2×2 stationary covariance matrix $\boldsymbol{\Gamma}_p$ models the BPS-related reactivity of the OU-specified process and is formulated as:

$$\boldsymbol{\Gamma}_p = \begin{bmatrix} \gamma_{1p} & \gamma_{12p} \\ \gamma_{21p} & \gamma_{2p} \end{bmatrix}. \tag{3.10}$$

Its diagonal elements (i.e., γ_{1p} and γ_{2p}) quantify the levels of fluctuation due to BPS input in the two dimensions, in our case valence and arousal. The off-diagonal $\gamma_{12p} = \gamma_{21p}$ quantifies how valence and arousal *covary within-person*.

The covariance can be decomposed into $\gamma_{12p} = \rho_{\gamma_p}\sqrt{\gamma_{1p}\gamma_{2p}}$, where ρ_{γ_p} is the contemporaneous (i.e., at the same time) correlation of the gamma parameters on the level of the latent process. The diagonal elements of the covariance matrix (γ_{1p} and γ_{2p}), that is, the variances, are constrained to be positive. We will model the square root of the variance, that is, the intraindividual standard deviation, and assign a lognormal (LN) population distribution to it to constrain it to the positive real line:

$$\sqrt{\gamma_{1p}} \sim LN\left(\boldsymbol{x}_p^{\mathrm{T}}\boldsymbol{\alpha}_{\gamma_1}, \sigma_{\gamma_1}^2\right).$$

The mean of this population-level distribution is modeled via the product of time-invariant covariates and their corresponding regression weights, in the same manner that was described for the regulatory force above. The vector $\boldsymbol{\alpha}_{\gamma_1}$ contains the (fixed) regression coefficients, belonging to the set of person-level covariates. The first element of $\boldsymbol{\alpha}_{\gamma_1}$ relates to the intercept and expresses the overall mean level of BPS-related reactivity. The parameter $\sigma_{\gamma_1}^2$ represents the unexplained interindividual variation in BPS-related reactivity in the first dimension. The specification and interpretation of the parameters for the second dimension follow the same logic.

The cross-correlation ρ_{γ_p} is bounded between -1 and 1. By taking advantage of the Fisher z-transformation $F(\rho_{\gamma_p}) = \frac{1}{2}\log\frac{1+\rho_{\gamma p}}{1-\rho_{\gamma p}}$, we can transform its distribution to the real line:

$$F(\rho_{\gamma_p}) \sim N\left(\boldsymbol{x}_p^{\mathrm{T}}\boldsymbol{\alpha}_{\rho_\gamma}, \sigma_{\rho_\gamma}^2\right).$$

Again, α_{ρ_γ} contains $K+1$ regression weights and $\boldsymbol{x}_p^{\mathrm{T}}$ the K covariate values for person p with 1 for the intercept, and $\sigma_{\rho_\gamma}^2$ quantifies the unexplained interindividual variation. The first coefficient of $\boldsymbol{\alpha}_{\rho_\gamma}$ belongs to the intercept and represents the overall population-level within-person (standardized) correlation between the two dimensions. We note that while the model can capture covariation across the two dimensions, the current implementation is limited in the sense that it does not capture how the processes may influence each other over time (i.e., the off-diagonal elements of B_p are not estimated).

3.4 Casting the Multilevel OU Process Model in the Bayesian Framework

Estimation of the full model and inference to both individuals and the population are facilitated by the Bayesian statistical framework. In brief, the Bayesian statistical inference framework entails using a full probability model that describes not only our uncertainty in the value of an outcome variable (y) conditional on some unknown parameter(s) (θ) but also the uncertainty about the parameter(s) themselves. The goal of Bayesian inference is to update our beliefs about the likely values of model parameters using the model and data. The relationship between our prior beliefs about the parameters (before observing the data) and our posterior beliefs about the parameters (after observing the data) is described by Bayes' theorem: $p(\theta|y) = p(y|\theta)p(\theta)/p(y)$, which states that the posterior probability distribution, $p(\theta|y)$, of parameter(s) θ given data y is equal to the product of a likelihood function $p(y|\theta)$ and prior distribution $p(\theta)$, scaled by the marginal likelihood of the data $p(y)$. With the posterior distribution, we can easily make nuanced and intuitive probabilistic inference. Since in Bayesian inference we obtain a posterior probability distribution over all possible parameter values, instead of merely point estimates, we can use the posterior distribution to make probabilistic statements about parameters (and functions of parameters) of interest. For example, we can easily derive the probability that the parameter value lies in any given interval.

To cast the described multilevel process model in Bayesian framework, we used Eqs. (3.6) and (3.7) as our likelihood function, and we specified non-informative prior distributions on all parameters. The posterior can be thought of as a compromise between the likelihood and the prior distributions and describes the relative plausibility of all parameter values conditional on the model being estimated. In the Bayesian framework, parameter estimation and inference focuses on the posterior distribution. For some simple models, posterior distributions can be calculated analytically, but for almost all nontrivial models, the posterior has to be approximated numerically. Most commonly, Bayesian software packages employ simulation techniques such as Markov chain Monte Carlo algorithms to obtain many draws from the posterior distribution. After a sufficiently large number of iterations, one obtains a Markov chain with the posterior distribution as its equilibrium distribution, and the generated samples can be considered as draws from the posterior distribution. Checks that the algorithm is behaving properly are facilitated by use of multiple chains of draws that are started from different initial values and should converge to the same range of values.

3.5 Investigating Core Affect Dynamics with the Bayesian Multilevel Ornstein-Uhlenbeck Process Model

3.5.1 A Process Model of Core Affect Dynamics Measured in an Ecological Momentary Assessment Study

Data used in the current illustration of the process modeling approach were collected at Pennsylvania State University in the United States from $N = 52$ individuals (35 female, mean age = 30 years, SD = 10) who participated in an EMA study of core affect and well-being. Participants were informed that the study protocol consisted of (1) filling out short web-based survey via their own smartphones, six times a day for 4 weeks, while going on with the course of their everyday life and (2) completing a battery of personality tests and demographics items during the introductory and exit sessions. After consent, the participants provided their phone number and were registered with a text messaging service. Over the course of the next month, participants received and responded to up to 168 text-message-prompted surveys. Compliance was high, with participants completing an average of 157 (SD = 15) of the surveys. Participants were paid proportional to their response rate, with maximum payment of $200. In addition to the core affect ratings of *valence* and *arousal* from the repeated surveys, we make use of two sets of covariates. Linear and quadratic representations of the *time of day* of each assessment, which ranged between 7 and 24 o'clock (centered at 12 noon), were used as TVCs. Select information from the introductory and exit batteries were used as TICs, namely, *gender* ($n = 35$ female) and two measures of emotional functioning: the *emotional well-being* ($M = 74$, SD = 18) and *role limitations due to emotional problems* ($M = 77$, SD = 36) scales from the 36-Item Short Form Health Survey (SF-36; Ware et al. 1993) (centered and standardized for analysis).

The parameter estimation for the multilevel, bivariate OU model described by Eqs. (3.6) and (3.7), was implemented in JAGS (Plummer 2003) and fitted to the above data.[2] Six chains of 65,000 iterations each were initiated from different starting values, from which the initial 15,000 were discarded (adaptation and burn-in). Convergence was checked using the Gelman-Rubin $\hat{R}$ statistic (for more information, see Gelman et al. 2013). All $\hat{R}$'s were under 1.1 which indicated no problems with convergence.

3.5.2 *Population-Level Summaries and Individual Differences of Core Affect Dynamics*

Table 3.1 summarizes the results on the population level by showing posterior summary statistics for the population mean values and interindividual standard deviation for each process model parameter. The posterior summary statistics are the posterior mean (column 2) and the lower and upper ends of the 90% highest probability density interval (HDI; columns 3 and 4), designating the 90% range of values with the highest probability density. We walk through each set of parameters in turn.

The valence and arousal baselines were allowed to vary as function of time of day (linear and quadratic). The displayed estimates for the population mean baselines therefore represent estimated baselines at 12 noon. As seen in the first line of Table 3.1, average core affect at noon is somewhat pleasant (59.60 on the 0–100 response scale) and activated (57.63). Posterior means for the linear and quadratic effects of time of day indicate practically no diurnal trend for valence (1.30 for

[2]Code and data used in this chapter are available as supplementary material at the book website http://www.springer.com/us/book/9783319772189.

Table 3.1 Population-level results

Model parameter		Posterior	90% HDI	
Description	Notation	mean	Low	High
Valence				
Baseline intercept	α_{1,μ_1}	59.60	54.54	64.75
Interindividual SD in baseline	σ_{μ_1}	12.05	9.51	14.43
Linear time effect	$\alpha_{\delta_{1,L}}$	1.30	0.92	1.74
Interindividual SD in linear time effect	$\sigma_{\delta_{1,L}}$	0.19	0.08	0.30
Quadratic time effect	$\alpha_{\delta_{1,Q}}$	−0.04	−0.05	−0.02
Interindividual SD in quadratic time	$\sigma_{\delta_{1,Q}}$	0.00	0.00	0.01
BPS-related reactivity intercept	$\alpha_{1,\sqrt{\gamma_1}}$	7.53	5.30	10.94
Interindividual SD in BPS-related reactivity	$\sigma_{\sqrt{\gamma_1}}$	4.94	2.50	9.46
Regulation intercept	α_{1,β_1}	0.17	0.09	0.33
Interindividual SD in regulation	σ_{β_1}	0.19	0.06	0.57
Arousal				
Baseline intercept	α_{1,μ_2}	57.63	51.07	64.23
Interindividual SD in baseline	σ_{μ_2}	16.13	12.70	19.32
Linear time effect	$\alpha_{\delta_{2,L}}$	6.32	5.73	6.90
Interindividual SD in linear time effect	$\sigma_{\delta_{2,L}}$	0.37	0.00	0.61
Quadratic time effect	$\alpha_{\delta_{2,Q}}$	−0.22	−0.23	−0.19
Interindividual SD in quadratic time effect	$\sigma_{\delta_{2,Q}}$	0.02	0.02	0.03
BPS-related reactivity intercept	$\alpha_{1,\sqrt{\gamma_2}}$	13.58	11.08	16.69
Interindividual SD in BPS-related reactivity	$\sigma_{\sqrt{\gamma_2}}$	5.17	3.20	7.91
Regulation intercept	α_{1,β_2}	0.44	0.25	0.80
Interindividual SD in regulation	σ_{β_2}	0.39	0.14	1.04
Cross-effects				
Within-person correlation intercept	α_{1,ρ_γ}	0.99	0.90	1.00
Between-person correlation intercept	$\sigma_{\mu_1\mu_2}$	0.68	0.53	0.84

linear time, −0.04 for quadratic time), but an inverted U-shaped pattern for arousal (6.32 for linear, −0.22 for quadratic). The interindividual SDs for baselines and the linear and quadratic time effects quantify the extent of between-person differences in baseline throughout the day. These differences are illustrated in Fig. 3.4. As can be seen, most of the interindividual differences are in the baseline intercepts (differences in level at 12 noon) and not in the shape of the trajectories. For example, the left panel shows that while there is remarkable inverted U-shaped trend in the daily arousal baselines, this pattern is quite similar across people. Similarly, the right panel shows the extent of between-person differences in level, and similarities in shape of the daily trends, for valence.

The population-level estimates of the biopsychosocial input-related reactivity are also summarized Table 3.1. The prototypical participant (male) had BPS input to valence of γ_1 7.53 and to arousal of $\gamma_2 = 13.58$, with the amount of perturbation

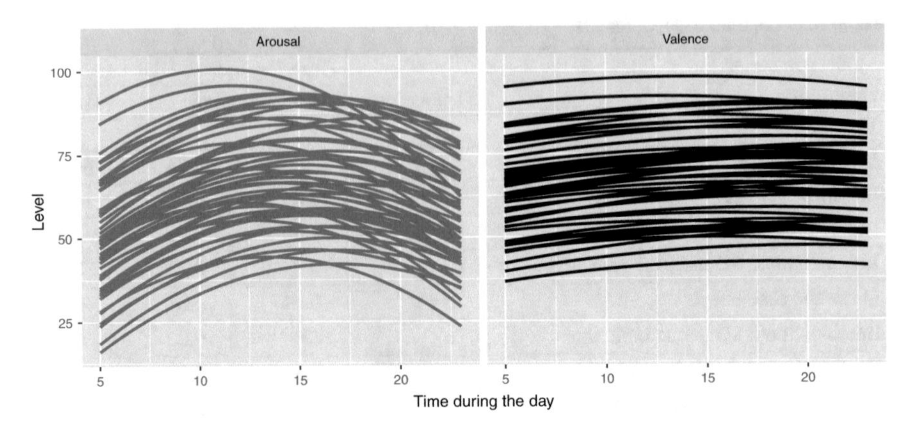

Fig. 3.4 Individual differences in changes in core affect over the day

differing substantially between-persons for both valence (SD = 4.94) and arousal (SD = 5.17).[3]

The population-level estimates of regulatory force (β_1 and β_2) are also shown in Table 3.1. These values quantify how quickly individuals' core affect is regulated back to baseline after perturbation. The prototypical participant had regulatory force of $\beta_1 = 0.17$ for valence and $\beta_2 = 0.44$ for arousal, with substantial between-person difference in how quickly the regulatory force acted (SD = 0.19 for valence, SD = 0.39 for arousal). As noted above, these parameters control the slope of the exponential continuous-time autocorrelation function of the process. Illustrations of the differences across persons for both valence and arousal are shown in Fig. 3.5.

Finally, the population-level estimates of the cross-effects, within-person and between-person covariation of valence and arousal, are shown in the bottom of Table 3.1. For the prototypical participant (male), valence and arousal are very strongly coupled within-person, $r = 0.99$, suggesting that valence and arousal change in similar ways when BPS-related perturbations come in (but see also covariate results regarding gender differences for this parameter in the next section). Across persons, individuals with higher baselines in valence also tend to have higher baselines in arousal, $r = 0.68$.

[3]Note that the means and standard deviations for this standard deviation parameter ($\sqrt{\gamma}$) are based on the first and second expectations of the lognormal distribution. Same applies for the regulation (β) parameters as well.

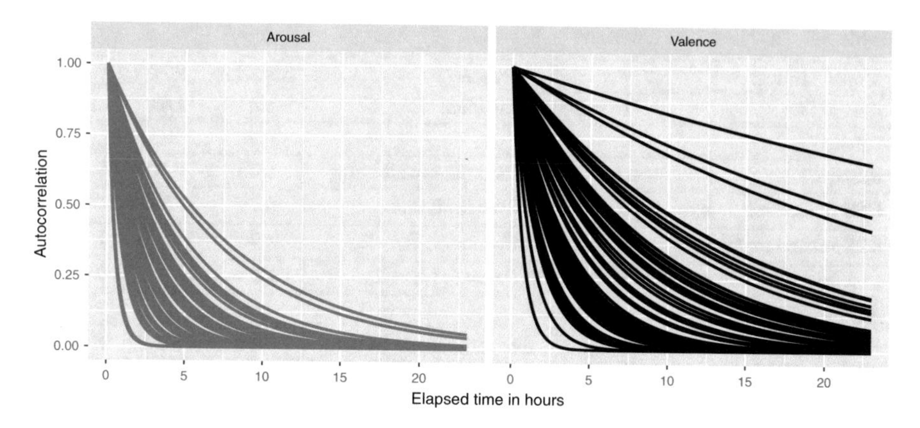

Fig. 3.5 Individual differences in the continuous-time autocorrelation functions for valence and arousal

3.5.3 Results on the Time-Invariant Covariates

Selected results on the time-invariant covariates are shown in Table 3.2. Only the TICs for which most of the posterior mass was positive or negative (90% highest density interval did not contain 0) were selected, as these are the coefficients that we consider remarkably different from 0. As mentioned before, γ (BPS-related reactivity) and β (regulation) were constrained to be positive; therefore their log-transformed values were regressed on the covariates; for these parameters relative effect sizes are reported in Table 3.2 (instead of the original regression coefficients that relate to the log scale). With regard to the covariates, gender was coded as 0 for male and 1 for female, and higher values on the emotional well-being scale indicate better well-being, while lower values on the role limitations due to emotional problems scale indicate more difficulties.

As expected, results show that higher levels of emotional well-being were associated with higher baseline levels of valence ($\alpha_{3,\mu_1} = 8.00$) and higher baseline levels of arousal ($\alpha_{3,\mu_2} = 6.54$). Higher levels of emotional well-being were also associated with greater BPS input into arousal: $\alpha_{3,\sqrt{\gamma_2}} = 1.14$. Again, the 1.14 value here is a relative effect size that relates to the original scale of the BPS-related reactivity and is interpreted the following way: consider a comparison point to be at 1, then the relative effect size expresses how many percent of change in the outcome (i.e., BPS-related reactivity) is associated with one standard deviation (or one point in case of gender) change in the covariate. For example, if we consider the association between emotional well-being and BPS-related reactivity, one standard deviation increase in emotional well-being is associated with 14% ($1.14 - 1 = 0.14$) increase in BPS-related reactivity. Relative effect sizes related to regulation are interpreted the same way.

Table 3.2 Results on the covariates

Model parameter		Person	Posterior	90% HDI	
Description	Notation	covariate	mean	Low	High
Valence					
Baseline	α_{3,μ_1}	Emotional well-being	8.00	3.95	12.22
BPS-related reactivity	$\alpha_{4,\sqrt{\gamma_1}}$	Role limitations	0.79	0.64	0.97
BPS-related reactivity	$\alpha_{2,\sqrt{\gamma_1}}$	Gender (1: female)	1.46	1.05	2.04
Regulation	α_{2,β_1}	Gender (1: female)	2.21	1.27	3.81
Arousal					
Baseline	α_{3,μ_2}	Emotional well-being	6.54	2.03	11.14
BPS-related reactivity	$\alpha_{3,\sqrt{\gamma_2}}$	Emotional well-being	1.14	1.01	1.29
Regulation	α_{4,β_2}	Role limitations	0.73	0.53	0.98
Cross-effects					
Synchronicity in changes	α_{2,ρ_γ}	Gender (1: female)	−0.89	−0.99	−0.34

Greater role limitation due to emotional problems was associated with higher levels of BPS-related valence reactivity ($\alpha_{4,\sqrt{\gamma_1}} = 0.79$; 21%) and stronger regulatory force on arousal (i.e., quicker return to baseline; $\alpha_{4,\beta_2} = 0.73$; 27%). There were also some notable gender differences: female participants tended to have higher levels of BPS-related valence reactivity ($\alpha_{2,\sqrt{\gamma_1}} = 1.46$; 46%) and stronger regulatory force for valence ($\alpha_{2,\beta_1} = 2.21$; 121%). This suggests that women participants experienced both greater fluctuations in pleasantness over time and regulated more quickly toward baseline. In line with the idea that females must contend with a more varied and less predictable set of perturbations, they also have less synchronicity between changes in valence and arousal ($\alpha_{2,\rho_\gamma} = -0.89$).

Note that this is an exploratory approach of looking at associations between-person traits and dynamical parameters. As can be seen, some coefficients in Table 3.2 represent very small effect sizes with 90% highest density interval (HDI) being close to 1 (the cutoff for the relative effect sizes). For more robust inference in a follow-up study, we would recommend using a stricter criterion (e.g., 99% HDI) or, more ideally, calculating a Bayes factor for these coefficients.

3.6 Discussion

In this paper we presented process modeling as a framework that can contribute to theory building and testing through mathematical specification of the processes that may produce observed data. We illustrated how the framework can be used to examine interindividual differences in the intraindividual dynamics of core affect. In conceptualization, core affect is a continuously changing neurophysiological blend of valence and arousal that underlies all emotional experiences (Barrett 2017)—a video. Measurement of core affect, however, requires that individuals consciously access their core affect and indicate their current level of pleasantness (valence) and activation (arousal)—a selfie. The inherent discrepancy between the continuous-time conceptualization of core affect and the moment-to-moment measurement of core affect requires a framework wherein the parameters of process models governing action in an individual's movie biopic can be estimated from a series of selfies that were likely snapped at random intervals (e.g., in an EMA study design).

Our illustration developed a mathematical description for the continuous-time conceptualization of core affect based on a mean-reverting stochastic differential equation, the OU model. From a theoretical perspective, this mathematical model is particularly useful because it explicitly maps key aspects of the hypothesized intraindividual regulatory dynamics (e.g., DynAffect model) of core affect, to three specific parameters, μ, β, and γ, that may differ across persons. Expansion into a bivariate model provides opportunity to additionally examine interrelations between affect valence and affect arousal. A key task in tethering the mathematical model to the psychological theory is "naming the betas" (Ram and Grimm 2015). Explicitly naming the parameters facilitates interpretation, formulation and testing of hypotheses, and potentially, theory building/revision. Here, we explicitly tethered μ to a baseline level of core affect—the "goal" or attractor point of the system; β to the strength of the "pull" of the baseline point—an internal regulatory force; and γ to the variability that is induced by affect-provoking biopsychosocial inputs. This content area-specific naming facilitated identification and inclusion into the model of a variety of time-varying and time-invariant covariates—putative "causal factors" that influence the intraindividual dynamics and interindividual differences in those dynamics. In particular, inclusion of time-of-day variables (linear and quadratic) provided for testing of hypotheses about how baseline valence and arousal change across the day in accordance with diurnal cycles, and inclusion of gender and indicators of emotional well-being provide for testing of hypotheses about how social roles and psychological context may influence affective reactivity and regulation. Generally, parameter names are purposively selected to be *substantively* informative in order that theoretical considerations may be easily engaged.

Statistical considerations come to the fore when attempting to match the model to empirical data, and particularly in situations like the one illustrated here, where a model with measurement and dynamic equations is placed within a multilevel framework that accommodates estimation of interindividual differences in intraindividual dynamics from *noisy* ILD that are unbalanced and time-unstructured. Our empirical illustration was constructed to highlight the utility of a multilevel Bayesian statistical estimation and inference framework (Gelman et al. 2013). The flexible Markov chain Monte Carlo methods provide for estimation of increasingly complex models that include multiple levels of nesting and both time-varying and time-invariant covariates. Specification of the full probability model, for both the observed outcomes and the model parameters, provides for robust and defensible probabilistic statements about the phenomena of interest. In short, the Bayesian estimation framework offers a flexible alternative to frequentist estimation techniques and may be particularly useful when working with complex multilevel models.

We highlight specifically the benefits of estimation for EMA data. Generally, these data are purposively collected to be time-unstructured. Intervals between assessments are kept inconsistent in order to reduce validity threats related to expectation bias in self-reports. This is problematic when a discrete-time mathematical model is used to describe the processes that may produce observed data, because the data cannot be mapped (in a straightforward way) to relations formulated in terms of $\theta(t)$ and $\theta(t-1)$. As such, when working with EMA data, the model is better formulated with respect to *continuous time*, $\theta(t)$ and $d\theta(t)$. Theoretically, continuous-time models may also be more accurate to describe any measured phenomena that do not cease to exist between observations: most processes in behavioral sciences unfold in continuous time and should be modeled as such, see more discussion in, for example, Oud (2002) and Oud and Voelkle (2014a). In our case, core affect is by definition *continuously* changing and thus requires a mathematical description based in a continuous-time model. The process modeling framework stands taller and with more confidence when the data, mathematical model, and theory are all continuously aligned.

Acknowledgment The research reported in this paper was sponsored by grant #48192 from the John Templeton Foundation, the National Institutes of Health (R01 HD076994, UL TR000127), and the Penn State Social Science Research Institute.

References

Blackwell, P. G. (1997). Random diffusion models for animal movements. *Ecological Modelling, 100*, 87–102. https://doi.org/10.1016/S0304-3800(97)00153-1

Bolger, N., & Laurenceau, J. (2013). *Intensive longitudinal methods: An introduction to diary and experience sampling research*. New York, NY: Guilford Press.

Chow, S.-M., Haltigan, J. D., & Messinger, D. S. (2010). Dynamic infant-parent affect coupling during the face-to-face/still-face. *Emotion, 10*(1), 101–114. https://doi.org/10.1037/a0017824

Chow, S.-M., Ram, N., Boker, S. M., Fujita, F., & Clore, G. (2005). Emotion as a thermostat: Representing emotion regulation using a damped oscillator model. *Emotion, 5*, 208–225. https://doi.org/10.1037/1528-3542.5.2.208

Deboeck, P. R., & Bergeman, C. (2013). The reservoir model: A differential equation model of psychological regulation. *Psychological Methods, 18*(2), 237–256. https://doi.org/10.1037/a0031603

Driver, C. C., Oud, J. H. L., & Voelkle, M. (2017). Continuous time structural equation modeling with R package ctsem. *Journal of Statistical Software, 77*(5), 1–35. https://doi.org/10.18637/jss.v077.i05

Driver, C. C., & Voelkle, M. (2018). Hierarchical Bayesian continuous time dynamic modeling. *Psychological Methods*. https://www.ncbi.nlm.nih.gov/pubmed/29595295

Dunn, J. E., & Gipson, P. S. (1977). Analysis of radio telemetry data in studies of home range. *Biometrics, 33*, 85–101. https://doi.org/10.2307/2529305

Barrett, L. F. (2017). The theory of constructed emotion: An active inference account of interoception and categorization. *Social Cognitive and Affective Neuroscience, 12*(1), 1–23. http://doi.org/10.1093/scan/nsw154

Gelman, A., Carlin, J. B., Stern, H. S., Dunson, D. B., Vehtari, A., & Rubin, D. B. (2013). *Bayesian data analysis* (3rd ed.). Boca Raton, FL: Chapman & Hall/CRC.

Gelman, A., & Hill, J. (2007). *Data analysis using regression and multilevel/hierarchical models*. Cambridge: Cambridge University Press.

Gross, J. J. (2002). Emotion regulation: Affective, cognitive, and social consequences. *Psychophysiology, 39*(03), 281–291. https://doi.org/10.1037/0022-3514.85.2.348

Hills, T. T., Jones, M. N., & Todd, P. M. (2012). Optimal foraging in semantic memory. *Psychological Review, 119*(2), 431–440. https://doi.org/10.1037/a0027373

Kuppens, P., Oravecz, Z., & Tuerlinckx, F. (2010). Feelings change: Accounting for individual differences in the temporal dynamics of affect. *Journal of Personality and Social Psychology, 99*, 1042–1060. https://doi.org/10.1037/a0020962

Oravecz, Z., Faust, K., & Batchelder, W. (2014). An extended cultural consensus theory model to account for cognitive processes for decision making in social surveys. *Sociological Methodology, 44*, 185–228. https://doi.org/10.1177/0081175014529767

Oravecz, Z., Tuerlinckx, F., & Vandekerckhove, J. (2011). A hierarchical latent stochastic differential equation model for affective dynamics. *Psychological Methods, 16*, 468–490. https://doi.org/10.1037/a0024375

Oravecz, Z., Tuerlinckx, F., & Vandekerckhove, J. (2016). Bayesian data analysis with the bivariate hierarchical Ornstein-Uhlenbeck process model. *Multivariate Behavioral Research, 51*, 106–119. https://doi.org/10.1080/00273171.2015.1110512

Oud, J. H. L. (2002). Continuous time modeling of the crossed-lagged panel design. *Kwantitatieve Methoden, 69*, 1–26.

Oud, J. H. L., & Voelkle, M. C. (2014a). Continuous time analysis. In A. C. E. Michalos (Ed.), *Encyclopedia of quality of life research* (pp. 1270–1273). Dordrecht: Springer. https://doi.org/10.1007/978-94-007-0753-5_561

Oud, J. H. L., & Voelkle, M. C. (2014b). Do missing values exist? Incomplete data handling in cross-national longitudinal studies by means of continuous time modeling. *Quality & Quantity, 48*(6), 3271–3288. https://doi.org/10.1007/s11135-013-9955-9

Plummer, M. (2003). JAGS: A program for analysis of Bayesian graphical models using Gibbs sampling. In *Proceedings of the 3rd International Workshop on Distributed Statistical Computing (DSC 2003)* (pp. 20–22).

Ram, N., & Grimm, K. J. (2015). Growth curve modeling and longitudinal factor analysis. In R. M. Lerner (Ed.), *Handbook of child psychology and developmental science* (pp. 1–31). Hoboken, NJ: John Wiley & Sons, Inc.

Ram, N., Shiyko, M., Lunkenheimer, E. S., Doerksen, S., & Conroy, D. (2014). Families as coordinated symbiotic systems: Making use of nonlinear dynamic models. In S. M. McHale, P. Amato, & A. Booth (Eds.), *Emerging methods in family research* (pp. 19–37). Cham: Springer International Publishing. https://doi.org/10.1007/978-3-319-01562-0_2

Ratcliff, R., & Rouder, J. N. (1998). Modeling response times for two-choice decisions. *Psychological Science, 9*, 347–356. https://doi.org/10.1111/1467-9280.00067

Raudenbush, S. W., & Bryk, A. S. (2002). *Hierarchical linear models: Applications and data analysis methods*. Newbury Park, CA: Sage.

Robert, C. P., & Casella, G. (2004). *Monte Carlo statistical methods*. New York, NY: Springer. https://doi.org/10.1007/978-1-4757-4145-2

Russell, J. A. (2003). Core affect and the psychological construction of emotion. *Psychological Review, 110*, 145–172. https://doi.org/10.1037/0033-295X.110.1.145

Shiffman, S., Stone, A. A., & Hufford, M. R. (2008). Ecological momentary assessment. *Annual Review of Clinical Psychology, 4*(1), 1–32. https://doi.org/10.1146/annurev.clinpsy.3.022806.091415 (PMID: 18509902).

Stone, A. A., & Shiffman, S. (1994). Ecological momentary assessment (EMA) in behavioral medicine. *Annals of Behavioral Medicine, 16*(3), 199–202.

Thomas, E. A., & Martin, J. A. (1976). Analyses of parent-infant interaction. *Psychological Review, 83*(2), 141–156. https://doi.org/10.1037/0033-295X.83.2.141

Uhlenbeck, G. E., & Ornstein, L. S. (1930). On the theory of Brownian motion. *Physical Review, 36*, 823–841. https://doi.org/10.1103/PhysRev.36.823

Walls, T. A., & Schafer, J. L. (2006). *Models for intensive longitudinal data*. New York, NY: Oxford University Press. https://doi.org/10.1093/acprof:oso/9780195173444.001.0001
Ware, I. E., Snow, K. K., Kosinski, M., & Gandek, B. (1993). *SF-36 Health Survey. Manual and interpretation guide*. Boston, MA: Nimrod Press.
Wetzels, R., Vandekerckhove, J., Tuerlinckx, F., & Wagenmakers, E.-J. (2010). Bayesian parameter estimation in the Expectancy Valence model of the Iowa gambling task. *Journal of Mathematical Psychology, 54*, 14–27. https://doi.org/10.1016/j.jmp.2008.12.001
Wiener, N. (1923). Differential-space. *Journal of Mathematics and Physics, 2*(1–4), 131–174. http://dx.doi.org/10.1002/sapm192321131
Wood, J., Oravecz, Z., Vogel, N., Benson, L., Chow, S.-M., Cole, P., ... Ram, N. (2017). Modeling intraindividual dynamics using stochastic differential equations: Age differences in affect regulation. *Journals of Gerontology: Psychological Sciences, 73*(1), 171–184. https://doi.org/10.1093/geronb/gbx013

Chapter 4
Understanding the Time Course of Interventions with Continuous Time Dynamic Models

Charles C. Driver and Manuel C. Voelkle

4.1 Introduction

Time has often been given only limited consideration when assessing experimental interventions or treatments, but as all such effects unfold over time, a genuine understanding of the system in question is only possible with the full consideration of the time course. How long does a treatment take to have maximum effect? Is the effect then maintained or does it dissipate or perhaps even reverse? Do certain sorts of people respond faster or stronger than others? Is the treatment more effective in the long run for those that respond quickly? These are the sorts of questions we should be able to answer if we truly understand an intervention and the system we apply it to.

The randomised controlled trial is recognised as something of a gold standard for the analysis of interventions, for good reason. Yet still, when it comes to the direct knowledge of the world such a trial can provide to us, we learn only about what happened to some subjects at a particular moment, or moments, of observation. Going beyond such experimental knowledge in order to produce useful, powerful and predictive inferences about the world requires some assumptions about regularity and stability of the universe.

Electronic Supplementary Material The online version of this article (https://doi.org/10.1007/978-3-319-77219-6_4) contains supplementary material, which is available to authorized users.

C. C. Driver (✉)
Centre for Lifespan Psychology, Max Planck Institute for Human Development, Berlin, Germany
e-mail: driver@mpib-berlin.mpg.de

M. C. Voelkle
Department of Psychology, Humboldt-Universität zu Berlin, Berlin, Germany
e-mail: manuel.voelkle@hu-berlin.de

K. van Montfort et al. (eds.), *Continuous Time Modeling in the Behavioral and Related Sciences*, https://doi.org/10.1007/978-3-319-77219-6_4

For some experimental manipulations, it may be sufficient that we learn that there tends to be an effect while the manipulation is occurring. In such cases we need only the most basic assumptions of universal regularity for the knowledge to be useful at later occasions—similar antecedents result in similar consequences. However there are many effects which are only of interest because we assume they persist in some sense outside the bounds of our observation window—a treatment for depression would not be very useful if it only improved a person's depression while being evaluated! As such, whether we are explicitly conscious of it or not, we typically rely on some model of temporal regularity.

In many cases such a model of temporal regularity may be vague and implicit—something like 'there is some sort of continuity in the observed effects even when we are not observing the subjects'. This stance may be adequate if we know enough about the nature of our system at the time scale we are interested in. For instance, if we assess the effect of a treatment on the level of depression 90 days later, we would also expect to know quite a lot about the effect of the treatment at 89 and 91 days, even though no observations were made on those days. But what about the effect after 30 days or 180 days? Probably, most people would agree that additional observations would be necessary. With multiple observations we could then interpolate between and beyond them, but how should this be done? Simple linear interpolation between the strength of an average treatment effect (across subjects) at various occasions can be adequate for some situations and research questions, but we can also do much better. In this work we endeavour to show that adopting a dynamic systems approach cannot only yield improved estimates of the effect of an intervention at unobserved times but can also help us better understand the nature of the intervention and system more fully and improve possibilities for personalised treatments.

Specifically, in this chapter we adopt a continuous time dynamic modelling approach to the problem, based on linear stochastic differential equations. With this approach, variability within a subject over time is partitioned into stochastic inputs at the system level (latent process variance), deterministic changes based on earlier states of the system, stochastic inputs at the measurement level (measurement error), deterministic inputs of unknown origin (i.e. trends) and, then finally, deterministic inputs of known cause—an intervention. In broad terms, the approach differs from what could be done using latent growth curves by the inclusion of the system noise and dynamics component. Thus, rather than sketching only a *description* of change over time, to the extent possible, the generating *process* is also considered, even if only in very broad terms. This can lead to more informative inferences, dependent of course on the quality of the data, assumptions and modelling. For an introduction to continuous time models in the field of psychology, see Voelkle et al. (2012), and for more background, see Oud and Jansen (2000). The text from Gardiner (1985) gives a detailed treatment of stochastic differential equations in general.

Interventions in a system dynamics context are already considered in fields such as pharmacodynamics and pharmacokinetics for drug discovery and testing. There they endeavour to describe processes on the causal path between blood plasma concentration and effect, for multiple subjects (see for general and estimation-focused overviews, respectively Danhof et al. 2008; Donnet and Samson 2013). Single-subject analysis, in which many measurements over time are required, has been more commonly undertaken with such an approach (see, for instance, the insulin dose and glucose monitoring work of Wang et al. 2014). In the realm of psychology, Boker et al. (2016) discuss how certain inputs can lead to changes in the equilibrium of a process. With respect particularly to continuous time models, Singer (1998, 1993) and Oud and Jansen (2000) have detailed model specifications that included time-varying parameters and exogenous inputs.

In the remainder of this work, we will first describe the continuous time dynamic model used here in more depth, and then consider why and how various deterministic input effects may be modelled, as well as mediation—how we may be able to use interventions to understand relations between the processes—and individual differences. We finish with a brief example demonstrating how, instead of being a main focus, interventions can also be used to aid in system identification. Throughout the work we will provide examples using the ctsem (Driver et al. 2017) software package for R (R Core Team 2014), which interfaces to both the OpenMx (Neale et al. 2016) and Stan (Carpenter et al. 2017) software.

4.2 The Model

The continuous time dynamic model we are interested in here is comprised of a latent dynamic model and a measurement model. We have previously described different approaches and software for estimating such models in either single- or multiple-subject contexts; for a maximum likelihood-based specification with mixed effects, see Driver et al. (2017), and for fully random-effects with a hierarchical Bayesian approach, see Driver and Voelkle (2018). Note that while various elements in the model depend on time, the fundamental parameters of the model as described here are time-invariant. Note also that while subject-specific subscripts on the parameters are possible, for simplicity they are not included at this point.

4.2.1 Latent Dynamic Model

The dynamic system is described by the following linear stochastic differential equation:

$$\mathrm{d}\boldsymbol{\eta}(t) = \left(\boldsymbol{A}\boldsymbol{\eta}(t) + \boldsymbol{b} + \boldsymbol{M}\boldsymbol{\chi}(t)\right)\mathrm{d}t + \boldsymbol{G}\mathrm{d}\boldsymbol{W}(t) \tag{4.1}$$

Vector $\boldsymbol{\eta}(t) \in \mathbb{R}^v$ represents the state of the latent processes at time t. The matrix $\boldsymbol{A} \in \mathbb{R}^{v \times v}$ is often referred to as the drift matrix, with auto effects on the diagonal and cross effects on the off-diagonals characterising the temporal dynamics of the processes. Negative values on the auto effects are typical of nonexplosive processes and imply that as the latent state becomes more positive, a stronger negative influence on the expected change in the process occurs—in the absence of other influences, the process tends to revert to a baseline. A positive cross effect in the first row and second column would imply that as the second process becomes more positive, the expected change in the first process also becomes more positive. The expected change for a given interval of time can be calculated and is shown in Eq. (4.4).

The continuous time intercept vector $\boldsymbol{b} \in \mathbb{R}^v$ provides a constant fixed input to the latent processes $\boldsymbol{\eta}$. In combination with $\boldsymbol{A}$, this determines the long-term level at which the processes fluctuate around. Without the continuous time intercept, the processes (if mean reverting) would simply fluctuate around zero.

Time-dependent predictors $\boldsymbol{\chi}(t)$ represent exogenous inputs (such as interventions) to the system that may vary over time and are independent of fluctuations in the system. Equation (4.1) shows a generalised form for time-dependent predictors that could be treated a variety of ways depending on the predictors assumed time course (or shape). We use a simple impulse form shown in Eq. (4.2), in which the predictors are treated as impacting the processes only at the instant of an observation occasion u. Such a form has the virtue that many alternative shapes are made possible via augmentation of the system state matrices, as we will describe throughout this work.

$$\boldsymbol{\chi}(t) = \sum_{u \in \boldsymbol{U}} \boldsymbol{x}_u \delta(t - t_u) \tag{4.2}$$

Here, time-dependent predictors $\boldsymbol{x}_u \in \mathbb{R}^l$ are observed at measurement occasions $u \in \boldsymbol{U}$, where $\boldsymbol{U}$ is the set of measurement occasions from 1 to the number of measurement occasions, with $u = 1$ treated as occurring at $t = 0$. The Dirac delta function $\delta(t - t_u)$ is a generalised function that is ∞ at 0 and 0 elsewhere yet has an integral of 1, when 0 is in the range of integration. It is useful to model an impulse to a system, and here is scaled by the vector of time-dependent predictors $\boldsymbol{x}_u$. The effect of these impulses on processes $\boldsymbol{\eta}(t)$ is then $\boldsymbol{M} \in \mathbb{R}^{v \times l}$. Put simply, the equation means that when a time-dependent predictor is observed at occasion u, the system processes spike upwards or downwards by $\boldsymbol{M}\boldsymbol{x}_u$.

$\boldsymbol{W}(t) \in \mathbb{R}^v$ represents v-independent Wiener processes, with a Wiener process being a random walk in continuous time. $\mathrm{d}\boldsymbol{W}(t)$ is meaningful in the context of stochastic differential equations and represents the stochastic error term, an infinitesimally small increment of the Wiener process. Lower triangular matrix $\boldsymbol{G} \in \mathbb{R}^{v \times v}$ represents the effect of this noise on the change in $\boldsymbol{\eta}(t)$. $\boldsymbol{Q}$, where $\boldsymbol{Q} = \boldsymbol{G}\boldsymbol{G}^\top$ represents the variance-covariance matrix of this diffusion process in continuous time. Intuitively, one may think of $\mathrm{d}\boldsymbol{W}(t)$ as random fluctuations and $\boldsymbol{G}$ as the effect of these fluctuations on the processes. $\boldsymbol{G}\mathrm{d}\boldsymbol{W}(t)$ then simply represents unknown changes in the direction of $\boldsymbol{\eta}$, which are distributed according to a multivariate normal with continuous time covariance matrix $\boldsymbol{Q}$.

4.2.2 Discrete Time Solution of Latent Dynamic Model

To derive expectations for discretely sampled data, the stochastic differential equation (4.1) may be solved and translated to a discrete time representation, for any observation $u \in \boldsymbol{U}$:

$$\boldsymbol{\eta}_u = \boldsymbol{A}^*_{\Delta t_u}\boldsymbol{\eta}_{u-1} + \boldsymbol{b}^*_{\Delta t_u} + \boldsymbol{M}\boldsymbol{x}_u + \boldsymbol{\zeta}_u \quad \boldsymbol{\zeta}_u \sim \mathrm{N}(\boldsymbol{0}_v, \boldsymbol{Q}^*_{\Delta t_u}) \tag{4.3}$$

The * notation is used to indicate a term that is the discrete time equivalent of the original, for the time interval Δt_u (which is the time at u minus the time at $u-1$). $\boldsymbol{A}^*_{\Delta t_u}$ then contains the appropriate auto and cross regressions for the effect of latent processes $\boldsymbol{\eta}$ at measurement occasion $u-1$ on $\boldsymbol{\eta}$ at measurement occasion u. $\boldsymbol{b}^*_{\Delta t_u}$ represents the discrete-time intercept for measurement occasion u. Since $\boldsymbol{M}$ is conceptualized as the effect of instantaneous impulses $\boldsymbol{x}$ (which only occur at occasions $\boldsymbol{U}$ and are not continuously present as for the processes $\boldsymbol{\eta}$), its discrete-time form matches the general continuous time formulation in Eq. (4.1). $\boldsymbol{\zeta}_u$ is the zero mean random error term for the processes at occasion u, which is distributed according to multivariate normal with covariance $\boldsymbol{Q}^*_{\Delta t_u}$. The recursive nature of the solution means that at the first measurement occasion $u = 1$, the system must be initialized in some way, with $\boldsymbol{A}^*_{\Delta t_u}\boldsymbol{\eta}_{u-1}$ replaced by $\boldsymbol{\eta}_{t0}$ and $\boldsymbol{Q}^*_{\Delta t_u}$ replaced by $\boldsymbol{Q}^*_{t0}$. These initial states and covariances are later referred to as T0MEANS and T0VAR, respectively.

Unlike in a purely discrete-time model, where the various discrete-time effect matrices described above would be unchanging, in a continuous time model, the discrete-time matrices all depend on some function of the continuous time parameters and the time interval Δt_u between observations u and $u-1$; these functions look as follows:

$$\boldsymbol{A}^*_{\Delta t_u} = e^{\boldsymbol{A}(t_u - t_{u-1})} \tag{4.4}$$

$$\boldsymbol{b}^*_{\Delta t_u} = \boldsymbol{A}^{-1}(\boldsymbol{A}^*_{\Delta t_u} - \boldsymbol{I})\boldsymbol{b} \tag{4.5}$$

$$\boldsymbol{Q}^*_{\Delta t_u} = \boldsymbol{Q}_\infty - \boldsymbol{A}^*_{\Delta t_u}\boldsymbol{Q}_\infty(\boldsymbol{A}^*_{\Delta t_u})^\top \tag{4.6}$$

where the asymptotic diffusion $\boldsymbol{Q}_\infty = \mathrm{irow}\big(-\boldsymbol{A}_\#^{-1}\ \mathrm{row}(\boldsymbol{Q})\big)$ represents the latent process variance as t approaches infinity, $\boldsymbol{A}_\# = \boldsymbol{A} \otimes \boldsymbol{I} + \boldsymbol{I} \otimes \boldsymbol{A}$, with $\otimes$ denoting the Kronecker product, row is an operation that takes elements of a matrix row wise and puts them in a column vector, and irow is the inverse of the row operation.

4.2.3 Measurement Model

While non-Gaussian generalisations are possible, for the purposes of this work, the latent process vector $\boldsymbol{\eta}(t)$ has the linear measurement model:

$$\boldsymbol{y}(t) = \boldsymbol{\Lambda}\boldsymbol{\eta}(t) + \boldsymbol{\tau} + \boldsymbol{\epsilon}(t) \quad \text{where } \boldsymbol{\epsilon}(t) \sim \mathrm{N}(\boldsymbol{0}_c, \boldsymbol{\Theta}) \tag{4.7}$$

$\boldsymbol{y}(t) \in \mathbb{R}^c$ is the vector of manifest variables, $\boldsymbol{\Lambda} \in \mathbb{R}^{c \times v}$ represents the factor loadings, and $\boldsymbol{\tau} \in \mathbb{R}^c$ represents the manifest intercepts. The residual vector $\boldsymbol{\epsilon} \in \mathbb{R}^c$ has covariance matrix $\boldsymbol{\Theta} \in \mathbb{R}^{c \times c}$.

4.2.4 Between Subjects Model and Estimation

We will not go into detail on between subjects and estimation aspects here, as they can be handled in various ways. A frequentist approach with random effects on the continuous time intercepts and manifest intercepts, and fixed effects on other model parameters, is presented in Driver et al. (2017), which also describes the R (R Core Team 2014) package ctsem which can be used for estimating these models. Driver and Voelkle (2018) extend the earlier work and ctsem software to fully random effects with a hierarchical Bayesian approach. The majority of this present work uses the frequentist approach for the sake of speed, but for an example of the Bayesian approach, see Sect. 4.4.5.

4.3 Shapes of Input Effects

When we speak of the effect of an input, we mean the effect of some observed variable that occurs at a specific moment in time. So, for instance, while a person's fitness is more easily thought of as constantly present and could thus be modelled as a latent process, an event such as 'going for a run' is probably easier to consider as a single event in time. We do not propose strict guidelines here; it would also be possible to model a 'speed of running' latent process, and it is also clear that all events take some time to unfold. However, we suggest it is both reasonable and simpler to model events occurring over short time scales (relative to observation intervals) as occurring at a single moment in time. Nevertheless, there may be cases when events should be thought of as persisting for some finite span of time, and this may also be approximated using the approaches we will outline.

So, although we speak of an input as occurring only at a single moment in time, the *effects* of such an input on the system of interest will persist for some time and may exhibit a broad range of shapes. While for the sake of clarity, we will discuss the shape of an input effect on a process that is otherwise a flatline; what we really mean by 'shape of the effect' is the *difference* between the expected value of the process conditional on an input and the expected value of the process without any such input.

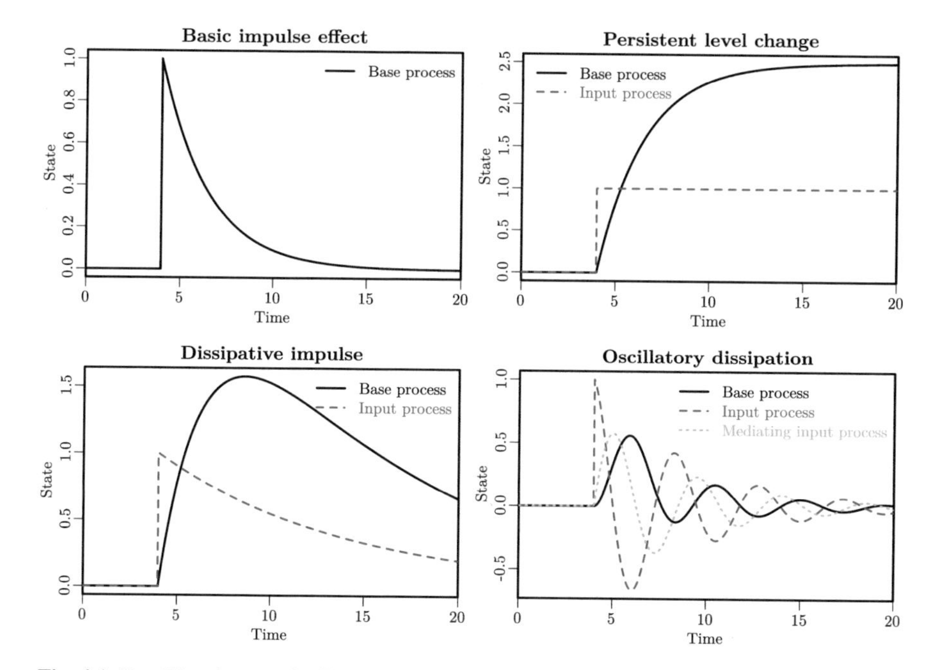

Fig. 4.1 Possible shapes of effects resulting from impulse input affecting a mean-reverting process, given various configurations of the state matrices. For the basic impulse effect, no additional processes need to be modelled, while the other examples require either one or two additional processes that are not directly observed

The shape of the effect could then be relative to a randomly fluctuating process, an oscillation, exponential trend, etc. Some examples of possible input effect shapes are shown in Fig. 4.1.

4.3.1 *Basic Impulse Effect*

As formalised by Eq. (4.2), the basic form of the effect, and fundamental building block for more complex effects, is that an input at a singular moment in time causes an impulse in the system at that moment, which then dissipates according to the temporal dependencies (drift matrix) in the system. The effect of such an impulse on a first-order, mean-reverting and non-stochastic process (i.e. a flatline with any changes driven by deterministic inputs) is shown in the top left of Fig. 4.1. The effect will take on a similar shape for any mean-reverting process; it is just not as easy to see when the process is stochastic and/or oscillating.

Effects we might plausibly model in such a way are those where the observed input is expected to have a sudden effect on the system, and that effect dissipates in a similar way to the other, unobserved, inputs on the system that are modelled via the stochastic term. An example of such could be the effect of encountering a friend in the street on ones' mood. In such a situation, mood may rapidly rise, then decline back to some baseline in much the same way as would occur for other random mood shifting events throughout the day.

Though the equivalence is perhaps not obvious at first, when time-varying covariates are included in a discrete-time cross-lagged or latent change score model, and no temporal dependencies in the covariates are modelled, it is just such an impulse effect that is being instantiated. As such, the approaches and thinking we outline in this paper can also be used in the discrete-time case, though care should be taken to consider divergences between the continuous and discrete-time approaches, particularly when models greater than first order (such as those with oscillations) are considered.

An example R script to simulate data from such a model, then fit, summarise, and plot results (remove hashes in front of plot lines to use), is as follows.[1]

```
install.packages("ctsem")
library(ctsem)
```

```
nlatent=1 #number of latent processes
nmanifest=1 #number of manifest variables
tpoints=30 #number of measurement occasions
ntdpred=1 #number of time dependent predictors
TDPREDMEANS=matrix(0,ntdpred*tpoints,1)
TDPREDMEANS[floor(tpoints/2)]=1 #input after 50% of observations

genm=ctModel(Tpoints=tpoints,
  n.latent=nlatent, n.manifest=nmanifest, n.TDpred=ntdpred,
  LAMBDA=matrix(c(1), nrow=nmanifest, ncol=nlatent),
  DRIFT=matrix(c(-.2),  nrow=nlatent, ncol=nlatent),
  DIFFUSION=matrix(c(.8), nrow=nlatent, ncol=nlatent),
  MANIFESTVAR=matrix(c(.8), nrow=nmanifest, ncol=nmanifest),
  TDPREDEFFECT=matrix(c(2), nrow=nlatent, ncol=ntdpred),
  CINT=matrix(c(0), nrow=nlatent, ncol=1),
  TDPREDMEANS=TDPREDMEANS,
  MANIFESTMEANS=matrix(c(0), nrow=nmanifest, ncol=1))

dat=ctGenerate(ctmodelobj=genm, n.subjects=50, burnin=50)
#ctIndplot(datawide=dat,n.subjects=10,n.manifest=1,Tpoints=tpoints)

fitm=ctModel(Tpoints=tpoints, type="omx",
  n.latent=nlatent, n.manifest=nmanifest, n.TDpred=ntdpred,
  LAMBDA=matrix(c(1), nrow=nmanifest, ncol=nlatent),
  DRIFT=matrix(c("drift11"), nrow=nlatent, ncol=nlatent),
  DIFFUSION=matrix(c("diffusion11"), nrow=nlatent, ncol=nlatent),
  MANIFESTVAR=matrix(c("merror11"), nrow=nmanifest, ncol=nmanifest),
  MANIFESTMEANS=matrix(c("mmeans_Y1"), nrow=nmanifest, ncol=nmanifest),
  TDPREDEFFECT=matrix(c("tdpredeffect21"), nrow=nlatent, ncol=ntdpred))

fit=ctFit(dat, fitm)
summary(fit)
ctKalman(fit, timestep=.1, subjects = 1:2, plot=TRUE,
  kalmanvec=c('y','ysmooth'))
```

[1]For the sake of simplicity, we generate and fit data without stable between-subjects differences, but in real-world analyses of multiple subjects, it may be advisable to account for such effects. With ctsem this can be done either via the MANIFESTTRAITVAR or TRAITVAR matrices in frequentist configuration or by allowing individually varying parameters with the Bayesian approach—discussed briefly in Sect. 4.4.5.

The matrix forms of the model equations for a basic impulse affecting a first-order process are as follows, with underbraced notations denoting the symbol used to represent the matrix in earlier formulas, and where appropriate also the matrix name in the R specification.

$$\underbrace{\mathrm{d}\left[\eta_1\right](t)}_{\mathrm{d}\eta(t)} = \left(\underbrace{\underbrace{\left[\text{drift11}\right]}_{A}}_{\text{DRIFT}}\underbrace{\left[\eta_1\right](t)}_{\eta(t)} + \underbrace{\underbrace{\left[0\right]}_{b}}_{\text{CINT}} + \underbrace{\underbrace{\left[\text{tdpredeffect21}\right]}_{M}}_{\text{TDPREDEFFECT}}\underbrace{\left[\chi_1\right]}_{\chi(t)}\right)dt$$

$$+\underbrace{\underbrace{\left[\text{diffusion11}\right]}_{G}}_{\text{DIFFUSION}}\underbrace{\mathrm{d}\left[W_1\right](t)}_{\mathrm{d}W(t)}$$

$$\underbrace{\left[Y_1\right](t)}_{Y(t)} = \underbrace{\underbrace{\left[1\right]}_{\Lambda}}_{\text{LAMBDA}}\underbrace{\left[\eta_1\right](t)}_{\eta(t)} + \underbrace{\underbrace{\left[\text{mmeans\_Y1}\right]}_{\tau}}_{\text{MANIFESTMEANS}} + \underbrace{\left[\epsilon_1\right](t)}_{\epsilon(t)} \quad \text{where} \underbrace{\left[\epsilon_1\right](t)}_{\epsilon(t)} \sim \mathrm{N}\left(\left[0\right], \underbrace{\underbrace{\left[\text{merror11}\right]}_{\Theta}}_{\text{MANIFESTVAR}}\right)$$

4.3.2 *Level Change Effect*

In contrast to the impulse effect, some inputs may result in a stable change in the level of a process. Such a change may occur near instantaneously or more gradually. The top right of Fig. 4.1 shows the more gradual change. Often, we would hope that treatments may generate such an effect. Consider, for instance, the effect of an exercise intervention on fitness. In the intervention condition, subjects are encouraged to increase the amount of exercise they do throughout the week. If the intervention is successful, we then wouldn't necessarily expect to see an immediate improvement in fitness but would hope that people had begun exercising more in general, which would slowly increase their fitness towards some new level.

There are various ways one could model such an effect. An intuitive approach that may spring to mind would be to code the input variable as 0 for both treatment and control prior to the intervention, then when treatment begins, code the input variable of the treated group as one for all further observations. This is somewhat problematic, however, as it will result in a series of impulses of the same strength occurring at the observations, with the process declining as normal after an observation, then jumping again with the next impulse. This is not at all representative of a consistent change in level when we consider the process at both observed and unobserved times, that is, *between the observations*. Nevertheless, if observations are equally spaced in time, it is potentially adequate for estimating the extent to which the treatment group exhibited a persistent change in their level. For instance, as illustrated in Fig. 4.2, in the case of unequal observation intervals, it is clear that both the underlying model and resulting predictions will be very wrong when taking such an input variable coding approach. With the equal

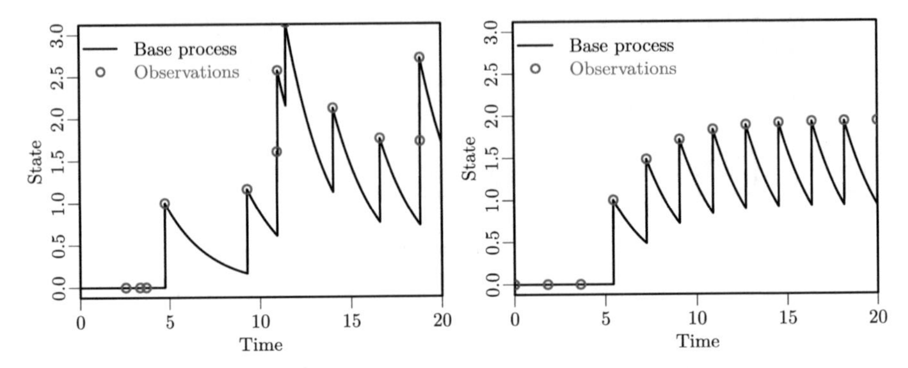

Fig. 4.2 Modelling a persistent level change via input variable coding. While strictly speaking the underlying model is incorrect, nevertheless with equal time intervals (right panel) a level change effect may be captured by a series of impulses, though this is only accurate at the exact moments of observation. With unequal time intervals (left panel), the approach is not even heuristically useful

observation time intervals on the right of Fig. 4.2, however, we see that although the underlying model is incorrect, predictions at the times when an observation occurs (and the input variable is coded as 1) do gradually rise towards a new level.

While possible in specific circumstances to model the level change via input variable coding, we would in general argue for a more adequate model specification, which is easily achievable. The approach we will take throughout this work is to model additional, *unobserved*, processes, which have no stochastic component. This might also be referred to as augmenting the state matrices. These additional processes are affected by the impulse of the input variable and in turn affect the actual process of interest, which we will refer to as the *base process*. What have we gained by including such an intermediary? This intermediary, or *input process*, is not restricted to an impulse shape but can exhibit a time course characterised by its own auto effect parameter.

One way we might think of such an input process is that of directly characterising the effect we are interested in. Consider, for instance, the effect of room temperature on attention and an experimental situation in which room temperature was changed at some point via a thermostat. We do not have direct experimental control or measurements of temperature, rather, we simply know that the thermostat was changed at some point. So we include 'change of thermostat' as our observed input variable, which generates a gradual rise in an unobserved 'temperature' process, which has some effect on attention that we are interested in.

Alternatively, we may have directly observed the manipulation we are interested in, but its effect on the process of interest may occur via some unobserved mediating process. Consider the example of fitness change in response to a cognitive behavioural intervention to motivate exercise. We have a direct observation of 'intervention that took place', but know that the subjects must actually exercise for any change in fitness to occur. Thus, we could include a dummy coded input variable of 'intervention', which generates an instantaneous and persistent change on some unobserved input process that we could think of as something like 'amount of weekly exercise', which in turn has an effect on our measure of 'fitness'. We do not propose any strong interpretation of the unobserved input process, as there will always be many more mediators at many more time scales than we can model. Rather, we propose to model the input process sufficiently such that the major features of the causal chain, at the time scale we are interested in, can be adequately captured. So, although we will not necessarily know what the causal links are, with such a model, we can aim at least to understand that (a) there appears to be some causal chain between our input and our process of interest, and (b) in the vicinity of the values of the context we have observed, the time course of the effect is likely to be similar to our estimated results.

Now, how to explicitly formulate such a level change model? Using ctsem, we configure our model just as for the impulse specification but include an additional latent process. For this process, we fix all input effects that do not stem from the input predictor to 0. These input effects we fix include the stochastic effects of the DIFFUSION and T0VAR matrices as well as intercept elements from the T0MEANS and CINT matrices—these are fixed to 0 because the input process should represent only the effect of our observed input variable. Because the process is unobserved, we must identify its scale in some way, which can be done either by fixing the effect of the input variable on the process (to any non-zero value, though one would seem sensible) or by similarly fixing the effect of the input process on our process of interest. The following script demonstrates simulating and fitting such a model. Note that because zeroes on diagonals can cause problems for matrix inversions, where necessary very small deviations from zero are used instead, which has a negligible effect on the model at the time scale of interest.

```
nlatent=2 #number of latent processes
nmanifest=1 #number of manifest variables
tpoints=30 #number of measurement occasions
ntdpred=1 #number of time dependent predictors
TDPREDMEANS=matrix(0,ntdpred*tpoints,1)
TDPREDMEANS[floor(tpoints/2)]=1 #intervention after 50% of observations

genm=ctModel(Tpoints=tpoints,
  n.latent=nlatent, n.manifest=nmanifest, n.TDpred=ntdpred,
  LAMBDA=matrix(c(1, 0), nrow=nmanifest, ncol=nlatent),
  DRIFT=matrix(c(-.4, 0, 1, -0.00001), nrow=nlatent, ncol=nlatent),
  DIFFUSION=matrix(c(.5, 0, 0, 0), nrow=nlatent, ncol=nlatent),
  MANIFESTVAR=matrix(c(.5), nrow=nmanifest, ncol=nmanifest),
  TDPREDEFFECT=matrix(c(0, .4), nrow=nlatent, ncol=ntdpred),
  CINT=matrix(c(0, 0), nrow=nlatent, ncol=1),
  TDPREDMEANS=TDPREDMEANS,
  MANIFESTMEANS=matrix(c(0), nrow=nmanifest, ncol=1))

dat=ctGenerate(ctmodelobj=genm, n.subjects=50, burnin=0)
#ctIndplot(datawide=dat,n.subjects=10,n.manifest=1,Tpoints=tpoints)

fitm=ctModel(Tpoints=tpoints, type="omx",
  n.latent=nlatent, n.manifest=nmanifest, n.TDpred=ntdpred,
  LAMBDA=matrix(c(1, 0), nrow=nmanifest, ncol=nlatent),
  DRIFT=matrix(c("drift11", 0, 1, -0.0001), nrow=nlatent, ncol=nlatent),
```

```
  DIFFUSION=matrix(c("diffusion11", 0, 0, 0), nrow=nlatent, ncol=nlatent),
  MANIFESTVAR=matrix(c("merror11"), nrow=nmanifest, ncol=nmanifest),
  MANIFESTMEANS=matrix(c("mmeans_Y1"), nrow=nmanifest, ncol=nmanifest),
  T0MEANS=matrix(c("t0m1",0), ncol=1),
  T0VAR=matrix(c("t0var11",0,0,0), nrow=nlatent, ncol=nlatent),
  TDPREDEFFECT=matrix(c(0, "tdpredeffect21"), nrow=nlatent, ncol=ntdpred))

fit=ctFit(dat, fitm)
summary(fit)
ctKalman(fit, timestep=.1, subjects = 1:2, plot=TRUE,
  kalmanvec=c('y','ysmooth'))
```

The matrix forms for a level change intervention affecting a first-order process are as follows: with underbraced notations denoting the symbol used to represent the matrix in earlier formulas and where appropriate also the matrix name in the ctsem specification.

$$\underbrace{\mathrm{d}\begin{bmatrix}\eta_1\\ \eta_2\end{bmatrix}(t)}_{\mathrm{d}\boldsymbol{\eta}(t)} = \left(\underbrace{\begin{bmatrix}\text{drift11} & 1\\ 0 & -1e-04\end{bmatrix}}_{\underbrace{\boldsymbol{A}}_{\text{DRIFT}}}\underbrace{\begin{bmatrix}\eta_1\\ \eta_2\end{bmatrix}(t)}_{\boldsymbol{\eta}(t)} + \underbrace{\begin{bmatrix}0\\ 0\end{bmatrix}}_{\underbrace{\boldsymbol{b}}_{\text{CINT}}} + \underbrace{\begin{bmatrix}0\\ \text{tdpredeffect21}\end{bmatrix}}_{\underbrace{\boldsymbol{M}}_{\text{TDPREDEFFECT}}}\underbrace{\begin{bmatrix}\chi_1\end{bmatrix}}_{\boldsymbol{\chi}(t)}\right)\mathrm{d}t$$

$$+\underbrace{\begin{bmatrix}\text{diffusion11} & 0\\ 0 & 0\end{bmatrix}}_{\underbrace{\boldsymbol{G}}_{\text{DIFFUSION}}}\underbrace{\mathrm{d}\begin{bmatrix}W_1\\ W_2\end{bmatrix}(t)}_{\mathrm{d}\boldsymbol{W}(t)}$$

$$\underbrace{\begin{bmatrix}Y_1\end{bmatrix}(t)}_{\boldsymbol{Y}(t)} = \underbrace{\begin{bmatrix}1 & 0\end{bmatrix}}_{\underbrace{\boldsymbol{\Lambda}}_{\text{LAMBDA}}}\underbrace{\begin{bmatrix}\eta_1\\ \eta_2\end{bmatrix}(t)}_{\boldsymbol{\eta}(t)} + \underbrace{\begin{bmatrix}\text{mmeans\_Y1}\end{bmatrix}}_{\underbrace{\boldsymbol{\tau}}_{\text{MANIFESTMEANS}}} + \underbrace{\begin{bmatrix}\epsilon_1\end{bmatrix}(t)}_{\boldsymbol{\epsilon}(t)} \quad \text{where} \underbrace{\begin{bmatrix}\epsilon_1\end{bmatrix}(t)}_{\boldsymbol{\epsilon}(t)} \sim \mathrm{N}\left(\begin{bmatrix}0\end{bmatrix}, \underbrace{\begin{bmatrix}\text{merror11}\end{bmatrix}}_{\underbrace{\boldsymbol{\Theta}}_{\text{MANIFESTVAR}}}\right)$$

4.4 Various Extensions

So far we have discussed two possible extremes in terms of effect shape, the sudden and singular impulse and the slower but constant level change. These are both somewhat restrictive in terms of the shape of the effects implied by the model, so in many cases, it may be worthwhile to further free the possible shape of effects, either as a comparison for the more restrictive model or directly as the model of interest.

4.4.1 Dissipation

The most obvious and simplest relaxation is to take the level change model and free the auto effect (diagonal of the drift matrix) for the input process. Then, the extent

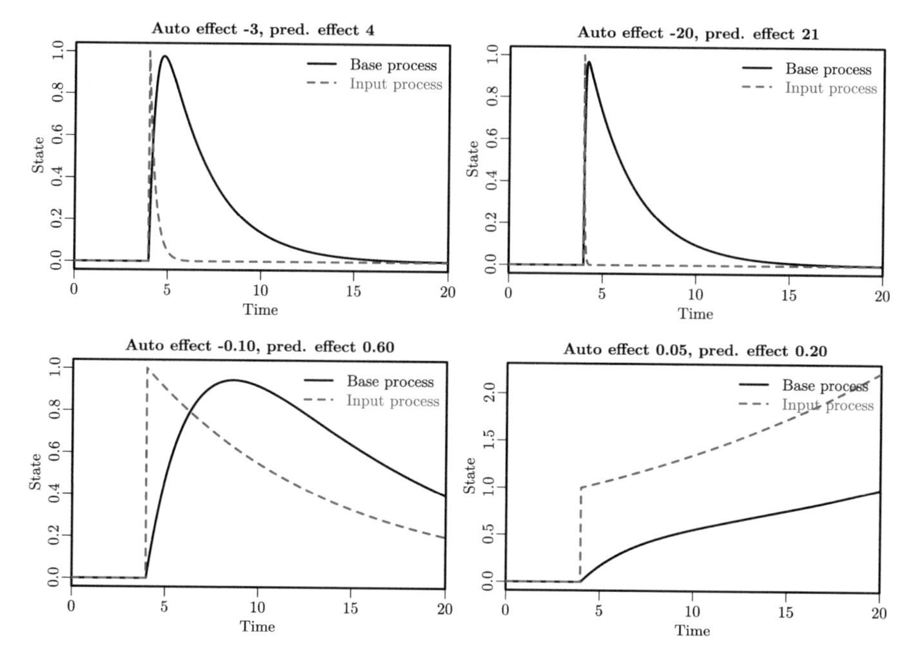

Fig. 4.3 Dissipating input effect, with different parameter values of the input process

to which the effect of the input persists is directly estimated, rather than assumed to persist forever. As the auto effect takes on more negative values, the input effect approaches the basic impulse type, and as the auto effect nears 0, the resulting input process approaches the level change type. Values greater than 0 would suggest that the input effect is compounding on itself with time, generating an explosive process. Such explosive processes tend to be highly unrealistic for forecasting much further into the future but may be an adequate characterisation over the time range considered. Note that, although with a highly negative auto effect the input process approaches the impulse shape, the value of the effect strength parameter will need to be much larger in order to match the impulse form with no mediating input process. This is shown in the top two plots of Fig. 4.3, while the bottom two show a slower dissipation on the left and an explosive effect on the right.

Such a dissipative input is probably a very reasonable starting point for modelling the effect of an intervention intended to have long-term consequences but where it is unclear if the consequences really do persist forever. Were we to use the simple impulse form, our estimated effect would only represent the magnitude of short-term changes and may not represent the bulk of the effect. With the level-change form of predictor, the estimated effect captures changes that persist for all later observations, and thus if the effect actually declines over time, it may underestimate the size of the effect at the short and medium term. Instead of these two extremes, it may instead be the case that the intervention was somewhat successful, with change persisting for some time but slowly dissipating. As the script for simulation and fitting of such a model is very similar to that for the level-change model (with the input process, auto effect simply given a parameter name rather than a fixed value of near zero), we do not repeat it here.

4.4.2 Multiple Time Scales and Oscillations

An additional flexibility that one could also consider is to take either the level-change or dissipation type model specification and free the direct effect of the input (via the TDPREDEFFECT matrix) on the base process itself. This will then allow for estimating both an instantaneous impulse that dissipates according to the base process and a more persistent input process, allowing for short-term effects to differ markedly from longer term, potentially even in the opposite direction. Some real-world examples of such could be the effect of various drugs, for which short-term effects are often very different to longer-term effects, or, at a different time scale, perhaps the effect of harshly disciplining a child—the child's behaviour may temporarily improve but with negative consequences later. A mild expansion of this approach could involve the specification of two independent input processes, each with a distinct effect on the base process (thus ensuring that dynamics of the base process in general are not confounded with dynamics in response to the intervention). Including parameters to try and tease apart different time-scale effects will likely make interpretation and hypothesis testing somewhat more complex, and empirical under identification may present an issue for convergence (at least when using the frequentist approach). An approach to mitigate these difficulties may be to revert to the level-change form (wherein the long-term persistence is fixed) rather than estimating the persistence—at least to attain initial estimates.

For more complex shapes of effects due to an intervention, we will need to change from a simple first-order input process to higher-order configurations. In the vector form used in this work and the relevant estimation software, higher-order processes are always modelled via additional first-order processes. Thus, whether one thinks of a single higher-order system or multiple first-order systems that interact makes no particular difference. A damped oscillation (as in Voelkle and Oud 2013) is probably the simplest higher-order model and could be used in similar circumstances to those of the above multiple time scales example, though the model will have somewhat different properties. In the multiple time scales model above, a strong interpretation of the model parameters would suggest that there were two distinct and independent effect processes unfolding, one short and one longer term. This is in contrast to the damped oscillation form, in which a single effect unfolds in a more complex fashion, requiring two unobserved processes that are coupled together.

Figure 4.4 plots an abstract example of the oscillation generating input effect, as well as an input effect comprised of two independent input processes—one fast and negative, the other slow and positive. The following R code generates and fits

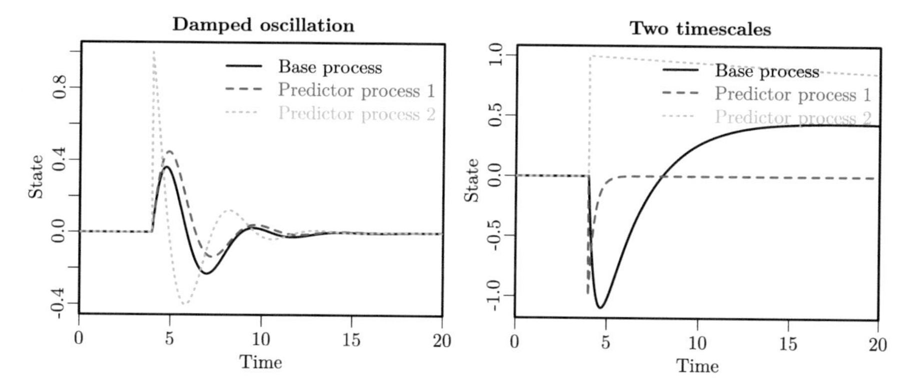

Fig. 4.4 Oscillating input effects. On the left the input process is a classic damped linear oscillator, involving two processes that are deterministically linked, and the direct effect on the base process occurring via input process 1. On the right, the two input processes are independent, with each having a direct effect on the base process. A simpler variant of the latter may neglect to include the short-term input process, instead having the input directly impact the base process

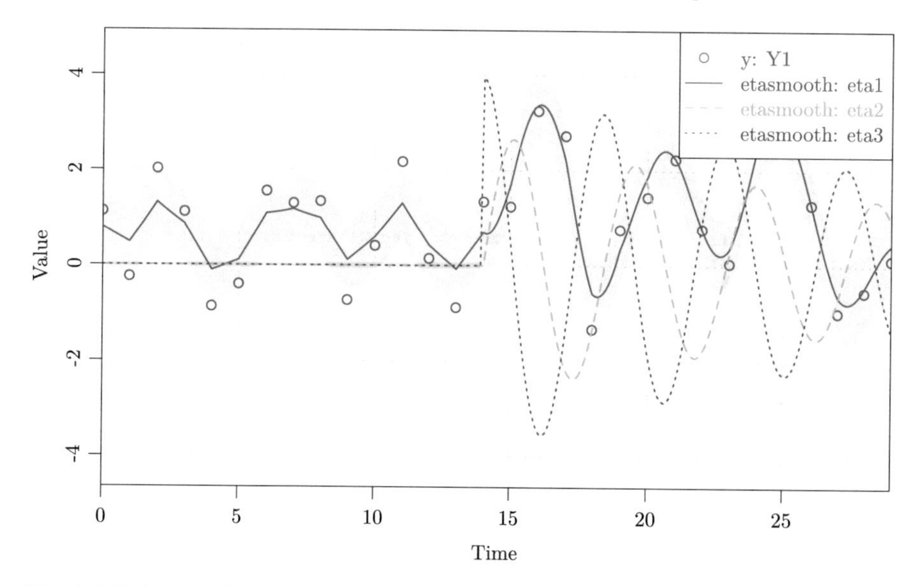

Fig. 4.5 Estimates of latent mean and covariance of base and input processes, for one subject, along with observed data. Estimates are from the Kalman smoother, so conditional on the fitted model and all time points. R code to obtain plots provided at top

a system with a single first-order process of interest (our base process), which is impacted upon by an input effect that generates a dissipating oscillation. Figure 4.5 shows the fit of this oscillation model for one of the generated subjects.

```
nlatent=3
nmanifest=1
tpoints=30
ntdpred=1
TDPREDMEANS=matrix(0,ntdpred*(tpoints),1)
TDPREDMEANS[floor(tpoints/2)]=1

genm=ctModel(Tpoints=tpoints,
  n.latent=nlatent,
  n.manifest=nmanifest,
  n.TDpred=ntdpred,
  LAMBDA=matrix(c(1, 0, 0), nrow=nmanifest, ncol=nlatent),
  DRIFT=matrix(c(
    -.3, 1, 0,
    0, 0,1,
    0,-2,-.1), byrow=TRUE, nrow=nlatent, ncol=nlatent),
  DIFFUSION=matrix(c(
    1, 0, 0,
    0,0,0,
    0,0,0), byrow=TRUE, nrow=nlatent, ncol=nlatent),
  MANIFESTVAR=matrix(c(.5), nrow=nmanifest, ncol=nmanifest),
  TDPREDEFFECT=matrix(c(0, 0, 4), nrow=nlatent, ncol=ntdpred),
  CINT=matrix(c(0), nrow=nlatent, ncol=1),
  TDPREDMEANS=TDPREDMEANS,
  MANIFESTMEANS=matrix(c(0), nrow=nmanifest, ncol=1))

dat=ctGenerate(ctmodelobj=genm, n.subjects=100, burnin=50)

#ctIndplot(datawide=dat,n.subjects=10,n.manifest=1,Tpoints=tpoints)

fitm=ctModel(Tpoints=tpoints,
  n.latent=nlatent,
  n.manifest=nmanifest,
  n.TDpred=ntdpred,
  LAMBDA=matrix(c(1, 0, 0), nrow=nmanifest, ncol=nlatent),
  DRIFT=matrix(c(
    "drift11", 1, 0,
    0, 0,1,
    0,"drift32","drift33"), byrow=TRUE, nrow=nlatent, ncol=nlatent),
  DIFFUSION=matrix(c(
    "diffusion11", 0, 0,
    0,0,0,
    0,0,0), byrow=TRUE, nrow=nlatent, ncol=nlatent),
  T0VAR=matrix(c(
    "t0var11", 0, 0,
    0,0,0,
    0,0,0), byrow=TRUE, nrow=nlatent, ncol=nlatent),
  TDPREDEFFECT=matrix(c(0, 0, "tdpredeffect31"), nrow=nlatent, ncol=ntdpred),
  CINT=matrix(c(0), nrow=nlatent, ncol=1),
  T0MEANS=matrix(c("t0mean1",0,0),nrow=nlatent,ncol=1),
  MANIFESTVAR=matrix(c("merror11"), nrow=nmanifest, ncol=nmanifest),
  MANIFESTMEANS=matrix(c("mmeans_Y1"), nrow=nmanifest, ncol=1))

fit=ctFit(dat, fitm)
summary(fit)
ctKalman(fit, timestep=.1, subjects=1, plot=TRUE, kalmanvec=c('y','etasmooth'))
```

The matrix forms for an intervention effect that first rises and then oscillates back to equilibrium, and which affects a first-order process, are as follows. Underbraced notations denoting the symbol used to represent the matrix in earlier formulas and where appropriate also the matrix name in the ctsem specification.

$$\underbrace{\mathrm{d}\begin{bmatrix}\eta_1\\ \eta_2\\ \eta_3\end{bmatrix}(t)}_{\mathrm{d}\boldsymbol{\eta}(t)} = \left(\underbrace{\begin{bmatrix}\text{drift11} & 1 & 0\\ 0 & 0 & 1\\ 0 & \text{drift32} & \text{drift33}\end{bmatrix}}_{\underbrace{\boldsymbol{A}}_{\text{DRIFT}}}\underbrace{\begin{bmatrix}\eta_1\\ \eta_2\\ \eta_3\end{bmatrix}(t)}_{\boldsymbol{\eta}(t)} + \underbrace{\begin{bmatrix}0\\ 0\\ 0\end{bmatrix}}_{\underbrace{\boldsymbol{b}}_{\text{CINT}}} + \underbrace{\begin{bmatrix}0\\ 0\\ \text{tdpredeffect31}\end{bmatrix}}_{\underbrace{\boldsymbol{M}}_{\text{TDPREDEFFECT}}}\underbrace{\begin{bmatrix}\chi_1\end{bmatrix}}_{\boldsymbol{\chi}(t)}\right)\mathrm{d}t$$

$$+\underbrace{\begin{bmatrix}\text{diffusion11} & 0 & 0\\ 0 & 0 & 0\\ 0 & 0 & 0\end{bmatrix}}_{\underbrace{\boldsymbol{G}}_{\text{DIFFUSION}}}\underbrace{\mathrm{d}\begin{bmatrix}W_1\\ W_2\\ W_3\end{bmatrix}(t)}_{\mathrm{d}\boldsymbol{W}(t)}$$

$$\underbrace{\begin{bmatrix}Y_1\end{bmatrix}(t)}_{\boldsymbol{Y}(t)} = \underbrace{\begin{bmatrix}1 & 0 & 0\end{bmatrix}}_{\underbrace{\boldsymbol{\Lambda}}_{\text{LAMBDA}}}\underbrace{\begin{bmatrix}\eta_1\\ \eta_2\\ \eta_3\end{bmatrix}(t)}_{\boldsymbol{\eta}(t)} + \underbrace{\begin{bmatrix}\text{mmeans\_Y1}\end{bmatrix}}_{\underbrace{\boldsymbol{\tau}}_{\text{MANIFESTMEANS}}} + \underbrace{\begin{bmatrix}\epsilon_1\end{bmatrix}(t)}_{\boldsymbol{\epsilon}(t)} \quad \text{where } \underbrace{\begin{bmatrix}\epsilon_1\end{bmatrix}(t)}_{\boldsymbol{\epsilon}(t)} \sim \mathrm{N}\left(\begin{bmatrix}0\end{bmatrix}, \underbrace{\begin{bmatrix}\text{merror11}\end{bmatrix}}_{\underbrace{\boldsymbol{\Theta}}_{\text{MANIFESTVAR}}}\right)$$

4.4.3 Trends

So far we have been using models where the base process is assumed to be stationary over the time window we are interested in (independent of any inputs). This means that given knowledge of only the observation times (and not any of the values), our expectations for the unobserved process states will always be the same, that is, neither expectation nor uncertainty regarding our processes is directly dependent on time. However, cases such as long-term development, as, for instance, when observing from childhood to adulthood, are likely to exhibit substantial trends. If unaccounted for, such trends are likely to result in highly non-Gaussian prediction errors, violating the assumptions of our model. Furthermore, there may be cases where influencing such a trend via an intervention is of interest, and we thus need to be able to incorporate a long-term trend in our model and include any potential effects of the input on the trend.

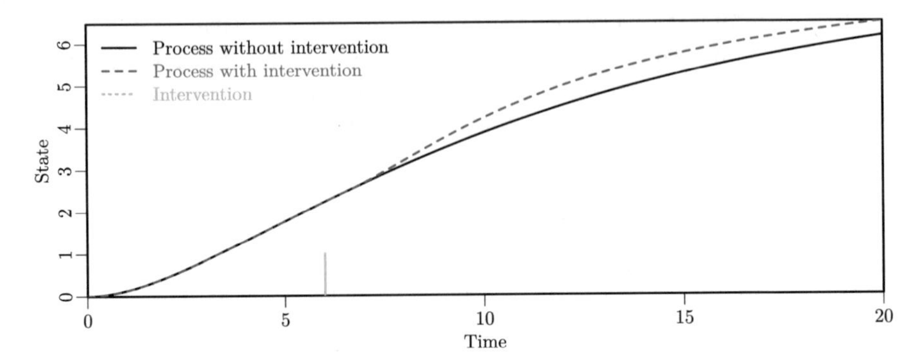

Fig. 4.6 Input effects on a developmental trend. The intervention at time 6 increases the trend slope temporarily

Let us consider an example of the influence of a health intervention on reading ability, in children of the developing world. Such a health intervention might consist of a short period involving health checks, treatment and education, with the idea that better health may facilitate learning, both at school and elsewhere. For an observation window of years with a limited observation frequency, the intervention period can reasonably be treated as a singular event (Fig. 4.6).

To specify such a model, we specify our base process as usual, capturing short-term fluctuations in our process of interest, reading ability. We then need to include an additional process that captures the slow trend component. The initial state (T0MEANS), temporal effects between the trend process and other processes and stochastic input (DIFFUSION) of this trend process are fixed to zero (or near zero), and in contrast to earlier models, we need to include a non-zero continuous-time intercept parameter, to capture the unknown trend size.[2] Other components are estimated as usual. Then we include an input effect onto some form of input process and have this input process affect the trend process. In this case, our measurement model must reflect that our measurements are a summation of both base and trend processes, so two elements of the LAMBDA matrix of factor loadings are now fixed to 1.00, in contrast to prior examples where only one latent process ever directly influenced the indicators (Fig. 4.7).

```
nlatent=3
nmanifest=1
tpoints=30
ntdpred=1
TDPREDMEANS=matrix(0,ntdpred*(tpoints),1)
TDPREDMEANS[floor(tpoints/2)]=1

genm=ctModel(Tpoints=tpoints,
```

[2]In a model with between subjects differences in the trend, variability in this parameter can be accommodated via the TRAITVAR matrix (for frequentist ctsem) or by simply setting the parameter to individually varying (in the Bayesian approach).

```
  n.latent=nlatent, n.manifest=nmanifest, n.TDpred=ntdpred,
  LAMBDA=matrix(c(1, 1, 0), nrow=nmanifest, ncol=nlatent),
  DRIFT=matrix(c(
    -.5, 0, 0,
    0, -.03,1,
    0,0,-.1), byrow=TRUE, nrow=nlatent, ncol=nlatent),
  DIFFUSION=matrix(c(
    3, 0, 0,
    0,0,0,
    0,0,.0001), byrow=TRUE, nrow=nlatent, ncol=nlatent),
  MANIFESTVAR=matrix(c(1), nrow=nmanifest, ncol=nmanifest),
  TDPREDEFFECT=matrix(c(0, 0, 4), nrow=nlatent, ncol=ntdpred),
  CINT=matrix(c(0,1,0), nrow=nlatent, ncol=1),
  TDPREDMEANS=TDPREDMEANS,
  MANIFESTMEANS=matrix(c(0), nrow=nmanifest, ncol=1))

dat=ctGenerate(ctmodelobj=genm, n.subjects=50, burnin=5)
#ctIndplot(datawide=dat,n.subjects=10,n.manifest=1,Tpoints=tpoints)

fitm=ctModel(Tpoints=tpoints,
  n.latent=nlatent, n.manifest=nmanifest, n.TDpred=ntdpred,
  LAMBDA=matrix(c(1, 1, 0), nrow=nmanifest, ncol=nlatent),
  DRIFT=matrix(c(
    "drift11", 0, 0,
    0, "drift22",1,
    0,0,"drift33"), byrow=TRUE, nrow=nlatent, ncol=nlatent),
  DIFFUSION=matrix(c(
    "diffusion11", 0, 0,
    0,0,0,
    0,0,.0001), byrow=TRUE, nrow=nlatent, ncol=nlatent),
  T0VAR=matrix(c(
    "t0var11", 0, 0,
    0,0,0,
    0,0,.0001), byrow=TRUE, nrow=nlatent, ncol=nlatent),
  TDPREDEFFECT=matrix(c(0, 0, "tdpredeffect31"), nrow=nlatent, ncol=ntdpred),
  CINT=matrix(c(0,"cint2",0), nrow=nlatent, ncol=1),
  T0MEANS=matrix(c("t0mean1",0,0),nrow=nlatent,ncol=1),
  MANIFESTVAR=matrix(c("merror11"), nrow=nmanifest, ncol=nmanifest),
  MANIFESTMEANS=matrix(c("mmeans_Y1"), nrow=nmanifest, ncol=1))

fit=ctFit(dat, fitm)
summary(fit)
```

```
ctKalman(fit, timestep=.1, subjects=1:2, plot=TRUE,
  plotcontrol=list(ylim=c(-5,50)), kalmanvec=c('y','etasmooth'),
  legendcontrol=list(x='topleft',bg='white'))
ctKalman(fit, timestep=.1, subjects = 1:2, plot=TRUE,
  kalmanvec=c('y','ysmooth'), plotcontrol=list(ylim=c(-5,50)),
  legend=FALSE)
```

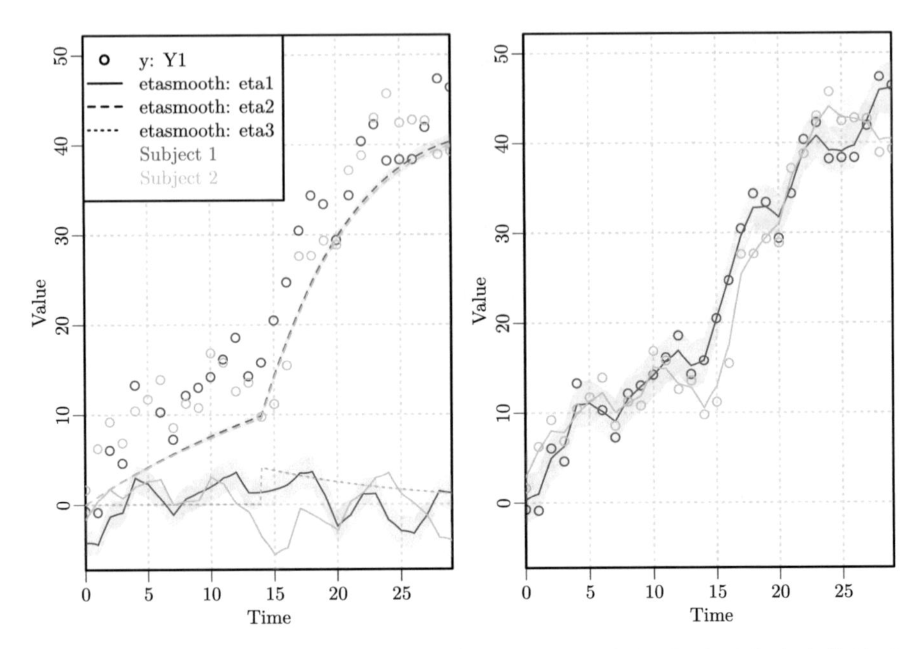

Fig. 4.7 Estimated trend with intervention model for two individuals. On the left, the individual latent processes—base, trend and predictor—and observed data points are shown. On the right, the latent processes are combined with the measurement model. Estimates are from the Kalman smoother, so conditional on the fitted model and all time points. R code to obtain plots provided at top

The matrix forms for an intervention effect on the long-term trend of a process are as follows. Underbraced notations denote the symbol used to represent the matrix in earlier formulas and where appropriate also the matrix name in the ctsem specification.

$$\underbrace{\mathrm{d}\begin{bmatrix}\eta_1\\ \eta_2\\ \eta_3\end{bmatrix}(t)}_{\mathrm{d}\boldsymbol{\eta}(t)} = \left(\underbrace{\begin{bmatrix}\text{drift11} & 0 & 0\\ 0 & \text{drift22} & 1\\ 0 & 0 & \text{drift33}\end{bmatrix}}_{\underbrace{\boldsymbol{A}}_{\text{DRIFT}}} \underbrace{\begin{bmatrix}\eta_1\\ \eta_2\\ \eta_3\end{bmatrix}(t)}_{\boldsymbol{\eta}(t)} + \underbrace{\begin{bmatrix}0\\ \text{cint2}\\ 0\end{bmatrix}}_{\underbrace{\boldsymbol{b}}_{\text{CINT}}} + \underbrace{\begin{bmatrix}0\\ 0\\ \text{tdpredeffect31}\end{bmatrix}}_{\underbrace{\boldsymbol{M}}_{\text{TDPREDEFFECT}}} \underbrace{\begin{bmatrix}\chi_1\end{bmatrix}}_{\boldsymbol{\chi}(t)} \right) \mathrm{d}t$$

$$+ \underbrace{\begin{bmatrix}\text{diffusion11} & 0 & 0\\ 0 & 0 & 0\\ 0 & 0 & 1e-04\end{bmatrix}}_{\underbrace{\boldsymbol{G}}_{\text{DIFFUSION}}} \underbrace{\mathrm{d}\begin{bmatrix}W_1\\ W_2\\ W_3\end{bmatrix}(t)}_{\mathrm{d}\boldsymbol{W}(t)}$$

$$\underbrace{\begin{bmatrix} Y_1 \end{bmatrix}(t)}_{Y(t)} = \underbrace{\underbrace{\begin{bmatrix} 1 & 1 & 0 \end{bmatrix}}_{\Lambda}}_{\text{LAMBDA}} \underbrace{\begin{bmatrix} \eta_1 \\ \eta_2 \\ \eta_3 \end{bmatrix}(t)}_{\eta(t)} + \underbrace{\underbrace{\begin{bmatrix} \text{mmeans\_Y1} \end{bmatrix}}_{\tau}}_{\text{MANIFESTMEANS}} + \underbrace{\begin{bmatrix} \epsilon_1 \end{bmatrix}(t)}_{\epsilon(t)} \quad \text{where } \underbrace{\begin{bmatrix} \epsilon_1 \end{bmatrix}(t)}_{\epsilon(t)} \sim \mathrm{N}\left(\begin{bmatrix} 0 \end{bmatrix}, \underbrace{\underbrace{\begin{bmatrix} \text{merror11} \end{bmatrix}}_{\Theta}}_{\text{MANIFESTVAR}} \right)$$

4.4.4 Mediation

Throughout this work, we have modelled a range of different shapes of input effects by including additional processes in our system model, and these processes have not been directly measured—regression strengths (i.e. elements of the LAMBDA matrix in ctsem) directly from these additional processes to data have been zero, in most cases. One possible interpretation of these unobserved processes is that they represent some aggregate over all mediating processes that occur between the measured input effect and our measured process of interest. While such processes can simply be left unobserved, understanding the mediators of effects is a common goal of psychological research, and the framework outlined here offers possibilities for such.

Let us consider again the example of an experimental intervention to improve fitness levels in a group of patients. A successful intervention is unlikely to generate a sudden increase in fitness, rather, it could be expected to gradually rise towards some new level, for which we have already discussed a modelling approach. However, suppose we had also observed some measure of the amount of daily exercise. Conditional on daily exercise, it seems unlikely that the intervention would have any further effect on fitness. That is, we assume that daily exercise mediates the effect of the intervention on fitness. We can test such a theory by comparing a model which includes both mediated and direct effects on fitness, with one that includes only the mediated effect. The following R script provides an example of such a model, and compares a model with effects from the input to all processes, to a restricted model where the input indirectly affects fitness via exercise. In order to provide a fair comparison, the full model for fitting contains an unobserved input process that is impacted by the input. In our example, however, the data does not support such a model. Such an over-specified model can make attaining convergence somewhat trickier, as multiple parameters must be estimated where only one would suffice. To minimise these difficulties, we fit the full model using the parameter estimates of the more restricted model as starting values. The mxCompare function from OpenMx is used to compare the two fits and will show that there is no significant difference between the model with the intervention directly affecting fitness and the restricted model that contains only the intervention effect on exercise rate.

```
set.seed(4)
nlatent=3
nmanifest=2
tpoints=30
ntdpred=1
TDPREDMEANS=matrix(0,ntdpred*(tpoints),1)
TDPREDMEANS[floor(tpoints/2)]=1

genm=ctModel(Tpoints=tpoints,
  n.latent=nlatent, n.manifest=nmanifest, n.TDpred=ntdpred,
  LAMBDA=matrix(c(1, 0, 0, 1, 0, 0), nrow=nmanifest, ncol=nlatent),
  DRIFT=matrix(c(
    -.2, .1, 0,
    0, -.3,1,
    0,0,-.0001), byrow=TRUE, nrow=nlatent, ncol=nlatent),
  DIFFUSION=matrix(c(
    1, 0, 0,
    3,2,0,
    0,0,.0001), byrow=TRUE, nrow=nlatent, ncol=nlatent),
  MANIFESTVAR=matrix(c(1,0,0,1), nrow=nmanifest, ncol=nmanifest),
  TDPREDEFFECT=matrix(c(0, 0, 5), nrow=nlatent, ncol=ntdpred),
  CINT=matrix(c(0), nrow=nlatent, ncol=1),
  TDPREDMEANS=TDPREDMEANS,
  MANIFESTMEANS=matrix(c(0), nrow=nmanifest, ncol=1))

dat=ctGenerate(ctmodelobj=genm, n.subjects=50, burnin=50)
#ctIndplot(datawide=dat,n.subjects=10,n.manifest=2,Tpoints=tpoints)

nlatent=4 #because for our fit we include extra input process

fullm=ctModel(Tpoints=tpoints, #type="stanct",
  n.latent=nlatent, n.manifest=nmanifest, n.TDpred=ntdpred,
  LAMBDA=matrix(c(1, 0, 0, 0, 0, 1, 0, 0), nrow=nmanifest, ncol=nlatent),
  DRIFT=matrix(c(
    "drift11", 1, "drift13", 0,
    0, "drift22", 0, 0,
    0, 0, "drift33", 1,
    0, 0, 0, "drift44"), byrow=TRUE, nrow=nlatent, ncol=nlatent),
  DIFFUSION=matrix(c(
    "diffusion11", 0, 0, 0,
    0, .0001, 0, 0,
    "diffusion31", 0, "diffusion33", 0,
    0, 0, 0, .0001), byrow=TRUE, nrow=nlatent, ncol=nlatent),
  T0VAR=matrix(c(
    "t0var11", 0, 0, 0,
    0, 0, 0, 0,
    "t0var31", 0, "t0var33", 0,
    0, 0, 0, 0), byrow=TRUE, nrow=nlatent, ncol=nlatent),
  TDPREDEFFECT=matrix(c("tdpredeffect11", "tdpredeffect21",
    "tdpredeffect31", "tdpredeffect41"), nrow=nlatent, ncol=ntdpred),
  CINT=matrix(c(0), nrow=nlatent, ncol=1),
  T0MEANS=matrix(c("t0mean1", 0, "t0mean3",0),nrow=nlatent,ncol=1),
  MANIFESTVAR=matrix(c("merror11",0,0,"merror22"), nrow=nmanifest, ncol=
      nmanifest),
  MANIFESTMEANS=matrix(c("mmeans_fit","mmeans_ex"), nrow=nmanifest, ncol=1))

mediationm=fullm
mediationm$TDPREDEFFECT[1:3,1]=0

mediationfit=ctFit(dat, mediationm)

fullfit=ctFit(dat, fullm,carefulFit=FALSE, #because we specify start values
omxStartValues = omxGetParameters(mediationfit$mxobj))

mxCompare(base = fullfit$mxobj, comparison = mediationfit$mxobj)
```

The matrix forms for the full (not yet restricted) mediation model, with an intervention affecting a measured first-order process that in turn affects another measured first-order process, are as follows. Underbraced notations denote the symbol used to represent the matrix in earlier formulas and where appropriate also the matrix name in the ctsem specification.

$$\underbrace{\mathrm{d}\begin{bmatrix}\eta_1\\ \eta_2\\ \eta_3\\ \eta_4\end{bmatrix}(t)}_{\mathrm{d}\boldsymbol{\eta}(t)} = \left(\underbrace{\begin{bmatrix}\text{drift11} & 1 & \text{drift13} & 0\\ 0 & \text{drift22} & 0 & 0\\ 0 & 0 & \text{drift33} & 1\\ 0 & 0 & 0 & \text{drift44}\end{bmatrix}}_{\underbrace{\boldsymbol{A}}_{\text{DRIFT}}} \underbrace{\begin{bmatrix}\eta_1\\ \eta_2\\ \eta_3\\ \eta_4\end{bmatrix}(t)}_{\boldsymbol{\eta}(t)} + \underbrace{\begin{bmatrix}0\\ 0\\ 0\\ 0\end{bmatrix}}_{\underbrace{\boldsymbol{b}}_{\text{CINT}}} + \underbrace{\begin{bmatrix}\text{tdpredeffect11}\\ \text{tdpredeffect21}\\ \text{tdpredeffect31}\\ \text{tdpredeffect41}\end{bmatrix}}_{\underbrace{\boldsymbol{M}}_{\text{TDPREDEFFECT}}} \underbrace{\begin{bmatrix}\chi_1\end{bmatrix}}_{\boldsymbol{\chi}(t)} \right) \mathrm{d}t$$

$$+ \underbrace{\begin{bmatrix}\text{diffusion11} & 0 & 0 & 0\\ 0 & 1e-04 & 0 & 0\\ \text{diffusion31} & 0 & \text{diffusion33} & 0\\ 0 & 0 & 0 & 1e-04\end{bmatrix}}_{\underbrace{\boldsymbol{G}}_{\text{DIFFUSION}}} \underbrace{\mathrm{d}\begin{bmatrix}W_1\\ W_2\\ W_3\\ W_4\end{bmatrix}(t)}_{\mathrm{d}\boldsymbol{W}(t)}$$

$$\underbrace{\begin{bmatrix}Y_1\\ Y_2\end{bmatrix}(t)}_{\boldsymbol{Y}(t)} = \underbrace{\begin{bmatrix}1\ 0\ 0\ 0\\ 0\ 0\ 1\ 0\end{bmatrix}}_{\underbrace{\boldsymbol{\Lambda}}_{\text{LAMBDA}}} \underbrace{\begin{bmatrix}\eta_1\\ \eta_2\\ \eta_3\\ \eta_4\end{bmatrix}(t)}_{\boldsymbol{\eta}(t)} + \underbrace{\begin{bmatrix}\text{mmeans\_fit}\\ \text{mmeans\_ex}\end{bmatrix}}_{\underbrace{\boldsymbol{\tau}}_{\text{MANIFESTMEANS}}} + \underbrace{\begin{bmatrix}\epsilon_1\\ \epsilon_2\end{bmatrix}(t)}_{\boldsymbol{\epsilon}(t)}$$

$$\text{where } \underbrace{\begin{bmatrix}\epsilon_1\\ \epsilon_2\end{bmatrix}(t)}_{\boldsymbol{\epsilon}(t)} \sim \mathrm{N}\left(\begin{bmatrix}0\\ 0\end{bmatrix}, \underbrace{\begin{bmatrix}\text{merror11} & 0\\ 0 & \text{merror22}\end{bmatrix}}_{\underbrace{\boldsymbol{\Theta}}_{\text{MANIFESTVAR}}}\right)$$

4.4.5 Individual Differences

While understanding how the *average* effect of an intervention develops over time is useful, it has long been observed that any such average may not be representative of the development in any single individual—the exercise intervention we have discussed may be more effective for those who live near a park or recreational space, for instance, as the barriers to following the intervention guidelines are lower. An extreme approach to such a problem is to treat individuals as entirely distinct, but this requires very many observations per subject if even moderately flexible

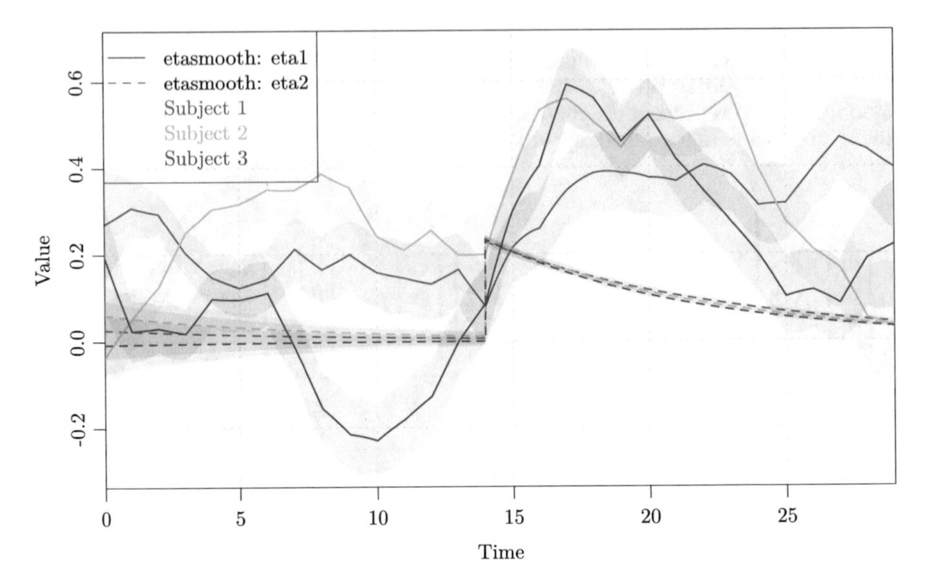

Fig. 4.8 Posterior estimates of the latent mean and covariance of base and input processes, for three subjects. Note the individual differences in the time spans of the input process (eta2). R code to obtain plots at top

dynamic models are to be fitted and also raises questions as to how one should treat individuals for which no observations exist. A benefit to this extreme view is that modelling is simplified, and the previously discussed approaches suffice.

A more flexible approach to individual differences is that of random effects (or hierarchical models). These approaches treat individuals as somewhat similar and estimate the extent of this similarity. This allows for situations where some individuals have been observed many times and others very few, with the resulting model for those observed only few times relying more on the average model across all individuals. For more extended discussion on such models in this context, see Driver and Voelkle (2018). Both frequentist and Bayesian approaches for random effects of observed input variables on latent processes are relatively straightforward; however, random effects on the parameters of any unobserved input processes are more complicated in the frequentist case. As such, we demonstrate the case of individual differences using the Bayesian formulation of the ctsem software, which can take longer to fit. In this case we have specified a minimal number of iterations, and it takes roughly 5–10 min on a modern PC—for many problems more iterations will be necessary (Fig. 4.8).

For this example, we will look how the strength and persistence of an intervention varies in our sample and relate this variation to an observed covariate. For this we will use the dissipative predictor model developed in Sect. 4.4.1 but allow for variation in the strength (parameters of the TDPREDEFFECT matrix) and persistence (the drift auto effect parameter of the unobserved input process).

```
nlatent=2 #number of latent processes
nmanifest=1 #number of manifest variables
tpoints=30 #number of measurement occasions
ntdpred=1 #number of time dependent predictors
nsubjects=30 #number of subjects
TDPREDMEANS=matrix(0,ntdpred*tpoints,1)
TDPREDMEANS[floor(tpoints/2)]=1 #intervention after 50% of observations

genm=ctModel(Tpoints=tpoints,
  n.latent=nlatent, n.manifest=nmanifest, n.TDpred=ntdpred,
  LAMBDA=matrix(c(1, 0), nrow=nmanifest, ncol=nlatent),
  DRIFT=matrix(c(-.2, 0, 1, -.2), nrow=nlatent, ncol=nlatent),
  DIFFUSION=matrix(c(.1, 0, 0, 0.00001), nrow=nlatent, ncol=nlatent),
  MANIFESTVAR=matrix(c(.1), nrow=nmanifest, ncol=nmanifest),
  TDPREDEFFECT=matrix(c(0, .2), nrow=nlatent, ncol=ntdpred),
  CINT=matrix(c(0, 0), nrow=nlatent, ncol=1),
  TDPREDMEANS=TDPREDMEANS,
  MANIFESTMEANS=matrix(c(0), nrow=nmanifest, ncol=1))

library(plyr)
#generate data w random parameter in DRIFT
dat=aaply(1:nsubjects, 1, function(x){
  tempm=genm
  stdage=rnorm(1)
  tempm$DRIFT[2, 2] = -exp(rnorm(1, -2, .5) + stdage * .5)
  cbind(ctGenerate(ctmodelobj=tempm, n.subjects=1, burnin=50), stdage)})

#convert to long format used by Bayesian ctsem
datlong=ctWideToLong(datawide = dat, Tpoints = tpoints, n.manifest = nmanifest,
  n.TDpred = ntdpred,n.TIpred = 1,manifestNames = c("Y1"),
  TDpredNames = c("TD1"), TIpredNames=c("stdage"))
#convert intervals to abs time
datlong=ctDeintervalise(datlong)

fitm=ctModel(Tpoints=tpoints, type="stanct",
  n.latent=nlatent, n.manifest=nmanifest,
  n.TDpred=ntdpred, n.TIpred=1, TIpredNames = "stdage",
  LAMBDA=matrix(c(1, 0), nrow=nmanifest, ncol=nlatent),
  DRIFT=matrix(c("drift11", 0, 1, "drift22"), nrow=nlatent, ncol=nlatent),
  DIFFUSION=matrix(c("diffusion11", 0, 0, 0.001), nrow=nlatent, ncol=nlatent),
  MANIFESTVAR=matrix(c("merror11"), nrow=nmanifest, ncol=nmanifest),
  MANIFESTMEANS=matrix(c("mmeans_Y1"), nrow=nmanifest, ncol=nmanifest),
  TDPREDEFFECT=matrix(c(0, "tdpredeffect21"), nrow=nlatent, ncol=ntdpred))

#only the persistence and strength of the predictor effect
#varies across individuals
fitm$pars$indvarying[-c(8,18)] = FALSE
#and thus standardised age can only affect those parameters
fitm$pars$stdage_effect[-c(8,18)] = FALSE

fit=ctStanFit(datlong, fitm, iter = 200, chains=3)
summary(fit)
```

```
ctKalman(fit, subjects = 1:3, timestep=.01, plot=TRUE,
  kalmanvec='etasmooth', errorvec='etasmoothcov',
  legendcontrol = list(x = "topleft"))
```

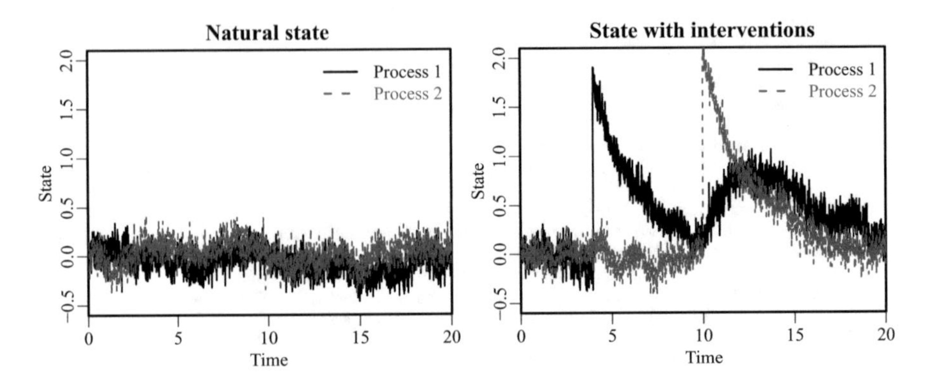

Fig. 4.9 Measurement error combined with limited natural variation in the system can make determining the relationships difficult. On the left, it is not clear which process is driving the other, but with interventions on the right, it is clear that changes in process 2 lead to changes in process 1, and not vice versa

Now, the summary of fit gives an estimated population standard deviation of the persistence and strength of the predictor effect (under *$popsd*) and also an estimate of the effect of standardised age on these parameters (under *$tipreds*). In this case there is no genuine effect of age on the effect strength, but it is important to allow for this effect because of the strong dependence between the strength and persistence parameters—it may be difficult to distinguish one rising from the other lowering in some datasets.

4.4.6 System Identification via Interventions

In addition to focusing on how interventions unfold over time, the concepts and modelling procedures we have discussed in this work can also be applied in a context wherein interventions are used primarily with the goal for developing an understanding of the underlying system—the intervention effect itself is not of primary interest. Figure 4.9 shows a simple case where this may be useful, where the natural state of the bivariate system exhibits limited variation and measurement error is relatively high. In this context, estimates of the system dynamics can have high uncertainty. By intervening on the system, first on one process and then later on the other, shown on the right of Fig. 4.9, the relation between the two processes becomes much more evident and more easily estimable. An R script is provided, first fit using data without intervention effects and then with intervention effects—in general the estimates of the dynamics should become more accurate, with lower standard errors, when the interventions are included.

```
set.seed(1)
nlatent=2 #number of latent processes
nmanifest=2 #number of manifest variables
tpoints=10 #number of measurement occasions
ntdpred=2 #number of time dependent predictors
TDPREDMEANS=matrix(0,ntdpred*tpoints,1)
TDPREDMEANS[c(3,tpoints+6)]=1 #intervention 1 at t3 and intervention 2 at t6

genm=ctModel(Tpoints=tpoints,
  n.latent=nlatent, n.manifest=nmanifest, n.TDpred=ntdpred,
  LAMBDA=diag(1,2),
  DRIFT=matrix(c(-.4, 0, .1, -.3), nrow=nlatent, ncol=nlatent),
  DIFFUSION=matrix(c(1, 0, 0, 1), nrow=nlatent, ncol=nlatent),
  MANIFESTVAR=matrix(c(1, 0, 0, 1), nrow=nmanifest, ncol=nmanifest),
  TDPREDEFFECT=matrix(c(0, 0, 0, 0), nrow=nlatent, ncol=ntdpred),
  CINT=matrix(c(0, 0), nrow=nlatent, ncol=1),
  TDPREDMEANS=TDPREDMEANS,
  MANIFESTMEANS=matrix(c(0), nrow=nmanifest, ncol=1))

dat=ctGenerate(ctmodelobj=genm, n.subjects=20, burnin=20)

fitm=ctModel(Tpoints=tpoints, type="omx", n.latent=nlatent,
  n.manifest=nmanifest, n.TDpred=ntdpred, LAMBDA=diag(1,2))

fit=ctFit(dat, fitm)
summary(fit)

#now with input effects
genm$TDPREDEFFECT[c(1,4)] = 10
interventiondat=ctGenerate(ctmodelobj=genm, n.subjects=50, burnin=20)
interventionfit=ctFit(interventiondat, fitm)
summary(interventionfit)
```

4.5 Discussion

The effect of some intervention or event on an ongoing process can manifest in many forms over a range of time scales. We have shown that using an impulse (the Dirac delta) as an exogenous input effect in continuous time dynamic models allows a wide variety of possible shapes to be estimated by including additional unobserved mediating processes. This allows for changes in subjects' baseline levels or trends that can happen either instantaneously or gradually. Such changes can dissipate very rapidly, persist for the entire observation window or even build on themselves over time. Changes at different time scales need not even be in the same direction, as we have shown how one may model oscillations back to equilibrium or alternatively an initial change with a slower recovery to a new equilibrium. Such an approach can be used both to formulate and test hypotheses regarding the response over time of individuals to some intervention or stimulus, or in a more exploratory approach using a flexible initial specification or iterative model development. We have demonstrated possibilities of formalising and testing mediation models, as well as for examining relations between individuals' specific traits and their response to an intervention—a key component for personalised approaches.

While the approach put forward here relies on an impulse input that may then be transformed to a range of shapes via the drift matrix, alternative treatments of input variables in continuous time have been proposed by Singer (1998) and Oud and Jansen (2000). In such approaches, the input is conceived as occurring over some time frame; a polynomial describes the temporal evolution, and some approximation of the integral over time must be computed. Because such an approach does not extend the state matrices, it may in some cases be computationally faster.

An aspect which has been implicit throughout the work so far is that of causal inference. We do not think it is controversial to state that, to the extent that the model is accurately specified and the observed input effects are exogenous to the system processes, causal interpretations of the input effects and their time course may be reasonable. While a fully explicated formal treatment of causality is beyond the scope of this work, we will briefly discuss exogeneity of effects:

When input effects are not exogenous, as, for instance, when they are used to model events in a person's life that the person can have some influence over—as, for example, when ending a relationship—it may still be helpful to model input effects as discussed, but interpretation is far less clear. In such a case, the observed response to a particular type of event may still be interesting, but the response cannot be assumed to be due to the event specifically, as it may instead be due to antecedents that gave rise to the event. Finding that those who choose to end a relationship become happier does not imply that ending relationships is a useful intervention strategy for people in general!

Amongst the benefits of the approach we have been discussing, there are also some limitations to be considered: the timing of the start of the input effect must be known, the value of the input variables must be known, non-linearities in the input effect are not directly modelled and the input variable only influences states of the processes, not parameters of the model. In the following paragraphs, we elaborate on these limitations.

In the approach we have put forward, the timing of all observations, including observations of input variables, are regarded as known and must be input as data. This could be troublesome if (a) the timing of the intervention or event in question is simply not well measured, or (b) there is some genuine lag time during which the effect of the input or intervention on the process of interest is truly zero, before suddenly taking effect. In both cases, when the possible error is small relative to the observation time intervals, there are unlikely to be substantial ramifications for model fitting and inference. The latter case may be somewhat trickier to determine, so far as we can imagine, should be limited to situations involving physical stimuli and very short time scales. For instance, a loud noise must travel from the source to the person, and the person must engage in some low-level processing of the noise, before any startle response would occur. For such situations, if the time-lag is reasonably well known, the timing of the input variable can simply be adjusted. In other cases, extending the model to include a measurement model of the observation timing would seem to make sense, though we have not experimented with such.

Just as we may be uncertain about the exact timing, we may also be uncertain about the exact values of the input variables. This latter case is more straightforward, however, simply requiring a measurement model on the input variables. This is not explicitly available within ctsem at time of writing but could be achieved by treating the input variable as an indicator of a regular process in the system and fixing the variance and covariance of system noise (diffusion) for this process to near zero. This configuration would result in a very similar model to the dissipating input effect model we have described, wherein the auto effect could be freely estimated or fixed if one wished explicitly for an impulse or level change effect. The strength of the intervention effect would then need to be captured either by the factor loading or the temporal effect in the drift matrix.

Throughout this paper, we have discussed a model in which the input effect is independent of the current system state. So, although the input effect parameters may depend in some way on the stable characteristics of the subject (either their model parameters or included covariates), this effect does not change depending on whether the subject is higher or lower on any processes. Taking the results from such a model at face value could imply that the joy of receiving a chocolate bar as a gift is independent of whether it is three o'clock in the afternoon or three in the morning. So, some care in application and interpretation is certainly warranted. This independence of effect from the system state may become more problematic in situations with repeated administration of an intervention. In these situations, one approach would be to include the repeated input effect as a distinct input variable with its own set of parameters, rather than another occurrence of the same input variable.

Such differential response to repeated inputs is similar to another potential issue, that of non-linear response to dosage. An example of this could be caffeine, wherein a moderate dose may improve performance on some tasks but too high a dose leads to decrements. This non-linearity is a common modelling problem and not specific to our continuous time approach to interventions. Possible ways of tackling the issue in this context include the addition of quadratic and higher-order polynomial versions of the input variables or binning together ranges of dosage levels and treating these as distinct input variables.

The final limitation we will discuss is that input variables only affect process states and may not alter the model parameters themselves. While a persistent level change effect is equivalent to a persistent change in the continuous intercept parameter matrix, there are no such analogues for the other parameter matrices such as the temporal dynamics or system noise. In the case that an intervention substantially changes the dynamics or measurement properties of the system, model parameters estimated using this approach will represent some average over the observation window. There is no cause (that we can see) to expect bias in either state estimates or the expectation given some intervention, but estimates of uncertainty may be inaccurate. In situations where one is concerned about the possibility of a change in the model parameters induced by an intervention, an approach to test for this could be to include a comparison model wherein the parameters are allowed to change in an instantaneous fashion when the intervention occurs, as with interrupted time-series approaches. This is relatively simple to implement and would correspond to an instantaneous level-change-type effect on the model parameters themselves. A more realistic though more complex approach is to include any relevant model parameters in the system state equations, which would require non-linear filtering techniques.

The continuous-time dynamic modelling approach allows for inference with regard to interventions in two important domains. The primary domain is that of the effect of the intervention on any processes of interest—how long do the effects take to manifest, do they vary in direction over time, and are there individual differences in the effects? The second domain is that of the processes themselves—by generating a substantial impulse of known cause, timing and quantity on the processes, this may enable us to better estimate any causal and mediating relations between processes.

So, just how long does a treatment take to reach maximum effect, how long does the effect last, and what sorts of people is it most effective for? By adopting a modelling framework such as we have proposed and developing an improved understanding of the timing of interventions and their effects, we may be able to better answer such questions.

References

Boker, S. M., Staples, A. D., & Hu, Y. (2016). Dynamics of change and change in dynamics. *Journal of Person-Oriented Research, 2*, 34–55. https://doi.org/10.17505/jpor.2016.05

Carpenter, B., Gelman, A., Hoffman, M. D., Lee, D., Goodrich, B., Betancourt, M., ... Riddell, A. (2017). Stan: A probabilistic programming language. *Journal of Statistical Software, 76*(1). https://doi.org/10.18637/jss.v076.i01

Danhof, M., de Lange, E. C. M., Della Pasqua, O. E., Ploeger, B. A., & Voskuyl, R. A. (2008). Mechanism-based pharmacokinetic-pharmacodynamic (PKPD) modeling in translational drug research. *Trends in Pharmacological Sciences, 29*(4), 186–191. https://doi.org/10.1016/j.tips.2008.01.007

Donnet, S., & Samson, A. (2013). A review on estimation of stochastic differential equations for pharmacokinetic/pharmacodynamic models. *Advanced Drug Delivery Reviews, 65*(7), 929–939. https://doi.org/10.1016/j.addr.2013.03.005

Driver, C. C., Oud, J. H. L., & Voelkle, M. C. (2017). Continuous time structural equation modeling with r package ctsem. *Journal of Statistical Software, 77*(5). https://doi.org/10.18637/jss.v077.i05

Driver, C. C., & Voelkle, M. C. (2018). Hierarchical Bayesian continuous time dynamic modeling. *Psychological Methods.*

Gardiner, C. W. (1985). *Handbook of stochastic methods* (Vol. 3). Berlin: Springer. Retrieved 2016, from http://tocs.ulb.tu-darmstadt.de/12326852.pdf

Neale, M. C., Hunter, M. D., Pritikin, J. N., Zahery, M., Brick, T. R., Kirkpatrick, R.M., ... Boker, S. M. (2016). OpenMx 2.0: Extended structural equation and statistical modeling. *Psychometrika, 81*(2), 535–549. https://doi.org/10.1007/s11336-014-9435-8

Oud, J. H. L., & Jansen, R. A. R. G. (2000). Continuous time state space modeling of panel data by means of SEM. *Psychometrika, 65*(2), 199–215. https://doi.org/10.1007/BF02294374

R Core Team. (2014). *R: A language and environment for statistical computing*. R Foundation for Statistical Computing. http://www.R-project.org/

Singer, H. (1993). Continuous-time dynamical systems with sampled data, errors of measurement and unobserved components. *Journal of Time Series Analysis, 14*(5), 527–545. https://doi.org/10.1111/j.1467-9892.1993.tb00162.x

Singer, H. (1998). Continuous panel models with time dependent parameters. *Journal of Mathematical Sociology, 23*(2), 77–98. https://doi.org/10.1080/0022250X.1998.9990214

Voelkle, M. C., & Oud, J. H. L. (2013). Continuous time modelling with individually varying time intervals for oscillating and non-oscillating processes. *British Journal of Mathematical and Statistical Psychology, 66*(1), 103–126. https://doi.org/10.1111/j.2044-8317.2012.02043.x

Voelkle, M. C., Oud, J. H. L., Davidov, E., & Schmidt, P. (2012). An SEM approach to continuous time modeling of panel data: Relating authoritarianism and anomia. *Psychological Methods, 17*(2), 176–192. https://doi.org/10.1037/a0027543

Wang, Q., Molenaar, P., Harsh, S., Freeman, K., Xie, J., Gold, C., ... Ulbrecht, J. (2014). Personalized state-space modeling of glucose dynamics for type 1 diabetes using continuously monitored glucose, insulin dose, and meal intake: An extended Kalman filter approach. *Journal of Diabetes Science and Technology, 8*(2), 331–345. https://doi.org/10.1177/1932296814524080

Chapter 5
Continuous-Time Modeling of Panel Data with Network Structure

Nynke M. D. Niezink and Tom A. B. Snijders

5.1 Introduction

In many statistical models, the assumption of independent observations is key for making inference. Such an assumption is likely to be valid in, for example, a survey study among a random sample of people from a large population. If a group of people fills out a survey every year, observations are no longer independent. They are nested within individuals.

Temporal dependence is merely one form of dependence between observations. A shared social context may be another reason to assume dependence, for example, for students in a classroom or employees in an organization. When the crime rate in neighboring areas is similar, we speak of spatial dependence. All these forms of dependence between observations are based on some notion of shared context: the individual, the social, and the location.

The interactions and relations between individuals within a social context introduce another layer of dependence. They are the object of study in the field of social network research (e.g., Kadushin 2012; Scott and Carrington 2011; Wasserman and Faust 1994). Examples of social relations include friendship (among students), collaboration (among firms), and advice seeking (among colleagues). The relations in a group of social actors (e.g., students, firms, or colleagues) constitute a social network. In this network, actors are represented by nodes and the relations between actors by ties (directed edges between pairs of nodes).

N. M. D. Niezink (✉)
Department of Statistics & Data Science, Carnegie Mellon University, Pittsburgh, PA, USA
e-mail: nniezink@andrew.cmu.edu

T. A. B. Snijders
Department of Sociology, University of Groningen, Groningen, The Netherlands
e-mail: t.a.b.snijders@rug.nl

K. van Montfort et al. (eds.), *Continuous Time Modeling in the Behavioral and Related Sciences*, https://doi.org/10.1007/978-3-319-77219-6_5

The definition of the group, the set of actors between whom the relation is studied, is part of the design in network research. It is assumed that relations outside this group may be ignored for the purpose of the analysis; this is of course always only an approximation. This is called the problem of network delineation or the "network boundary problem" (cf. Marsden 2005), and it is considered to have been solved before embarking upon the analysis.

Social ties can change over time. They affect and are affected by the characteristics of the actors. An employee can increase his or her performance by seeking advice. At the same time, advice is more likely to be sought from high-performing colleagues. As such, the advice-seeking network and the performance of a group of colleagues may well develop interdependently (coevolve) over time.

A major reason for the fruitfulness of a network-oriented research perspective is this entwinement of networks and individual behavior, performance, attitudes, etc. of social actors. The effect of peers on individual behavior is a well-studied issue. The composition of the social context of individuals influences their attitudes and behaviors, and the choice of interaction partners is itself dependent on these attitudes and behaviors. Studying the entwinement of networks and actor-level outcomes is made difficult because of this endogeneity: the network affects the outcomes, while the outcomes affect the network. One way to get a handle on this is to model these dynamic dependencies in both directions in studies of the coevolution of networks and nodal attributes. This is called the coevolution of networks and behaviors, where for the nodal attributes in the role of dependent variables we use the term "behavior" as a catchword that also can represent other outcomes such as performance, attitudes, etc.

The coevolution of a social network and the attributes of network actors can be studied based on network-attribute panel data. For the analysis of such data, continuous-time models are a natural choice. Many social relations and individual outcomes do not change at fixed-time intervals. Social decisions can be made at any point in time. Moreover, continuous individual outcomes often reflect the consequence of many decisions (Hamerle et al. 1991). Continuous-time models allow us to model gradual change in networks and behavior and have the additional advantage that their results do not depend on the chosen observation interval, as for discrete-time models (Gandolfo 1993; Oud 2007).

In general, continuous-time models are fruitful especially for systems of variables connected by feedback relations and for observations taken at moments that are not necessarily equidistant. Both issues are relevant for longitudinal data on networks and individual behavior. Network mechanisms such as reciprocity and transitive closure ("friends of friends becoming friends") are instances of feedback processes that do not follow a rhythm of regular time steps. The same holds for how actors select interaction partners based on their own behavior and that of others and for social influence of interaction partners on an actor's own behavior. For this reason, continuous-time models are eminently suitable for the statistical modeling of network dynamics.

The dependence structure in social network data is complex. Neither the actors in the network, nor the ties between them are independent. The stochastic actor-oriented model has been developed for the study of social network dynamics (Snijders 2001) and deals with these intricate dependencies. It is a continuous-time Markov chain model on the state space of all possible networks among a set of actors. The model represents network dynamics by consecutive tie change decisions made by actors. The network change observed in panel data is considered the aggregate of many individual tie changes. The model is developed in the tradition of network evolution models by Holland and Leinhardt (1977). Earlier models in this tradition assumed that the pairs of actors (dyads) in a network evolve independently (e.g., Wasserman 1980). The stochastic actor-oriented model does not impose this assumption.

This chapter discusses a model for network-attribute panel data, in which the attributes are measured on a continuous scale. Earlier extensions of the stochastic actor-oriented model for network-attribute coevolution were suited for actor attributes measured on an ordinal categorical scale (Snijders et al. 2007) or on a binary scale representing whether or not an actor has adopted an innovation (Greenan 2015). While those models can be entirely specified within the continuous-time Markov chain framework for discrete (finite) outcome spaces, the model presented here integrates the stochastic actor-oriented model for network dynamics and the stochastic differential equation model for attribute dynamics. The probability model is a combination of a continuous-time model on a discrete outcome space (Kalbfleisch and Lawless 1985) and one on a continuous outcome space.

The model presented in this chapter was first discussed by Niezink and Snijders (2017). While in that paper, we introduced the mathematical framework of the model, here we focus on its applicability. In the next section, we elaborate the type of research questions that the model can help to answer and describe the data necessary to answer the questions. In Sect. 5.3, we formulate the model for the coevolution of a social network and continuous actor attributes. Section 5.4 outlines the method of moment procedure for parameter estimation.

The statistical procedures developed for the analysis of longitudinal network and behavior data are implemented in the R package RSiena (Ripley et al. 2018). Section 5.5 describes how this package can be used to estimate the class of models described in this chapter. In Sect. 5.5, we also present a study of the coevolution of friendship and school performance, in terms of mathematics grades, based on the data from 33 school classes. We simultaneously consider the role of school performance in how students select their friends and how friends influence each other's performance. The study is the first application of the coevolution model for networks and continuous behavior based on multiple groups (networks). Section 5.6 concludes with a summary and a discussion of some potential directions for future research.

5.2 Background Context

Friends have been found to be more similar to each other than non-friends. For example, adolescent friends are often similar in their cigarette and drug use (Kandel 1978) and show similar delinquent behavior (Agnew 1991). Formally, this type of association with respect to individual characteristics of related social actors is referred to as network autocorrelation. It can be caused by peer influence and by homophilous selection, e.g., an adolescent could start smoking, because his friend smokes, and a smoking adolescent could select a fellow smoker as his friend.

Steglich et al. (2010) review the network autocorrelation puzzle from a sociological point of view. They present the stochastic actor-oriented model for network-attribute coevolution as a way to disentangle selection and influence, based on longitudinal data. Applications of the model include studies on the spread of delinquent behavior (Weerman 2011) and moral disengagement (Caravita et al. 2014) in adolescent friendship networks and on the relationship between countries' democracy and their formation of trade agreements (Manger and Pickup 2016).

While the model described by Steglich et al. (2010) assumes actor attributes to be measured on an ordinal categorical scale, in this chapter, we focus on the model for the coevolution of networks and continuous actor attributes (Niezink and Snijders 2017). Many attributes of social actors, such as the performance of an organization or the health-related characteristics of a person, are naturally measured on a continuous scale. Apart from selection and influence processes, the stochastic actor-oriented model can be used to study the effects of local network structures on such attributes.

Vice versa, the effects of individual attributes on the dynamics of social networks can be studied. We can also take into account the effect of covariates, measured on the individual level (e.g., sex of a student) or on the dyad level (e.g., geographic distance between two organizations). Finally, but very essentially, in the stochastic actor-oriented model, the dynamic state of the network itself can affect network change. Transitivity (the tendency to befriend the friends of one's friends) is but one endogenous network mechanism that has been shown to play an important role in the evolution of social networks.

5.2.1 Data

To study a coevolution process as described above, we need repeated observations of a complete social network among a set of actors $\mathscr{I} = \{1, \ldots, n\}$ and of the attributes of these actors. The social networks that are studied using stochastic actor-oriented models typically include between 20 and 400 actors. Networks of this size often still describe a meaningful social context for a group of actors. For very large networks, such as online social networks, this is no longer true.

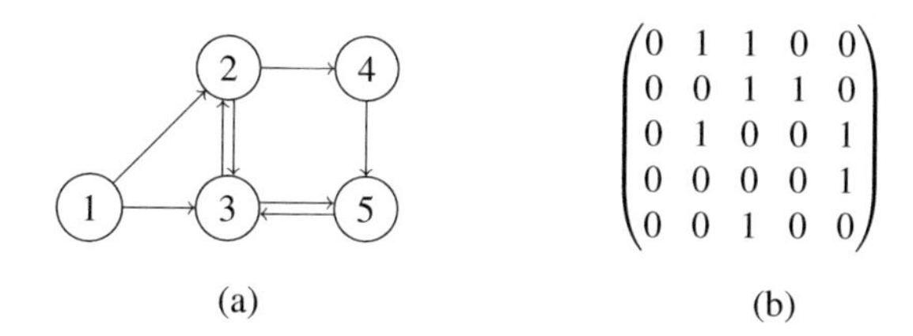

Fig. 5.1 Two representations of the same relational data (**a**) The network. (**b**) The adjacency matrix

Complete social network data contains the information about all $n(n-1)$ tie variables x_{ij} between the actors. The presence of a tie from actor i to actor j is indicated by $x_{ij} = 1$ and its absence by $x_{ij} = 0$. Adjacency matrix $x = (x_{ij}) \in \{0, 1\}^{n \times n}$ summarizes all tie variable information. Figure 5.1 shows an example of a network and the corresponding adjacency matrix. Ties are assumed to be nonreflexive, so $x_{ii} = 0$ for $i \in \mathscr{I}$. Reflexive social ties would be either conceptually different from the ties actors have with others ("being your own friend") or would make no sense ("asking yourself for advice"). We also assume ties to be directed, so x_{ij} and x_{ji} are not necessarily equal. Friendship, advice seeking, and bullying are examples of directed relations; even though i calls j his friend, j may not call i his friend. Undirected relations, such as collaboration, can be studied using the stochastic actor-oriented model as well (Snijders and Pickup 2016) but are not the focus of this chapter.

The attribute values of all actors can be summarized in a matrix $z \in \mathbb{R}^{n \times p}$. In Sect. 5.3, we present the network-attribute coevolution model for the case that actor i has several coevolving attributes $z_i = (z_{i1}, \ldots, z_{ip})$. The implementation in the package RSiena currently only allows for one coevolving attribute, i.e., $p = 1$.

The network $x(t_m)$ and attributes $z(t_m)$ are measured at several points in time $t_1, \ldots, t_M$, not necessarily equidistant. For most longitudinal studies of networks among people, the number of measurements M ranges between two and five. Between the measurement moments, changes in the network and actor attributes take place without being directly observed. Therefore, we assume the measurements to be the discrete-time realizations of a continuous-time process. More specifically, the networks $x(t_m)$ and attributes $z(t_m)$ are assumed to be realizations of stochastic networks $X(t_m)$ and attributes $Z(t_m)$, embedded in a continuous-time stochastic process $Y(t) = (X(t), Z(t))$, with $t_1 \leq t \leq t_M$. Non-stochastic individual and dyadic covariates may also be part of this process but are left implicit in the notation.

The network changes between consecutive measurements provide the information for parameter estimation and therefore should be sufficiently numerous. At the same time, the number of network changes should not be too large. A very large number of changes would contradict the assumption that the change process under study is gradual or, in case the change is gradual, would mean that the measurements are too far apart (Snijders et al. 2010b).

The Jaccard index quantifies the amount of network change between two measurements. Let N_{11} denote the number of ties present at both measurements, N_{10} those present only at the first, and N_{01} those present only at the second measurement.

The Jaccard index is defined as the proportion of stable ties

$$\frac{N_{11}}{N_{11} + N_{10} + N_{01}}. \tag{5.1}$$

For a stochastic actor-oriented model analysis, Jaccard indices should preferably be higher than 0.3. If Jaccard indices are very low while the average degree of the network actors is not strongly increasing, this indicates that the change process is not gradual, compared to the measurement frequency (Ripley et al. 2018).

Collecting complete social network data is a considerable effort. Since complete social network data are especially sensitive to missing data due to their complex dependence structure (Huisman and Steglich 2008), a high response rate is very important. Respondents in complete network studies often form a meaningful group (e.g., a school class or the employees of an organization), which provides a natural choice for the network boundary. Missing responses can therefore not be compensated by the addition of a few randomly selected other individuals to the study. At the same time, for the participants, answering multiple network questions such as "who in this school is your friend/do you study with/do you dislike?" can be a wearying task. We refer to Robins (2015) for guidance on collecting network data and doing social network research.

5.3 Stochastic Actor-Oriented Model

The change observed in network ties and actor attributes between two measurements, at time t_m and t_{m+1}, is the accumulation of the gradual change that has taken place in the meantime. We model the gradual change in network-attribute state $Y(t) = (X(t), Z(t))$ by a continuous-time Markov process: the continuous attribute evolution is represented by a stochastic differential equation (Sect. 5.3.1) and the network evolution by a continuous-time Markov chain (Sect. 5.3.2). In Sect. 5.3.3, we present a simulation scheme for the network-attribute coevolution process, which integrates both components. The simulation scheme is used for parameter estimation (Sect. 5.4).

The $M - 1$ periods between consecutive observations t_m and t_{m+1} do not necessarily have equal length. Moreover, even in periods of equal length, the amount of observed change can differ. For example, in the first month of a school year, friendship ties among students may evolve more rapidly than in the fifth month. In the specification of the stochastic actor-oriented model below, we take this into account by including period-specific parameters that differentiate rates of change per period (cf. Snijders 2001). We do assume that the change is governed by the same social mechanisms over all periods. Therefore, the parameters indicating the strength of these mechanisms are equal across periods.

5.3.1 Attribute Evolution

The evolution of the attributes $Z_i(t)$ of actor i is modeled by the linear stochastic differential equation

$$dZ_i(t) = \tau_m[A\, Z_i(t) + B\, u_i(t)]\, dt + \sqrt{\tau_m}\, G\, dW_i(t) \tag{5.2}$$

for the period between two observation moments t_m and t_{m+1}. The drift matrix $A \in \mathbb{R}^{p\times p}$ specifies how the change in the attributes $Z_i(t)$ depends on the state of the attributes. The elements of input vector $u_i(t) \in \mathbb{R}^r$ are called effects and can depend on the state $Y(t)$ of the coevolution process. Parameter matrix $B \in \mathbb{R}^{p\times r}$ indicates the strength of the effects in $u_i(t)$.

An example of an effect that depends on the network state $X(t)$ is the number of outgoing ties of actor i:

$$\sum_j X_{ij}(t). \tag{5.3}$$

Figure 5.2a shows the consequence of a positive outdegree effect: actor i has three outgoing ties and therefore his attribute value increases. A corresponding hypothesis could be "having friends has a positive effect on a person's mood." Figure 5.2b visualizes a social influence process, a dependence of the attribute dynamics of actor i on the attribute values of the actors to whom actor i has outgoing ties (his "alters"). A corresponding hypothesis could be "a person's mood is affected by the mood of his friends." In the figure, actor i takes on the average value of the attribute values of his alters. Generally, social influence processes are modeled by the combination of network and attribute information. The average alter effect of the kth attribute, defined as

$$\textstyle\sum_j X_{ij}(t)(Z_{jk}(t) - \bar{z}_k)\, /\, \sum_j X_{ij}(t), \tag{5.4}$$

is one way to operationalize social influence. Here, $\bar{z}_k$ denotes the overall observed mean of attribute k. If actor i has no outgoing ties, the effect is set to 0. The effects in $u_i(t)$ may also depend on individual and dyadic covariates.

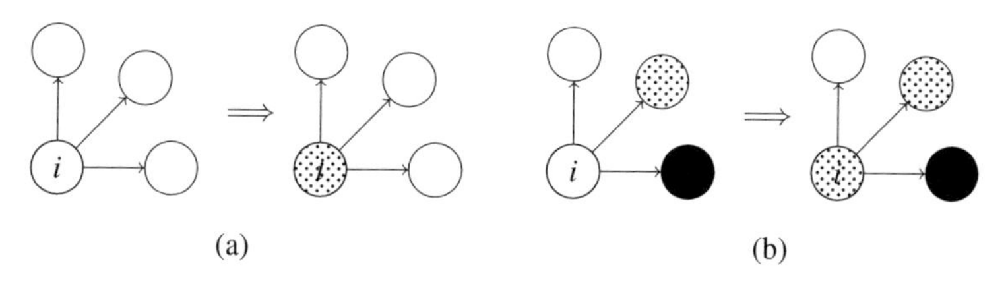

Fig. 5.2 Two examples of network-related effects in input vector $u_i(t)$. Darker nodes represent actors with a higher attribute value. (**a**) Outdegree effect. (**b**) Average alter effect

Parameter τ_m in stochastic differential equation (5.2) is a period-specific parameter and accounts for potential heterogeneity in period length. The parameter indicates how the equation scales in time. For $s = \tau\, t$, a standard Wiener process $W(s)$ transforms as

$$W(s) = W(\tau\, t) = \sqrt{\tau}\, W_\tau(t), \tag{5.5}$$

where $W_\tau(t)$ is again a standard Wiener process. This explains the $\sqrt{\tau_m}$ in Eq. (5.2). Parameter τ_m in the deterministic part of the equation follows from $\mathrm{d}s/\mathrm{d}t = \tau$. We can assume that the time parameter t in our model runs between 0 and 1 for each period m, as the free—and to be estimated—parameter τ_m will absorb the true time scale of the period.

Finally, to make model (5.2) identifiable, we assume that matrix $G \in \mathbb{R}^{p\times p}$ is a lower diagonal matrix with upper left entry equal to 1 (Niezink and Snijders 2017). In case the number of observations is two, we can set $\tau_1 = 1$ instead.

5.3.2 Network Evolution

A characteristic assumption of the stochastic actor-oriented model (Snijders 2001) is that actors control their outgoing ties. Changes in the network are modeled as choices made by actors. We assume that, at any given moment, all actors act conditionally independently of each other given the current state of the network and attributes of all actors. Moreover, actors are assumed not to make more than one change in their ties to other actors at any given time point. These assumptions exclude, for example, the possibility for actors to coordinate their decisions but have the advantage that they keep the model parsimonious and relatively simple. Depending on the application, the assumptions will make more or less sense.

The network evolution model, given the attribute values, is a continuous-time Markov chain with a discrete state space (Norris 1997). This means that at discrete moments, the state "jumps" to a different value. The time between these jump moments must have the exponential distribution because of the Markov property; the assumptions made in the preceding paragraph imply that the new state differs in exactly one tie variable from the old state.

In the network evolution model, the rate of network change or, more precisely, the rate at which each actor in the network gets the opportunity to change one of his outgoing ties is period-specific. In period m, the waiting time until the next change opportunity is, for each actor, exponentially distributed with rate λ_m. The minimum of these waiting times, i.e., the time until the next change opportunity for any actor, is exponentially distributed with rate $\lambda_+ = n\lambda_m$. The probability that a particular actor i gets the opportunity to make a change is $1/n$. The parameters λ_m play a role similar to that of the parameters τ_m in the stochastic differential equation. They account for heterogeneity in period length and allow us to model each period as having unit duration. For models with variable change rates across actors, see Snijders (2001).

An actor i with the opportunity to make a tie change can choose to create a new tie to one of the other actors, to dissolve one of his existing ties, or to make no change. The probability distribution of the change made by actor i is determined by objective function $f_i(x, z)$. This function can be interpreted as the value actor i assigns to a particular network-attribute state (x, z), when making a network change.

Let $x^{(\pm ij)}$ denote the adjacency matrix equal to x, in which entry x_{ij} is changed into $1 - x_{ij}$, and let $x^{(\pm ii)} = x$. Given that x is the current network state, the choice probabilities for actor i are given by

$$\mathrm{P}\left\{x \text{ changes to } x^{(\pm ij)}\right\} = p_{ij}(x, z) = \frac{\exp(f_i(x^{(\pm ij)}, z))}{\sum_{h=1}^{n} \exp(f_i(x^{(\pm ih)}, z))}. \tag{5.6}$$

These probabilities can be obtained as the result of actor i selecting the network state $x^{(\pm ij)}$ $(j = 1, \ldots, n)$ for which

$$f_i(x^{(\pm ij)}, z) + \epsilon_i$$

is highest, where the ϵ_i are independently generated for each next actor choice and follow a standard Gumbel distribution (McFadden 1974).

The objective function is defined as a weighted sum of effects $s_{ik}(x, z)$,

$$f_i(x, z) = \sum_k \beta_k s_{ik}(x, z). \tag{5.7}$$

Parameter β_k indicates the strength of the kth effect, controlling for all other effects in the model. The effects represent the actor-level mechanisms governing network change, as the effects in $u_i(t)$ in Eq. (5.2) do for attribute change. Ripley et al. (2018) provide an overview of the many effects that are currently implemented for stochastic actor-oriented models. Examples are discussed in Sect. 5.5. As the effects $s_{ik}(x, z)$ are allowed to depend on the actor attributes z, we can model selection based on attribute similarity (homophilous selection). We can also model the differential tendency of actors with high attribute values to send (ego effect) or receive (alter effect) more network ties.

For given actor attribute values z, the network evolution model reduces to a continuous-time Markov chain (Norris 1997) with transition rate matrix or infinitesimal generator:

$$q(x, x') = \begin{cases} \lambda_m \, p_{ij}(x) & \text{if } x' = x^{(\pm ij)} \text{ and } i \neq j, \\ -\sum_{i \neq j} \lambda_m \, p_{ij}(x) & \text{if } x' = x, \\ 0 & \text{else.} \end{cases}$$

In the following section, we specify how the evolution of a network and the continuous attributes of network actors can be simulated, in case their dynamics are interdependent.

5.3.3 Coevolution

For linear stochastic differential equations, the transition density from one measurement to the next can be expressed exactly in the so-called exact discrete model (EDM) (e.g., Hamerle et al. 1991; Oud and Jansen 2000). This model is the solution of the stochastic differential equation. For most specifications of the stochastic actor-oriented model, it is not possible to express the transition density of a network-attribute state explicitly. Therefore, we specify the coevolution model as a simulation model.

In the coevolution model, we use the EDM to model the attribute change not between measurements but between the (unobserved) tie changes. The EDM expresses the discrete-time consequences of continuous-time model (5.2). Between tie changes at times t and $t + \Delta t$, the attribute change for actor i is given by

$$z_{i,t+\Delta t} = A_{\Delta t}\, z_{i,t} + B_{\Delta t}\, u_{i,t} + w_{i,\Delta t}, \tag{5.8}$$

where $w_{i,\Delta t}$ is multivariate normally distributed with zero mean and covariance $Q_{\Delta t}$ (Bergstrom 1984; Hamerle et al. 1991). The EDM exactly relates the continuous-time parameters in (5.2) to the discrete-time parameters in (5.8) via

$$\begin{aligned} A_{\Delta t} &= e^{\tau_m A \Delta t} \\ B_{\Delta t} &= A^{-1}(A_{\Delta t} - I_p)B \\ Q_{\Delta t} &= \text{ivec}[(A \otimes I_p + I_p \otimes A)^{-1}(A_{\Delta t} \otimes A_{\Delta t} - I_p \otimes I_p)\, \text{vec}(GG^\top)], \end{aligned} \tag{5.9}$$

in case the input $u_i(t)$ is constant between t and $t + \Delta t$. This is true if $u_i(t)$ only contains effects that are a function of the covariates and the network state—the network is constant between tie changes. However, social influence effects, such as average alter effect (5.4), also depend on the attribute states of the actors. For this case, Niezink and Snijders (2017) show in a simulation study that the approximation error in the attribute values induced by setting $u_{i,t}$ constant between tie changes is negligible.

Intuitively, the negligible approximation error can be understood by considering Fig. 5.3. The periods Δt between the unobserved tie changes are much smaller than the time period between the measurements. In practice, in the coevolution scheme presented hereafter, the number of simulated tie changes within one period often easily tops a hundred. Repeatedly assuming $u_i(t)$ to be constant over a very small

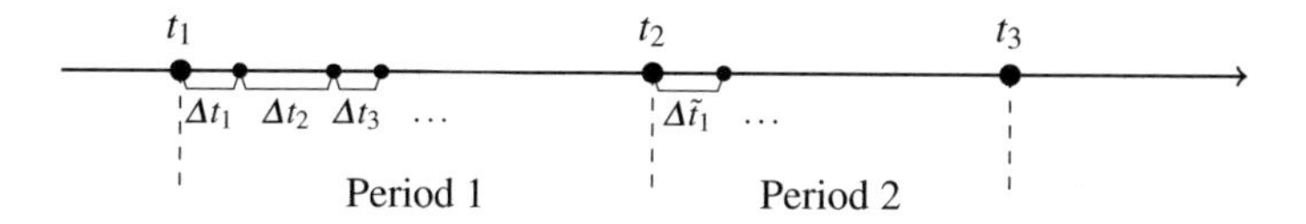

Fig. 5.3 Time scales in network-attribute coevolution. The times Δt between consecutive tie changes are very small compared to the lengths of the two observations periods (the Δt are depicted larger for sake of visualization)

time span does not accumulate into large deviations in the actor attributes (Niezink and Snijders 2017).

By combining the EDM with the network evolution model defined in the previous section, we specify the network-attribute coevolution model in a simulation scheme (see Algorithm 1). In this scheme, the following steps are repeated: after a waiting time Δt until a next network change is drawn, the actor attributes are updated over the Δt period, and an actor i makes a tie change to an actor j.

The simulation time length for the coevolution process between observation times t_m and t_{m+1} is set to 1. While this is an arbitrary choice, the actual duration of period m is accounted for by the network change rates λ_m and by the parameters τ_m of the stochastic differential equation.

We condition the simulation of the coevolution process in period m on the initial observation of that period, the network $x(t_m)$ and attributes $z(t_m)$. Estimation is carried out conditional on these initial states.

Data: network $x(t_m)$, attributes $z(t_m)$, and covariates.
Input: values for parameter matrices A, B, G and for τ_m, λ_m, and all β_k.

set $t = 0$, $x = x(t_m)$, $z = z(t_m)$ and $u_i = u_i(x, z)$ for all $i \in \mathscr{I}$
sample Δt from an exponential distribution with rate $n\lambda_m$
while $t + \Delta t < 1$ **do**
  for all $i \in \mathscr{I}$: sample c_i from a $\mathscr{N}(A_{\Delta t} z_i + B_{\Delta t} u_i, Q_{\Delta t})$ distribution
  for all $i \in \mathscr{I}$: set $z_i = c_i$
  select actor $i \in \mathscr{I}$ with probability $1/n$
  select alter $j \in \mathscr{I}$ with probability $p_{ij}(x, z)$
  set $t = t + \Delta t$ and $x = x^{(\pm ij)}$, and $u_i = u_i(x, z)$ for all $i \in \mathscr{I}$
  sample Δt from an exponential distribution with rate $n\lambda_m$
end
for all $i \in \mathscr{I}$: sample c_i from a $\mathscr{N}(A_{(1-t)} z_i + B_{(1-t)} u_i, Q_{(1-t)})$ distribution
for all $i \in \mathscr{I}$: set $z_i = c_i$

Algorithm 1: Simulating the network-attribute coevolution in period m

5.4 Estimation

The parameters in a stochastic actor-oriented model represent the strengths of the effects included in the model. To estimate the parameters, the method of moment procedure (Snijders 2001) is most commonly used. Bayesian and maximum likelihood estimation methods also have been proposed (Koskinen and Snijders 2007; Snijders et al. 2010a) but require much more computation time.

Let θ denote the vector containing all parameters in the model, and $S(Y)$ a vector of statistics, one for each parameter, such that statistic S_k is sensitive to changes in parameter θ_k. For example, for the reciprocity parameter, we select the number of reciprocated ties at the observation moments as our statistic. The method of moment estimate $\hat{\theta}$ is that value of θ for which the expected value of $S(Y)$ equals its observed value $s = S(y)$,

$$\mathrm{E}_{\hat{\theta}}(S(Y)) = s. \tag{5.10}$$

Snijders (2001) proposed statistics for the network evolution parameters, and Niezink and Snijders (2017) derived statistics for the attribute evolution parameters. The latter are based on the sufficient statistics of an autoregression model, the parameters of which can be linked to the continuous-time parameters via the EDM.

It follows from the delta method (e.g., Lehmann 1999, p. 315) that the covariance matrix of $\hat{\theta}$ can be approximated as

$$\mathrm{cov}(\hat{\theta}) \approx D_\theta^{-1} \,\mathrm{cov}_\theta(S(Y))(D_\theta^{-1})^\top, \tag{5.11}$$

where

$$D_\theta = \frac{\partial}{\partial \theta} \mathrm{E}_\theta(S(Y)) \tag{5.12}$$

and $\mathrm{cov}_\theta(S(Y))$ are evaluated at the estimate $\hat{\theta}$.

The values of $\mathrm{E}_\theta(S(Y))$ and $\mathrm{cov}_\theta(S(Y))$ generally cannot be evaluated analytically, but they can be estimated based on simulations. To this end, for each $m = 1, \ldots, M - 1$ the coevolution process $Y(t)$ is simulated from time t_m to t_{m+1}, given the observed initial value $Y(t_m) = y(t_m)$ and parameters θ. We can estimate $\mathrm{E}_\theta(S(Y))$ by the average of the statistics computed based on a large random sample of simulated coevolution processes and $\mathrm{cov}_\theta(S)$ by their covariance. Also D_θ can be estimated based on simulated coevolution trajectories, using the score function method described by Schweinberger and Snijders (2007) and elaborated for continuous dependent variables by Niezink and Snijders (2017).

To find the solution $\hat{\theta}$ to Eq. (5.10), we use the multivariate version of the Robbins-Monro (1951) stochastic approximation algorithm, a stochastic equivalent of the Newton-Raphson method. Let S_N denote the values of the statistics based on the Nth simulation of the coevolution process with $\theta = \theta_N$. The central iteration step in the algorithm is

$$\theta_{N+1} = \theta_N - a_N \operatorname{diag}(\hat{D})^{-1}(S_N - s), \tag{5.13}$$

where $\operatorname{diag}(\hat{D})$ is a an estimate of derivative matrix (5.12), in which the outer diagonal entries have been multiplied by a constant between 0 and 1, and a_N is a sequence of positive numbers converging to zero at rate N^{-c}, with $0 < c < 0.5$, controlling the convergence of the algorithm (Polyak 1990; Ruppert 1988).

The estimation procedure, implemented in the R package RSiena (Ripley et al. 2018), consists of three phases.

Phase 1. Given an initial parameter θ_0, matrix $\hat{D}$ is estimated based on a small number n_1 of simulations. Then the parameter value is updated once using (5.13) with $S_0 = \overline{S}$, the mean of the statistics values based on the n_1 simulations.

Phase 2. The value of θ is estimated in n_2 subphases of increasing length and with a_N constant per subphase. At the end of each subphase, θ is estimated by the average of the θ_N values generated in the subphase, and a_N is reduced by a factor 0.5.

Phase 3. Given the final estimate of θ in Phase 2, a large number n_3 of coevolution processes is simulated. These are used to estimate covariance (5.11) and to check the convergence of the algorithm.

For more details about the estimation procedure and convergence checking, see Snijders (2001, 2005) and Ripley et al. (2018). After parameter estimation, the goodness of fit of a model can be checked by evaluating whether, apart from the network configurations and attribute effects explicitly fitted, also other aspects of the network structure and attributes are adequately represented by the model (Ripley et al. 2018). Examples of network structures that are often considered in goodness of fit analyses are the in- and outdegree distribution and the triad census (all possible network configurations on three actors) of a network.

5.5 Example: Coevolution of Friendship and Grades

To illustrate the method discussed in this chapter, we present a study of the coevolution of the friendship ties between adolescents at the start of their secondary education and their academic performance, in terms of mathematics grades. We explore whether academic performance plays a role in the formation of friendship between students and whether the performance of a student is influenced by the performance of his friends. Although stereotypes about female inferiority in mathematics still persist, actual performance differences favoring males are generally not found (Hyde et al. 1990; Lindberg et al. 2010). In this study, we also assess gender differences in the evolution of mathematical performance.

The analysis is based on school class data collected by Knecht (2004). Parts of these data were studied by Knecht (2010) and Knecht et al. (2011). In these references, detailed information about the data and procedure is provided. In this section, we analyze a subset of the classes in the original data, using the stochastic actor-oriented model. Throughout the text, code snippets provide guidance on how to analyze a single class using the R package RSiena. We aggregate the classroom results in a meta-analysis (Cochran 1954; Snijders and Baerveldt 2003).

5.5.1 Data

The original sample (Knecht 2004) included over 3000 students from 120 classrooms in 14 secondary schools in the Netherlands, who were followed over their first year at secondary school (2003/04). Data were collected four times, every 3 months. Friendship ties were assessed by asking students to nominate up to 12 best friends in class. Mathematics grade information was only available at the last three measurements. We will refer to these time points as time 1, time 2, and time 3.

The numbers of students in the classrooms ranged from 14 to 33. These numbers challenge the feasibility of the method of moment estimation procedure. For a class of size n, with no missing data, only $n \times 3$ grade observations are available, based on which the grade dynamics have to be inferred. To avoid convergence issues due to small class sizes or missing data, we selected the classes with grade data available for more than 25 students at least two waves. The 39 selected classes ranged in size from 27 to 33. By selecting classes of similar size, we also ensure that the results based on the different classes are comparable. The Jaccard indices of the friendship networks in the 39 classes ranged from 0.34 to 0.71. These values are sufficiently high to consider the observed networks as part of an evolution process (see Sect. 5.2.1).

The selected sample includes 1160 students, aged 11–13 at the beginning of the school year. Table 5.1 presents some descriptives of the friendship networks and mathematics grades at the three time points. Network density is defined as the ratio of the observed number of ties to the total number of possible ties $n(n-1)$. The networks in our selected sample have a low density, which is expected for networks of best friendship nominations. Reciprocity in a network is defined as the number of reciprocated (mutual) ties over the total number of ties. The reciprocity in the best friendship networks is high.

The theoretical range of the mathematics grades is 10 to 100, and the minimum passing grade is 55. Table 5.1 shows that the differences in average mathematics grades between boys and girls are very small, especially given the large variation within the groups. For the stochastic actor-oriented model analyses, the grades are centered on the average grade 69.7 of the 1160 students and divided by 10. This transformation does not affect the substantive conclusions of the analyses but does ensure that the scale of the parameters in the stochastic differential equation and the grade-related network parameters are not too small.

Table 5.1 Descriptive statistics of the friendship networks and mathematics grades

	Time 1 mean (sd)	Time 2 mean (sd)	Time 3 mean (sd)
Friendship			
Density	0.13 (0.020)	0.14 (0.015)	0.12 (0.019)
Reciprocity	0.88 (0.032)	0.88 (0.035)	0.88 (0.026)
Missing (%)	16.7 (10.3)	13.2 (6.2)	16.4 (8.5)
Mathematics grades			
Boys	69.6 (11.2)	70.2 (10.7)	70.1 (11.6)
Girls	69.2 (11.5)	69.6 (10.7)	69.9 (11.0)
Missing (%)	19.1	10.9	11.8

The network descriptives are computed on the network level (39 networks) and the grade descriptives on the individual level (time 1, 938 students; time 2, 1033 students; time 3, 1023 students)

A network-attribute data set on 30 students can be read into the RSiena framework as described below. The function `sienaDependent` defines the coevolving dependent variables, based on the three observed 30×30 adjacency matrices and the 30×3 matrix with mathematics grades. Gender is a constant covariate (`coCovar`).

```
friends <-
sienaDependent(array(c(friend1, friend2, friend3),
dim=c(30,30,3), type="oneMode")
grades <- sienaDependent(gradeData, type="continuous")
sexM <- coCovar(sexData, center=FALSE)
mydata <- sienaDataCreate(friends, grades, sexM)
```

5.5.2 Model Specification

The specification of an actor-based model is given by the list of effects included in the objective function (5.7) and the input vector $u_i(t)$ in stochastic differential equation (5.2). The basic specification of the model, obtained from the command

```
myeff <- getEffects(mydata)
```

accounts for the density of the friendship network and the tendency of students to reciprocate friendship ties,

$$f_i(x, z) = \beta_1 \sum_j x_{ij} + \beta_2 \sum_j x_{ij} x_{ji}, \tag{5.14}$$

and includes an intercept term in the stochastic differential equation,

$$dZ_i(t) = \tau_m[a\, Z_i(t) + b_0]\, dt + \sqrt{\tau_m}\, dW_i(t). \tag{5.15}$$

For data collected at two measurement moments, the alternative specification of differential equation (5.2), fixing τ_1 to 1 instead of G, is obtained by including the option `onePeriodSde=TRUE` in the function `getEffects`.

We add to this default model a selection of effects which represent social mechanisms governing the friendship and grade dynamics. The manual for the RSiena package provides an overview of all effects that are implemented for the stochastic actor-oriented model (Ripley et al. 2018). It also provides some guidelines on the practice of model specification.

In this study, we are interested in the interdependence between the dynamics of friendship and mathematics grades. We model social influence on grades by the average alter effect, specified in expression (5.4). We also assess the effect of gender on the grade dynamics.

```
myeff <- includeEffects(myeff, name="grades", avAlt,
interaction1="friends")
myeff <- includeEffects(myeff, name="grades", effFrom,
interaction1="sexM")
```

For the function `includeEffects`, the parameter `name` specifies the dependent variable, and `interaction1` specifies the explanatory variable. The effects to be included are designated by their short names `avAlt` and `effFrom`; the short names are listed in the manual (Ripley et al. 2018).

The specification of the dynamics of the friendship network is done in line with the usual recommendations for network dynamics, as given in Snijders et al. (2010b) and Ripley et al. (2018). The effects of grades on friendship are modeled by a grade ego and a grade alter effect, representing the differential tendency for students with high grades to send or receive more friendship nominations. We also include an interaction effect of ego's and alter's grade, measuring the differential attractiveness of students with high grades for students with high grades. A positive grade ego × grade alter effect indicates that students with high grades tend to nominate friends with high grades (i.e., homophilous selection based on grades). The three effects are added to the objective function (5.14),

$$\cdots + \beta_3\, z_i \sum_j x_{ij} + \beta_4 \sum_j x_{ij} z_j + \beta_5\, z_i \sum_j x_{ij} z_j. \tag{5.16}$$

```
myeff <- includeEffects(myeff, name="friends",
egoX, altX, egoXaltX, interaction1="grades")
```

Reciprocity is not the only endogenous network mechanism playing a role in friendship dynamics. In friendship, people have a tendency toward transitive closure, i.e., to befriend the friends of their friends. As a tendency against reciprocation is generally found in transitive groups (Block 2015), we also account for the interaction effect of reciprocity and transitivity. Moreover, we account for degree-related effects on friendship. The outdegree activity effect represents the differential tendency for students with a high outdegree to send friendship nominations. The outdegree (indegree) popularity effect represents the tendency for students with a high outdegree (indegree) to receive friendship nominations. A homophilous selection effect based on gender is also included in the model. Let v_i denote the gender of student i (0 = female, 1 = male) and $I\{C\}$ the indicator function for condition C. We complete objective function (5.14) with

$$\cdots + \beta_6 \sum_{j,h} x_{ij}x_{ih}x_{hj} + \beta_7 \sum_j x_{ij}x_{ji} \sum_h x_{ih}x_{hj} + \beta_8 \Big(\sum_j x_{ij}\Big)^2$$

$$+ \beta_9 \sum_j x_{ij} \sum_h x_{jh} + \beta_{10} \sum_j x_{ij} \sum_h x_{hj} + \beta_{11} I\{v_i = v_j\}. \tag{5.17}$$

```
myeff <- includeEffects(myeff, name="friends",
transTrip, transRecTrip, outAct, outPop, inPop)
myeff <- includeEffects(myeff, name="friends", sameX,
interaction1="sexM")
```

5.5.3 *Analysis*

We estimate the model parameters using the method of moments in RSiena. The estimation procedure requires little further implementation by the user. The function `sienaAlgorithm` specifies an estimation algorithm with `nsub` subphases in phase 2 and `n3` iterations in phase 3. The default number of phase 2 subphases is 4. Increasing this number can help in obtaining converged models but makes the estimation procedure lasts longer. The default number of iterations in phase 3 is 1000. How to select the right number of iterations in phase 3 to obtain accurate standard error estimates is the subject of ongoing research. Here we do 30,000 iterations. Several other parameters can be passed to the algorithm (Ripley et al. 2018). We set the random number seed to 123, to be able reproduce our results. As students were asked to nominate up to 12 best friends, we set the maximum possible outdegree in the friendship network to 12.

```
myalg <- sienaAlgorithmCreate(projname="analysis",
nsub=5, n3=30000, seed=123, MaxDegree=c(friends=12))
ans <- siena07(myalg, data=mydata, effects=myeff)
```

For 2 of the 39 classrooms, the estimation procedure did not reach convergence. For four of the remaining classrooms, the models converged, but the standard errors were estimated inadequately. Table 5.2 presents the results of the meta-analysis based on the finally retained sample of 33 classes. In the meta-analysis, it is assumed

Table 5.2 Results of meta-analysis of stochastic actor-oriented model analyses

	Mean parameter		Variation	
	$\hat{\mu}_\theta$	s.e.	$\hat{\sigma}_\theta$	p
Friendship dynamics				
Rate period 1	5.09	0.30[a]	1.36	<0.001
Rate period 2	4.86	0.25[a]	1.02	<0.001
Density	−1.70	0.11*	0.35	0.058
Reciprocity	2.25	0.082*	0.19	0.162
Transitivity	0.58	0.028*	0.10	0.025
Transitivity × reciprocity	−0.28	0.031*	<0.001	0.980
Outdegree activity	−0.020	0.008*	0.026	0.011
Outdegree popularity	−0.33	0.022*	0.085	<0.001
Indegree popularity	0.011	0.013	0.039	0.105
Same gender	0.65	0.054*	0.14	0.142
Grades ego	0.040	0.028	0.070	0.118
Grades alter	−0.024	0.024	<0.001	0.671
Grades ego × grades alter	0.094	0.030*	<0.001	0.726
Grade dynamics				
Scale period 1	0.57	0.056[a]	0.19	0.001
Scale period 2	0.47	0.051[a]	0.20	<0.001
Feedback	−0.58	0.047[a]	0.16	0.018
Intercept	0.099	0.070	0.27	0.001
Average alter	0.069	0.095	<0.001	0.991
Male	0.023	0.063	<0.001	0.948

$^*p < 0.01$
[a]These effects are not tested, as the hypothesis $H_0 : \theta = 0$ is irrelevant. The mean parameter tests are based on the t-ratio of the estimated mean parameter ($\hat{\mu}_\theta$) to its standard error (s.e.). Estimate $\hat{\sigma}_\theta$ denotes the estimated true between-classroom standard deviation; the p-value is based on a chi-squared test ($df = 32$) for testing $\sigma_\theta^2 = 0$

that parameters may be different across classrooms and that they are the sum of a mean parameter value μ_θ and a random deviation with mean 0 and standard deviation σ_θ. Table 5.2 indicates the estimated mean parameter estimates $\hat{\mu}_\theta$ (with standard errors) and the estimated between-classroom standard deviations $\hat{\sigma}_\theta$ (with p-values of the tests that variance $\hat{\sigma}_\theta^2$ is 0).

5.5.3.1 Results for Friendship Dynamics

The results indicate no evidence for a differential effect of grades on sending friendship ties (grades ego effect), nor an effect of grades on the students' popularity (grades alter). We do find a significant effect of grade-based selection (grades ego × grades alter), indicating that students with high grades are more likely to select friends with high grades.

The grade-related parameters can be interpreted as follows. When a student i makes a single tie change, and x^a and x^b are two possible results of this change, then

$$\exp(f_i(x^a, z) - f_i(x^b, z)) \tag{5.18}$$

is the odds ratio for choosing between x^a and x^b. So when a student who scores 20 points above the average creates a friendship tie, his odds ratio for nominating a classmate who scores 20 points above the average (alter1), compared to one who scores 10 points below the average (alter2), is

$$\exp\Big(- 0.024 \times (\underset{\text{alter1}}{2} - \underset{\text{alter2}}{-1}) + 0.094 \times \underset{\text{ego}}{2} \times (\underset{\text{alter1}}{2} - \underset{\text{alter2}}{-1})\Big) \approx 1.64, \tag{5.19}$$

when all else is equal. The latter means that in the two networks, induced by selecting the high-performing and the low-performing classmate, respectively, the contributions of all other effects to the objective function $f_i(x, z)$ are equal. We also find strong evidence of gender homophily. Everything else being equal, students are almost twice as likely (odds ratio = 1.92) to select same gender classmates as friends.

As expected, Table 5.2 indicates strong evidence for reciprocity and transitivity. Moreover, reciprocity is less likely to occur in transitive groups. Students who already nominate many friends do not tend to nominate even more friends (negative outdegree activity) and are less popular (negative outdegree popularity).

The right half of Table 5.2 provides insight in the differences across classrooms. For the endogenous network effects of transitivity, outdegree activity, and outdegree popularity, we find significant ($p < 0.05$) differences across classes. The substantive conclusions drawn earlier for transitivity and outdegree popularity remain generally valid, given the size of the variation in the parameters. For the outdegree activity effect, we consider the approximate 95% prediction interval of the parameter β_8, given by

$$\hat{\mu}_{\beta_8} \pm t_{0.975,k-2}\sqrt{\left(\text{s.e.}(\hat{\beta}_8)\right)^2 + \left(\hat{\sigma}_{\beta_8}\right)^2}, \tag{5.20}$$

where k is the number of classrooms in the meta-analysis (Borenstein et al. 2009). This interval ranges from −0.076 to 0.035, thus containing both positive and negative values.

5.5.3.2 Results for Grade Dynamics

Considering the grade dynamics, we do not find evidence of an influence effect of peers on mathematics grades (average alter). Also, girls and boys do not significantly differ in their grade evolution.

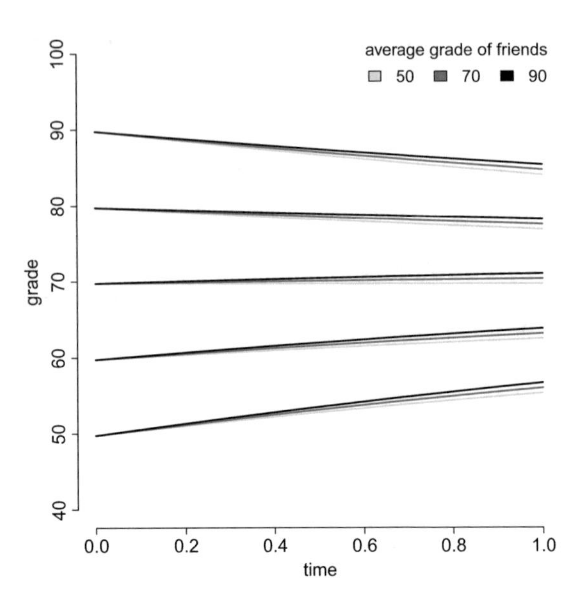

Fig. 5.4 Visualization of the grade dynamics: the mean trajectories for boys with initial grades ranging from 50 to 90 and average friends' grades 50, 70, or 90

Based on the estimated parameters, Fig. 5.4 visualizes the size of the nonsignificant social influence effect, by showing the average grade dynamics for boys between the first two measurements for various initial grade values and various average friends' grades. The analogous figure for girls (not depicted here) is very similar. We can conclude from Fig. 5.4 that the performance level of a student's friends matters little for his or her grade evolution.

Table 5.2 shows a significant and substantive variation in the intercept parameter across classrooms; the 95% prediction interval for the intercept parameter ranges from −0.46 to 0.66. This indicates classroom-level variation in the development of class averages.

5.6 Discussion

In this chapter, we have shown how the interdependent dynamics of a social network and the continuous attributes of network actors can be modeled based on panel data. The stochastic actor-oriented model presented here employs a continuous-time approach, which allows us to model the unobserved network-attribute dynamics between the measurement moments. The model combines a Markov chain model for the network evolution and a stochastic differential equation for the attribute evolution.

Continuous-time models are a natural representation of processes that occur continuously in time. In our case, continuous-time modeling has the advantage that it allows us to specify the coevolution model as a generative model. For non-network panel data, autoregressive and cross-lagged models, with their limitations (e.g., Voelkle et al. 2012), constitute a classic alternative for stochastic differential equations. For complete network panel data, similar regression-type alternatives, representing the complex dependence structure of complete network data, did not exist before the introduction of the stochastic actor-oriented model (Snijders 1996).

We account for irregularly spaced measurement moments by period-specific parameters in the model. These parameters allow us to model each period as having unit duration. This may not seem to be in the spirit of continuous-time modeling, where the true time intervals do their work without the insertion of extra parameters (e.g., Oud and Jansen 2000). However, in social network panel data, the measurements are usually months apart, and the expected amount of change in equal time intervals is not necessarily the same. In the first month of a school year, friendship ties among high school students may evolve more rapidly than in the fifth month. Therefore, instead of letting the amount of change depend on the product of the true interval lengths and a single fixed change rate, we estimate the change rates per period. The period-specific parameters thus also account for the heterogeneity in change activity between periods.

Depending on their attributes or on positional characteristics such as network outdegree, actors might change their social ties at different rates. For example, girls may change their friendship ties more frequently than boys. Such differences can be investigated by the parametrization of the rate into a so-called rate function (Snijders 2001). The stochastic actor-oriented model also allows for the differentiation between tie creation and tie termination (Snijders 2001). In this chapter, we have assumed that both processes are governed by the same social mechanisms to the same extent. This assumption may not always be valid. For example, the difference between a reciprocated and a non-reciprocated friendship may be larger in the choice frame of losing a friend than in the frame of making a new friend.

The stochastic actor-oriented model is mostly applied to panel data with two to five measurements. Theoretically, it could also be applied to time series data. In case that all network changes and the time points at which these changes occur are known, as well as the attribute values of the actors at these time points, the likelihood corresponding to the model presented in this chapter can be formulated explicitly. Parameters could then be estimated by maximizing this likelihood. Unfortunately, fine-grained information of this sort about network evolution is hard to collect and rarely available.

The stochastic actor-oriented model aims to model the change of a network *state* over time, based on "snapshots" of this state. Stadtfeld (2012; see also Stadtfeld et al. 2017) generalized the model to time-stamped *event* stream data. When people make phone calls, send e-mails, or visit each other, these actions can be considered as directed dyadic relational events. Stadtfeld (2012) models such events from an actor-oriented perspective. The methodology has been applied on private message communication in an online question and answer community of around 88,000 people over a 3-year time span (Stadtfeld and Geyer-Schulz 2011).

Stochastic actor-oriented coevolution models are often used with the aim to disentangle selection and social influence processes. We studied these processes in a secondary school context, exploring the relation between the friendship and academic performance, in terms of mathematics grades. Based on the data from 33 classrooms, we find evidence of friendship selection on grade similarity, but not of peer influences on grades.

We combined the results of the 33 classrooms in a meta-analysis. For the study of network dynamics based on multiple groups, a random coefficient model has been implemented that combines the actor-oriented network model with the ideas of hierarchical linear modeling (Ripley et al. 2018). Integrating the coevolving attributes in this framework would be a natural extension of the method presented in this chapter. This approach could mean an increase in statistical power compared to the simple meta-analysis. This would make the method suitable for the analysis of a collection of small networks (e.g., with fewer than 20 actors).

References

Agnew, R. (1991). The interactive effects of peer variables on delinquency. *Criminology, 29*, 47–72. https://doi.org/10.1111/j.1745-9125.1991.tb01058.x

Bergstrom, A. R. (1984). Continuous time stochastic models and issues of aggregation over time. In Z. Griliches & M. D. Intriligator (Eds.), *Handbook of econometrics, volume II* (pp. 1146–1212). Amsterdam: Elsevier Science Publishers BV.

Block, P. (2015). Reciprocity, transitivity, and the mysterious three-cycle. *Social Networks, 40*, 163–173. https://doi.org/10.1016/j.socnet.2014.10.005.

Borenstein, M., Hedges, L. V., Higgins, J. P. T., & Rothstein, H. R. (2009). *Introduction to meta-analysis*. Chichester: Wiley. https://doi.org/10.1002/9780470743386.

Caravita, S. C. S., Sijtsema, J. J., Rambaran, A. J., & Gini, G. (2014). Peer influences on moral disengagement in late childhood and early adolescence. *Journal of Youth and Adolescence, 43*, 193–207. https://doi.org/10.1007/s10964-013-9953-1.

Cochran, W. G. (1954). The combination of estimates from different experiments. *Biometrics, 10*(1), 101–129. https://doi.org/10.2307/3001666.

Gandolfo, G. (1993). Continuous-time econometrics has come of age. In G. Gandolfo (Ed.), *Continuous-time econometrics* (pp. 1–11). London: Chapman & Hall.

Greenan, C. C. (2015). Diffusion of innovations in dynamic networks. *Journal of the Royal Statistical Society, Series A, 178*, 147–166. https://doi.org/10.1111/rssa.12054.

Hamerle, A., Nagl, W., & Singer, H. (1991). Problems with the estimation of stochastic differential equations using structural equation models. *Journal of Mathematical Sociology, 16*(3), 201–220. https://doi.org/10.1080/0022250X.1991.9990088.

Holland, P. W., & Leinhardt, S. (1977). A dynamic model for social networks. *Journal of Mathematical Sociology, 5*(1), 5–20. https://doi.org/10.1080/0022250X.1977.9989862.

Huisman, M., & Steglich, C. (2008). Treatment of non-response in longitudinal network studies. *Social Networks, 30*, 297–308. https://doi.org/10.1016/j.socnet.2008.04.004.

Hyde, J. S., Fennema, E., & Lamon, S. J. (1990). Gender differences in mathematics performance: A meta-analysis. *Psychological Bulletin, 107*(2), 139–155. https://doi.org/10.1037/0033-2909.107.2.139.

Kadushin, C. (2012). *Understanding social networks: Theories, concepts, and findings*. New York: Oxford University Press.

Kalbfleisch, J. D., & Lawless, J. F. (1985). The analysis of panel data under a Markov assumption. *Journal of the American Statistical Association, 80*, 863–871. https://doi.org/10.1080/01621459.1985.10478195.

Kandel, D. B. (1978). Similarity in real-life adolescent friendship pairs. *Journal of Personality & Social Psychology, 36*(3), 306–312. https://doi.org/10.1037/0022-3514.36.3.306.

Knecht, A. (2004). *Network and actor attributes in early adolescence*. DANS. https://doi.org/10.17026/dans-z9b-h2bp

Knecht, A., Burk, W. J., Weesie, J., & Steglich, C. E. G. (2011). Friendship and alcohol use in early adolescence: A multilevel social network approach. *Journal of Research on Adolescence, 21*, 475–487. https://doi.org/10.1111/j.1532-7795.2010.00685.x.

Knecht, A., Snijders, T. A. B., Baerveldt, C., Steglich, C. E. G., & Raub, W. (2010). Friendship and delinquency: Selection and influence processes in early adolescence. *Social Development, 19*(3), 494–514. https://doi.org/10.1111/j.1467-9507.2009.00564.x.

Koskinen, J. H., & Snijders, T. A. B. (2007). Bayesian inference for dynamic social network data. *Journal of Statistical Planning and Inference, 137*, 3930–3938. https://doi.org/10.1016/j.jspi.2007.04.011.

Lehmann, E. L. (1999). *Elements of large sample theory*. New York: Springer. https://doi.org/10.1007/b98855.

Lindberg, S. M., Hyde, J. S., & Petersen, J. L. (2010). New trends in gender and mathematics performance: A meta-analysis. *Psychological Bulletin, 136*(6), 1123–1135. https://doi.org/10.1037/a0021276.

Manger, M. S., & Pickup, M. A. (2016). The coevolution of trade agreement networks and democracy. *Journal of Conflict Resolution, 60*(1), 164–191. https://doi.org/10.1177/0022002714535431.

Marsden, P. V. (2005). Recent developments in network measurement. In P. Carrington, J. Scott, & S. Wasserman (Eds.), *Models and methods in social network analysis* (pp. 8–30). New York: Cambridge University Press. https://doi.org/10.1017/CBO9780511811395.002.

McFadden, D. (1974). Conditional logit analysis of qualitative choice behavior. In P. Zarembka (Ed.), *Frontiers in economics* (pp. 105–142). New York: Academic Press.

Niezink, N. M. D., & Snijders, T. A. B. (2017). Co-evolution of social networks and continuous actor attributes. *The Annals of Applied Statistics, 11*(4), 1948–1973. https://doi.org/10.1214/17-AOAS1037.

Norris, J. R. (1997). *Markov chains*. New York: Cambridge University Press. https://doi.org/10.1017/CBO9780511810633.

Oud, J. H. L. (2007). Continuous time modeling of reciprocal effects in the cross-lagged panel design. In S. M. Boker & M. J. Wenger (Eds.), *Data analytic techniques for dynamical systems* (pp. 87–129). Mahwah: Lawrence Erlbaum Associates.

Oud, J. H. L., & Jansen, R. A. R. G. (2000). Continuous time state space modeling of panel data by means of SEM. *Psychometrika, 65*(2), 199–215. https://doi.org/10.1007/BF02294374

Polyak, B. T. (1990). New method of stochastic approximation type. *Automation and Remote Control, 51*, 937–946.

Ripley, R. M., Snijders, T. A. B., Boda, Z., Vörös, A., & Preciado, P. (2018). Manual for RSiena (version May 2018) [Computer software manual]. Oxford: University of Oxford, Department of Statistics, Nuffield College.

Robbins, H., & Monro, S. (1951). A stochastic approximation method. *Annals of Mathematical Statistics, 22*, 400–407. https://doi.org/10.1214/aoms/1177729586.

Robins, G. (2015). *Doing social network research: Network-based research design for social scientists*. London: Sage.

Ruppert, D. (1988). *Efficient estimation from a slowly convergent Robbins-Monro process* (Tech. Rep.). Cornell University, School of Operations Research and Industrial Engineering.

Schweinberger, M., & Snijders, T. A. B. (2007). Markov models for digraph panel data: Monte Carlo-based derivative estimation. *Computational Statistics & Data Analysis, 51*, 4465–4483. https://doi.org/10.1016/j.csda.2006.07.014.

Scott, J., & Carrington, P. J. (Eds.). (2011). *The SAGE handbook of social network analysis*. London: SAGE Publications Ltd.

Snijders, T. A. B. (2001). The statistical evaluation of social network dynamics. In M. Sobel & M. Becker (Eds.), *Sociological methodology* (pp. 361–395). Boston: Basil Blackwell.

Snijders, T. A. B. (1996). Stochastic actor-oriented models for network change. *Journal of Mathematical Sociology, 21*, 149–172. https://doi.org/10.1080/0022250X.1996.9990178.

Snijders, T. A. B. (2005). Models for longitudinal network data. In P. J. Carrington, J. Scott, & S. S. Wasserman (Eds.), *Models and methods in social network analysis*. New York: Cambridge University Press.

Snijders, T. A. B., & Baerveldt, C. (2003). A multilevel network study of the effects of delinquent behavior on friendship evolution. *Journal of Mathematical Sociology, 27*, 123–151. https://doi.org/10.1080/00222500305892.

Snijders, T. A. B., Koskinen, J., & Schweinberger, M. (2010a). Maximum likelihood estimation for social network dynamics. *The Annals of Applied Statistics, 4*(2), 567–588. https://doi.org/10.1214/09-AOAS313.

Snijders, T. A. B., & Pickup, M. (2016). Stochastic actor-oriented models for network dynamics. In J. N. Victor, A. M. Montgomery, & M. Lubell (Eds.), *The Oxford handbook of political networks*. Oxford: Oxford University Press.

Snijders, T. A. B., Van de Bunt, G. G., & Steglich, C. E. G. (2010b). Introduction to actor-based models for network dynamics. *Social Networks, 32*, 44–60. https://doi.org/10.1016/j.socnet.2009.02.004.

Snijders, T. A. B., Steglich, C. E. G., & Schweinberger, M. (2007). Modeling the co-evolution of networks and behavior. In K. van Montfort, J. H. L. Oud, & A. Satorra (Eds.), *Longitudinal models in the behavioral and related sciences* (pp. 41–71). New York: Cambridge University Press.

Stadtfeld, C. (2012). *Events in social networks: A stochastic actor-oriented framework for dynamic event processes in social networks*. Phd dissertation, Karlsruher Institut für Technologie.

Stadtfeld, C., & Geyer-Schulz, A. (2011). Analyzing event stream dynamics in two-mode networks: An exploratory analysis of private communication in a question and answer community. *Social Networks, 33*, 258–272. https://doi.org/10.1016/j.socnet.2011.07.004.

Stadtfeld, C., Hollway, J., & Block, P. (2017). Dynamic network actor models: Investigating coordination ties through time. *Sociological Methodology, 47*, 1–40. https://doi.org/10.1177/0081175017709295.

Steglich, C. E. G., Snijders, T. A. B., & Pearson, M. (2010). Dynamic networks and behavior: Separating selection from influence. *Sociological Methodology, 40*, 329–392. https://doi.org/10.1111/j.1467-9531.2010.01225.x

Voelkle, M. C., Oud, J. H. L., Davidov, E., & Schmidt, P. (2012). An SEM approach to continuous time modeling of panel data: Relating authoritarianism and anomia. *Psychological Methods, 17*, 176–192. https://doi.org/10.1037/a0027543.

Wasserman, S. S. (1980). A stochastic model for directed graphs with transition rates determined by reciprocity. *Sociological Methodology, 11*, 392–412. https://doi.org/10.2307/270870.

Wasserman, S. S., & Faust, K. (1994). *Social network analysis: Methods and applications*. Cambridge: Cambridge University Press. https://doi.org/10.1017/CBO9780511815478.

Weerman, F. (2011). Delinquent peers in context: A longitudinal network analysis of selection and influence effects. *Criminology, 49*, 253–286. 10.1111/j.1745-9125.2010.00223.x

Chapter 6
Uses and Limitation of Continuous-Time Models to Examine Dyadic Interactions

Joel S. Steele, Joseph E. Gonzales, and Emilio Ferrer

6.1 Introduction

An important aspect of psychological research is a focus on how individuals change. Modeling change has a long history in psychology (Cronbach and Furby 1970; Meredith and Tisak 1990; Tucker 1958), and most experts would agree on two major views of change: either discrete or continuous. This distinction is most relevant to the theory of how a process evolves over time. While theoretically separate from any specific computational method, ultimately, this distinction reflects a necessary decision made through the selection of an appropriate method for modeling data. However, when limited or unsuitable methods are the only approaches available, researchers run the risk of augmenting their designs and hypotheses to conform to the limits of a selected method. Often the process of data collection in psychology

Electronic Supplementary Material The online version of this article (https://doi.org/10.1007/978-3-319-77219-6_6) contains supplementary material, which is available to authorized users.

J. S. Steele (✉)
Department of Psychology, Portland State University, Portland, OR, USA
e-mail: j.s.steele@pdx.edu

J. E. Gonzales
Department of Psychology, University of Massachusetts, Lowell, MA, USA
e-mail: Joseph_Gonzales@uml.edu

E. Ferrer
Department of Psychology, University of California, Davis, Davis, CA, USA
e-mail: eferrer@ucdavis.edu

K. van Montfort et al. (eds.), *Continuous Time Modeling in the Behavioral and Related Sciences*, https://doi.org/10.1007/978-3-319-77219-6_6

limits the density of measures that can reasonably be collected, making it seemingly impossible to ask more theoretically sophisticated questions about continuously evolving processes. The result is an extensive, perhaps over-reliance on panel studies that comprise discrete snapshots of a process collected at either weekly, monthly, or even yearly intervals.

Beyond an over-reliance on panel data, there is an equal over-reliance on panel based methods for modeling change (e.g., Box 1950; Davidson 1972; O'Brien and Kaiser 1985). Here time is often treated simply as a factor- or interval-based measure—little attention is paid to the size of the interval so long as it is consistent. While ostensibly a reasonable approach, there is a recognition in the field that not all treatments of time are equal (Miller and Ferrer 2017). Rather, estimates from models of the same process can differ dramatically depending on whether the process is modeled in continuous or discrete time (e.g., Oud 2004; Singer 1991a; Voelkle 2017). This raises a genuine concern that, if the process is thought to evolve in continuous time, then any statistical model fit to the data should reflect continuous-time parameters. The remaining challenge then is in how to use discrete time panel data for examining continuous-time phenomena?

In what follows we explore the use of the exact discrete model (EDM; Bergstrom 1988), a method for modeling continuous-time processes that are measured at discrete time points, and apply this method to the study of change in daily affect in romantic couples. In an attempt to present a broad and accessible view of continuous-time methods, we organize what follows around three main goals: (a) To examine some of the benefits and limitations of continuous-time models. (b) To present the details of fitting continuous-time models to data. (c) To illustrate the specification and interpretation of a continuous-time model through the use of an empirical example. For this last point, we fit a model based on a theoretical specification of dyadic interactions originally introduced by Felmlee and Greenberg (1999) and present results from both simulated and empirical data.

6.2 The Promise of Continuous-Time Models

As methods and measures continue to advance, more examples emerge that incorporate a continuous-time perspective for studying change in social and behavioral sciences (e.g., Boker 2001; Ferrer and Steele 2014; Singer 1991a; Steele et al. 2014; Voelkle 2017). The basic assumption is similar to earlier approaches, in that time is not viewed as the driving force of change, but rather it is seen as a requisite component, or context, for observing change (Baltes et al. 1988; Coleman 1968). Unique to continuous-time models, however, is that change can be specified in such a way that expected differences can be computed for any interval of time. In particular, through the use of differential equations, change can be modeled instantaneously by considering smaller and smaller increments of time, until the limit of change is considered when the increment is infinitely small. Equation (6.1) below illustrates this limit,

$$\frac{df}{dt} = \lim_{\Delta t \to 0} \frac{f(t + \Delta t) - f(t)}{\Delta t}. \tag{6.1}$$

In Eq. (6.1), $f(t)$ represents the original function for which instantaneous change is to be computed, $\frac{df}{dt}$ represents the process of differentiation, and Δt represents the interval of time considered. Metaphorically, this interval serves as the stage upon which differences in the function are allowed to take shape. The difference in the numerator, $f(t + \Delta t) - f(t)$, represents change in the output of the function on two occasions separated by the interval Δt. The ratio specified in Eq. (6.1) is literally the rise over the run, or slope, of the function over the interval. As the interval becomes infinitely small, this slope describes a line tangent to the original function. This line describes the direction and rate of change of the function at that instant. Therefore, the result of differentiation is another function that specifies how the original function changes. If the observable shape of change is known, differentiation can be used to describe how this trajectory behaves at any point in its path. This result is illustrated in Fig. 6.1.

In Fig. 6.1 the function $f(t) = t^2$ is plotted, and lines tangent to the function at three different points are indicated. The slopes of these tangent lines represent the result of the first derivative, or the instantaneous rate of change in the function, which is equal to $\frac{df}{dt} = 2t$. Depending on the function and the time point selected, this change can be negative or positive. Importantly, once the derivative is known, change in the original function can be computed for any amount of time. In practice,

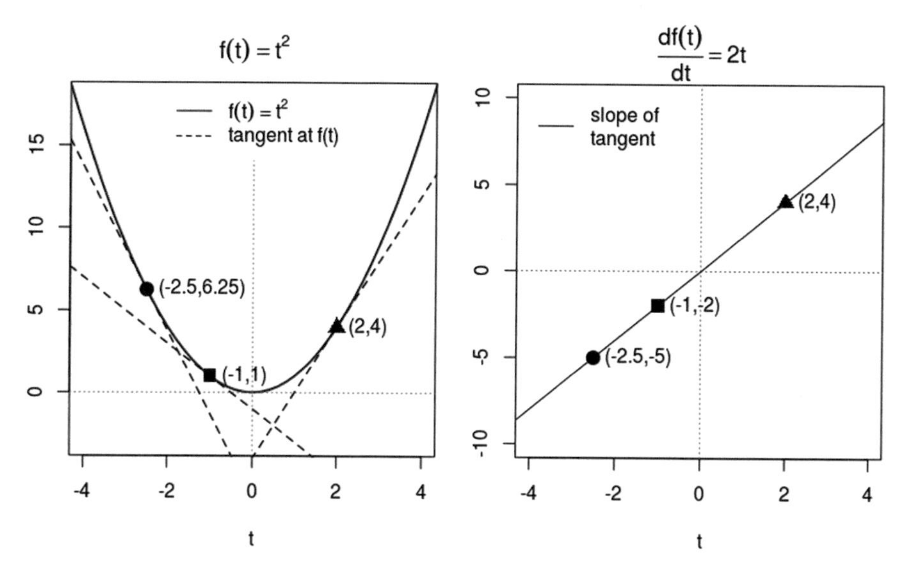

Fig. 6.1 Plot of a function and the function of its first derivative. *Note*: In the figure $f(t) = t^2$ was evaluated at points $t \in \{-2.5, -1, 2\}$

this means that the amount of time that elapses between measurement occasions can be whatever suits the investigation.

One beneficial result of differentiation for applied research is that longitudinal studies of the same process that use different time metrics become comparable. For example, a study that collects measures daily can be compared to a study that collects measures hourly, assuming the process can be adequately captured at each interval.

6.3 Building an Intuitive Understanding of Continuous-Time Methods

We shift the focus here toward building an understanding of how continuous-time models are estimated in practice. We begin by presenting the state-space model which explicitly separates the analytic specification of a continuously evolving process into the smoothly evolving state and the discretely observed measurements generated by it. This separation results in two levels of equations, referred to from here on as the state equation and the measurement equation. Error can be added to each of these equations, implying that estimation may be fallible in capturing the true change of the process as well as in accurately accounting for its measurement. A specification based on the work of Jazwinski (1970) is provided below:

$$dx_t = [Ax_t + Bu_t]dt + GdW_t, \tag{6.2}$$

$$y_t = Hx_t + Du_t + \epsilon_t. \tag{6.3}$$

Equation (6.2) defines change in the process through state vector x_t. This vector contains all of the components of the continuous-time process and is determined by a linear model involving previous states x, exogenous inputs u, and random error W_t. The term A represents a matrix containing the direct influences of each state variable on all other state variables. The B term is another matrix that permits the influence of exogenous inputs into the system dynamics. These two components of the equation combine to define the deterministic behavior of the process. The final term, GdW_t, by contrast, permits the inclusion of stochastic input or the state-level error. In practice, stochastic influences are modeled by GdW_t, the change in continuous-time random walk or Wiener process W_t, scaled by matrix G.

Equation (6.3) represents the measurement equation with observed values contained in a vector y_t. This specification defines the observed values as a combination of the state-level process, scaled through the Hx_t term, and exogenous influences through Du_t. Error at this level is represented through the ϵ_t term. Important to note is that the errors at both the state and measurement levels are assumed to be Gaussian distributed. Figure 6.2 presents a schematic path diagram of the state-space model. This is similar to the latent and manifest models common in structural equations models (Chow et al. 2010; MacCallum and Ashby 1986).

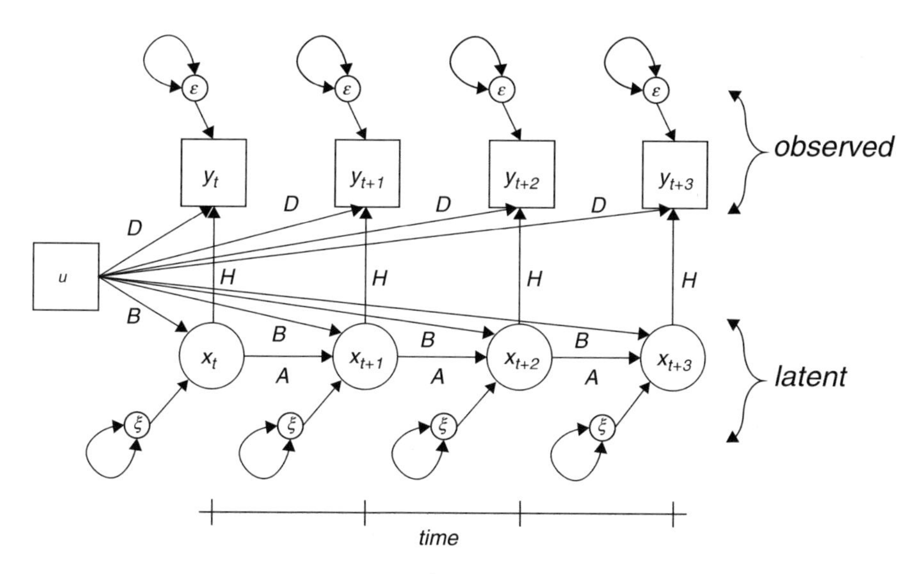

Fig. 6.2 Diagram of state-space model. *Note*: In the diagram circles represent latent variables, squares represent observed variables Paths are denoted according the representation in Eqs. (6.2) and (6.3)

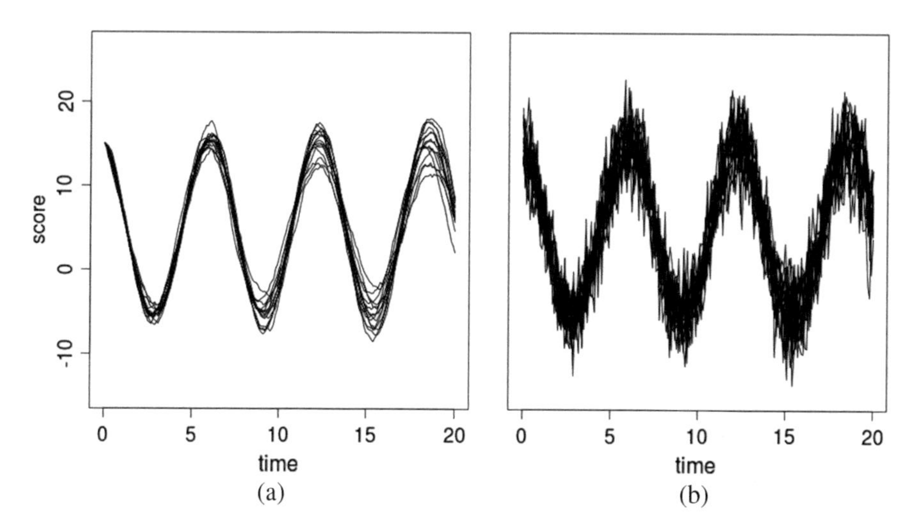

Fig. 6.3 Simulated SDE. (**a**) Process noise. (**b**) Process and measurement noise

The inclusion of state-level noise is noteworthy in that it specifies a source of error in the smoothly evolving differential equation mentioned previously. The inclusion of noise at this level results in a stochastic differential equation (SDE), wherein changes are no longer determined solely by the differential equation. Rather, the expected smooth trajectory may be perturbed over time. The influence of this random source of input specified in an SDE is illustrated in Fig. 6.3a.

Figure 6.3b presents the same SDE trajectory with the additional inclusion of measurement level error. Both Fig. 6.3a, b present multiple realizations of the same process. The only differences are due to the inclusion of state-level noise and measurement noise.

6.3.1 The Kalman Filter Algorithm

Different approaches are available for estimating the parameter of the state-space model (see Oud 2004, 2007; Singer 1991a). Among them, the Kalman filter (Kalman 1960) is perhaps the most flexible and most widely used. The Kalman filter algorithm estimates parameters of the state-space model by minimizing the difference between where the observations are expected to occur versus their actual values once observed, a process known as prediction error decomposition (PED). The PED approach is common in time series analysis and is attractive since it takes into account the fact that repeated measures of the same process are not independent of each other but rather they are serially related (see Canova 2007; Singer 1991b, for a thorough discussion of PED.)

A simplified explanation of the Kalman filter algorithm begins with a projection of the state-level process forward in time. This prediction is based on the current value of the state along with the corresponding system parameters that determine how the state changes. The projected point is then compared against an actual observation, recorded through the measurement equation, and the difference is computed. The error between expected and observed values is used to update the entire process before the next step of the system is projected. If the error is large, the algorithm adapts to more closely match the observed measures. However, if the forecast was close to the observed measures and the error is small, the filter relies more on the process equation that produces each prediction.

By analogy, the algorithm makes an informed guess about where the system is headed; it then compares this guess to an actual observation. If the initial guess was close to what was observed, the process of informed guessing is more trusted and used again to produce another guess. If not, the algorithm tunes the guessing process to rely more heavily on the collected measures.

This trade-off is reflected through a term called the Kalman gain, which is updated at each step and is used by the algorithm to tune the forecasting process. Its value represents how much the algorithm relies on either the state model dynamics or on the values of the collected observations. Mathematically, the Kalman gain can be conceptualized as a measure of reliability of the state-level process, in which state-level variance is divided by variance in both the state and measurement levels. Values close to one represent a reliance on the state equation; values closer to zero represent a more heavy reliance on the collected observations.

Through use of the PED approach, the state-space model parameters can be estimated recursively via the Kalman filter algorithm. This algorithm can be used for real-time forecasting as well as for parameter estimation. In short, the steps of the algorithm include: (1) the computation of a forecast prediction of the future state, (2) an update of the expected error based on this forecast, (3) the collection of an observation, (4) and the comparison of the observation to the forecast in order to update the gain. This fairly intuitive process of forecast, observe, update, and repeat is both robust and flexible.

6.3.2 The Discrete-Continuous Connection

The final step in connecting discrete time measures to continuous-time dynamics is accomplished through the exact discrete model (Bergstrom 1988; Oud 2004; Singer 1991a). The EDM accomplishes this connection by combining the analytic solution of linear differential equations, with the formulation of stochastic process error together in one framework. One major benefit of this approach is that it can readily be accomplished using a state-space formulation.

The EDM provides an explicit solution to the state-level formulation in Eq. (6.2), which can be formulated as a difference equation,

$$x_{t+\Delta t} = A^* x_t + B^* u_t + \upsilon_t, \tag{6.4}$$

for $t = \{0, \ldots, T-1\}$ and $\upsilon_t \sim \mathcal{N}(0, \Omega^*)$. The analytic solutions of the parameters from Eq. (6.4) are

$$\begin{aligned} A^* &= e^{A\Delta t}, \\ B^* &= \int_0^{\Delta t} e^{A(\Delta t - s)} Bx(t_i + s) ds, \\ \Omega^* &= \int_0^{\Delta t} e^{As} \Omega e^{A^T s} ds, \quad \Omega = GG^T. \end{aligned} \tag{6.5}$$

The term $B^* u_t$ represents the exogenous input, and, when not measured continuously, its influence must be approximated if the model is to be estimated. This approximation can be varied in form, yet it is most convenient to assume a step function, in which the exogenous inputs have a constant level of impact for the entire interval (see Singer 1991b). This leads to

$$B^* = A^{-1}(A^* - I)B. \tag{6.6}$$

The terms in Eqs. (6.5) and (6.6) provide the mean vector and covariance matrix of the multivariate normal distribution of the latent variables at time t. They are used to express the likelihood of the data for parameter estimation via the Kalman filter (see Singer 1991b, for more detail on the estimation process).

The innovation of the EDM is the use of nonlinear constraints on the A^* matrix to provide an analytic solution to the parameters of the state-space model. In practice these nonlinear constraints mirror the general solution to linear differential equations which take the form e^{zt}, with Euler's number e providing the model of continuous growth, z the rate of change, and t representing time (cf. Coleman 1964, 1968).

As shown in Eq. (6.5), the same exponentiation process is applied to the A matrix from Eq. (6.2). Recall that the A matrix is a central component of the deterministic portion of the state-space equation, in that it specifies how the prior values of the state influence future values. It is often referred to as the transition matrix in that it represents how the system transitions from time point to time point. This exact analytic solution is where the model gets its name and provides a means for integration of the system at any conceivable time.

Further inspection of Eq. (6.5) reveals the dependence of the system on the matrix A^*. This dependence reflects the fact that the estimation of instantaneous change through the A matrix is scaled in the available data by a factor of Δt. As can be seen, both the impact of exogenous inputs and the scaling of the stochastic input depend on this formulation of the A^* matrix.

6.3.3 Suitable State-Level Error

There remains a question of how to account for noise or randomness in the process of change. This randomness is reflected in process level by the term W_t in Eq. (6.2). Measurement error is reflected in the term ϵ_t in Eq. (6.3). This noise may not reflect only aspects of poor measurement reliability but may also reflect sources of error that are not accounted for in the model. Such differences, in the context of differential equations, require that the infinitesimal accumulations of change should also reflect randomness.

Integration of a random process in the classical way is not possible. However, a solution was specified by Norbert Wiener and was later generalized by Koyashi Itô in 1944. The Itô integral provides a mathematical means for integrating a random process and has been widely used in such fields as physics, economics, and biology, to model changes that are subject to error. The theoretical benefit of the Itô integral is the specification of error in the change process.

It is important at this point to highlight some concerns that should be acknowledged when fitting continuous-time models. Specifically, analysts and researchers should consider the following: (a) the nature of the parameter estimation techniques—either from continuous measures or discrete measures like those common in psychological research, (b) the nature of the theoretical model that specifies how changes progress—whether or not changes arise from linear or nonlinear combinations of parameter, whether the model requires any parameter constraints during estimation in order to accurately reflect the theory, and (c) whether or not to include process-level error. We believe that these questions are necessary before researchers select a method for fitting a continuous-time model.

6.4 Application of SDEs and the EDM

Next we present the application of the EDM to model dyadic interactions. We begin with the presentation of the theoretical model and proceed to outline how this model can be specified in the EDM framework. The model specification is verified with simulated data before it is fit to empirical data from romantic couples.

6.4.1 *Theoretical Model of Dyadic Interactions*

The model selected for the analyses represents a dynamic system of dyadic interaction proposed by Felmlee and Greenberg (1999). Given the dyadic focus of this model, there are two differential equations that make up the system describing each partner's changes. To help contextualize the presentation, the construct under investigation is daily affect specific to one's romantic relationship. The actual measures to be used represent the average of eighteen ratings made specifically about the relationship, nine were for positive affect and nine were for negative affect. Since ratings were collected for each partner in the relationship each day, we distinguish female and male ratings using f and m, respectively. The full system of equations is presented below:

$$\begin{aligned}\frac{df}{dt} &= a_1(f^* - f_t) + b_1(m_t - f_t),\\ \frac{dm}{dt} &= a_2(m^* - m_t) + b_2(f_t - m_t).\end{aligned} \tag{6.7}$$

Here the terms $\frac{df}{dt}$ and $\frac{dm}{dt}$ represent the instantaneous change in levels of affect, f and m, over time, for the female and male, respectively. The terms f^* and m^* represent the ideal levels of affect for each individual and are explained in more detail below. The terms containing a t subscript represent levels at time t.

In the original theory of the model, the terms f^* and m^* represented ideal goals set by each individual. Conceptually, these values represent how high or low on a given dimension, either positive or negative affect, a participant would like to feel about their relationship. As an example, a high level of f^* for positive affect would reflect a goal of the female in the dyad to feel highly positive about her relationship. These values could also be viewed as the emotional set point around which each individual fluctuates. These goals were not explicitly measured in the data used for these analyses, but their values can be estimated from the data.

From Eq. (6.7) we see that in this model, change is determined by the summation of weighted differences, one between the individual's ideal and their daily rating and the other between the daily ratings of each partner. Assuming that all coefficients in the model are positive, a daily rating lower than the ideal (e.g., $f^* > f_t$) results in an upward driving force in affect for that individual. This is also the case for situations where the partner is expressing a level higher than the individual (e.g., $m_t > f_t$). Although both weighted differences contribute to change in overall affect over time in Eq. (6.7), the components can be viewed as distinct contributions representing self-regulation, in the first term, and co-regulation between partners, in the second term. The combination of these self-regulatory and co-regulatory processes can result in highly varied trajectories, though each is governed by the same underlying process (see Felmlee 2006; Felmlee and Greenberg 1999; Ferrer and Steele 2014, for full discussion).

6.5 The SDE Representation of the Theoretical Model

Let us now turn attention to the translation of the model in Eq. (6.7) into the SDE/EDM framework. To start off, the model is reexpressed using the vector θ to define the deterministic model parameters; thus $\theta = (a_1, a_2, b_1, b_2)$. This is done mainly to avoid confusion with the matrices of the EDM. The reexpressed model takes the form of

$$\begin{aligned} \frac{df}{dt} &= \theta_1(f^* - f_t) + \theta_2(m_t - f_t), \\ \frac{dm}{dt} &= \theta_3(m^* - m_t) + \theta_4(f_t - m_t). \end{aligned} \tag{6.8}$$

Given some algebraic distribution of terms and simplification, we arrive at the following specification:

$$\begin{aligned} df &= [-(\theta_1 + \theta_2) f_t + \theta_2 m_t + \theta_1 f^*]dt, \\ dm &= [\theta_4 f_t - (\theta_3 + \theta_4) m_t + \theta_3 m^*]dt. \end{aligned} \tag{6.9}$$

It is important to note that the terms $\theta_1 f^*$ and $\theta_3 m^*$ are constant influences that represent the individual ideal levels of affect unique to each member in the dyad. One assumption of the model is that these ideal levels do not change over time. Equation (6.9) is easily translated into matrix form as

$$d\begin{bmatrix} f \\ m \end{bmatrix} = \begin{bmatrix} -(\theta_1 + \theta_2) & \theta_2 \\ \theta_4 & -(\theta_3 + \theta_4) \end{bmatrix} \begin{bmatrix} f \\ m \end{bmatrix} dt + \begin{bmatrix} \theta_1 f^* \\ \theta_3 m^* \end{bmatrix} [1]dt. \tag{6.10}$$

This matrix specification of the model mirrors the terms from the state-space equations (6.2) and (6.3). In particular, the direct influences of affect on change in affect are represented in the first term to the right of the equal sign which equates to the A matrix. This is followed by the B matrix with again allows for exogenous influences to impact the progression of the state process.

Notice here that the constant terms are moved from the A matrix and placed in the B matrix, which permits exogenous input to the system. The placement of these terms in the B matrix enables the estimation of these terms solely from the observed data. This differs from the theoretical definition of these terms presented originally, in which these terms are assumed to be measured separately (Felmlee 2006). Their estimation here requires an assumed constant influence $u = 1$ (this term is omitted from Eq. (6.11)).

This specification fully characterizes the deterministic components of the model. Note that the parameters in both the A matrix and the B matrix are specified such that their estimation is constrained, specifically, as either a sum or a product with other parameters in the model. The specification of these constraints may be overcome analytically; however, within the modeling framework selected for these analyses, these constraints can be directly specified in the model syntax. This is illustrated in the syntax provided in Appendix 3.

Below we present the full specification with the addition of stochastic influences dW_t^f and dW_t^m,

$$d\begin{bmatrix} f \\ m \end{bmatrix} = \left(\begin{bmatrix} -(\theta_1 + \theta_2) & \theta_2 \\ \theta_4 & -(\theta_3 + \theta_4) \end{bmatrix} \begin{bmatrix} f \\ m \end{bmatrix} + \begin{bmatrix} \theta_1 f^* \\ \theta_3 m^* \end{bmatrix} \right) dt + \begin{bmatrix} \Omega_f & 0 \\ 0 & \Omega_m \end{bmatrix} \begin{bmatrix} dW_t^f \\ dW_t^m \end{bmatrix}. \tag{6.11}$$

Thus, the model above represents an SDE. The distribution of the dt term to only the deterministic portion of the equation (on the right hand side) is necessary because, in continuous time, the first derivative of a Wiener process, such as dW_t^f or dW_t^m, does not exist. The final step in the translation involves the specification of the observation-level equation given below:

$$\begin{bmatrix} z_f \\ z_m \end{bmatrix} = \begin{bmatrix} 1 & 0 \\ 0 & 1 \end{bmatrix} \begin{bmatrix} f \\ m \end{bmatrix} + \begin{bmatrix} \phi_f & 0 \\ 0 & \phi_m \end{bmatrix} \begin{bmatrix} \xi_f \\ \xi_m \end{bmatrix}. \tag{6.12}$$

Notice that the D matrix and exogenous inputs u are lacking from this equation. The omission of the inputs from the measurement equation was deliberate, based on a theoretical interpretation of the model's original presentation (Felmlee 2006; Felmlee and Greenberg 1999). That is, change in affect is determined by a combination of these constant terms with the daily measures, as in Eq. (6.7). The inclusion of additional exogenous inputs has not been fully explored regarding where in the model these influences would enter, either at the process or the observation level.

6.6 Methods

Historically, few frameworks were available to estimate parameters using the EDM (see Oud and Jansen 2000; Oud and Singer 2008); however, with advancement in software and methods, this is no longer the case. Numerous packages are available for modeling dynamics systems as SDEs including freely available packages written in R such as CTSMr (Juhl 2015), CTSEM (Driver et al. 2017), dse (Gilbert 2006), OpenMx (Boker et al. 2015; Neale et al. 2016) via the SSCT expectation routine, and many more. For this presentation, however, the linear stochastic differential equations package (LSDE; Singer 1991b) written in SAS/IML® (SAS Institute Inc. 2002–2008) was selected. The reasoning behind this choice has to do with the ease with which estimation could be accomplished even in the presence of parameter constraints based on the model. Undoubtedly, similar constraints are possible in many of the packages listed above. However, the LSDE package makes these constraints fairly explicit, and thus it is possible to directly reflect the mathematical specification of the model within the LSDE syntax. Appendix 3 provides example syntax in LSDE for one of the models.

6.6.1 Verification of the Model Specification with Simulated Data

To this point a specification that mathematically represents the hypothesized model has been developed, and the technical aspects of the syntax needed have been explained. Due to the complexity of the model and the package used to fit the model, the specification was tested by fitting various sets of simulated data. Though functions are available in the LSDE package for simulating data, the simulated data used for verification was generated using the SAS/ETS® MODEL procedure (SAS Institute Inc. 2002–2008). The simulated data were then fit using the LSDE package. This was done to ensure that the translation of the model in Eqs. (6.11) and (6.12) into the EDM framework was accurate.

Our goal in performing these simulation steps was simply to confirm the correct specification of the model rather than to examine the behavior of the estimates; therefore, differing sets of simulation values were not explored. The simulation consisted of 100 replications of the same trajectory over 50 time points. Table 6.1 contains simulated and estimated values for a strictly deterministic system, that is, a system without error.

As is evident in Table 6.1, the LSDE procedure clearly recaptures the true parameter values. Given the constrained nature of the parameters that make up the system, the recapturing of the simulated values was evidence of the correct specification of the model. Therefore, the next step in the analysis was to fit the model to the empirical data.

Table 6.1 Simulated data, no error model

Simulated		Fitted values
Parameter	True value	Estimated
a_1, θ_1	0.001	0.001
b_1, θ_2	1.000	1.000
a_2, θ_3	1.000	1.000
b_2, θ_4	−2.000	−2.000
f^*, β_1	10.000	10.009
m^*, β_2	5.000	5.000
$f(0), \mu_1$	15.000	15.000
$m(0), \mu_2$	12.000	12.000

Note: $N = 100$ replications, $T = 50$

6.6.2 *Empirical Data*

6.6.2.1 Sample and Measures

Participants

The data used are part of a project focused on the development of models to analyze dyadic interactions (Ferrer and Steele 2011). Participants include couples involved in a romantic relationship. As part of the overall project, participants were asked to complete a daily questionnaire about their affect for up to 90 consecutive days. This report presents data from 108 couples who had at least 50 days of complete data.

Relationship-Specific Affect

The questionnaire was intended to examine day-to-day fluctuations in affect specific to one's relationship. As part of this questionnaire, 18 items were intended to tap into the participants' positive and negative emotional experiences specific to their relationship. Participants were asked to complete these items responding to the instructions "Indicate to what extent you have felt this way about your relationship today." The nine positive items included (1) *emotionally intimate*, (2) *trusted*, (3) *committed*, (4) *physically intimate*, (5) *free*, (6) *loved*, (7) *happy*, (8) *loving*, and (9) *socially supported* , whereas the nine negative mood items included (1) *sad*, (2) *blue*, (3) *trapped*, (4) *argumentative*, (5) *discouraged*, (6) *doubtful*, (7) *lonely*, (8) *angry*, and (9) *deceived*.

For all items, participants were asked to respond using a five-point Likert-type scale ranging from one (very slightly or not at all) to five (extremely). For all analyses, daily averages for positive and negative affect, for each person, were computed using all of the items in the scales. The reliability of change was computed within person using generalizability analysis (Cranford et al. 2006). The reliability

Table 6.2 DDIP summary statistics

	Mean	SD	Min	Max
f_+	3.542	0.858	1.000	5.000
f_-	1.299	0.495	1.000	4.889
m_+	3.566	0.802	1.000	5.000
m_-	1.316	0.501	1.000	4.778

Note: $N = 108$ dyads, $t = 50$ days

coefficient was computed as

$$\rho_\Delta = \frac{\sigma^2_{p\times d}}{\sigma^2_{p\times d} + \frac{\sigma^2_e}{m}}$$

Where $p \times d$ represents the person by day interaction, e represents error, and m represents the number of indicators. The resulting reliability coefficients for positive and negative affect were 0.85 and 0.87 for females and 0.82 and 0.85 for males. Summary statistics for relationship affect measures are reported in Table 6.2.

6.7 Results

Investigation of Model Starting Values

Through the simulation process, the sensitivity of the LSDE procedure to starting values became apparent. Deviations from the true values of the simulated data, which may be considered small in other modeling frameworks, led to large deviations in parameter estimates; at times such deviations resulted in a lack of model convergence. This sensitivity is a by-product of the nonlinear estimation procedure and can result in estimation difficulties including the production of a nonpositive definite Hessian matrix.[1] In such instances, extensive prior investigation may be necessary to identify suitable starting values for the model. When fitting the empirical data, estimation of suitable starting values became necessary for the positive affect model specifically. We estimated starting values for this model using the PROC MODEL procedure, which fit a strictly deterministic version of the model from Eq. (6.7).

Additionally, alternative model specifications were not investigated in these analyses; therefore model comparisons are lacking. For this reason, the likelihood values for the models are omitted, but parameters estimates, standard errors, and the ratio of the estimate to the standard error, which represents the t value of the parameters, are reported. The standard error values are available only if the model

[1] This matrix is used to estimate the Fisher information matrix which is in turn used to estimate standard errors of the model parameters.

Hessian matrix is positive definite, which implies that a maximum of the likelihood function was found.

6.7.1 *Positive Affect*

Results from fitting the model to positive and negative affect are listed in Table 6.3. For positive affect, all parameter estimates were greater than zero and at least twice the magnitude of their standard errors. Theoretically, this may represent a cooperative system of dyadic interactions in which discrepancies between partners as well as between ideal and observed levels for the individual are minimized (Felmlee 2006; Steele et al. 2014). Among the control parameters, the self-regulation controls were slightly smaller in magnitude than the co-regulatory controls, $a_1 = 0.190$ versus $b_1 = 0.290$ for females; and $a_2 = 0.236$ versus $b_2 = 0.250$ for males. Also of interest are the estimates of the ideal levels of affect. These were similar to the average values for each measure, $\hat{f}^*_+ = 3.502$, compared to $\bar{f}_+ = 3.542$; and $\hat{m}^*_+ = 3.590$, compared to $\bar{m}_+ = 3.566$. In previous work, these values were not estimated but were set to the mean level of affect to represent the idealized level (Steele et al. 2014). The estimates obtained here serve to confirm the previous selection. Also evident from the model estimates is an appreciable amount of process-level variance for both measures ($\epsilon_{df} = 0.702$ and $\epsilon_{dm} = 0.663$). This variance may reflect the exclusion of separate estimates for observation-level noise. Observation-level noise is not included in the models due to lack of convergence. Additional estimates are reported in Table 6.3.

Table 6.3 Relationship-specific affect estimates

	Positive			Negative		
	Est	SE	Est/SE	Est	SE	Est/SE
a_1, θ_1	0.190	0.015	12.939	0.592	0.031	19.168
b_1, θ_2	0.290	0.018	15.819	0.546	0.034	16.037
a_2, θ_3	0.236	0.014	16.751	0.556	0.030	20.070
b_2, θ_4	0.250	0.016	15.489	0.373	0.032	11.519
f^*, β_1	3.502	0.052	67.958	1.278	0.016	78.924
m^*, β_2	3.589	0.039	91.648	1.320	0.016	82.934
ϵ_{df}, Ω_1	0.702	0.010	69.254	0.654	0.012	54.478
ϵ_{dm}, Ω_3	0.663	0.009	71.112	0.618	0.010	60.496
$f(0), \mu_1$	3.615	0.072	50.362	1.413	0.060	23.799
$m(0), \mu_2$	3.621	0.067	53.881	1.424	0.051	28.052
$SD_{f(0)}, \Sigma_1$	0.556	0.076	7.347	0.381	0.052	7.347
$SD_{m(0)}, \Sigma_2$	0.488	0.066	7.347	0.278	0.038	7.346

Note: $N = 108$, $T = 50$

6.7.2 Negative Affect

For negative affect, all parameter estimates were greater than zero and over two times as large as the estimated standard errors, as was the case for positive affect. The estimates of these control parameter were higher in absolute magnitude than those obtained from the positive affect models (e.g., $a_{1_+} = 0.190$ vs. $a_{1_-} = 0.592$ and $b_{1_+} = 0.290$ vs. $b_{1_-} = 0.546$). This might reflect a tendency for couples to reconcile differences between their own feelings and those of their partners more quickly regarding negative affect as compared to positive affect.

Within negative affect, the control parameters of the self-regulation terms were higher in magnitude than for the co-regulation terms, $a_1 = 0.592$ versus $b_1 = 0.546$ for females; and $a_2 = 0.556$ versus $b_2 = 0.373$ for males. This difference was greatest for males in the sample and may indicate a fairly autonomous system of negative affect regulation for males. Within this dimension, the estimates of the ideal point also resembled the mean values reported in Table 6.2, $\hat{f}^*_- = 1.278$ compared to $\bar{f}_- = 1.299$ and $\hat{m}^*_- = 1.320$ compared to $\bar{m}_- = 1.316$. These values were considerably lower than those from positive affect. This was expected because the means of negative affect ratings tend to be lower than those for positive affect, and the negative affect ratings tend to be less variable overall (Ferrer and Steele 2011; Steele and Ferrer 2011). These items may be more difficult to endorse than the positive affect items. This trend was also evident in the estimates of the variability in initial conditions, $\widehat{SD_{f(0)_+}} = 0.556$ compared to $\widehat{SD_{f(0)_-}} = 0.381$ and $\widehat{SD_{m(0)_+}} = 0.488$ compared to $\widehat{SD_{m(0)_-}} = 0.278$. These parameters represent one important aspect of interindividual variability. Not all dyads follow the exact same trajectory. For negative affect, the estimates were smaller than those for positive affect, indicating that, overall, there was less variability in the negative affect measures. Additional estimates are reported in Table 6.3.

6.7.3 Expected System Behavior

The comparison of parameter magnitudes is somewhat informative, but the interpretability of these parameters independently is limited. The system described in Eq. (6.7) represents a model of change in affect rather than a description of the absolute levels. Thus, the interpretation of the model estimates does not mirror the interpretation in traditional regression. Specifically, the effects of a given parameter are not independent of the influences of other parameters in the model (see Nielsen and Rosenfeld 1981). Restrictions imposed on the parameter estimates further alter the interpretation of the parameters, as described in Eq. (6.10). For example, the combined scaling of the a_1 and b_1 parameters on previous values of f determines their effect on the instantaneous change. Put another way, the influence of f on change in f over time is scaled by the combined effects of self-regulation and co-regulation.

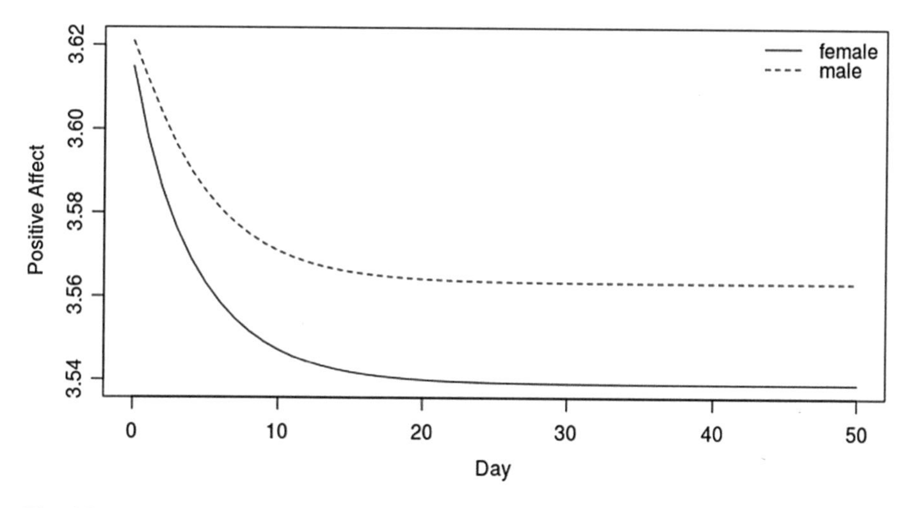

Fig. 6.4 Positive affect solution, deterministic portion. *Note*: Dashed line represents male affect; solid line represents female affect

The examination of the different systems that arise from unique sets of the parameters in this model has been presented previously (see Felmlee 2006; Felmlee and Greenberg 1999; Ferrer and Steele 2011). These presentations were constructed to simplify the interpretation of the model and illustrate salient classes of possible outcomes that can arise from this specification of dyadic interactions. However, the examples may not represent real-life behaviors. Based on these presentations, one may be led to think that the reported estimates in Table 6.3 are indicative of cooperative systems. Theoretically, this seems reasonable; however, the deterministic solutions to both positive and negative affect are indicative of independent systems dynamics. This is evident when the solutions are plotted. Simulated results are presented in Figs. 6.4 and 6.5.

In Fig. 6.4, the trajectories for both males and females quickly approach an asymptotically stable solution. Although this solution is stable, it does not match the expected trajectories presented previously (see Felmlee 2006; Ferrer and Steele 2011; Steele et al. 2014), in which both partners approached a common asymptote due to their cooperation.

Figure 6.5 presents a trajectory that converges quickly to a stable solution. It should be noted that the coefficients reported in Table 6.3 are all positive, which according to the theory should reflect a cooperative systems. Notably, however, the expected trajectories based on the model estimates appear to be independent.

The difference between the simplified expected behavior and the resultant trajectories may be reconciled when the solution is investigated analytically. It is important to note that the following is relevant only to the deterministic portion

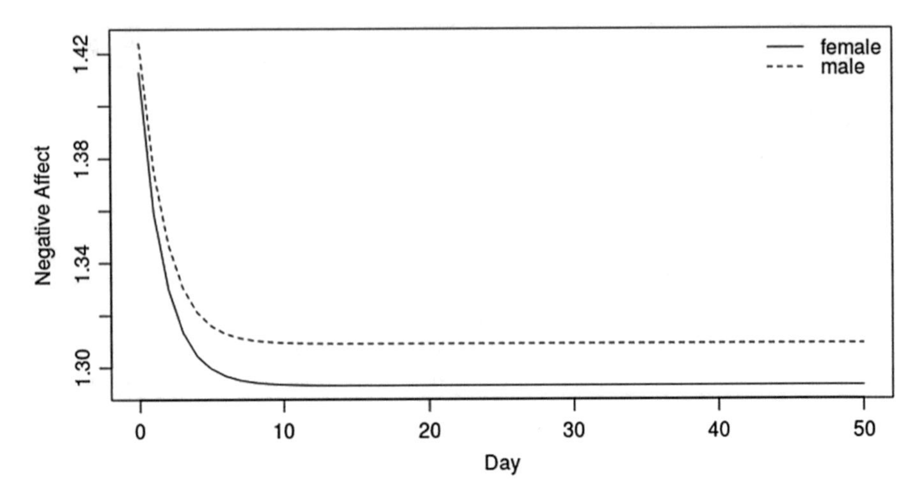

Fig. 6.5 Negative affect solution, deterministic portion. *Note*: Dashed line represents male affect; solid line represents female affect

of the system. The dynamic solution to the system in Eq. (6.7) takes the following form:

$$f_t = \alpha e^{\lambda_1 t} + \beta e^{\lambda_2 t}$$
$$m_t = \delta e^{\lambda_1 t} + \gamma e^{\lambda_2 t} \tag{6.13}$$

where the terms α, β, δ, and γ represent constants of integration. The term λ_1, λ_2 represent eigenvalues of the system and take on the form of the solution to the quadratic equation:

$$\lambda = \frac{-b \pm \sqrt{b^2 - 4(ac)}}{2}.$$

In the context of the model in Eq. (6.7), this solution can be expressed as

$$\frac{-(a_1 + b_1 + a_2 + b_2) \pm \sqrt{(a_1 + b_1 + a_2 + b_2)^2 - 4[(a_1 + b_1)(a_2 + b_2) - b_1 b_2]}}{2} \tag{6.14}$$

and represents the eigenvalues of the transition matrix from the deterministic portion of the system specification presented in Eq. (6.10) (see Felmlee and Greenberg 1999).

The values of λ represent eigenvalues of the equation and determine the behavior of the resultant solutions. If they are both real and negative, the solution converges to an equilibrium point. If, however, one or both are positive, the solution explodes. In the situation that the value under the radical sign is negative, complex roots result.

This can lead to additional types of solutions showing various patterns of oscillation over time.

For positive affect, the tendency is for both males and females to approach independent equilibrium points. This is also illustrated in the computed eigenvalues of this solution, $\lambda_1^+ = -0.214$ and $\lambda_2^+ = -0.752$, which are both real and negative indicating an asymptotically stable solution. Again, the equation above applies to the eigenvalues of a deterministic form of the original model and does not account for the influence of stochastic input at the process level. Therefore, these values and plotted solution in Fig. 6.4 are for illustrative purposes only.

This last point is also evident when examining the negative affect solution relative to the systems eigenvalues and trajectory depicted in Fig. 6.5. For negative affect, the computed roots are $\lambda_1^- = -0.570$ and $\lambda_2^- = -1.497$, which indicate an asymptotic approach to equilibrium according to the theory. Appendix 4 provides syntax in the R statistical software for plotting these trajectories.

6.7.4 Assessing Model Fit

Naturally, as with any statistical model, some indication of how well the model fit the data is desired. The assessment of model fit from the LSDE package mirrors methods commonly found in time series analysis. Namely, the residuals of the model are inspected for some indication of a remaining structure which would imply that some aspect of the data generation process was not included in the model specification. This is accomplished either through computation of the auto-covariance function to determine whether lagged residuals show significant dependence for small lags, visual inspection of residual plots, or calculation of formal statistics to test for independence among the model residuals.

For univariate tests, the most common test on the residuals is either the Box-Pierce test or the more robust Ljung-Box test (Box and Pierce 1970; Ljung and Box 1978), both of which are occasionally referred to as portmanteau tests. These tests provide an assessment of independence in the residual of a fitted model. If the statistic is significant, the residuals are not strictly independent. Figure 6.6 provides a plot of the proportion of significant results of a series of univariate Box-Pierce test run for each partner, male and female, on each dimension, positive and negative separately. In Fig. 6.6, it is apparent that few of the models produce residuals with remaining structure. These results present an encouraging, perhaps overly rosy, view of model fit. Whether these tests are appropriate for multivariate systems is questionable. Below we present a multivariate-based assessment of the model residuals.

A multivariate extension of the portmanteau test, proposed by Hosking (1980), also was fit to the residuals of each dyadic bivariate model, male and female, for each dimension of affect, positive and negative, respectively. Figure 6.7 presents the proportion of the multivariate portmanteau tests that were significant at different levels of lag. The test is used to determine if the residuals of a multivariate time

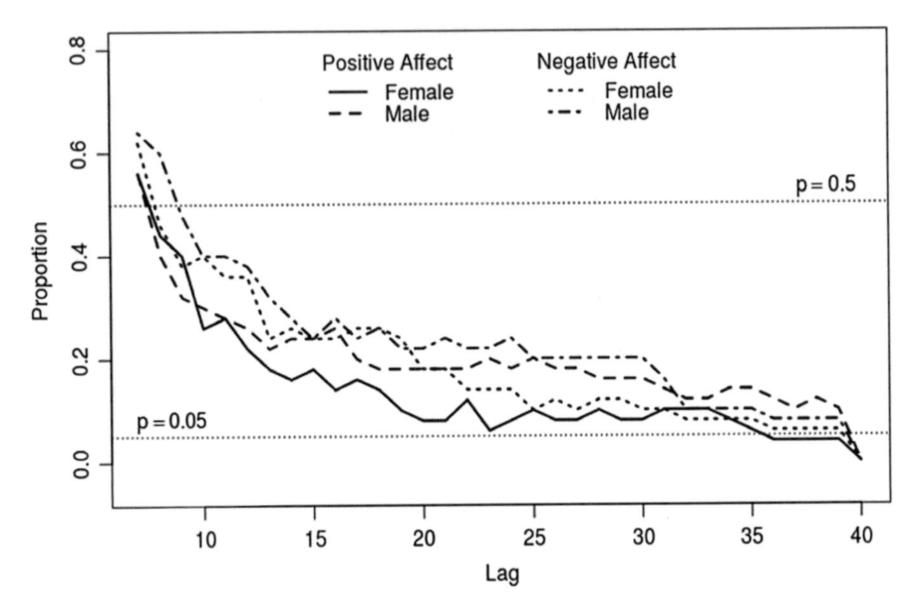

Fig. 6.6 Proportion of significant Box-Pierce test statistics by lag value. *Note*: Univariate Box-Pierce tests were run for each partner, male and female, on each dimension, positive and negative separately

series model differ from a white-noise process. This is accomplished by computing the sum of covariances among the residuals for a specified number of lags referred to as the MaxLag. The resulting sum is compared against a χ^2 distribution with degrees of freedom equal to $k^2(l - o)$, where k represents the dimension of the process, l represents the lag, and o the order of the model fit.[2] There are few guidelines available for selecting the value of the MaxLag, but one rule of the thumb is that MaxLag should be set to the lesser of either half the length of the series minus 2 or a value of 40. In Fig. 6.7, the test was run with MaxLag set from 7 to 40. As can be seen in Fig. 6.7 when the MaxLag term is equal to half the series, less than half of the dyadic models produced residuals that differ from white noise. While not a traditional assessment of model fit, this does indicate that the SDE is adequately capturing the trajectories of the majority of dyadic time series. When researchers are interested in building a model in practice, numerous model specifications could be compared via likelihood ratio tests to determine the best fitting model to the data. In our example no such alternative specifications were explored. In this situation we believe a direct examination of the residuals as well as a summary of the portmanteau test is a sufficient approach for assessing model fit. Were this model to be used for forecasting purposes, more work would be necessary

[2]For the Felmlee-Greenberg specification $o = 6$.

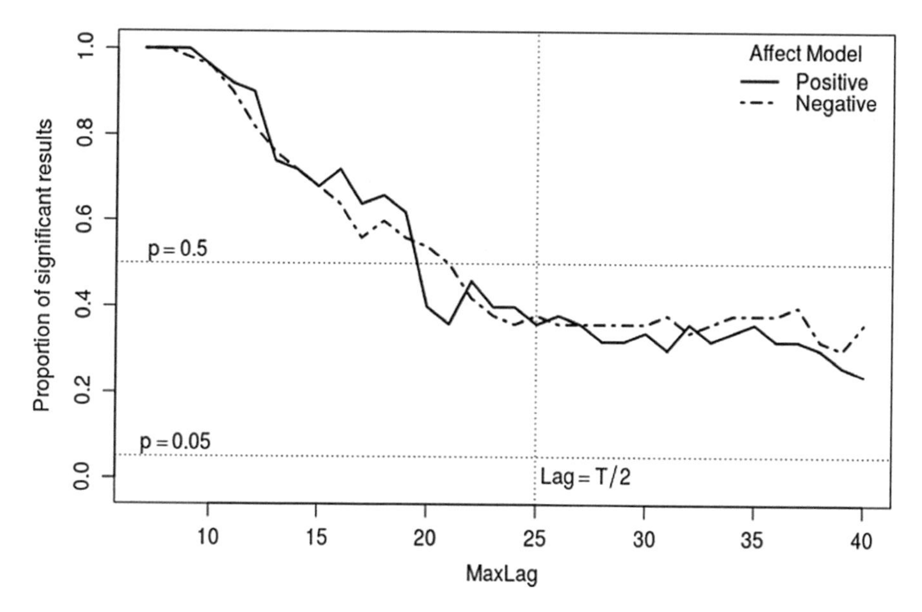

Fig. 6.7 Proportion of significant Portmanteau test statistics by MaxLag value. *Note*: Dashed line represents negative affect; solid line represents positive affect

to identify poorly fitting series, and alternative model specifications may need to be explored.

6.8 Discussion

Much is to be gained from modeling dynamical systems as stochastic differential equations using the exact discrete model. In the present work, the benefits emerged through the estimation of the parameters governing change as well as estimation of idealized levels specified in the original model. Additionally, the estimation of initial conditions and their variability is very beneficial, as this variability represents interindividual differences and conveys useful information about the system.

What is lacking in this application, however, is a straightforward means to interpret the estimated parameters. These parameters determine change in the observed values rather than differences in relative levels. To this end we encourage users to combine visualization of the solution with analytic examination of the stability of the system that is estimated. Related to this point is the formulation of the model used for parameter estimation, in which the parameters are combined in the drift matrix. The estimates from this matrix influence every other aspect of the model, including the levels of stochastic input.

Additionally, we would like to reiterate the promise of continuous-time methods, that being that once the parameters are adequately estimated, the interval of discrete observations can be whatever suits the study. This assumes that the process can be adequately captured. To this point, the importance of continuous-time methods can be illustrated by plotting the behavior of the estimates from the drift matrix A^* from Eq. (6.5) for different time intervals Δt.

The impacts of the time interval on the magnitude and direction of the estimates from the drift matrix may be more pronounced depending on the system under study. For example, assume a model is fit that results in the following values of the estimated A matrix:

$$A = \begin{bmatrix} -0.65 & 0.00 & 1.7 \\ 0.00 & -0.85 & 1.0 \\ -3.50 & -1.50 & 0.0 \end{bmatrix}.$$

If the system were fit using discrete time methods, these parameter estimates would depend on the size of the interval between measures. Changes in the diagonal elements of this matrix dependent on the interval size are illustrated in Fig. 6.8. As can be seen in Fig. 6.8, both the magnitude and the sign of the elements of the drift matrix change depending on the interval adopted for measurement. This presents an important caveat for researchers conducting longitudinal analysis and highlights the importance of continuous-time methods.

Not all estimates will show such dramatic fluctuations based on interval size. Figures 6.9 and 6.10 present the magnitude of each element of the A^* matrix from the models estimated for our empirical examples. These plots trace the magnitude of the expected effects that would be estimated from the system for a given time interval of observation.

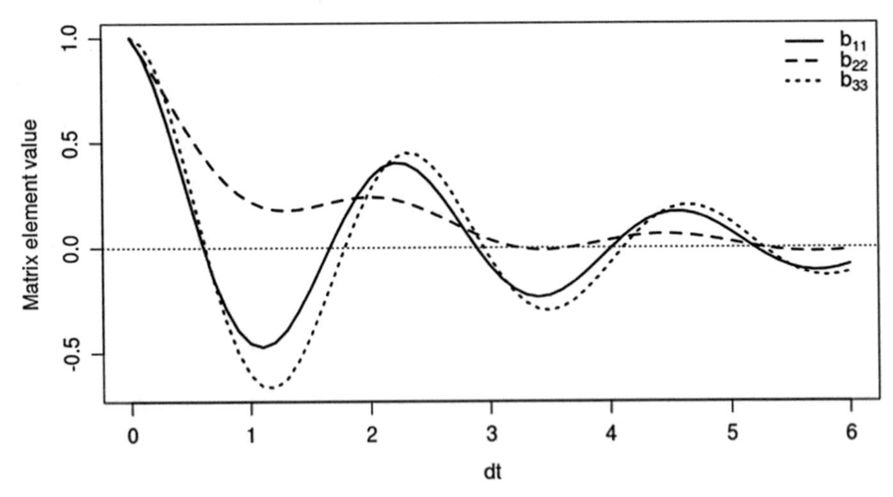

Fig. 6.8 Expected discrete time coefficients from the EDM drift matrix

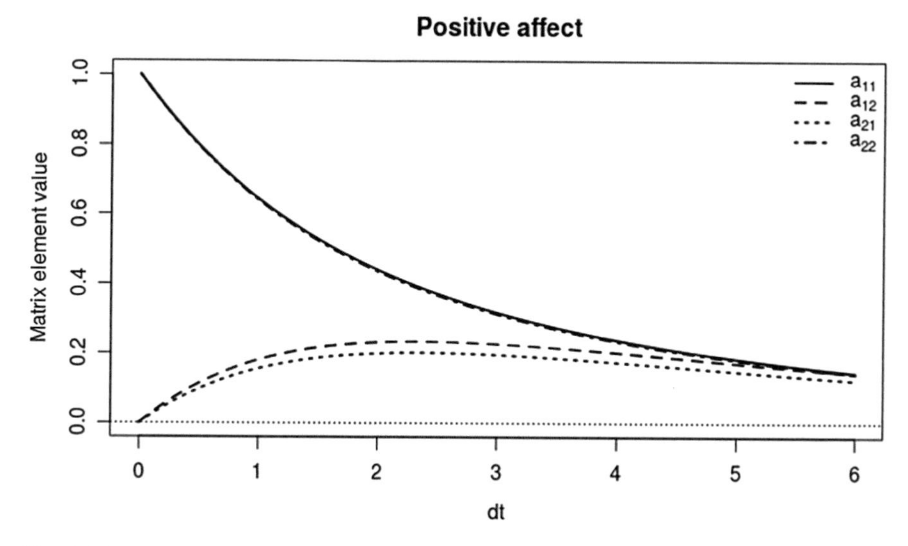

Fig. 6.9 Expected discrete time coefficients from the EDM drift matrix for positive affect

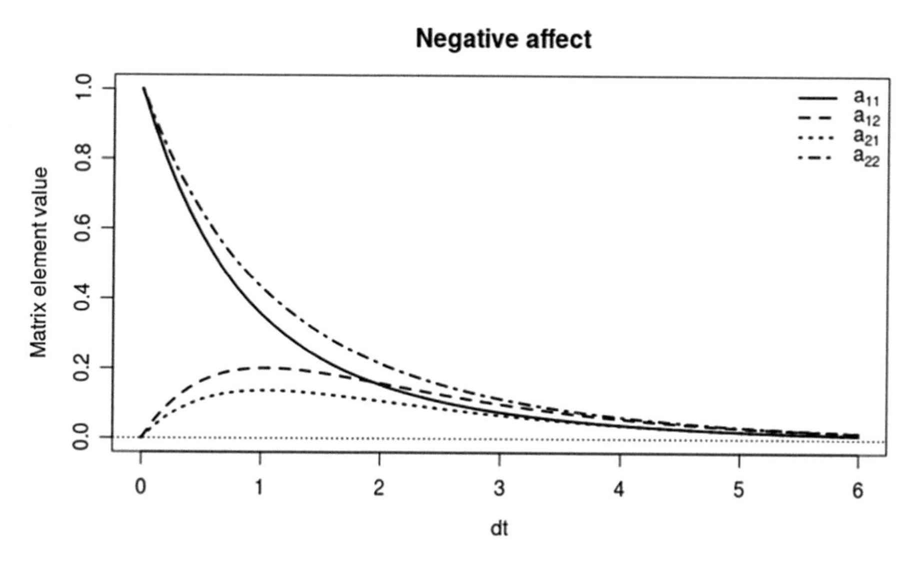

Fig. 6.10 Expected discrete time coefficients from the EDM drift matrix for negative affect

The trajectories presented in Figs. 6.9 and 6.10 appear to converge rather quickly. It is important to keep in mind that these plots represent the magnitude of the estimated parameters that would result if the same system were measured discretely at different intervals. These plots provide ostensible evidence that the system of interactions may not be identifiable at larger intervals between measurements since some of the parameters approach zero. Again, this effect may be more pronounced for different systems; our estimates presented here are indicative of largely independent dynamics.

The main benefit of the use of the EDM concerns the connection of the discrete measurements to the underlying continuous-time process. Affective dynamics are most certainly constantly evolving. This important aspect of these social processes is not easily captured by other methods. The use of the analytic solution provided by the EDM accomplishes discrete to continuous translation. While not a trivial task, the specification of a differential equation model and the use of the LSDE package for modeling in continuous time can be worthwhile. The preceding represents a methodological step in translating theory into practice.

Appendix 1: McDonald-Swaminathan Matrix Differentiation

The specification of an LSDE model in the SAS/IML® framework requires the specification of matrix derivatives. Matrix differentiation is specified for each of the matrices that make up the state-space specification and is evaluated with respect to the parameters that make up each element. The implementation of matrix differentiation used by the LSDE package is based on the work of McDonald and Swaminathan (1973) which outlines how matrix differentiation can be performed. Given a matrix Y that is $n \times m$ in size, with elements that represent some function of the elements of another matrix X, which is $p \times q$ in size, the differentiation of Y with respect to X in the McDonald-Swaminathan rules would result in a matrix of partial derivatives $\frac{\partial Y}{\partial X}$ that is $n(m) \times p(q)$ in size.

As an example, imagine Y is a 2×2 matrix and is to be differentiated base on another matrix X that is also 2×2. The result of differentiation is a 4×4 matrix of partial derivatives. Specifically, each of the rows of Y are arranged into a single row vector that is $1 \times n(m)$ in size,

$$Y = \begin{bmatrix} y_{11} & y_{12} \\ y_{21} & y_{22} \end{bmatrix} \xrightarrow{row} \begin{bmatrix} y_{11} & y_{12} & y_{21} & y_{22} \end{bmatrix}. \tag{6.15}$$

The same process is performed for the matrix X and transposed to produce a column vector of elements that is $p(q) \times 1$ in size,

$$X = \begin{bmatrix} x_{11} & x_{12} \\ x_{21} & x_{22} \end{bmatrix} \xrightarrow{row-transpose} \begin{bmatrix} x_{11} \\ x_{12} \\ x_{21} \\ x_{22} \end{bmatrix}. \tag{6.16}$$

The differentiation is arranged as the outer product of these vectors, resulting in a $p(q) \times n(m)$ matrix of partial derivatives,

$$\frac{\partial Y}{\partial X} = \begin{bmatrix} \frac{\partial y_{11}}{\partial x_{11}} & \frac{\partial y_{12}}{\partial x_{11}} & \frac{\partial y_{21}}{\partial x_{11}} & \frac{\partial y_{22}}{\partial x_{11}} \\ \frac{\partial y_{11}}{\partial x_{12}} & \frac{\partial y_{12}}{\partial x_{12}} & \frac{\partial y_{21}}{\partial x_{12}} & \frac{\partial y_{22}}{\partial x_{12}} \\ \frac{\partial y_{11}}{\partial x_{21}} & \frac{\partial y_{12}}{\partial x_{21}} & \frac{\partial y_{21}}{\partial x_{21}} & \frac{\partial y_{22}}{\partial x_{21}} \\ \frac{\partial y_{11}}{\partial x_{22}} & \frac{\partial y_{12}}{\partial x_{22}} & \frac{\partial y_{21}}{\partial x_{22}} & \frac{\partial y_{22}}{\partial x_{22}} \end{bmatrix}. \tag{6.17}$$

An example drawn directly from the original paper is presented below. These equations are a reproduction of equations 3–5 in McDonald and Swaminathan (1973). Assume we are given a matrix Y,

$$Y = \begin{bmatrix} e^{x_{11}x_{12}} & sin(x_{11} + x_{12}) \\ log(x_{11} + x_{12} + x_{21}) & x_{11}x_{12}x_{21}x_{22} \end{bmatrix} \tag{6.18}$$

with elements that are a function of another matrix X,

$$X = \begin{bmatrix} x_{11} & x_{12} \\ x_{21} & x_{22} \end{bmatrix}. \tag{6.19}$$

The derivative of Y with respect to the elements of X would result in

$$\frac{\partial Y}{\partial X} = \begin{bmatrix} x_{12}e^{x_{11}x_{12}} & cos(x_{11} + x_{22}) & \frac{1}{x_{11}+x_{12}+x_{21}} & x_{12}x_{21}x_{22} \\ x_{11}e^{x_{11}x_{12}} & 0 & \frac{1}{x_{11}+x_{12}+x_{21}} & x_{11}x_{21}x_{22} \\ 0 & 0 & \frac{1}{x_{11}+x_{12}+x_{21}} & x_{11}x_{12}x_{22} \\ 0 & cos(x_{11} + x_{22}) & 0 & x_{11}x_{12}x_{21} \end{bmatrix}. \tag{6.20}$$

Appendix 2: Matrix Differentiation of the Felmlee–Greenberg Model

Next we illustrate how the steps outlined above are performed using the two matrices A and B, from our model from Eq. (6.11). These matrices represent the deterministic portion of the state equation for Felmlee–Greenberg model Felmlee and Greenberg (1999). Below we differentiate these matrices with respect to a parameter vector Θ.

To begin, we reparameterize the model and place the f^* and m^* terms in the parameter vector; thus $\Theta = (a_1, b_1, a_2, b_2, f^*, m^*)$. This allows us to express the A and B matrices as

$$A = \begin{bmatrix} -(\theta_1 + \theta_2) & \theta_2 \\ \theta_4 & -(\theta_3 + \theta_4) \end{bmatrix}, B = \begin{bmatrix} \theta_1\theta_5 \\ \theta_3\theta_6 \end{bmatrix}.$$

The derivative of the matrix A with respect to the parameter vector Θ is

$$\frac{\partial A}{\partial \Theta} = \begin{bmatrix} -1 & 0 & 0 & 0 \\ -1 & 1 & 0 & 0 \\ 0 & 0 & 0 & -1 \\ 0 & 0 & 1 & -1 \\ 0 & 0 & 0 & 0 \\ 0 & 0 & 0 & 0 \end{bmatrix}.$$

For the matrix B, the result is

$$\frac{\partial B}{\partial \Theta} = \begin{bmatrix} \theta_5 & 0 \\ 0 & 0 \\ 0 & \theta_6 \\ 0 & 0 \\ \theta_1 & 0 \\ 0 & \theta_3 \end{bmatrix}.$$

This example is only a portion of what is required in the LSDE syntax for this model. Please see Appendix 3 for a complete listing of the LSDE syntax required to fit the model for positive affect.

References

Baltes, P. B., Reese, H. W., & Nesselroade, J. R. (1988). *Life-span developmental psychology: Introduction to research methods* (reprint of 1977 edition). Mahwah: Erlbaum.

Bergstrom, A. R. (1988). The history of continuous-time econometric models. *Econometric Theory, 4*, 365–383. https://doi.org/10.1017/S0266466600013359

Boker, S. M. (2001). Differential structural equation modeling of intraindividual variability. In L. M. Collins & A. G. Sayer (Eds.), *New methods for the analysis of change* (pp. 5–27). Washington: American Psychological Association. https://doi.org/10.1037/10409-001.

Boker, S. M., Neale, M. C., Maes, H. H., Wilde, M. J., Spiegel, M., Brick, T. R., ... Driver, C. (2015). OpenMx 2.3.1 user guide [Computer software manual].

Box, G. E. P. (1950). Problems in the analysis of growth and wear curves. *Biometrics, 6*(4), 362–389. https://doi.org/10.2307/3001781.

Box, G. E. P., & Pierce, D. A. (1970). Distribution of residual correlations in autoregressive-integrated moving average time series models. *Journal of the American Statistical Association, 65*, 1509–1526.

Canova, F. (2007). *Methods for applied macroeconomic research* (vol. 13). Princeton: Princeton University Press.

Chow, S.-M., Ho, M. H. R., Hamaker, E. J., & Dolan, C. V. (2010). Equivalences and differences between structural equation and state-space modeling frameworks. *Structural Equation Modeling, 17*, 303–332. https://doi.org/10.1080/10705511003661553

Coleman, J. S. (1964). *Introduction to mathematical sociology*. New York: Free Press.

Coleman, J. S. (1968). The mathematical study of change. In H. M. Blalock & A. Blalock (Eds.), *Methodology in social research* (pp. 428–478). New York: McGraw-Hill.

Cranford, J. A., Shrout, P. E., Iida, M., Rafaeli, E., Yip, T., & Bolger, N. (2006). A procedure for evaluating sensitivity to within-person change: Can mood measures in diary studies detect change reliably? *Personality and Social Psychology Bulletin, 32*(7), 917–929.

Cronbach, L. J., & Furby, L. (1970). How we should measure change—or should we? *Psychological Bulletin, 74*, 68–80.

Davidson, M. L. (1972). Univariate versus multivariate tests in repeated measures experiments. *Psychological Bulletin, 77*(6), 446–452. https://doi.org/10.1037/h0032674.

Driver, C. C., Oud, J. H. L., & Voelkle, M. C. (2017). Continuous time structural equation modelling with R package ctsem. *Journal of Statistical Software, 77*(5), 1–35. https://doi.org/10.18637/jss.v077.i05.

Felmlee, D. H. (2006). Application of dynamic systems analysis to dyadic interactions. In A. Ong & M. V. Dulmen (Eds.), *Oxford handbook of methods in positive psychology* (pp. 409–422). Oxford: Oxford University Press.

Felmlee, D. H., & Greenberg, D. F. (1999). A dynamic systems model of dyadic interaction. *The Journal of Mathematical Sociology, 23*(3), 155–180. https://doi.org/10.1080/0022250X.1999.9990218.

Ferrer, E., & Steele, J. S. (2011). Dynamic systems analysis of affective processes in dyadic interactions using differential equations. In G. R. Hancock & J. R. Harring (Eds.), *Advances in longitudinal methods in the social and behavioral sciences* (pp. 111–134). Charlotte: Information Age Publishing.

Ferrer, E., & Steele, J. S. (2014). Differential equations for evaluating theoretical models of dyadic interactions. In *Handbook of developmental systems theory and methodology* (pp. 345–368). New York: Guilford Press.

Gilbert, P. D. (2006). Brief user's guide: Dynamic systems estimation [Computer software manual]. Retrieved from http://cran.r-project.org/web/packages/dse/vignettes/Guide.pdf.

Hosking, J. R. M. (1980). The multivariate portmanteau statistic. *Journal of the American Statistical Association, 75*, 602–608. https://doi.org/10.1080/01621459.1980.10477520

Jazwinski, A. H. (1970). *Stochastic processes and filtering theory*. New York: Academic Press.

Juhl, R. (2015). Ctsmr: Ctsm for R [Computer software manual]. R package version 0.6.8-5.

Kalman, R. E. (1960). A new approach to linear filtering and prediction problems. *Journal of Basic Engineering, 82*(1), 35–45.

Ljung, G. M., & Box, G. E. P. (1978). On a measure of lack of fit in time series models. *Biometrika, 65*, 297–303. https://doi.org/10.1093/biomet/65.2.297.

MacCallum, R., & Ashby, F. G. (1986). Relationships between linear systems theory and covariance structure modeling. *Journal of Mathematical Psychology, 30*(1), 1–27.

McDonald, R. P., & Swaminathan, H. (1973). A simple matrix calculus with application to multivariate analysis. *General Systems, XVIII*, 37–54.

Meredith, W., & Tisak, J. (1990). Latent curve analysis. *Psychometrika, 55*(1), 107–122. https://doi.org/10.1007/BF02294746.

Miller, M. L., & Ferrer, E. (2017). The effect of sampling-time variation on latent growth curve model fit. *Structural Equation Modeling, 24*, 831–854. https://doi.org/10.1080/10705511.2017.1346476.

Neale, M. C., Hunter, M. D., Pritikin, J. N., Zahery, M., Brick, T. R., Kickpatrick, R. M., ... Boker, S. M. (2016). OpenMx 2.0: Extended structural equation and statistical modeling. *Psychometrika, (81)*, 535–549. https://doi.org/10.1007/s11336-014-9435-8.

Nielsen, F., & Rosenfeld, R. A. (1981). Substantive interpretations of differential equation models. *American Sociological Review, 46*(2), 159–174. https://doi.org/10.2307/2094976.

O'Brien, R. G., & Kaiser, M. K. (1985). MANOVA method for analyzing repeated measures designs: An extensive primer. *Psychological Bulletin, 97*(2), 316. https://doi.org/10.1037/0033-2909.97.2.316.

Oud, J. H. L. (2004). SEM state space modeling of panel data in discrete and continuous time and its relationship to traditional state space modeling. In K. van Montfort, J. H. L. Oud, & A. Satorra (Eds.), *Recent developments on structural equation models: Theory and applications* (pp. 13–40). Dordrecht: Kluwer.

Oud, J. H. L. (2007). Comparison of four procedures to estimate the damped linear differential oscillator for panel data. In K. van Montfort, J. H. L. Oud, & A. Satorra (Eds.), *Longitudinal models in the behavioral and related sciences* (pp. 19–40). Mahwah: Lawrence Erlbaum.

Oud, J. H. L., & Jansen, R. A. R. G. (2000). Continuous time state space modeling of panel data by means of SEM. *Psychometrika, 65*(2), 199–215. https://doi.org/10.1007/BF02294374.

Oud, J. H. L., & Singer, H. (2008). Continuous time modeling of panel data: SEM versus filter techniques. *Statistica Neerlandica, 62*, 4–28.

SAS Institute Inc. (2002–2008). SAS 9.2 Help and documentation [Computer software manual]. Cary, NC: SAS Institute Inc.

Singer, H. (1991a). Continuous-time dynamical systems with sampled data, errors of measurement and unobserved components. *Journal of Time Series Analysis*, *14*, 527–545.

Singer, H. (1991b). *LSDE- A program package for the simulation, graphical display, optimal filtering and maximum likelihood estimation of Linear Stochastic Differential Equations*. Meersburg: Author.

Steele, J. S., & Ferrer, E. (2011). Latent differential equation modeling of self-regulatory and coregulatory affective processes. *Multivariate Behavioral Research*, *46*(6), 956–984.

Steele, J. S., Ferrer, E., & Nesselroade, J. R. (2014). An idiographic approach to estimating models of dyadic interactions with differential equations. *Psychometrika*, *79*(4), 675–700. https://doi.org/10.1007/s11336-013-9366-9.

Tucker, L. R. (1958). Determination of parameters of a functional relation by factor analysis. *Psychometrika*, *23*(1), 19–23. https://doi.org/10.1007/BF02288975.

Voelkle, M. C. (2017). A new perspective on three old methodological issues: The role of time, missing values, and cohorts in longitudinal models of youth development. In A. C. Petersen, S. H. Koller, F. Motti-Stefanidi, & S. Verma (Eds.), *Positive youth development in global contexts of social and economic change* (pp. 110–136). New York: Routledge.

Chapter 7
Makes Religion Happy or Makes Happiness Religious? An Analysis of a Three-Wave Panel Using and Comparing Discrete and Continuous-Time Techniques

Heiner Meulemann and Johan H. L. Oud

7.1 Introduction

Application of continuous-time methods in behavioural science is still rare. The analysis of longitudinal data almost always takes place in discrete time. In this chapter we explain in detail the serious problems connected with a discrete-time analysis and how to solve these problems by continuous-time modelling. We do this by means of an empirical example, the effect of religiosity on life satisfaction, which has been the subject of several discrete-time analyses in the past. In the next paragraphs, we discuss the theoretical background of the example, previous discrete-time studies and the discrete-time model adapted from Meulemann (2017). It is this model that will first be estimated in discrete time and next criticized and improved from a continuous-time perspective.

Electronic Supplementary Material The online version of this article (https://doi.org/10.1007/978-3-319-77219-6_7) contains supplementary material, which is available to authorized users.

H. Meulemann (✉)
Institute for Sociology and Social Psychology, University of Cologne, Cologne, Germany
e-mail: meulemann@wiso.uni-koeln.de

J. H. L. Oud
Behavioural Science Institute, University of Nijmegen, Nijmegen, The Netherlands
e-mail: j.oud@pwo.ru.nl

K. van Montfort et al. (eds.), *Continuous Time Modeling in the Behavioral and Related Sciences*, https://doi.org/10.1007/978-3-319-77219-6_7

As human beings die and are aware of being mortal, they must distinguish between This World and the World Beyond. Every human being has to get along with what this world is for and whether and what is beyond this world. Traditionally, religion promises to provide answers to these kinds of questions. It explains bad luck and injustice showing up in every human life within an overarching order. It provides a "sacred canopy" which "nomizes" life (Berger 1967). It is a resource to cope with life. The religiosity of a person is supposed to increase his or her life satisfaction. This is called the *nomization hypothesis*.

However, the more one is satisfied with life, the more one looks optimistically at it and takes it as it is. The more one is inclined to take the answers given by religion for granted, the more one will choose a religious belief that justifies and a religious community that reinforces one's satisfaction. In brief, the more one will become religious. This self-selection of satisfied people into religiosity is called the *optimism hypothesis*.

The nomization hypothesis has been examined longitudinally in four panel studies. First, in 16 yearly waves of the German Socio-Economic Panel (GSOEP), a fixed effects regression—that is, a regression of the change of the dependent variable on changes of the independent variables—shows a positive impact of the frequency of church attendance on the general life satisfaction (Headey et al. 2008, p. 18). Second, in a 1-year panel in the USA, regressions controlling for the former level of the dependent variables (yet not for the former levels of the independent variables) show a positive impact of the frequency of church attendance on general life satisfaction. However, a real effect is doubtful, because in the short time interval of 1 year, church attendance and life satisfaction change only slightly (Lim and Putnam 2010, p. 924). Third, in a further 1-year panel in the USA, not church membership but assessment of the importance of religion in life increases a specific life satisfaction—namely, with the family (Regnerus and Smith 2005, pp. 39–40). Fourth, in a 12-year panel study controlling for the former dependent variables, neither public nor private religious practices have an effect on general life satisfaction (Levin and Taylor 1998). We conclude that a positive impact of religiosity on life satisfaction has been confirmed strictly—over a longer time span and by the appropriate means of a fixed effects regression—only once: In the GSOEP study. As plausible as the nomization hypothesis seems to be, it is not yet strongly founded empirically. To our knowledge the optimism hypothesis as a causal hypothesis has never been examined empirically.

At first sight the nomization and optimism hypotheses seem to contradict each other. However, both could also be operating simultaneously in a reciprocal relationship across time. Whether the effect is in one of the two directions, in both directions or in none, and whether the sign of the effect is positive or negative can only be tested, if both are measured more than once—that is, longitudinally. In the following, how religiosity and life satisfaction measured at age 30 affect each other at age 43 and how religiosity and life satisfaction measured at age 43 affect each other at age 56 are examined. Thus, stabilities and cross-effects of life satisfaction and religiosity are analysed across time. At all three ages, 30, 43 and 56, religiosity is split up in its two main dimensions, practice and belief—measured as church

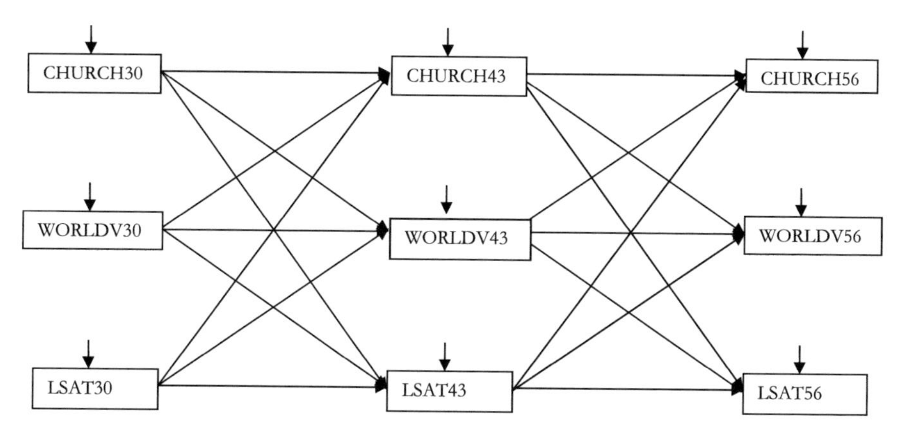

Fig. 7.1 Religiosity as church attendance (CHURCH) and Christian world view (WOLRDV) and life satisfaction (LSAT) at ages 30, 43 and 56

attendance and Christian worldviews. Thus, there are three variables measured three times, and the path diagram of the hypothesized causal system is given in Fig. 7.1.

In this so-called Markov chain model, causal hypotheses connect neighbour time points only. We suppose that at each time point, only experiences of the preceding period can have an impact. Impacts of earlier periods are supposed to be mediated and controlled by the immediately preceding ones. LSAT30, for example, can have no impact on LSAT56 that has not been already taken up by its impact on LSAT43 and the two stabilities in between. Stated differently, at each time point, the variables contain all information relevant to predict the future, and previous time points do not improve prediction.

As a structural equation model, the model has the rather simple recursive structure. The model structure would have been nonrecursive (interdependent), if in addition to lagged reciprocal effects (e.g. CHURCH30 to WORLDV43 and WORLDV30 to CHURCH43) also instantaneous reciprocal effects would have been specified (e.g. CHURCH43 to WORLDV43 and WORLDV43 to CHURCH43). In cross-sectional research, such a nonrecursive structure would be the only possibility to analyse reciprocal effects. In longitudinal research, often both lagged and instantaneous effects (instead of correlated residuals) are specified, in particular when the observation intervals between waves are big (e.g. Abele et al. 2011). We did not do this, because it leads to special problems that can be avoided by continuous-time analysis techniques.

The causal system in Fig. 7.1 has been analysed by Meulemann (2017). He used discrete-time (DT) structural equation modelling (Bollen and Brand 2010), controlled for unobserved heterogeneity by a random person factor, and used full information maximum likelihood (FIML) estimation, which takes care of arbitrary missing values based on the so-called missing-at-random (MAR) assumption

(Wothke 2009). In the following, we reanalyse the same data using continuous-time (CT) structural equation modelling (Oud and Delsing 2010; Oud and Jansen 2000) in order to show exemplarily the added value of CT over DT. In Meulemann (2017) the effects of some exogenous variables were also part of the model. To ease comparison, in this chapter, we skip both in DT and CT exogenous effects and limit the model to the relationships between the three endogenous variables. However, as in Meulemann's DT model, a single random person factor or "trait" for all three endogenous variables simultaneously was added to the CT model. A single trait not only saves degrees of freedom, but controlling for a general level in these variables with similar standardized score scale (see next section) was considered sufficient.

After presenting in the next section more specific information about the data and measurement, we report in Sect. 7.3 detailed results of the DT model. Section 7.4 explains the essence and advantages of CT over DT, while the CT results are given in Sect. 7.5. The DT analysis has been done by SAS-CALIS (SAS Institute Inc. 2013), the CT analysis by the R package CT-SEM (Driver et al. 2017).[1]

7.2 Data and Measurement

The sample is the Cologne High School Panel (*Kölner Gymnasiastenpanel, KGP*). Tenth grade high school students in the German federal state *Northrhine-Westfalia* have been first interviewed in written form in their classroom about their life plans in 1969 and reinterviewed orally three times in 1984, 1997 and 2010 about their life career between early and late midlife. The modal ages of the respondents in the reinterviews were 30, 43 and 56. Of the 3240 respondents in 1969, 1301 (40.1%) have been reinterviewed at ages 30, 43 and 56. However, time intervals between interviews were not for all respondents exactly equal to 13 years. The exact time intervals for each individual subject were known and will be used in the CT analysis. The sample is socially selective, because it has been drawn from students from the highest stratum of the German tripartite secondary school system.

The frequency of church attendance (CHURCH) has been measured by a single question with six options. The Christian worldview (WORLDV) has been ascertained by three statements of an inventory of Felling et al. (1987): "Life has meaning for me only because there is a God", "Life has a meaning because there is something after death" and "I believe that human existence has a clear meaning and follows a specific plan". For each statement, there were five response options; responses have been averaged. Life satisfaction (LSAT) has been measured on a scale from 0 to 10 as follows: "How satisfied are you nowadays altogether with your life" with a numbered response scale from 0 to 10, the extremes of which were labelled "totally unsatisfied" and "totally satisfied". Means were considerably

[1]The programming code of the analyses is available as supplementary material at the book website http://www.springer.com/us/book/9783319772189.

above the middle point and increased between the first and third observation time point slightly from 7.6 to 7.7; standard deviations decreased from 1.5 to 1.4. Detailed information about the data set is given in Weber (2017). For all three variables, values have been transformed into the standardized scores of the cumulative frequencies under the normal distribution (z-scores). This was done for the cumulative distribution of the three time points combined, so that the differences between them have been kept.

7.3 DT Model Results

As shown in Fig. 7.1, the model has 9 variables, that is, 54 nonidentical elements in the augmented moment matrix. Of our 1301 subjects, 1262 have complete information; 39 subjects have in total 107 missing values, which are taken care of by the missing value procedure of FIML. The model starts off at age 30 with the three initial means and six (co)variances, in total nine parameters, of three so-called predetermined variables. At age 43, the same variables are endogenous, and each depends on every other one at age 30. Thus, 9 regression parameters are needed; and, just as at age 30, 9 parameters for intercepts and residual (co)variances—resulting in altogether 18 parameters. At age 56, the model is exactly the same as at age 43 such that in a DT analysis, one normally would use again 18 parameters. It is customary in DT to test for equality of each of the corresponding parameters and set them equal if the test is passed. From a CT perspective, this is already questionable in the frequent case of unequal intervals, because then significant differences would show up in DT even if the underlying parameters are equal. In CT, however, time-varying parameters would not be handled stepwise at each of the discrete time points separately but by a function covering the whole time range of the model. In order to make the DT analysis comparable to the CT analysis, therefore, we set all parameters at age 56 equal to the respective ones at age 43. That is, there are no additional parameters: The model specified in both cases is time-invariant. Finally, one parameter for the trait variance is added which shows up in each of the six equations.

Altogether, the model has 28 parameters—leaving 26 degrees of freedom. The DT model has a $-2\times$Log-Likelihood (-2LL) of 26,563.86. The Chi-square value of 176.04 for testing the model against the saturated model is for 26 degrees of freedom significant.[2] This is expected, however, for such a big sample. The popular fit measure RMSEA (Browne and Cudeck 1993) with value 0.067 indicates that the model fits "reasonably". The results are presented in Table 7.1.

[2]The fact that the degrees of freedom left is a positive number, so making the Chi-square test possible, is only a necessary condition for identification of the model. A sufficiency proof of the identification of both the DT and the CT version of the model for as few as three observation time points is given in Appendix B of Angraini et al. (2014).

Table 7.1 Church attendance, Christian worldview and life satisfaction at ages 30, 43 and 56 in DT model

Initial parameters (age 30)				
	Means	(Co)variances		
		CHURCH	WORLDV	LSAT
CHURCH	0.0819***	0.4751***		
WORLDV	−0.0436	0.1903***	0.5024***	
LSAT	−0.0541	−0.1076***	−0.0437*	0.8847***
Dynamic parameters (ages 43, 56)				
	Intercepts	Regressions		
		CHURCH	WORLDV	LSAT
CHURCH	−0.0516**	0.4892***	−0.0279	−0.1188***
WORLDV	0.0235	−0.0244	0.4432***	−0.1117***
LSAT	0.0318	−0.2049***	−0.1410***	0.2591***
		Residual (co)variances		
		CHURCH	WORLDV	LSAT
	CHURCH	0.3074***		
	WORLDV	0.0468***	0.4679***	
	LSAT	−0.0589***	−0.0741***	0.6339***
	Trait variance	0.2051***		

−2LL = 26,563.86; Chi-square = 176.04 with df = 26; RMSEA = 0.067
*** $p<0.001$, ** $p<0.01$, * $p>0.05$

The results are similar to the ones in Meulemann (2017), although no exogenous variables are taken into consideration. First, all autoregressions are significantly positive, but church attendance and Christian worldviews turn out to be more persistent properties than life satisfaction. They can be seen as internalized early in life and held upright fairly well against shocks from outside. Also, there is substantial autoregression of life satisfaction, but this seems to be more susceptible to outside influences such as success and failure in life than religiosity.

Contrary to expectation, neither church attendance nor Christian worldviews have a positive effect on life satisfaction, nor life satisfaction on church attendance or Christian worldviews. Rather, all cross-regressions between religiosity and life satisfaction are significantly negative. On the one hand, religiosity seems to be rather a cost than a benefit in terms of life satisfaction. It costs time to go to church, and it may cost self-actualization to believe. On the other hand, neither is there a tendency of satisfied people to become religious. Moreover, the two cross-regressions between the two dimensions of religiosity are also negative but only slightly negative and not significantly. Practice and belief do not buttress each other.

The residual variances are found in conformity with the autoregressions to be smaller for the two religiosity dimensions than for life satisfaction. Finally, there is a strong trait variance representing unobserved heterogeneity. This may stem from private and occupational life success but also from personality factors and from socialization in family and school.

7.4 Discrete-Time Modelling Problems Solved by Continuous-Time Modelling

The first problem of a discrete-time model is that it ignores the processes taking place in continuous time between the measurement time points. The path diagram in Fig. 7.1 makes this clear. The arrows jump from one point in time to the next one, assuming that nothing happens between measurements. In fact, the discrete-time autoregression coefficients (horizontal arrows) and cross-regression coefficients (nonhorizontal arrows) in Fig. 7.1 are complicated mixtures of underlying continuous-time auto- and cross-effects in a constant interchange and dependent on the chosen observation interval. So, the true underlying auto- and cross-effects in continuous-time (CT) coefficients should be differentiated from the resulting auto- and cross-regressions in discrete-time (DT) coefficients. A variable with a high auto-effect, meaning that there is a strong tendency to sustain its value over time, tends also to retain the influence of other variables better and over a longer time interval than a variable with a low auto-effect. So, even a relatively small CT cross-effect can result in a relatively high DT cross-regression coefficient, if the variable influenced has a high auto-effect. But the converse can also be true: A relatively strong CT cross-effect having only small impact over a discrete-time interval (low DT cross-regression coefficient) because of a rather low auto-effect in the dependent variable.

Oud and Delsing (2010) show that, going in this way from DT to the underlying CT results, paradoxical changes can take place: Equal DT coefficients become different in CT, the strength order of coefficients reverses from DT to CT (e.g. if in DT the effect of CHURCH on WORLDV is larger than in the opposite direction, it becomes the other way around in CT), and nonzero coefficients in DT vanish or even change sign in CT. So, the first contribution of a CT analysis is to disentangle the true underlying CT auto- and cross-effects from the DT mixtures. One finds these mixtures in the autoregression matrix $\mathbf{A}_{\Delta t}$ (autoregression coefficients on the diagonal and cross-regression coefficients off-diagonally) in discrete-time equation (7.1), which further contains the DT intercepts $\mathbf{b}_{\Delta t}$ and prediction errors $\mathbf{e}_{t-\Delta t}$. The observation interval Δt as a subscript of $\mathbf{A}_{\Delta t}$ and $\mathbf{b}_{\Delta t}$ reminds that the discrete-time mixtures may differ for different observation intervals.

$$\mathbf{x}_t = \mathbf{A}_{\Delta t}\mathbf{x}_{t-\Delta t} + \mathbf{b}_{\Delta t} + \mathbf{e}_{t-\Delta t} \tag{7.1}$$

$$\frac{\mathrm{d}\mathbf{x}(t)}{\mathrm{d}t} = \mathbf{A}\mathbf{x}(t) + \mathbf{b} + \mathbf{e}(t) \tag{7.2}$$

The so-called drift matrix $\mathbf{A}$ in CT equation (7.2) analogously has the underlying CT auto-effects on its diagonal and the underlying CT cross-effects off-diagonally. Differential equation (7.2) explains the derivative $\mathrm{d}\mathbf{x}(t)/\mathrm{d}t$ or change in $\mathbf{x}$ at t for the interval Δt going to zero: $\Delta t \to 0$. Because of $\Delta t \to 0$, $\mathbf{A}$ and the CT intercepts $\mathbf{b}$ do not depend on the interval Δt any more. Oud and Delsing (2010) show how the DT mixtures $\mathbf{A}_{\Delta t}$ and $\mathbf{b}_{\Delta t}$ arise from the underlying CT $\mathbf{A}$ and $\mathbf{b}$ and explain in detail how a CT analysis gets the true underlying $\mathbf{A}$ and $\mathbf{b}$ back again from the mixtures. Autoregression matrix $\mathbf{A}_{\Delta t}$ in Eq. (7.1) and drift matrix $\mathbf{A}$ in Eq. (7.2) relate by the matrix exponential function as given in Eq. (7.3). It is by the highly nonlinear character of this relation that most of the paradoxical differences between DT and CT results can be explained. For the relation between DT and CT intercepts $\mathbf{b}_{\Delta t}$ and $\mathbf{b}$ as well as between errors $\mathbf{e}_{t-\Delta t}$ and $\mathbf{e}(t)$, we refer to Oud and Delsing (2010).

$$\mathbf{A}_{\Delta t} = \mathrm{e}^{\mathbf{A}\cdot\Delta t} \tag{7.3}$$

CT modelling can also be approached from the perspective of the lagged and instantaneous effects mentioned earlier. A big time lag is expected to result in relatively low values of the lagged coefficients (in any case in the autoregression coefficients but after some time interval also in the cross-regression coefficients), leading for many analysts to the correct feeling that lagged coefficients alone are not sufficient and to the decision to add the corresponding instantaneous effects to give a more appropriate picture of the underlying effects. However, Oud and Delsing (2010) describe a second problem of DT modelling and call it the lagged and instantaneous effects dilemma. It consists in the fact that the corresponding lagged and instantaneous coefficient values give quite different results and that in general the longer the lag, the higher the instantaneous coefficients become in comparison to the lagged ones. In the study by Vuchinich et al. (1992), for example, the dilemma was whether to choose for instantaneous or lagged effects between parental disciplinary behaviour and child antisocial behaviour. The authors found significant instantaneous effects but no significant lagged effects between these variables. Another example is in Becker et al. (2017), who related church attendance and education. When relating these variables instantaneously, they found a highly significant positive effect of education on church attendance. However, the lagged effects of education on church attendance were significantly negative. The dilemma is solved by Bergstrom (1966) in a nonrecursive model that imposes such restrictions on both types of coefficients—lagged and instantaneous—that one combined set of values results which approximates the true underlying CT coefficients (see details in Oud and Delsing 2010). In fact, Bergstrom presented his DT approximation of underlying differential equation (7.2) as a justification of the very use of nonrecursive models in economics. Instead of Bergstrom's approximate procedure, we use an exact procedure, but both solve the lagged and instantaneous effects dilemma of DT modelling.

The consequence of CT modelling to solve both problems described is that we should not causally interpret the DT autoregressions and cross-regressions in $\mathbf{A}_{\Delta t}$ and intercepts in $\mathbf{b}_{\Delta t}$ of Eq. (7.1) nor the instantaneous coefficients discussed in the previous paragraph but the CT auto-effects and cross-effects in drift matrix $\mathbf{A}$ and the intercepts in $\mathbf{b}$ of differential equation (7.2). This does not mean that the autoregression and cross-regression coefficients are useless. They tell a quite important but different story. Autoregressions and cross-regressions give the response over specific intervals for a unit impulse at the starting point. In particular, autoregression tells what after a specific interval Δt is left from a unit quantity in the variable at the starting time point. A cross-regression tells what after specific interval Δt the increase is in the dependent variable as a result of a unit increase in the independent variable at the starting point. However, a DT analysis gives these resulting quantities only for one specific interval Δt.

A third contribution of a CT analysis is that it provides the entire autoregression and cross-regression functions over the whole continuous time scale, that is, for all intervals, by modelwise interpolating between and predicting after the observation time points. An autoregressive function enables to answer, for example, after which interval only half of the unit value is left. A cross-regression function starts at zero for a zero interval (a causal effect needs some time to operate), then goes to a maximum at some point on the time scale and finally converges to zero again in a stable model. It enables to answer, for example, at what interval the maximal effect of the independent variable is reached and at what interval the effect becomes virtually zero. In addition to the autoregression and cross-regression functions, CT also provides the mean and covariance functions and so displays the means and variances/covariances not only for the discrete observation time points in the study but for all points in continuous time.

One main problem is the dependence of DT results on the chosen time interval. This leads to incomparability of results over different observation intervals within and between studies. If unaccounted for, it can easily lead to contradictory conclusions. In a multivariate model with three or more variables, one researcher could find a positive effect between two variables x and y, while another researcher, again in DT, finds no or a negative effect between the same variables, just because of a different observation interval length. This might well be the case, for example, in our study. After 13 years an originally strong effect of religiosity on life satisfaction might have faded away. Because results depend on the specific length of the chosen observation interval, even the use of equal intervals in DT studies does not solve the problem (Oud and Delsing 2010). Another interval might have given different results to both researchers as we discussed in the first problem of a discrete-time model.

A fourth contribution of CT analysis is therefore making the different and possibly contradictory effects in DT independent of the interval for equal as well as unequal intervals. So, by reporting the CT results instead of or in addition to the DT results, one enables other researchers with different or equal intervals to make a correct comparison with one's own results.

The last but not the least important contribution of CT analysis is in missing data handling (Oud and Voelkle 2014). In its attempt to limit the quantity of missing data, a DT analysis classifies the data in a restricted number of equidistant time groups: 1, 2, 3,.... The implication is that all data in one such group come from exactly the same time point. This is seldom the case. Measurements differ almost always in time, be it hours or even minutes. By putting data actually coming from different time points in the same group, the results of the analysis will become at least inaccurate and possibly unacceptable. Some missing value patterns can be handled in DT by so-called phantom variables, but this approach is limited to rather simple cases. For example, suppose one has a panel data set with four waves, 2 years between wave 1 and 2 and between wave 2 and 3 but only 1 year between wave 3 and 4. In DT one could choose time groups 1, 2, 3, 4, 5 and 6 and use phantom variables for totally missing groups 2 and 4. In CT the missing data problem is translated into an unequal measurement interval problem, and the missing data vanish. Each datum gets exactly the treatment it needs by combining it with its exact time interval. In this way, even a data set with all subjects having different measurement time points and different intervals is unproblematic. The use of different intervals is advocated by Voelkle and Oud (2013). While the previously mentioned advantages and solutions of CT do not lead to a different model fit, if no extra restrictions are imposed, giving data their exact time intervals in CT for each subject separately instead of the approximate equidistant ones in DT will change the data and therefore also lead to a different model fit in CT.

7.5 CT Model Results

The CT model results are reported in Table 7.2. The table contains also the model implied DT dynamic parameter values. The small differences of those values as well as of the initial parameters and $-2LL$ with the ones in Table 7.1 are exclusively caused by the fact that CT inserts for all subjects individually the exact measurement intervals, while DT assumes equal intervals for all subjects.

When interpreting the values of the dynamic parameters in Table 7.2, it should be kept in mind that the scale range of autoregression from 1 (maximum autoregression in a stable model, no decay) to 0 (minimum autoregression, no predictability at all) translates to a range from 0 to $-\infty$ for the auto-effect in the CT drift matrix. So, the autoregression of 0.4892 for CHURCH in Table 7.1, which is highly significantly deviating from minimum 0 ($p < 0.001$), corresponds to the auto-effect -0.0647, also deviating significantly from 0 but which in this case is from the maximum value in a stable model. The story to be told for the autoregressions/auto-effects in general turns out similar in DT and CT. CHURCH is the most persistent and predictable variable, followed by WORLDV and LSAT, respectively. As in Table 7.1 also, all cross-effects are negative. Of course, the implied DT dynamic parameters in Table 7.2, which are easily calculated by the matrix exponential in Eq. (7.3) for $\Delta t = 13$, do not differ much from the ones in Table 7.1, because they only improve on the inexact measurement time points used in Table 7.1. If formally tested, the differences between the results in Table 7.1 and the implied DT results in Table 7.2 would in this case probably not be significant. Also the CT auto-effects and the sign of the CT cross-effects resemble those in Table 7.1. But, different from Table 7.1, the relatively low negative cross-effects between CHURCH and WORLDV turn out to be significant in CT ($p < 0.05$). Different also from Table 7.1 is that the strength order of reciprocal effects between CHURCH and WORLDV reverses in CT, the effect of WORLDV on CHURCH becoming more negative than in the opposite direction. Interesting is that all diffusion (co)variances

Table 7.2 Church attendance, Christian worldview and life satisfaction at ages 30, 43 and 56 with exact time intervals in CT model

Initial parameters				
	Means	(Co)variances		
		CHURCH	WORLDV	LSAT
CHURCH	0.0802***	0.4750***		
WORLDV	−0.0453	0.1903***	0.5024***	
LSAT	−0.0558	−0.1076***	−0.0437**	0.8447***
Dynamic parameters				
	Intercepts	Drift coefficients		
		CHURCH	WORLDV	LSAT
CHURCH	−0.0054*	−0.0647***	−0.0110*	−0.0313***
WORLDV	0.0029	−0.0128*	−0.0704***	−0.0316***
LSAT	0.0037	−0.0538***	−0.0403***	−0.1293***
		Diffusion (co)variances		
		CHURCH	WORLDV	LSAT
	CHURCH	0.0482***		
	WORLDV	0.0123***	0.0776***	
	LSAT	0.0179***	0.0174***	0.1630***
	Trait variance	0.2083***		
Implied DT dynamic parameters for Δt = 13 years				
	Intercepts	Regressions		
		CHURCH	WORLDV	LSAT
CHURCH	−0.0541	0.4813	−0.0276	−0.1189
WORLDV	0.0229	−0.0255	0.4363	−0.1136
LSAT	0.0315	−0.2068	−0.1425	0.2474
		Residual (co)variances		
		CHURCH	WORLDV	LSAT
	CHURCH	0.3089		
	WORLDV	0.0454	0.4761	
	LSAT	−0.0636	−0.0785	0.6377
	Trait variance	0.2083		

−2LL = 26,579.13
$^{***}p<0.001$, $^{**}p<0.01$, $^{*}p>0.05$

are positive, but because of the effects in the rest of the model, this results in negative values for the covariances of CHURCH and WORLDV with LSAT in DT (−0.0589 and −0.0741 in Table 7.1 and −0.0636 and −0.0785 in Table 7.2). Again, differences in interpretation between DT and CT in this case should not be exaggerated. Nevertheless, it is important to realize that, being independent of any specific interval, it is more reliable to interpret CT results than DT or implied DT results.

Beyond the more fundamental model specification and especially its independence of a specific DT time interval, CT has the advantage over DT of clearly depicting the process over the total period. Figures 7.2, 7.3, 7.4 and 7.5 display for

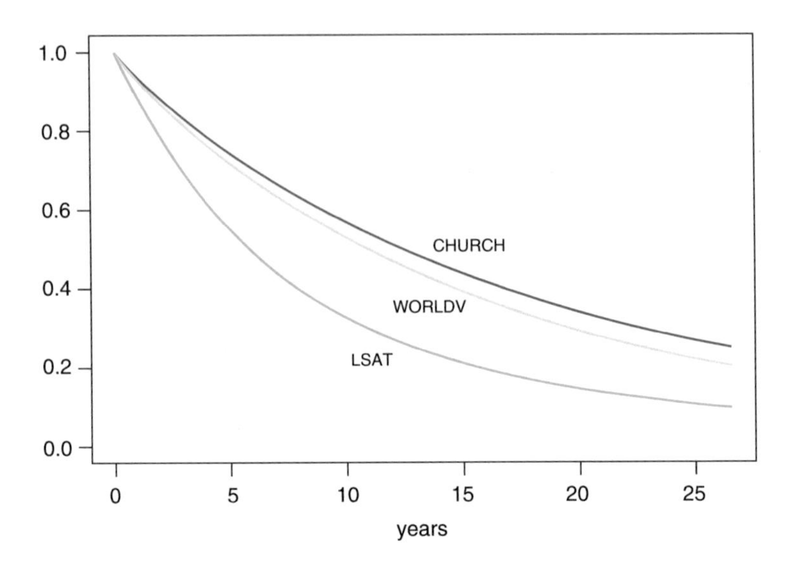

Fig. 7.2 Autoregression functions

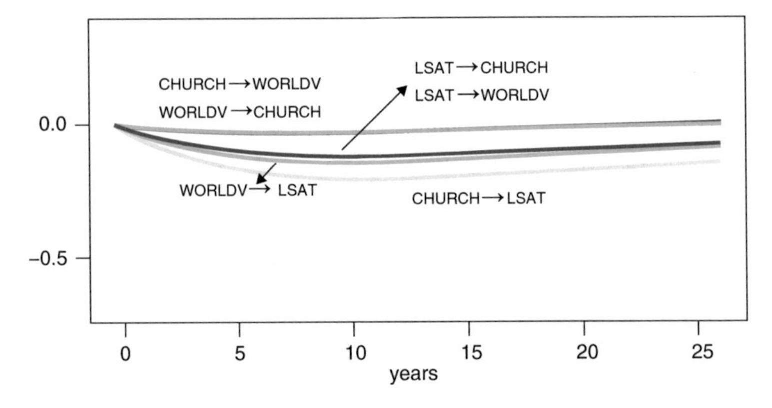

Fig. 7.3 Cross-regression functions

increasing intervals in continuous time over 26 years the estimated autoregressions and cross-regressions as well as the expected means and (co)variances in the studied group of subjects. Figure 7.2 shows that the autoregression is for CHURCH highest over the whole CT time scale and such that after 12 years still half of its value is left and after 26 years still more than 20%. Predictability of Christian worldview (WORLDV) on the basis of its previous value is at all intervals less, but the difference with CHURCH is small: after 11 years half of its value and about 20% after 26 years. Predictability of LSAT is considerably lower: half of its value only after 6 years and about 10% after 26 years.

All cross-regressions between the three variables in Fig. 7.3 turn out to be negative until the final interval of 26 years. However, not much is happening

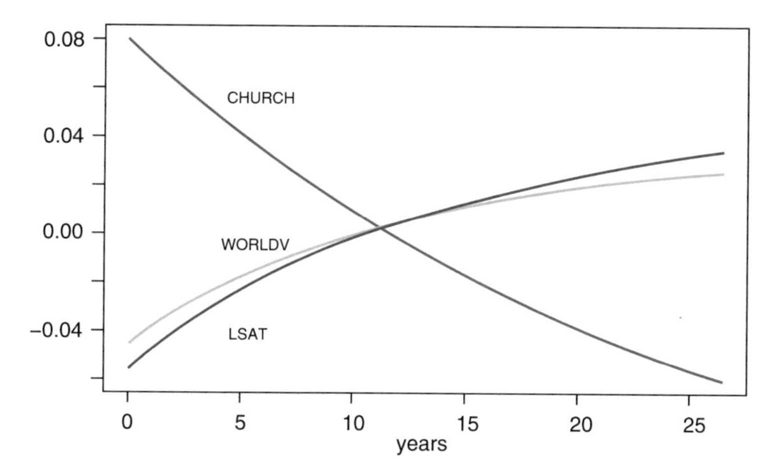

Fig. 7.4 Means across continuous time

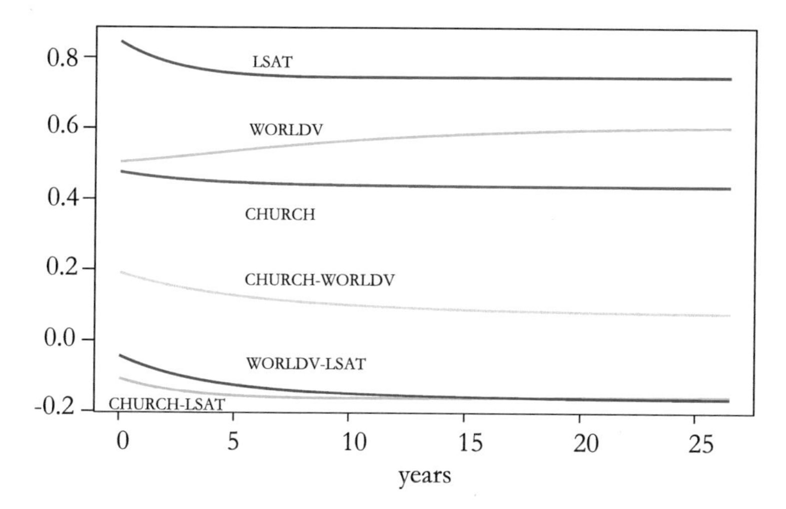

Fig. 7.5 Variances/covariances across continuous time

between CHURCH and WORLDV, neither in the short term nor in the long term. There are more substantial negative contributions from LSAT to these variables and from these variables to LSAT. Biggest is the negative contribution from CHURCH to LSAT, which reaches its maximum of −0.208 after 11 years, followed by the negative contribution from WORLDV to LSAT with maximum −0.146 after 10 years. The cross-regression functions in the opposite direction from LSAT to CHURCH and WORLDV, respectively, are almost identical and reach both their negative maxima of −0.120 and −0.116 after 11 years. It should be noted that these substantial contributions do not fade away rapidly, because after 26 years none of them is less negative than −0.075.

It will be no surprise that the CHURCH mean in Fig. 7.4 decreased over the 26 years period. However, WORLDV and LSAT showed a small increase over the same period. While in Fig. 7.5 the variances of CHURCH and LSAT kept being rather stable across time, the variance in WORLDV increased somewhat over the 26 years. There is a small positive correlation between CHURCH and WORLDV, which decreases somewhat across time. The other variables show small negative correlations across time.

7.6 Conclusion

We all lead our lives without really noticing how time passes by. Yet time and again we stop and look back. We register what has happened in the meantime—in our lives as well as in the lives of our fellow men. We notice the differences between some former and some current state; but we do not notice what has happened in between. We wonder how a difference could come up. If we are untroubled, we stick to contrasting former and current states. If we are pensive, we try to imagine a process that has led step by step from there to here. We live in continuous time, but we reflect upon our lives in discrete time.

Research on life histories, which is a reflection upon other people's life, cannot but do the same. It asks what people think and do at different times, it notices stabilities or differences and it tries to explain them. If it follows conventional wisdom, it takes the differences as given and looks for factors that may have determined them; time is split up into a sequence of discrete snapshots. If it is sophisticated, it assumes a process that has led from this to that value and constructs the values in between according to the rules of the process. In this chapter, we have compared two modelling approaches with the very same question and data. The question how religiosity and life satisfaction determine each other over a life span of 26 years of 1301 persons has been modelled by structural equations in discrete and in continuous time.

If we compare the results of the actual discrete-time analysis, supposing the data are collected at exactly the same discrete time points for all people, with the discrete-time results as implied by the parameters of continuous-time analysis, which accounts of the true individual measurement time points and intervals, the differences are small. Both analyses agree in three points. First, church attendance and Christian worldviews have stronger autoregressions than life satisfaction; the former seem to be habits of action and thought internalized early in life and the latter more easily subject to shocks from outside. Second, there are—contrary to expectation—negative rather than positive impacts of religiosity on life satisfaction and vice versa. In fact, both questions in the title should be answered negatively, on the basis of the discrete time as well as the continuous-time analysis. Religiosity, on the one hand, behaves rather as a cost than as a benefit in terms of life satisfaction. On the other hand, there is no self-selection of satisfied people into religiosity. Third, there is also a small negative reciprocal impact between the two dimensions of religiosity. Practice and belief interfere rather with one another than that they reinforce each other. Obviously, inconsistent patterns of both dimensions in cross-sectional population surveys which have often been interpreted as "believing without belonging" (Davie 2010) also show up in the longitudinal perspective on life histories.

Over and above these common results, the continuous-time analysis provides insights from the drift and the diffusion matrix that have no counterpart in discrete time; moreover, continuous-time analysis enables to depict the course of parameters, including means and variances/covariances, over time instead of only at a few specific discrete time points. The drift matrix in Table 7.2 reveals the underlying process and its realization across time in Figs. 7.2 and 7.3. The drift and diffusion matrix provided two main new insights into our specific data set. First, while the discrete-time dynamic parameters in Table 7.1 as well as in Table 7.2 showed that worldviews have a stronger negative impact on church attendance than vice versa, the corresponding drift parameters in Table 7.2 reversed this order. In addition, while both effects in Table 7.1 are nonsignificant, in Table 7.2 both are significant. So, practice seems to precede belief in a sense. Second, while the discrete-time residual covariances in Table 7.1 as well as in Table 7.2 between the two dimensions of religiosity and life satisfaction are negative, the respective diffusion covariances in Table 7.2 were positive. In both cases, the causal system as displayed in Fig. 7.1 is controlled for by unobserved heterogeneity—a person factor or trait. So, omitted effects seem to impact religiosity and life satisfaction in the same direction.

References

Abele, A., Spurk, D., & Volmer, J. (2011). The construct of career success: Measurement issues and an empirical example. *Zeitschrift für Arbeitsmarktforschung, 43*, 195–206. https://doi.org/10.1007/s12651-010-0034-6

Angraini, Y., Toharudin, T., Folmer, H., & Oud, J. H. L. (2014). The relationships between individualism, nationalism, ethnocentrism, and authoritarianism in Flanders: A continuous time-structural equation modeling approach. *Multivariate Behavioral Research, 49*, 41–53. https://doi.org/10.1080/00273171.2013.836621

Becker, S. O., Nagler, M., & Woessmann, L. (2017). Education and religious participation: city-level evidence from Germany's secularization period 1890–1930. *Journal of Economic Growth, 22*, 273–311. https://doi.org/10.1007/s10887-017-9142-2

Berger, P. (1967). *The sacred canopy: Elements of a theory of religion*. New York: Doubleday.

Bergstrom, A. R. (1966). Nonrecursive models as discrete approximations to systems of stochastic differential equations. *Econometrica, 34*, 173–182. https://doi.org/10.2307/1909861

Bollen, K. A., & Brand, J. E. (2010). A general panel model with random and fixed effects: A structural equations approach. *Social Forces, 89*, 1–34. https://doi.org/10.1353/sof.2010.0072

Browne, M. W., & Cudeck, R. (1993). Alternative ways of assessing model fit. In K. A. Bollen & J. Scott Long (Eds.), *Testing structural equation models*. Newbury Park, CA: Sage.

Davie, G. (2010). Vicarious religion: A response. *Journal of Contemporary Religion, 25*, 261–266. https://doi.org/10.1080/13537901003750944

Driver, C. C., Oud, J. H. L., & Voelkle, M. C. (2017). Continuous time structural equation modelling with R-Package ctsem. *Journal of Statistical Software, 77*(5), 1–35. https://doi.org/10.18637/jss.v077.i05

Felling, A., Peters, J., & Schreuder, O. (1987). Religion im Vergleich: Bundesrepublik Deutschland und Niederlande. Frankfurt am Main: Peter Lang.

Headey, B., Scup, J., Gucci, I., & Wagner, G. G. (2008). *Authentic Happiness Theory Supported by Impact of Religion on Life Satisfaction: A Longitudinal Analysis for Germany.* IZA Discussion Paper No. 3915.

Levin, J. S., & Taylor, R. J. (1998). Panel analyses of religious involvement and well-being in African Americans: Contemporaneous vs. longitudinal effects. *Journal for the Scientific Study of Religion, 37*, 695–709. https://doi.org/10.2307/1388151

Lim, C., & Putnam, R. D. (2010). Religion, social networks and life satisfaction. *American Sociological Review, 75*, 914–933. https://doi.org/10.1177/0003122410386686

Meulemann, H. (2017). Macht Religion glücklich-oder Glück religiös? In K. Birkelbach & H. Meulemann (Eds.), Lebensdeutung und Lebensplanung in der Lebensmitte (pp. 49–72). Wiesbaden: Springer. https://doi.org/10.1007/978-3-658-15362-5_3

Oud, J. H. L., & Delsing, M. J. M. H. (2010). Continuous time modeling of panel data by means of SEM. In K. van Montfort, J. H. L. Oud, & A. Satorra (Eds.), *Longitudinal research with latent variables* (pp. 201–244). New York: Springer.

Oud, J. H. L., & Jansen, R. A. R. G. (2000). Continuous time state space modeling of panel data by means of SEM. *Psychometrika, 65*, 199–215. https://doi.org/10.1007/BF02294374

Oud, J. H. L., & Voelkle, M. C. (2014). Do missing values exist? Incomplete data handling in cross-national longitudinal studies by means of continuous time modeling. *Quality & Quantity, 48*, 3271–3288. https://doi.org/10.1007/s11135-013-9955-9

Regnerus, M. D., & Smith, C. (2005). Selection effects in studies of religious influence. *Review of Religious Research, 47*, 23–50. https://doi.org/10.2307/4148279

SAS Institute Inc. (2013). *SAS/STAT®13.1 User's Guide: The CALIS procedure*. Cary, NC: SAS Institute Inc.

Voelkle, M. C., & Oud, J. H. L. (2013). Continuous time modelling with individually varying time intervals for oscillating and non-oscillating processes. *British Journal of Mathematical and Statistical Psychology, 66*, 103–126. https://doi.org/10.1111/j.2044-8317.2012.02043.x

Vuchinich, S., Bank, L., & Patterson, G. R. (1992). Parenting, peers, and the stability of antisocial behavior in preadolescent boys. *Developmental Psychology, 38*, 510–521. https://doi.org/10.1037/0012-1649.28.3.510

Weber, A. (2017). Lebenszufriedenheit im Lebensverlauf: Allgemein, privat und beruflich. In K. Birkelbach & H. Meulemann (Eds.), *Lebensdeutung und Lebensplanung in der Lebensmitte* (pp. 19–48). Wiesbaden: Springer. https://doi.org/10.1007/978-3-658-15362-5_2

Wothke, W. (2009). Longitudinal multigroup modeling with missing data. In T. D. Little, K. H. Schnabel, & J. Baumert (Eds.), *Modelling longitudinal and multilevel data: Practical issues, applied approaches, and specific examples* (pp. 219–240). Mahwah, NJ: Erlbaum.

Chapter 8
Mediation Modeling: Differing Perspectives on Time Alter Mediation Inferences

Pascal R. Deboeck, Kristopher J. Preacher, and David A. Cole

8.1 Introduction

This chapter proceeds in three parts. In Part I we discuss and differentiate discrete and continuous time perspectives on time using structural equation modeling (SEM) diagram conventions. In Part II we introduce a substantive example, in which we demonstrate a mediation model in continuous time and contrast it with a more familiar longitudinal mediation model in discrete time. In Part III we build on the prior sections to consider the implications of continuous time models for mediation modeling and provide an agenda for future work in mediation research.

8.2 Part I: Discrete and Continuous Time Perspectives on Data

8.2.1 Discrete Longitudinal Mediation

In line with a growing trend in the mediation literature, we assume that longitudinal data are necessary for making the types of inferences common to mediation analyses and that the same inferences are rarely (if ever) justified using cross-sectional data (Cole and Maxwell 2003; Gollob and Reichardt 1991; Preacher 2015; Selig

P. R. Deboeck (✉)
Department of Psychology, University of Utah, Salt Lake City, UT, USA
e-mail: pascal@psych.utah.edu

K. J. Preacher · D. A. Cole
Department of Psychology and Human Development, Vanderbilt University, Nashville, TN, USA
e-mail: kris.preacher@vanderbilt.edu; david.cole@vanderbilt.edu

K. van Montfort et al. (eds.), *Continuous Time Modeling in the Behavioral and Related Sciences*, https://doi.org/10.1007/978-3-319-77219-6_8

and Preacher 2009). Imagine, therefore, the collection of data on three variables (X, M, Y), over three waves of measurement, on n subjects. In the mediation literature, the most common approaches to modeling such data seem to be variations of a *cross-lagged panel model* (CLPM; Cole and Maxwell 2003; Gollob and Reichardt 1987, 1991; MacKinnon 2008; Maxwell and Cole 2007).

All CLPMs are defined by two necessary elements. First, repeated measures on the same variable are regressed on themselves at earlier occasions. It is common practice for an observation X_T to be regressed on the prior observation X_{T-1}, but one could also consider additionally regressing X_T onto X_{T-2}. The paths depicting regressions of variables on the same variables at different times, or *autoregression*, are often repeated for all variables (Fig. 8.1, black dashed arrows). Second, one or more variables are regressed on different variables at a prior observation; these effects are called *cross-lags* (Fig. 8.1, gray arrows). The number of these cross-lags differs in CLPMs, but in a mediation model, there would be a minimum of two regressions: Y_T onto M_{T-1} and M_{T-1} onto X_{T-2}. Taken together, they represent the *indirect* effect of X on Y via the mediator M. In addition, a *direct effect* of X on Y is often included by regressing Y_T onto either X_{T-1} or X_{T-2}. For the latter regression, methodologists disagree as to whether one should regress Y_T on the prior observation (X_{T-1}), as would be commonly done in related time series models, or two observations prior (X_{T-2}), based on the thinking that the indirect effect requires two observation intervals for X to have an effect on Y through M

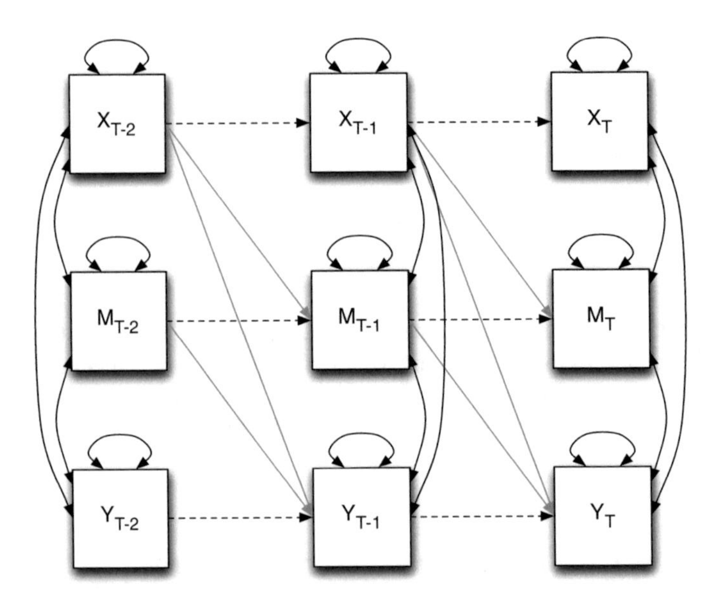

Fig. 8.1 Example of a common cross-lagged panel model used for testing mediation hypotheses. Later values (T) of each variable (X, M, Y) are predicted by proceeding values $(T - 1)$ of the variables. In this model, X affects Y both directly (path $X \rightarrow Y$) and indirectly $(X \rightarrow M \rightarrow Y)$. Thus, M is a mediator of the relation between X and Y

(i.e., $X_{T-2} \rightarrow M_{T-1} \rightarrow Y_T$). Variations of the CLPM can include additional cross-lags, a measurement model, and random intercepts to account for time-invariant differences in variables (Hamaker et al. 2015).

8.2.2 *A Continuous Time Perspective*

Figure 8.1 is incomplete from a continuous time perspective, as it represents only those discrete observations that were actually made. From the continuous time perspective, time does not exist in discrete quantities, and therefore Fig. 8.1 fails to give attention to the times that occur between sampling occasions. One way to see this perspective more clearly is to represent the unobserved occasions by depicting a large number of latent variables (Fig. 8.2, circles) between the observation occasions (Fig. 8.2, squares). Whereas latent occasions are depicted for only four additional times, the ellipses indicate that an infinite number of latent occasions actually exist. Contrasting Figs. 8.1 and 8.2 throws into relief the principal difference between common CLPMs and the continuous time perspective: that the continuous time perspective assumes that constructs that underlie variables and people continue to exist even when not being observed and could (in theory) have been sampled. Because of this ongoing existence of constructs, the two figures differ in terms of their relations between variables. In Fig. 8.1, M_T is regressed onto X_{T-1}, where one unit of time separates waves. This lag may constitute weeks, months, or even years. This regression requires the researcher to make the implicit assumption that

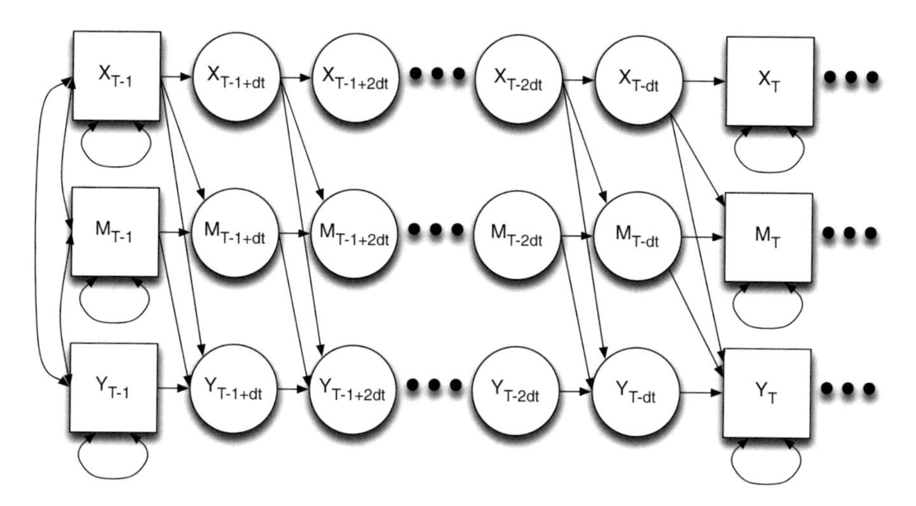

Fig. 8.2 A continuous time perspective on Fig. 8.1. Occasions at which no observations are being made are represented by latent variables (circles). Ideally, an infinite number of latent variables would be drawn between Times 1 and 2; for clarity, the infinite number of circles is represented by ellipses

the variables do not continue to have effects on one another during the intervening interval. This may be a tenable assumption when the rate at which observations are sampled is high relative to the rate at which the constructs underlying the variables are changing. In much of the social, behavioral, and medical sciences, however, constructs may change at rates much faster than the rate of observation. In Fig. 8.2, M_T is regressed on X_{T-dt}, where dt is a period of time much smaller than the observation interval. Smaller and smaller intervals of dt could be used, such that dt approaches (but never reaches) zero. That is, M_T depends on the value of X that exists an infinitesimally small moment (dt) before T, denoted with subscript $T-dt$. The variables at the prior moment are themselves dependent on the variables at a prior moment $T-2dt$, which are dependent on the variables at a prior moment, and so forth. This assumes not only the continued existence of constructs but ongoing effects between them from one moment to the next. The two figures, and perspectives, can be differentiated in that they make different assumptions about whether or not one can aggregate the momentary changes of variables over time.

The CLPM in Fig. 8.1 can be labeled a discrete time model because the progression of time is only implicitly expressed based on the ordering of the observations (Voelkle et al. 2012). Neither the autoregressive nor cross-lagged relations express precisely the interval occurring between observations. Although model diagrams may include indicators of time, these time indices exist only on the diagrams and are not incorporated into the mathematical model. Whether the first and second occasions of measurement are separated by 6 months or 12 months, the regressions are performed the same way although the resulting parameter estimates will differ.

The regressions in Fig. 8.1 do not specify how variables change from one observation to the next; many different patterns of change can produce the same regression coefficients. The model in Fig. 8.2 is not yet a truly continuous time model. To complete the continuous time model started in Fig. 8.2, we must specify a model that describes how variables change from one moment to the next and how they will affect each other, even when not being observed. That is, rather than allowing many possible patterns of change by fitting the model in Fig. 8.1, a very specific change dynamic is specified within the continuous time perspective. Figure 8.3 builds upon Fig. 8.2 by including (1) latent variables dX/dt, dM/dt, and dY/dt which express the change in each of the variables of interest with respect to time (first derivative) and (2) paths labeled dt, expressing an elapsed period of time. With these additions, time becomes explicit in the model, as each subsequent value of a variable (e.g., X_T) is equal to the prior value (X_{T-dt}), plus the rate at which change was occurring (dX/dt) multiplied by the elapsed time dt.

The question then becomes, what affects the latent change variables dX/dt, dM/dt, and dY/dt? That is, how do the unobserved constructs change from one moment to the next, and how do they affect each other? Figure 8.3 depicts one differential equation model that is relatively common, a first-order stochastic differential equation. Each of the latent change variables depicted has three components: (1) a path from the level of the variable to the velocity of the same variable (e.g.,

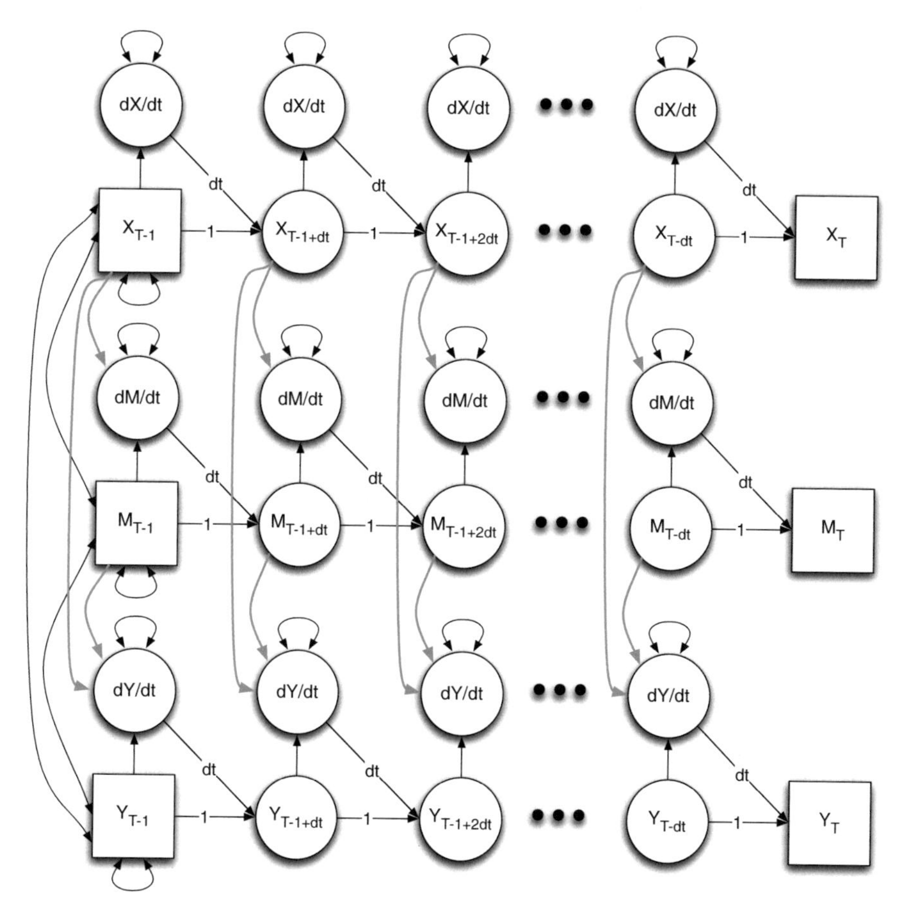

Fig. 8.3 A continuous time model. The change (first derivative) for each variable ($dX/dt, dM/dt, dY/dt$) is represented, in addition to the level (zeroth derivative) for each construct (X, M, Y). The gray arrows allow the level of variables to affect how other variables are changing; this is one of many possible mediation models that could be tested. Time becomes explicit in the model due to the dt path. The variables at each subsequent moment in time are equal to the prior values (path $= 1$) plus the rate of change (e.g., dY/dt) times the elapsed time (dt). Covariances between first derivative are omitted to reduce visual clutter

$X \rightarrow dX/dt$), (2) a path from the level of other variables (e.g., $X \rightarrow dY/dt$), and (3) a double-headed arrow ($dX/dt \leftrightarrow dX/dt$).

The first component, the level-velocity path within the same variable, allows for the velocity (change) of a variable to be related to the momentary level of the same variable. This may occur, for example, when individuals regulate toward some homeostatic level (e.g., levels of X negatively correlated with dX/dt). The second component, the level-velocity relations between variables, allows the level of X to be related to the velocity of Y (i.e., dY/dt). This is similar to the CLPM which allows variance in Y not accounted for by the prior observation of Y (which will be related to changes in Y) to be predicted by the level of X.

These relations can be more formally expressed as

$$\frac{d\mathbf{X}}{dt} = \mathbf{AX} + \boldsymbol{\varepsilon}$$

where $\mathbf{X}$ is a matrix representing all three variables (X, M, and Y), $\mathbf{A}$ is a 3×3 matrix expressing the level-velocity relations, and $\boldsymbol{\varepsilon}$ represents error. The within-variable regressions fall along the diagonal of $\mathbf{A}$ and are called *autoeffects* (Voelkle et al. 2012). The between-variable regressions fill the off-diagonal and are called *cross-effects*. The resemblance of the naming of these effects to the autoregressive and cross-lagged effects of the CLPM is because of the mathematical relation that exists between the first-order stochastic differential equation model and the CLPM. If autoregressive and cross-lagged effects are arranged in the same manner as $\mathbf{A}$, such that

$$\mathbf{X}_T = \mathbf{A}_\Delta \mathbf{X}_{T-1} + \boldsymbol{\varepsilon}_\Delta$$

then

$$\mathbf{A}_\Delta = \mathbf{e}^{\mathbf{A}*\Delta},$$

where the subscript Δ is used to indicate a dependency of the values on the interval between occasions of measurement and Δ also represents the interval between measurement occasions. This equation states that the autoregressive and cross-lagged effects in the CLPM ($\mathbf{A}_\Delta$) are equal to the matrix exponential ($\mathbf{e}$) of the continuous time auto- and cross-effects ($\mathbf{A}$) multiplied by the interval between observations (Δ). Because of this correspondence between the discrete and continuous time models, the path $X \rightarrow dY/dt$ can be defined as the *direct effect* of X on Y. Similarly, the combination of the paths $X \rightarrow dM/dt$ and $M \rightarrow dY/dt$ constitutes the *indirect effect* of X on Y (Deboeck and Preacher 2015).

The third component, the double-headed arrow, allows the variance of the velocity to be nonzero. If only the first two paths were included in the model, the model would be *deterministic*, as if there were no perturbations to the system; later observations would be perfectly predicted by prior observations. The double-headed arrows represent residual variances that allow for *stochastic* perturbations to the system. These perturbations differ from measurement error, as they perturb a system from deterministic dynamics, and affect future observations. Conceptually, the perturbations can be thought of as representing the unpredictable events that occur constantly in social, behavioral, and medical systems and are true changes in the latent variables. The combination of differential equations with stochastic perturbations, *stochastic differential equations*, represents a landmark opportunity. As stochastic differential equations include a stochastic process, integration of stochastic differential equations produces a distribution of possible trajectories over time. This is unlike the integration of more common (deterministic) differential equations, which produce only a single trajectory over time.

8.3 Part II: Substantive Example

A continuous time approach requires a specific mathematical model of how people change from one time to the next, even if that path is probabilistic rather than deterministic. Ideally, a series of competing models would be explored to consider different ways in which X affects Y and different ways in which M could mediate those relations. In this section, a substantive example is presented to facilitate comparisons of this modeling framework to more common longitudinal mediation approaches such as the CLPM.

8.3.1 Data

Drawn from a larger data set, the data consist of a longitudinal sample of 291 children in grades 3, 4, and 5 (Cole et al. 1999a, 1997, 1999b). In this sample, children were assessed on anxiety, depression, and social competence. Each of the three variables was assessed by self-report of the children at each occasion. The anxiety scale was the Revised Children's Manifest Anxiety Scale (RCMAS: Reynolds and Richmond 1985), a self-report measure of the frequency and severity of anxiety symptoms including worry, oversensitivity, physiological reactivity, and concentration problems. The depression measure was the Children's Depression Inventory (CDI: Kovacs 1981, 1982) a 27-item self-report measure of cognitive, affective, and behavioral symptoms of depression. The social competence measure was the social acceptance subscale of Harter's (1985) Self-Perception Profile for Children (SPPC). In this example, a continuous time mediation model is built on literature suggesting that both anxiety (X) and depression (M) affect social competence (Y) but that anxiety (X) is often a precursor to depression (M) (Cole et al. 1998; Dobson 1985; Kendall and Brady 1995). Within the sample, 33.6% of observations were missing (880 observations missing out of 291 participants by 9 indicators). Full information maximum likelihood was utilized to make use of all available data. Possible predictors of missing observations, which could improve adherence to the missing-at-random assumption, were not explored for these analyses.

8.3.2 Model: Discrete Time

To provide a basis for comparison, a more commonly used discrete time mediation model was fit to the data. This model, shown in Fig. 8.4, is a variation of the CLPM. The CLPM can be seen in the time-varying level of each variable and includes cross-effects from anxiety to depression, anxiety to social competence, and depression to social competence. Per recommendations in the literature (Hamaker et al. 2015), and the expectation that the variables consist of both time-varying and time-invariant components, a time-invariant latent variable (i.e., random intercept) was included for each of the variables.

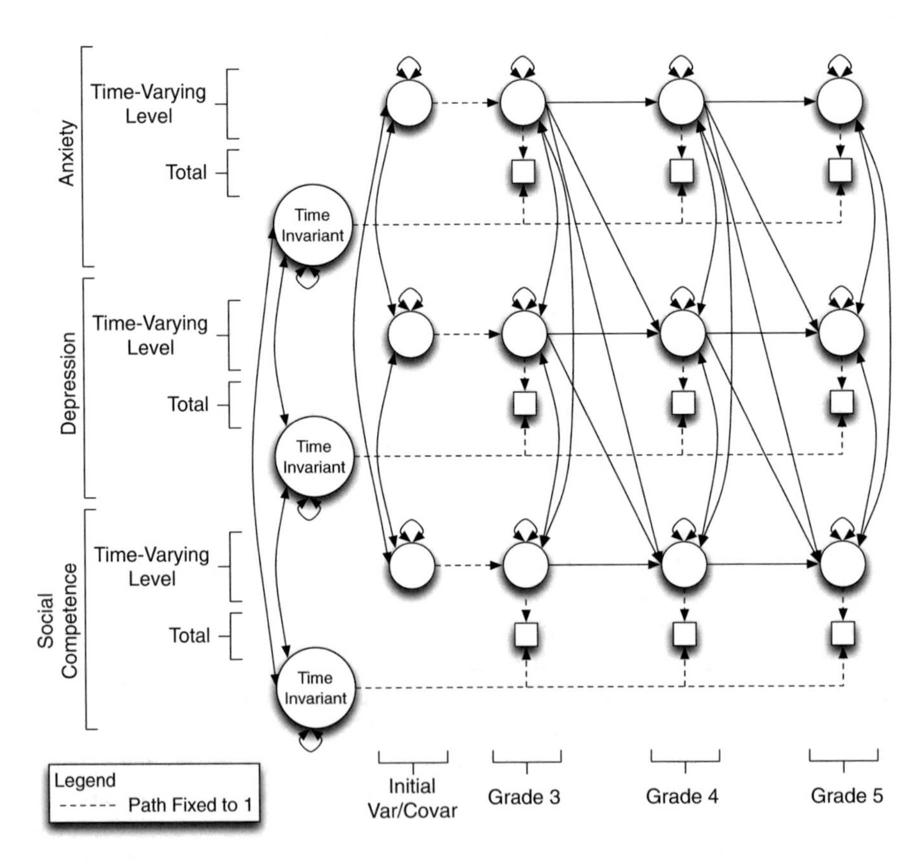

Fig. 8.4 Cross-lagged panel model. In this model, variables are regressed onto prior observations of the same variables (autoregressive effects) and prior observations of other variables (cross-lagged effects). Time-invariant components are included for each variable (random intercepts), to account for differences in mean values for each individual. The cross-effects allow for a direct effect of anxiety onto social competence, as well as an indirect effect through depression

8.3.3 Models: Continuous Time Model

Within-Variables To specify a continuous time model, we first consider how people change from one time to the next *within* each of the variables: anxiety, depression, and social competence. How are later observations on variable X_T dependent (or not) on one or more prior observations of X? For the present example, we began by specifying that each of the variables is equal to its prior value, plus some linear (first derivative) change multiplied by dt:

$$X_T = X_{T-dt} + (dt)\frac{dX_{T-dt}}{dt}.$$

This is similar to latent difference score models (McArdle 2009) and is also depicted in Fig. 8.3. In this model, however, the inclusion of additional latent states (ellipses) between observations allows the interval dt to be smaller than the observed interval; when a large number of latent states is introduced, dt will approach zero, and the model parameters will approach an analytic continuous time model solution (Deboeck and Boulton 2016). The first derivative $\frac{dX_{T-dt}}{dt}$ was allowed to have a nonzero residual variance, $\frac{dX_{T-dt}}{dt} \sim N(0, \sigma^2)$, to allow for stochastic perturbations. Although not explored here, a model with higher-order derivatives could also be considered.

Two additional sets of parameters are specified for each of the three variables. First, a random intercept was included for each variable, so as to allow variables to consist of a composite of a time-invariant component (i.e., random intercepts) over the interval of the study, and a time-varying component. The time-invariant components were allowed to correlate. Second, the velocity (first derivative) of each variable was regressed on the level (zeroth derivative, position relative to one's intercept) of the same variable. These autoeffects allow for modeling parts of the time-varying variance. The autoeffects are typically negative, as in such a case the velocity is in a direction opposite to the current level of the variable. Consequently, when negative, this relation allows for regulation toward one's time-invariant level. That is, people who are above their intercept will tend to have decreasing scores and people who are below their intercept will tend to have increasing scores—if the autoeffects are negative.

Between-Variables By allowing for correlated random intercepts, part of the relation between variables will already be explained. Above and beyond those relations, questions can be posed about how the time-varying levels (zeroth derivative) and velocity (first derivative) of the variables are related at any given moment. These questions include: Do higher levels of anxiety relate to the rate of change (velocity) of depression? Are the velocities correlated such that change in anxiety also leads to change in depression, regardless of the levels of each? Or is it perhaps some combination of both? (Deboeck et al. 2015) Admittedly, the literature is sparse concerning the momentary dynamics between variables and the relations between derivatives of variables. Like other modeling frameworks, competing models can be established, but strong theories become increasingly essential to explore the parameter space. In this example, we have greatly simplified the parameter space by assuming that higher-order derivatives are not required, but it is difficult to support this assumption with the present literature.

Admittedly, the choices for between-variable relations tested in this chapter are more exploratory and are primarily informed by a common first-order stochastic differential equation model. First, a set of level-to-velocity relations were examined. Informed by prior research, the level of anxiety was allowed to affect the velocities of both depression and social competence. The level of depression was also allowed to affect the velocity of social competence. Second, the velocities of each of the three variables were allowed to correlate. This will allow the stochastic perturbations to have nonzero correlations. This could occur if there are events or unmeasured variables that affect the rate of change of more than one variable.

Full Model In the full model, the observed variables at each occasion (Fig. 8.5, Total rows) consist of the sum of the time-invariant and time-varying contributions of the model. The *time-varying level* and *velocity* rows depict the within- and between-variable relations specified above. The model was fit by introducing a number of latent steps between observation occasions, so as to approximate continuous time parameter estimates (Deboeck and Boulton 2016); the additional latent steps are replaced by ellipses in Fig. 8.5.

8.3.4 Analysis

The full model was fit to the data using OpenMx (version 3.2.4; Neale et al. 2016), a package available in R (version 3.2.1; R Core Team 2015). Twelve theoretically plausible variations of the full model were run, each removing either a within- or between-variable parameter previously discussed. Models *Within1*, *Within2*, and *Within3* sequentially tested fixed intercepts, rather than random intercepts for anxiety, depression, and social competence, respectively. Related covariances between random intercepts were also removed as appropriate. Models *Within4*, *Within5*, and *Within6* sequentially removed the autoeffect parameters for each of the three variables. Models *Between1*, *Between2*, and *Between3* removed the effect of the level of one variable on the velocity of another ($\text{A} \rightarrow \Delta\text{Dep}$, $\text{A} \rightarrow \Delta\text{Soc}$, $\text{D} \rightarrow \Delta\text{Soc}$, respectively, where Δ is a symbol indicating "change in"). Models *Between4*, *Between5*, and *Between6* removed the covariances between velocities ($\Delta\text{Anx} \leftrightarrow \Delta\text{Dep}$, $\Delta\text{Anx} \leftrightarrow \Delta\text{Soc}$, $\Delta\text{Dep} \leftrightarrow \Delta\text{Soc}$, respectively).

8.3.5 Results and Discussion

Table 8.1 presents parameter estimates, and selected significance tests, for the discrete and continuous time models. Starting from the top of the table, it can be

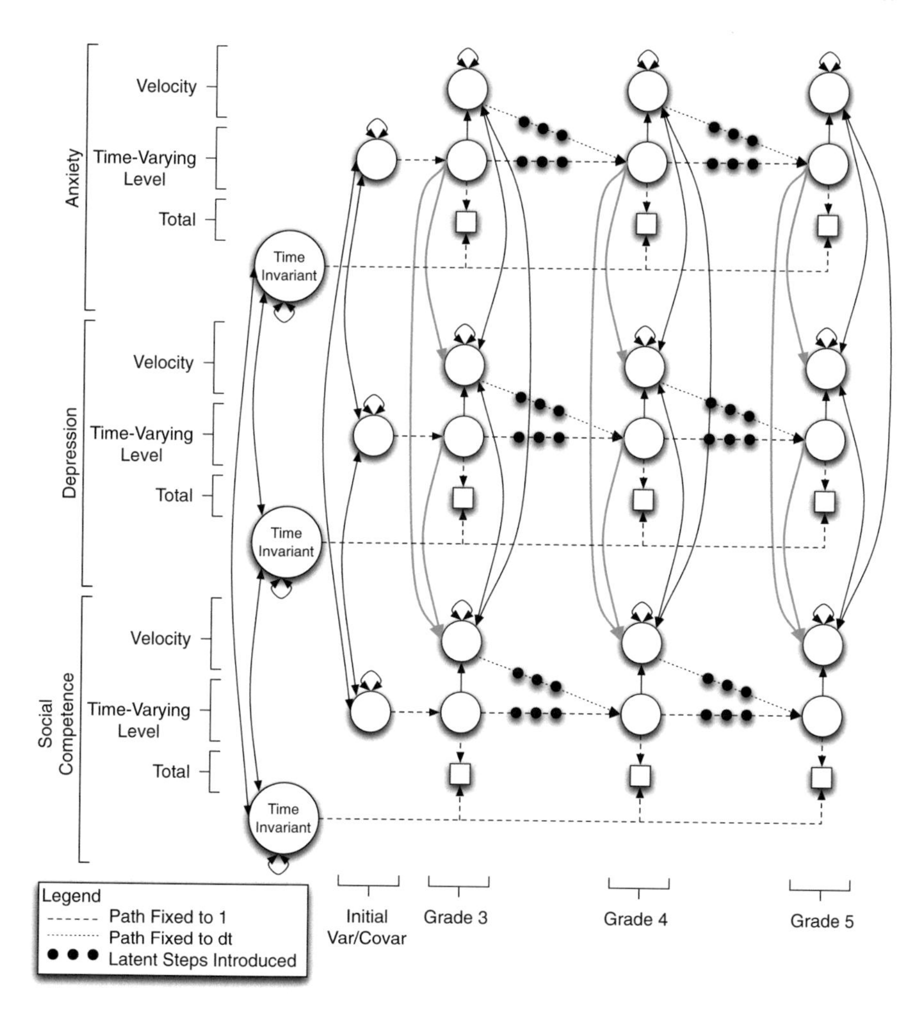

Fig. 8.5 Full model. This model is a variation of a first-order stochastic differential equation model. As with the CLPM, a time-invariant component is included for each of the variables. The time-varying component consists of a first-order stochastic differential equation, where the velocity at each time is related to the level of the same variable, the levels of other variables, and stochastic perturbations (velocity residual variance). The variance of the stochastic perturbations and relations between latent variables were constrained to be equal over time

seen that the estimated means, variances, and covariances for the time-invariant effects were essentially the same for the two models. The covariances between the time-invariant effects were not significantly different from zero.

The next set of results presents the initial variance/covariance estimates for the time-varying portion of the data. These parameter estimates were also essentially the same in both discrete and continuous time. The similarity of the time-invariant effects and initial time-varying variances/covariances can be expected based on the

Table 8.1 Results from model fitting

Parameter	Discrete time model			Continuous time model		
	Estimate	−2LL change (1df)	*p*-Value	Estimate	−2LL change (1df)	*p*-Value
Mean TI Anx	2.000			2.000		
Mean TI Dep	0.761			0.761		
Mean TI Soc	1.222			1.222		
Variance TI Anx	0.100			0.102		
Variance TI Dep	0.435			0.425		
Variance TI Soc	0.048			0.047		
Covariance TI Anx-Dep	0.301	0.6	$p = 0.437$	0.306	0.6	$p = 0.431$
Covariance TI Anx-Soc	−0.112	0.6	$p = 0.457$	−0.114	0.6	$p = 0.451$
Covariance TI Dep-Soc	−0.121	2.0	$p = 0.161$	−0.119	2.0	$p = 0.162$
Initial TV variance Anx	1.685			1.682		
Initial TV variance Dep	0.334			0.336		
Initial TV variance Soc	0.166			0.166		
Initial TV covariance Anx-Dep	0.521	32.3	$p < 0.001$	0.511	31.4	$p < 0.001$
Initial TV covariance Anx-Soc	−0.168	6.2	$p = 0.013$	−0.165	6.0	$p = 0.015$
Initial TV covariance Dep-Soc	−0.098	14.0	$p < 0.001$	−0.098	13.8	$p < 0.001$
TV residual variance Anx	0.76			–	–	–
TV residual variance Dep	0.237			–	–	–
TV residual variance Soc	0.137			–	–	–
TV residual covariance Anx-Dep	0.252	42.1	<0.001	–	–	–
TV residual covariance Anx-Soc	−0.098	81.0	<0.001	–	–	–
TV residual covariance Dep-Soc	−0.051	10.7	0.001	–	–	–
TV stochastic variance Anx	–	–	–	1.068		
TV stochastic variance Dep	–	–	–	3.419		
TV stochastic variance Soc	–	–	–	0.249		

TV stochastic covariance Anx-Dep	–	–	–	0.715	0.3	$p=0.619$
TV stochastic covariance Anx-Soc	–	–	–	−0.081	−0.8	$p=1.000$
TV stochastic covariance Dep-Soc	–	–	–	−0.282	0.7	$p=0.422$
Autoregressive Anx	0.681	15.1	$p<0.001$	–	–	–
Autoregressive Dep	−0.106	0.9	$p=0.353$	–	–	–
Autoregressive Soc	0.304	5.3	$p=0.022$	–	–	–
Autoeffect Anx	–	–	–	−0.380	66.0	$p<0.001$
Autoeffect Dep	–	–	–	−13.139	113.0	$p<0.001$
Autoeffect Soc	–	–	–	−1.153	86.9	$p<0.001$
Cross-lag Anx-Dep	0.231	6.3	$p=0.012$	–	–	–
Cross-lag Anx-Soc	−0.062	2.7	$p=0.099$	–	–	–
Cross-lag Dep-Soc	−0.017	0.1	$p=0.802$	–	–	–
Cross-effect Anx-Dep	–	–	–	3.688	5.5	$p=0.019$
Cross-effect Anx-Soc	–	–	–	0.240	0.0	$p=0.903$
Cross-effect Dep-Soc	–	–	–	−1.298	0.2	$p=0.630$

Note: TI=time-invariant; TV=time-varying; Anx=anxiety; Dep=depression; Soc=social competence. A subset of parameters was tested for statistical significance. The column "−2LL change" indicates the change in −2LL when the parameter was constrained to equal zero; the column "*p*-value" reports the likelihood ratio test *p*-value. Some parameters that appear to have similar interpretations are not listed in the same rows because transformation is required for direct comparisons to be meaningful similarities in the models specified in Figs. 8.4 and 8.5. The covariances between the time-invariant effects were all significantly different from zero, with anxiety and depression being positively correlated and social competence being negatively correlated with both anxiety and depression. Although it is tempting to interpret these relations, these covariances offer only a snapshot at the initial wave of observations, and do not actually reflect whether changes in the variables are related.

The key differences between Fig. 8.4 (the discrete time model) and Fig. 8.5 (the continuous time model) become more apparent in the remaining sets of results in Table 8.1. The results for the residuals (discrete time) and stochastic perturbations (continuous time) show little resemblance. The same is true for the autoregressive effects (discrete time) and autoeffects (continuous time) and the cross-lags (discrete time) and cross-effects (continuous time). These dissimilarities are because the two models partition the time-varying variance differently. The results have been presented in groups, however, because these pairs are related. Actually, these models are transformations of each other under specific conditions, a fact that becomes apparent when examining the fit of the models (Deboeck and Preacher 2015). The discrete time model showed good fit to the data (RMSEA $= 0.069$, CFI $= 0.966$, $\chi^2(27) = 64.3$), as did the continuous time model (RMSEA $= 0.070$, CFI $= 0.965$, $\chi^2(27) = 65.1$). The small discrepancy is related to the CLPM, in this instance, producing a negative estimated value of the autoregressive effect of depression that cannot be replicated by the current continuous time model.[1]

The relation between the discrete time residual variances and stochastic perturbations (change residual variances) is beyond the scope of this chapter (see Oud and Jansen 2000). However, it should be noted that from the continuous time perspective, the covariances of the stochastic perturbations represent correlations among the velocities of variables. In the discrete time model, the equivalents of stochastic perturbations get the unflattering name of "residuals"; however, in the continuous time model, the covariances between the latent change variables suggest whether stochastic perturbations to the variables are correlated. In the present example, the stochastic covariances were not significantly different from zero. Had they been different from zero, we would have been unable to disentangle whether the velocity of one of these variables results in a corresponding velocity for the other variable or whether there is a mutual third variance (e.g., events) that resulted in both variables changing.

The remaining estimates in Table 8.1 represent the autoregressive and cross-lagged effects for the discrete time model and the autoeffects and cross-effects for the continuous time model. The discrete time model and the autoregressive and cross-lagged effects represent the expected relations for a particular interval between observations—in this example, 1 year. These estimates can vary for different

[1]In the current continuous time model, the autoeffects (continuous time) typically range from minus infinity to zero, which corresponds to an autoregressive effect (discrete time) of 0-1. The continuous time model specified does not allow for the equivalent of a negative autoregressive effect. When the CLPM is constrained to produce autoeffects between 0 and 1, the -2LL for the constrained CLPM and continuous time model are equivalent to at least three decimal places.

intervals of time. The continuous time parameters, on the other hand, represent the instantaneous (i.e., as dt goes to zero) effects between the derivatives of the variables. Although interpreting continuous time parameters can be challenging, as presented in Table 8.1, through integration, the effects can be summed for any desired interval of time. This produces estimates of parameters such as the autoregressive and cross-lagged effects in a CLPM. The primary difference is that rather than estimating these parameters for only a single interval between observations present in the data, the continuous time model can yield estimates of these effects for any observation interval of interest (lags), to give an impression of how the variables affect each other over differing time intervals.

Figure 8.6 depicts the autoregressive and cross-lagged effects expected from the continuous time model for lags varying from 0 to 2 years; the maximum interval of 2 years was selected based on the maximal interval observed in the data. The lines, corresponding to the results of the continuous time model, are produced by integrating the differential equation model[2] for a range of lags, which produces the expected discrete time parameters for a range of lags. Solving for the indirect and direct effects requires constraining the model parameters, prior to integration, so as to allow only for one effect at a time. Details on these calculations are available in Deboeck and Preacher (2015). It should be noted that one can integrate the differential equation model for any lag, including lags that are not well supported by the data. It is consequently important when looking at this figure to be cognizant of the fact that data are present to support the results at lags of 1 and 2 years, but estimates of relations at lags of much less than 1 year or much greater than 2 years extrapolate beyond the available data.

The points in the plots represent the 1-year interval parameter estimates produced by the discrete time model. Figure 8.6 panels a and b represent the results produced by the discrete and continuous time models. If these models were perfect transformations of each other, the points would fall on the lines. In this example, there is some misfit because the continuous time model cannot produce negative autoregressive estimates as currently specified (see depression in panel a), but the discrete and continuous time models still produced similar estimates for a 1-year interval for Anx→Dep, Dep→Soc, and the total effect of Anx→Soc. Figure 8.6 panels c and d show the CLPM results when the autoregressive effects are constrained to be between 0 and 1; this, and the equivalent log-likelihoods (see footnote 1), highlight that under some conditions, the discrete and continuous time models examined here are transformations of each other. Comparing the results from the two models, one difference that is immediately apparent is that the discrete time model provides only a single "snapshot" of the relations among the variables, as the

[2]The full model parameters are used, as removal of the nonsignificant relations began producing continuous time parameter estimates outside the typically expected range. For example, the social competence continuous time autoeffect became equivalent to a CLPM autoregressive effect greater than 1. This may suggest that the effect of anxiety on social competence may be important to incorporate into the model, although it was not significantly different from zero.

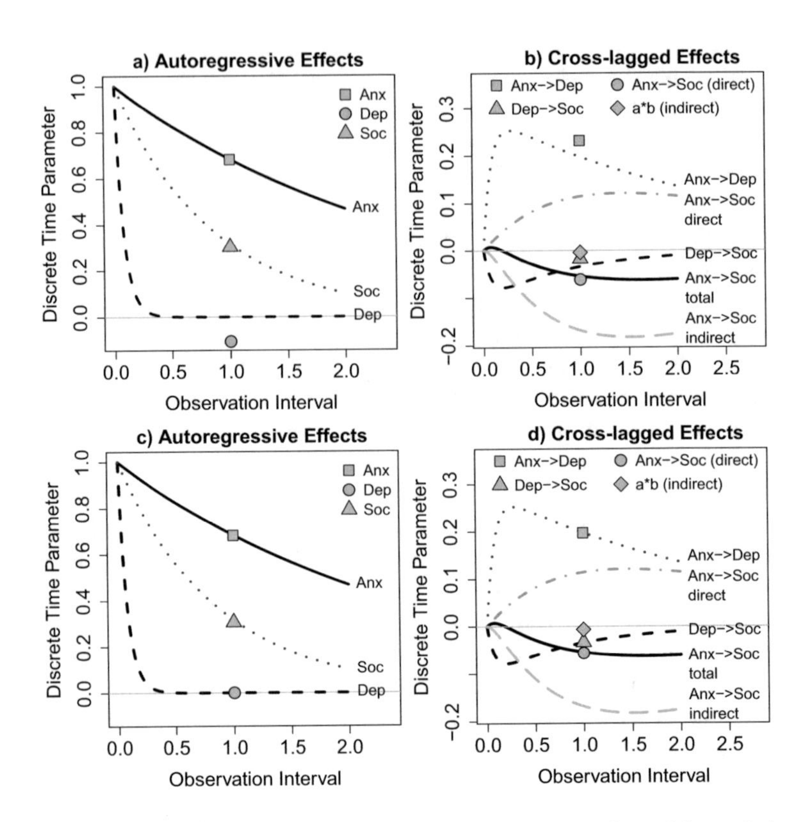

Fig. 8.6 Plots of the relations between variables expected for a variety of intervals between observations. The continuous lines are based on integrating the continuous time differential equation model for differing intervals of time. The gray points are the single interval estimates produced by the discrete time CLPM when unconstrained (panels (**a**) and (**b**)) and when autoregressive effects are constrained to be between 0 and 1 (panels (**c**) and (**d**)). The Anx→Soc relation plotted for the CLPM is the cross-lag that is interpreted as the direct effect of anxiety on social competence. Differences in the discrete and continuous time model estimates occur in panels (**a**) and (**b**), as the autoregressive effect of depression is less than zero, although not significantly different from zero (estimate=−0.106, s.e.=0.108). The differences in panel (**b**) are also much smaller than the standard errors of the CLPM estimates (0.080, 0.037, 0.070 for Anx→Dep, Anx→Soc, and Dep→Soc, respectively). The discrete and continuous time models provide an equivalent fit to the data, panels c and d, when autoregressive effects of the CLPM are constrained to be between 0 and 1. The models differ, however, in the effect that would be considered the direct effect (panels (**b**) and (**d**)); contrast the circle representing the discrete time direct effect of Anx → Soc with the continuous line labeled 'Anx → Soc direct.'

parameter estimates are specific to a single observation interval (e.g., 1 year, as in the running example).

Whereas the specific results require further investigation, this substantive example was intended to contrast discrete and continuous time modeling. These models began with consideration of ways to model the dynamics of within-variable change, followed by specification of the dynamics of between-variable change. The possible ways of specifying each part are numerous and were limited in this example due only to the assumption that derivatives beyond velocity were unnecessary (i.e., that a first-order model would be a sufficient initial step with these data). We discussed three different ways to consider the between-variable relations: (1) the covariances between time-invariant random intercepts, (2) coupled stochastic perturbations (correlated change), and (3) level-velocity relations. The latter are the source of the relations plotted in Fig. 8.6, which gave an impression of how the effects between variables may change for different intervals of observation. Moreover, this potential wealth of information is available from a relatively modest data set, consisting of only three waves of observation.

8.4 Part III: Implications, Questions, and the Future of Mediation

Having introduced the idea of continuous time modeling and having seen how a substantive example differs in this modeling framework relative to a more traditional approach, we now consider three topics common to mediation.

8.4.1 Direct and Indirect Effects

It would seem that direct and indirect effects should be concepts that are defined independent of the method or model being used to assess these effects and should transcend particular methodologies. Returning to Fig. 8.1, the effect from X_{T-1} to Y_T is commonly considered a direct effect, whereas X_{T-2} to M_{T-1} to Y_T is an indirect effect of X on Y.

Figure 8.2 calls into question this interpretation, however, as between $T - 1$ and T, there are ongoing transactions between the variables. The consequence is that from the continuous time perspective, the purported direct effect in Fig. 8.1 actually represents a summation of both direct and indirect effects. Figure 8.6 from the substantive example represents the effects expected from the continuous time model for a variety of intervals; CLPM typically provides us only a single vertical slice of Fig. 8.6, unless a researcher fits the CLPM for several differing intervals. The line in this figure labeled "Anx→Soc Total" represents the results produced by the CLPM, which in discrete time models would be labeled the direct effect (Deboeck and Preacher 2015). From a continuous time perspective, however, the discrete time interpretation of direct and indirect effects is fundamentally flawed, as the discrete time direct effect is actually a blending of direct and indirect effects.

This blending of direct and indirect effects may be better understood using the diagraming techniques in Part I and tracing rules. In Fig. 8.7, we can take the

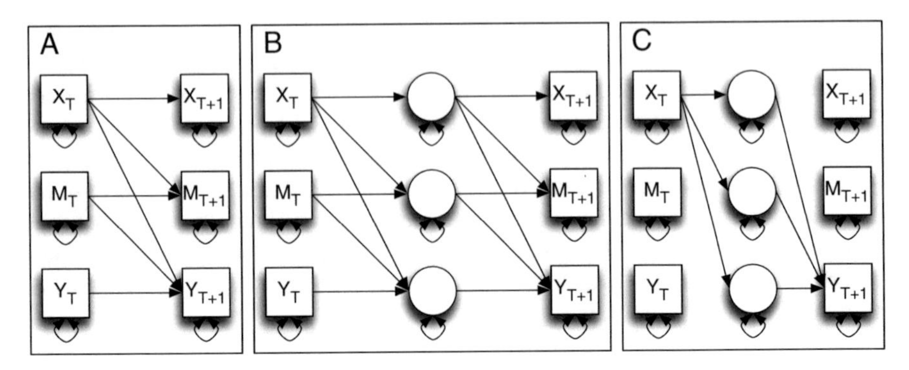

Fig. 8.7 The direct effect in the CLPM (panel (**a**)) bears an interpretation different from a direct effect in a continuous time perspective. The continuous time perspective can be understood by inserting a set of latent variables between observations (panel (**b**)). The paths in panel (**c**) represent the paths one would combine when using tracing rules on panel (**b**) to solve for the direct effect in panel (**a**). Thus, from a continuous time perspective, the direct effect in the CLPM is a combination of both direct and indirect effects

CLPM (panel a) and insert an unobserved set of latent variables between the two observations (panel b). Using Fig. 8.7 panel b, tracing rules can be applied to solve for the direct effect of X_{T-1} on Y_T, depicted in Fig. 8.7 panel a. If such a latent observation exists, the purported direct effect in Fig. 8.7 panel a will consist of the paths depicted in Fig. 8.7 panel c, which include indirect effects through the occasions at which X, M, and Y are not observed. Thus, the direct effect in CLPMs, from the continuous time perspective, is a combination of direct and indirect effects; this will lead the CLPM estimate of the indirect effect to be attenuated. The continuous time model can be conceptualized as including an infinite number of these additional latent variables, such that the moment-to-moment accumulation of the direct effect can be separated from the indirect effect.

For the continuous time model, the equivalents of direct and indirect effects are represented in the paths from level to velocity. The equivalent for the direct path is the $X \rightarrow dY/dt$ relation, and the equivalent for the indirect path is the combination of the $X \rightarrow dM/dt$ and $M \rightarrow dY/dt$ paths. The substantive example highlights that, based on the *Between2* model, there was no instantaneous effect of levels of anxiety on changes in social competence, which could be considered the lack of a significant direct effect. The use of the term "direct effect," however, perhaps should be discouraged. Although the parallel to direct and indirect effects seems close, these effects are dependent on the differential equation model that is selected. To the degree that a model is approximately correct, labeling the level-velocity relations "direct" and "indirect" effects may be useful, but clearly, the opposite is true for poorly specified models. Thus, the degree to which continuous time models capture these effects depends, like models of all real-world processes, on unknowable assumptions that can be explored through the testing of competing models.

Taking both discrete and continuous time perspectives into consideration provides no clear resolution for how to disentangle direct and indirect effects in a particular model. The continuous time perspective suggests that the labeling of direct and indirect effects in the CLPM is flawed, but the use of the terms for continuous time parameters requires a strong assumption about the degree to which the correct model has been selected. Rather, the dual perspectives highlight the need for definitions of these terms that supersede any particular model and that applying the label in the context of a particular model may be as inaccurate as when a label is applied to a latent variable. Concepts of "direct effects" and "indirect effects" are necessarily conceptual abstractions, representing all possible (linear, nonlinear, accelerating, etc.) effects of a particular sort. Given that any estimable model will likely neglect some aspects of these effects, they all run the risk of inaccurately estimating these effects to some degree or another.

8.4.2 Complete Mediation

The term "complete mediation," as used in discrete time models, has questionable value regardless of whether it is used to describe the importance of the effect or to suggest the lack of room for additional mediators (Preacher and Kelley 2011). Many desirable properties of effect size measures are not conveyed by the use of terms like "complete" and "partial." The continuous time perspective offers yet another set of reasons why these terms are of dubious value. Focusing on the level-velocity relations, the substantive example presents two ways of thinking about the relations between variables. One pertains to the continuous time parameters that are estimated and tested in Table 8.1. The second pertains to the effects that are expected to be observed for specific observation intervals, as shown in Fig. 8.6.

First, the continuous time parameters that are estimated and tested offer one natural way to conceptualize complete mediation. One could examine the level-velocity relation between anxiety and social competence with and without depression and demonstrate whether the relation is completely mediated by depression. Is it sufficient, however, for the level-velocity relations to be equal to zero to demonstrate complete mediation, or would all possible relations have to be equal to zero? Would the *level→level*, *level→velocity*, and *velocity→velocity* relations all have to demonstrate the desired pattern to be considered complete mediation, or would it be sufficient for only some of these to equal zero?

For a model to meet the requirements of mediation, it seems necessary for one or more derivatives representing change to be incorporated into a model (e.g., velocity, acceleration, etc.). If, for instance, the *level→velocity* and *velocity→velocity* relations were equal to zero, this would indicate a lack of related change scores but could still allow for correlated levels of the variables. While this may seem a relatively plausible task, a Pandora's box of possible change relations exists, as higher-order derivatives could be considered. This is not to say that models with higher-order derivatives lack utility, are beyond substantive theory, or cannot be fit to data (Deboeck et al. 2015). However, the existence of higher-order derivative models presents a problem for creating a definition of complete mediation that transcends a specific model. How many higher-order derivatives must be considered before mediation can be considered complete? Demonstrating a lack of all change relations in order to demonstrate complete mediation may be a task worthy of Sisyphus.

Second, the effects that are expected for a variety of observation intervals (as in Fig. 8.6) also raise problems for the definition of *complete mediation*. Although not the case in our substantive example, with relatively simple continuous time models, the relation between the same variables can produce effects that are negative for some intervals, indistinguishable from zero for other intervals, and positive for yet other intervals. It seems undesirable for the term *complete mediation* to apply only to specific observation intervals. It can be anticipated that for many possible differential equation models, at very short and very long lags, the total (i.e., unmediated) effect of X on Y will almost certainly be indistinguishable from zero, as in one case, not enough time is allowed for a detectable effect to accumulate and, in the other, so much time occurs that the influence of many other factors would contribute enough additional variance to render the effect undetectable. One could judiciously apply “complete mediation” only when the direct effect becomes zero, after inclusion of the mediator, for all possible lags. But this is not how the concept is commonly employed. Applications of the CLPM typically produce relations for only a single lag, rather than all (or even many) possible lags. Even then, such judicious application does not address the issue raised in the prior paragraph; in combination, the issues raised by the continuous time perspective suggest that “complete mediation” is a widely misapplied term in the current application of CLPMs.

8.4.3 What Is a Mediation Model?

Since Baron and Kenny’s (1986) classic paper, mediation has often been conceptualized as a single model applied to cross-sectional data. The mediation literature is moving from cross-sectional data toward emphasizing the need for longitudinal data. Articles concerning longitudinal mediation often replace the single Baron and Kenny mediation model with another relatively limited set of models, variations of the CLPM. In the present chapter, the substantive example began by discussing the need to consider modeling dynamic temporal variations in the relations between variables. Some variation of models was considered, including combinations of both deterministic and stochastic model components, as well as multiple ways in which observed variables could be related. The present chapter explored a relatively small subset of models, running the risk of replacing one narrow set of models with another. The full set of continuous and discrete time models that could be specified is much wider than has been expressed here. Naturally, models with higher-order lags could be explored in discrete time, and models with higher-order derivatives could be explored in continuous time. Even with models that incorporate higher-order relations, however, there would still be differences between the discrete and continuous time models paralleling those presented in this chapter with the CLPM and first-order stochastic differential equation models.

For some variables, models will reveal very small stochastic components, akin to the developmental trajectories specified by growth curves. For other variables, models may reveal substantial stochastic components that describe frequent perturbations to variables. Although some variables may consist entirely of additive stochastic perturbations (e.g., a *random walk*), others may regulate toward some homeostatic value. Such regulation can take different forms, such as the exponential decay toward homeostasis of a first-order differential equation or the sinusoidal curving toward homeostasis of a second-order differential equation. Moreover, these characteristics may occur around a changing homeostatic value, which could be described using any number of variations on growth curve models. This barely begins to outline the possible variations that can be considered.

Between variables, there are similar considerations. Many relations can be considered, but a first step might be to consider the linear relations between the first 2–3 derivatives of pairs of variables. The first three derivatives of each variable would constitute the 0th (level, a horizontal line describing a single value over time), 1st (velocity, straight-line change), and 2nd (acceleration, quadratic curve change). For a given variable (e.g., Y), the question could be posed whether it is the level, velocity, or acceleration of X that results in acceleration (changes in the rate of change) of Y. In many cases, it may be not only the level of X that is a predictor of Y but also X's recent rate of change, or it could be that the rate of change in Y is changing (Deboeck et al. 2015); one potential practical example is that acceleration in a child's externalizing symptomatology may be related not only to the level of parental marriage problems but also to recent changes in marriage problems (velocity). Consideration of multiple possible relations between derivatives of variables is uncharted territory for many domains of research. But the exploration of these derivative relations would help to nuance what is coupled when two variables are "related" while also highlighting variables that have merely correlated levels rather than correlated changes.

8.5 Future Directions

In presenting the reader with both discrete and continuous time perspectives, models, and results in a book about continuous time modeling, it may appear that we aim to reignite the often heated debate about whether discrete or continuous time models are better. The future of mediation analyses, however, is not dependent on universal agreement about the usage of discrete and continuous time models. Rather, its future depends to a large extent on the understanding that these processes *could* be modeled using different perspectives on time. As highlighted in the discussion of Part III, the perspective one takes (through the model that is used) has implications for fundamental concepts in mediation.

The first fundamental step will be to create more precise, consensual definitions of mediation concepts that transcend the selection of specific models. For *direct* and *indirect* effects consensual definitions may readily be obtainable, as these terms are dependent on the variables present in a model and the relations specified between three or more variables. The *direct* effect could continue to be defined as the "causal effects that are not mediated by any other variables in the model" (Cohen et al. 2003). Building consensus on the definition of complete mediation, however, may be more challenging and of questionable value, as the term "complete mediation" refers to something that is *not* present. If testing for mediation required only the fitting of a small number of prespecified models, demonstrating evidence for the nonexistence of an effect would be straightforward; but with an expansive number of possible models and observation intervals, establishing consensus on what constitutes reasonable proof for a nonexistent effect seems a quixotic pursuit.

It will also be necessary to develop widespread consensus on what constitutes a mediation model. Many descriptions of mediation directly state the need for causal relations between variables. Mathematically, this could be understood as a requirement that one observe not simply two variables changing in tandem, but one variable producing change in another variable—that the *level* of one variable results in the *change in level* of another variable, the *velocity* of one variable results in the *change in velocity* of another variable, and so forth. This suggests that, for a model to be a *mediation model*, the derivatives of the purported cause should be related to a higher order of derivative in the variable that is being affected. This idea is already present in the mediation literature on the CLPM, which highlights that by regressing a variable on itself (e.g., $Y_{T-1} \rightarrow Y_T$), the levels of other variables (e.g., X_T, M_T) are logically limited to explaining the variance that is not stable—that is, the portion of the dependent variable that is changing with respect to time (Preacher 2015). Both the discrete time and continuous time models presented in this chapter would qualify as mediation models under this definition. Models with only single measures of X, M, and Y over time, even if individually measured at differing times, would not meet this definition as they would be unable to demonstrate that antecedent variables are related to changes in the consequent variables.

The second fundamental step is to realize the limitations of the models used to test mediation. Although it is clear that no model ever fully describes real processes, researchers rarely consider how well a model captures the ideal definition of a *direct* or *indirect* effect. This chapter has highlighted that different models, perspectives on time, and observation intervals all affect the simple concept of a direct effect. From the continuous time perspective, there are many cases in which the CLPM will overestimate the purported direct effect and return an attenuated indirect effect. This bias, however, depends on factors such as the rate at which samples are collected over time. Conversely, continuous time models are predicated on an assumption about processes that occur when observations are not being made. No doubt there are conditions which will bias estimates of direct effects for these models. The limitations of our models go beyond saying that "all models are wrong" (Box 1976). Even when all necessary variables and the directionality of their relations are known, the data and model selected can lead to limitations on the generalizability of the inference that is made. Inferences are limited to the interval of observations in the data, and some models (e.g., CLPM) may offer only biased approximations of ideals such as direct and indirect effects.

The final step is to realize the undiscovered that already exists in our literature and many, many existing data sets. By adopting both discrete and continuous time perspectives in this chapter, we show that (1) much of our "mediation" literature may not actually be supported with "mediation models" because of the lack of consideration of change, (2) common longitudinal mediation models have offered estimates of effects that are biased from a continuous time perspective, (3) many of our estimates are limited to specific observation intervals and do not consider how these effects may change as a function of observation intervals, and (4) unlike continuous time models, the CLPM does not define a specific model of change over time. This is not intended to be discouraging. Instead, we have intended to show that the understanding of mediation processes is advancing and to highlight the potential for new ways of understanding old data. The substantive example, and Fig. 8.6 in particular, highlights the potential for understanding data in a completely novel way. We hope that file drawers of old data are reexamined using the perspectives offered in this chapter.

8.6 Conclusion

This chapter was not so much about the time points in our studies when data are collected. It was more about the in-between time points when data are *not* collected. Most longitudinal approaches in the mediation literature are implicitly based on an assumption of discrete time. This assumption was contrasted with a perspective in which time is sampled from a continuous set of possible times. In Parts I and II, a foundation for discussion was presented, primarily aimed at introducing continuous time concepts and demonstrating a continuous time model that can be considered a mediation model. Part III presented three questions which are easily addressed when mediation is considered with cross-sectional data or using a discrete time model such as the CLPM: *What are direct and indirect effects? What is complete mediation? What is a mediation model?* All were intended to be fundamental and simple enough questions that their answers would involve concepts that transcend specific modeling approaches. The answers began with difficulty in labeling parameters from either discrete or continuous time model "direct" effects. The difficulty continued with confronting the fact that the continuous time perspective further undermines the use of terms such as "complete" mediation because of the way relations between variables can vary with the observation interval. Finally, we raised more questions than answers with respect to "What is a mediation model," as mediation analysis can never be thought of as fitting a particular model or set of models. The possibilities are as varied as the possible ways in which change can occur. The questions raised in Part III serve as a reminder that time is unlike any other variable that is collected, and our perspective is limited if time is considered only implicitly.

References

Baron, R. M., & Kenny, D. A. (1986). The moderator-mediator variable distinction in social psychological research: Conceptual, strategic, and statistical considerations. *Journal of Personality and Social Psychology, 51*, 1173–1182. https://doi.org/10.1037/0022-3514.51.6.1173

Box, G. E. P. (1976). Science and statistics. *Journal of the American Statistical Association, 71*, 791–799. https://doi.org/10.1080/01621459.1976.10480949

Cohen, J., Cohen, P., West, S. G., & Aiken, L. S. (2003). *Applied multiple regression/ correlation analysis for the behavioral sciences*. New York: Routledge.

Cole, D. A., Martin, J. M., Peeke, L. A., Seroczynski, A. D., & Fier, J. (1999a). Children's over- and underestimation of academic competence: A longitudinal study of gender differences, depression, and anxiety. *Child Development, 70*(2), 459–473. https://doi.org/10.1111/1467-8624.00033

Cole, D. A., Martin, J. M., & Powers, B. (1997). A competency-based model of child depression: A longitudinal study of peer, parent, teacher, and selfevaluations. *Journal of Child Psychology and Psychiatry, 38*(5), 505–514. https://doi.org/10.1111/j.1469-7610.1997.tb01537.x

Cole, D. A., & Maxwell, S. E. (2003). Testing mediational models with longitudinal data: Questions and tips in the use of structural equation modeling. *Journal of Abnormal Psychology, 112*, 558–577. https://doi.org/10.1037/0021-843X.112.4.558

Cole, D. A., Peeke, L., Dolezal, S., Murray, N., & Canzoniero, A. (1999b). A longitudinal study of negative affect and self-perceived competence in young adolescents. *Journal of Personality and Social Psychology, 77*(4), 851. https://doi.org/10.1037/0022-3514.77.4.851

Cole, D. A., Peeke, L. G., Martin, J. M., Truglio, R., & Seroczynski, A. D. (1998). A longitudinal look at the relation between depression and anxiety in children and adolescents. *Journal of Consulting and Clinical Psychology, 66*(3), 451–460. https://doi.org/10.1037/0022-006X.66.3.451

Deboeck, P. R., & Boulton, A. J. (2016). Integration of stochastic differential equations using structural equation modeling: A method to facilitate model fitting and pedagogy. *Structural Equation Modeling, 23*, 888–903. https://doi.org/10.1080/10705511.2016.1218763

Deboeck, P. R., Nicholson, J. S., Kouros, C., Little, T. D., & Garber, J. (2015). Integrating developmental theory and methodology: Using derivatives to articulate change theories, models, and inferences. *Applied Developmental Science, 19*(4), 217–231. https://doi.org/10.1080/10888691.2015.1021924

Deboeck, P. R., & Preacher, K. J. (2015). No need to be discrete: A method for continuous time mediation analysis. *Structural Equation Modeling, 23*, 61–75. https://doi.org/10.1080/10705511.2014.973960

Dobson, K. S. (1985). The relationship between anxiety and depression. *Clinical Psychology Review, 5*, 307–324. https://doi.org/10.1016/0272-7358(85)90010-8

Gollob, H. F., & Reichardt, C. S. (1987). Taking account of time lags in causal models. *Child Development, 58*, 80–92. https://doi.org/10.2307/1130293

Gollob, H. F., & Reichardt, C. S. (1991). Interpreting and estimating indirect effects assuming time lags really matter. In L. M. Collins & J. L. Horn (Eds.), *Best methods for the analysis of change: Recent advances, unanswered questions, future directions* (pp. 243–259). Washington, DC: American Psychological Association. https://doi.org/10.1037/10099-015

Hamaker, E. L., Kuiper, R. M., & Grasman, R. P. (2015). A critique of the cross-lagged panel model. *Psychological Methods, 20*(1), 102–116. https://doi.org/10.1037/a0038889

Harter, S. (1985). *Manual for the self-perception profile for children (revision of the perceived competence scale for children)*. University of Denver.

Kendall, P. C., & Brady, E. U. (1995). Comorbidity in the anxiety disorders of childhood: Implications for validity and clinical significance. In K. D. Craig & K. S. Dobson (Eds.), *Banff international behavioral science series. Anxiety and depression in adults and children* (pp. 3–36). Thousand Oaks, CA: Sage Publications.

Kovacs, M. (1981). Rating scales to assess depression in school-aged children. *Acta Paedopsychiatrica: International Journal of Child & Adolescent Psychiatry, 46*, 305–315.

Kovacs, M. (1982). *The Children's Depression Inventory: A self-rating depression scale for school-aged youngsters*. Unpublished manuscript, University of Pittsburgh.

MacKinnon, D. P. (2008). *Introduction to statistical mediation analysis*. Mahwah, NJ: Erlbaum.

Maxwell, S. E., & Cole, D. A. (2007). Bias in cross-sectional analyses of longitudinal mediation. *Psychological Methods, 12*, 23–44. https://doi.org/10.1037/1082-989X.12.1.23

McArdle, J. J. (2009). Latent variable modeling of differences and changes with longitudinal data. *Annual Review of Psychology, 60*, 577–605. https://doi.org/10.1146/annurev.psych.60.110707.163612

Neale, M. C., Hunter, M. D., Pritikin, J. N., Zahery, M., Brick, T. R., Kirkpatrick, R. M., et al. (2016). OpenMx 2.0: Extended structural equation and statistical modeling. *Psychometrika, 80*(2), 535–549. https://doi.org/10.1007/s11336-014-9435-8

Oud, J. H. L., & Jansen, R. A. R. G. (2000). Continuous time state space modeling of panel data by means of SEM. *Psychometrika, 65*, 199–215. https://doi.org/10.1007/BF02294374

Preacher, K. J. (2015). Advances in mediation analysis: A survey and synthesis of new developments. *Annual Review of Psychology, 66*, 825–852. https://doi.org/10.1146/annurev-psych-010814-015258

Preacher, K. J., & Kelley, K. (2011). Effect size measures for mediation models: Quantitative strategies for communicating indirect effects. *Psychological Methods, 16*, 93–115. https://doi.org/10.1037/a0022658

R Core Team. (2015). R: *A language and environment for statistical computing*. Vienna: R Foundation for Statistical Computing. Retrieved from http://www.R-project.org/

Reynolds, C. R., & Richmond, B. O. (1985). *Revised children's manifest anxiety scale*. Los Angeles: Western Psychological Services.

Selig, J. P., & Preacher, K. J. (2009). Mediation models for longitudinal data in developmental research. *Research in Human Development, 6*, 144–164.

Voelkle, M. C., Oud, J. H. L., Davidov, E., & Schmidt, P. (2012). An SEM approach to continuous time modeling of panel data: Relating authoritarianism and anomia. *Psychological Methods, 17*, 176–192. https://doi.org/10.1037/a0027543

Chapter 9
Stochastic Differential Equation Models with Time-Varying Parameters

Meng Chen, Sy-Miin Chow, and Michael D. Hunter

9.1 Introduction

The world we live in is full of complexities, and we as humans are no exception. Our bodies are composed of trillions of cells, and yet we each function as an integrated unit. We move, think, learn, socialize, and thrive. Our actions could reflect a conscious effort, or an emergent reaction to—or interaction with—changes in the environment (Gottlieb et al. 1996; Newtson 1993; Newtson et al. 1987). In other words, our minds and bodies are capable of *self-organization*.

Self-organization is a process through which orderliness emerges from seeming disorderliness (Bosma and Kunnen 2011; Kelso 1995a; Lewis and Ferrari 2001; Magnusson and Cairns 1996). For instance, a simple motion of the human body involves approximately 10^2 muscles, 10^3 joints, and 10^{14} cells, yet the physical movement of the human body can be characterized by relatively few dimensions

Electronic Supplementary Material The online version of this article (https://doi.org/10.1007/978-3-319-77219-6_9) contains supplementary material, which is available to authorized users.

M. Chen (✉) · S.-M. Chow
Department of Human Development and Family Studies, Pennsylvania State University, University Park, PA, USA
e-mail: mxc681@psu.edu; quc16@psu.edu

M. D. Hunter
School of Psychology, Georgia Institute of Technology, Atlanta, GA, USA
e-mail: mhunter43@gatech.edu

K. van Montfort et al. (eds.), *Continuous Time Modeling in the Behavioral and Related Sciences*, https://doi.org/10.1007/978-3-319-77219-6_9

(Bertenthal 2007; Turvey 1990). Thus, our body organizes complex systems into simple patterns described by comparatively few elements. Mathematically, the notion of self-organization may be represented as a process through which a system transitions through qualitatively distinct patterns through continuous changes in the parameters that govern the system. One possibility for modeling self-organization in a dynamic system is to allow critical parameters that lead to disparate dynamics in the system to show continuous variations over time (e.g., continuous changes in the angle of an individual's elbows to allow the individual to lift a heavy object). Kitagawa (1998) used the term "self-organizing" models to describe models displaying such properties.

9.1.1 Continuous-Time Models with Time-Varying Parameters (TVPs) as a Way to Represent Self-Organization and Multi-Time Scale Processes

Dynamical systems modeling offers a framework for studying processes that undergo self-organization. The key characteristic of any dynamical system is that the complex behaviors that the system manifests can be captured with relatively simple rules or patterns (Nowak and Lewenstein 1994). This line of thinking has considerable appeal to social and behavioral scientists. In this chapter, we propose a stochastic differential equation (SDE) modeling framework with time-varying parameters (TVPs) as a way to capture self-organizational dynamical systems in continuous time.

Incorporating TVPs into a dynamic model of choice provides one way of representing self-organization as well as multi-time scale processes in that the TVPs generally change at distinct time scales from the time scales of other key variables (i.e., endogenous variables) in the system. The notion of a multi-time scale dynamic process is by no means novel to researchers in the physical, social, and behavioral sciences. Indeed, changes happening in the human body can be captured at multiple levels and across multiple time scales: from neural networks that fire every few milliseconds to cognitive and behavioral processes that progress in seconds, to even longer-term processes like personality traits that develop through the life span. To this end, Newell (1990) proposed a time scale of human actions that organized topics of interest in the field of psychology into a hierarchy of levels of analysis (see Table 9.1). According to Newell (1990), the biological band consists of neural activities, which happen on the scale of milliseconds. In the cognitive band, cognitive operations—such as recognizing a person approaching in order to engage in a conversation—happen on the scale of seconds. The next hierarchy consists of the rational band with tasks such as decision-making, which span minutes or hours. At the highest level, the social band encompasses social processes such as interactions with others and forming relationships, which may unfold over days, weeks, months, or even longer (Bertenthal 2007). Individual and

Table 9.1 Newell's table: time scale of human action

Scale (s)	Time units	System	World (theory)
10^7	Months		Social band
10^6	Weeks		
10^5	Days		
10^4	Hours	Task	Rational band
10^3	10 min	Task	
10^2	Minutes	Task	
10^1	10 s	Unit task	Cognitive band
10^0	1 s	Operations	
10^{-1}	100 ms	Deliberate act	
10^{-2}	10 ms	Neural circuit	Biological band
10^{-3}	1 ms	Neuron	
10^{-4}	100 μs	Organelle	

life span development, too, can be conceived as a multi-time scale process that interweaves more gradual developmental changes and short-term fluctuations or intraindividual variability (Baltes and Nesselroade 1979; Kelso 1995b; Ram and Gerstorf 2009)—a process described by Nesselroade (1991) as the warp and woof of the developmental fabric.

Due in part to the scarcity of data that span multiple bands and methodological difficulties in integrating multiple processes with distinct time scales, applications of dynamical systems concepts in the behavioral sciences have hitherto been limited to the representation of phenomena and activities within a particular band in Newell's table (see, e.g., Vallacher and Nowak 1994). Data availability, however, is no longer an insurmountable hurdle. The past decade has evidenced an increased popularity of studies aimed at collecting ecological momentary assessments. In these studies, participants are measured in real time in their natural environment and often for many repeated occasions (Shiffman et al. 2008). This, along with advances in data collection technology such as wearable assessment tools and smartphone-based surveys, affords renewed opportunities for modeling multi-time scale processes as they unfold in *continuous time*.

In continuous-time models, time is denoted by real numbers, as opposed to integers. Although conceptualizing and representing dynamic phenomena in discrete time has many practical advantages from a computational standpoint, modeling a dynamical process in continuous time offers several other advantages (Molenaar and Newell 2003; Oud and Singer 2008; Voelkle and Oud 2013). One of these advantages is that a continuous representation of time mirrors the phenomena of interest: time and human behaviors that unfold in time are continuous in nature. That is, even though empirical measurements of human dynamics are always observed at discrete time points, human behaviors do not cease to exist between successive measurements. Changes are unfolding at any particular time point even when observed data may not be available at those time points. Another important advantage of continuous-time modeling is that it can readily accommodate irregularly spaced observations (Chow et al. 2016; Oud 2007b; Voelkle and Oud 2013).

Extant applications involving models with TVPs have been restricted to formulation in discrete time. Examples of models with TVPs in the discrete-time modeling framework include time-varying autoregressive moving average model (Bringmann et al. 2017; Tarvainen et al. 2006; Weiss 1985), the local linear trend model (Harvey 2001), stochastic regression model (Pagan 1980), time-varying cyclic models (Chow et al. 2009), and dynamic factor analysis models with TVPs (Chow et al. 2011; Del Negro and Otrok 2008; Molenaar 1994; Molenaar et al. 2009; Stock and Watson 2008). In the continuous-time realm, Oud and Jansen (2000) illustrated possibilities of incorporating TVPs into continuous-time models formulated as structural equation models. However, the TVPs are constrained to show deterministic (i.e., predictable), piecewise functions of time, as opposed to stochastic changes with uncertainties, as assumed in the present chapter. Overall, work for representing TVPs in continuous time remains scarce.

In sum, our work thus extends previous work on dynamical systems modeling in a number of ways. First, we extend the damped oscillator model—a benchmark dynamical systems model of human behaviors with time-invariant parameters (Boker and Graham 1998; Oud 2007a)—by allowing the parameters to be time-varying. Second, we propose and discuss several approximation functions that can be used to represent the dynamics of TVPs even when their underlying change mechanisms are unknown. Third, we present a continuous-time framework and provide several illustrative examples of how TVPs may be represented in continuous time. Fourth, we present a set of algorithms based on the continuous-discrete Kalman filter (Bar-Shalom et al. 2001; Chow et al. 2017; Kulikov and Kulikova 2014; Kulikova and Kulikov 2014) that allows the simultaneous estimation of the dynamics of a system and all associated TVPs and time-invariant parameters. Finally, we illustrate how our proposed models can be fitted using a statistical package, `dynr` (Dynamic Modeling in R; Ou et al. 2016, 2017), in `R`.

The remainder of this chapter is organized as follows. We begin by describing the SDE framework on which our proposed continuous-time models with TVPs are based. We then present several time-varying extensions of the stochastic damped linear oscillation model (Oud and Singer 2008; Voelkle and Oud 2013), with some hypothetical scenarios of why and how TVPs may be used to capture multi-time scale oscillatory processes in continuous time. Illustrative simulation results are then used to highlight the importance of incorporating TVPs and the utility of the proposed approach. We then present results from a formal Monte Carlo simulation study in which the performance of the proposed estimation approach is examined in further detail and the usefulness of information criterion measures such as the Bayesian information criterion (Schwarz 1978) and the Akaike information criterion (Akaike 1973) as model selection tools in models with TVPs is investigated. Finally, we include a brief illustration of the code for specifying the proposed TVP models in `dynr`.

9.2 A Stochastic Differential Equation Framework

Differential equation models capture a system's change processes in terms of rates of change involving variables and their derivatives. Our hypothesized system of SDEs is assumed to be in the general form of:

$$d\boldsymbol{\eta}_i(t) = \boldsymbol{f}\Big(\boldsymbol{\eta}_i(t),\ t,\ \boldsymbol{x}_i(t)\Big)dt + \mathbf{G}d\mathbf{W}_i(t). \tag{9.1}$$

This system is represented by a deterministic part, $\boldsymbol{f}\left(\boldsymbol{\eta}_i(t), t, \boldsymbol{x}_i(t)\right)dt$, and a stochastic part, $\mathbf{G}d\mathbf{W}(t)$, where i is the smallest independent unit of analysis (e.g., person, dyad; $i = 1, \ldots, n$), t is an index for time that can take on any real number, $\boldsymbol{\eta}_i$ is a vector of latent system variables, $\boldsymbol{x}_i$ represents the set of covariates that may influence the system, and $\boldsymbol{f}$ is a vector of drift functions that governs the deterministic (i.e., predictable given perfect knowledge of the previous states of the system and the true parameters) portion of the changes in the system variables. Without the stochastic part, the system reduces to an ordinary differential equation (ODE) model. The term $\mathbf{W}_i$ denotes a vector of standard Wiener processes (referred to herein as the process noise) whose differentials, $d\mathbf{W}_i(t)$, over the time interval, dt, are Gaussian distributed and are characterized by variances that increase linearly with increasing dt. $\mathbf{G}$ is a diffusion matrix containing matrix square root of the process noise variance-covariance matrix. Thus, the SDE model in (9.1) can be used to represent latent processes that develop and fluctuate over time.

The initial conditions (ICs) for the SDEs are defined explicitly to be the latent variables at an initial time point, t_1 (e.g., the first observed time point), denoted as $\boldsymbol{\eta}_i(t_{i,1})$ and are specified to be normally distributed with means $\boldsymbol{\mu}_{\eta_1}$ and covariance matrix, $\boldsymbol{\Sigma}_{\eta_1}$. The time interval between time $t_{i,j}$ and $t_{i,j+1}$ is denoted by $\Delta t_{i,j} = t_{i,j+1} - t_{i,j}$.

The corresponding measurement model, which specifies the relations between the underlying latent processes and their observed manifestations measured at discrete and person-specific occasion j, $t_{i,j}$ ($j = 1, \ldots, T_i$; $i = 1, \ldots, n$) is expressed as:

$$\begin{aligned}\mathbf{y}_i(t_{i,j}) &= \boldsymbol{\tau} + \boldsymbol{\Lambda}\boldsymbol{\eta}_i(t_{i,j}) + \mathbf{A}\boldsymbol{x}_i(t_{i,j}) + \boldsymbol{\epsilon}_i(t_{i,j})\\ \boldsymbol{\epsilon}_i(t_{i,j}) &\sim N\left(\mathbf{0}, \mathbf{R}\right),\end{aligned} \tag{9.2}$$

where $\boldsymbol{\tau}$ is a vector of intercepts, $\boldsymbol{\Lambda}$ is the factor loading matrix, $\mathbf{A}$ is the regression matrix associated with the vector of covariates $\boldsymbol{x}_i$, and $\boldsymbol{\epsilon}_i(t_{i,j})$ is a vector of Gaussian distributed measurement errors.

Although both follow a Gaussian distribution, the process noise in $\mathbf{W}$ is different from the measurement noise in $\boldsymbol{\epsilon}$ in several important ways. As noted, the former contains "dynamical" or "process" noise, and it captures stochastic shocks to the system variables that continue to influence the underlying states of these variables over time. The latter comprises "measurement" errors whose influences are restricted to one and only one measurement occasion, namely, at time $t_{i,j}$. As such, the process noise—but not the measurement noise—helps capture additional sources of uncertainties that affect the latent dynamics of the system at the current as well as future time points. In the study of individual development and more broadly in behavioral sciences as a whole, it is unreasonable to assume that we have collected information about all the factors that influence a certain dynamic process. Referring back to Table 9.1, any change on any level of analysis is subjected to the influence of all the other levels in the system. The additional process noise component under the SDE framework, compared to the ODE framework, allows for the possibility of incorporating the influence of other unknown and unmodeled (by the deterministic drift functions, $\boldsymbol{f}(.)$) sources of disturbances to the system.

9.3 Estimation Procedures

The proposed SDE models with TVPs were fitted using a set of algorithms based on the continuous-discrete extended Kalman filter (CDEKF; Kulikov and Kulikova 2014) available in `dynr` (Chow et al. 2017; Ou et al. 2017). The model fitting procedure can be summarized into three steps: a filtering step, a parameter estimation step, and a smoothing step. Estimates of $\boldsymbol{\eta}_i(t_{i,j})$ obtained with observations from up to time $t_{i,j-1}$, up to time $t_{i,j}$, and the entire time string $t_{i,T}$ are referred to as the predicted, filtered, and smoothed latent variable estimates. We define the following terms: (1) $\mathbf{Y}_i(t_{i,j}) \triangleq \{\boldsymbol{y}_i(t_{i,1}), \ldots, \boldsymbol{y}_i(t_{i,j})\}$, all observed measurements from $t_{i,1}$, $\ldots$, up to $t_{i,j}$; (2) $\boldsymbol{\eta}_i(t_{i,j}|t_{i,j-1}) \triangleq E(\boldsymbol{\eta}_i(t_{i,j})|\mathbf{Y}_i(t_{i,j-1}))$, the predicted means of the latent variables at time $t_{i,j}$ conditional on $\mathbf{Y}_i(t_{i,j-1})$; (3) $\mathbf{P}_i(t_{i,j}|t_{i,j-1}) \triangleq Cov[\boldsymbol{\eta}_i(t_{i,j})|\mathbf{Y}_i(t_{i,j-1})]$, the covariance matrix of the latent variables at time $t_{i,j}$ conditional on $\mathbf{Y}_i(t_{i,j-1})$; (4) $\boldsymbol{\eta}_i(t_{i,j}|t_{i,j}) \triangleq E(\boldsymbol{\eta}_i(t_{i,j})|\mathbf{Y}_i(t_{i,j}))$, the filtered mean of the latent variables at time $t_{i,j}$ conditional on $\mathbf{Y}_i(t_{i,j})$; and (5) $\mathbf{P}_i(t_{i,j}|t_{i,j}) \triangleq Cov[\boldsymbol{\eta}_i(t_{i,j})|\mathbf{Y}_i(t_{i,j})]$, filtered covariance matrix of the latent variables at time $t_{i,j}$ conditional on $\mathbf{Y}_i(t_{i,j})$. If the entire time series of observations is available prior to model fitting, as is often the case in the study of human behavior, then all observations from each time string across all time points, $\mathbf{Y}_i(T_i)$, can be used to perform smoothing on the latent variable estimates to yield estimates of the means and covariance matrix of $\boldsymbol{\eta}_i(t_{i,j})$ conditional on all observations in $\mathbf{Y}_i(T_i)$, namely, $E(\boldsymbol{\eta}_i(t_{i,j})|\mathbf{Y}_i(T_i))$ and $\mathbf{P}_i(t_{i,j}|T_i) = Cov[\boldsymbol{\eta}_i(t_{i,j})|\mathbf{Y}_i(T_i)]$.

Step 1. Filtering The CDEKF is a nonlinear, continuous-time counterpart of the Kalman filter which assumes that the underlying dynamic processes unfold in continuous time but the associated empirical measurements are taken at discrete time points, and are used to estimate the latent variables $\boldsymbol{\eta}$ in Eqs. (9.1)–(9.2). For this step, the parameter values are assumed to be known and fixed ($\hat{\boldsymbol{\theta}}$). Parameter estimation for $\boldsymbol{\theta}$ through optimizing a log-likelihood function is described in Step 2. Following the general procedure of the Kalman filter, the CDEKF involves a series of prediction steps and correction steps over all subjects and time points to yield filtered estimates $\boldsymbol{\eta}_i(t_{i,j}|t_{i,j})$ and $\mathbf{P}_i(t_{i,j}|t_{i,j})$. Unlike the traditional prediction step in the discrete-time extended Kalman filter (EKF), the first step in the CDEKF integrates the SDE model from time $t_{i,j-1}$ to the next measured time $t_{i,j}$ by means of numerical integration, which in `dynr`, the specific R software package used for model estimation in the present chapter, is done with a fourth-order Runge–Kutta method. This is an approximation process in which the predicted means of the latent variables of interest in continuous time are obtained by "interpolating" changes at several intermediate, discrete, and equally spaced time points. That is, to deduce the changes that occur from time $t_{i,j-1}$ to the next measured time at $t_{i,j}$, interpolations (i.e., or intermediate values of the latent variables) are obtained at four equally spaced time intervals and then smoothing over the changes that occur over these four time intervals (Chow et al. 2007). This process yields numerical solutions to the following ODEs, with $\boldsymbol{\eta}_i(t_{i,j-1}|t_{i,j-1})$ and $\mathbf{P}_i(t_{i,j-1}|t_{i,j-1})$ as the respective initial conditions:

$$\frac{d\hat{\boldsymbol{\eta}}_i(t)}{dt} = \boldsymbol{f}\left(\hat{\boldsymbol{\eta}}_i(t), t, \boldsymbol{x}_i(t)\right), \tag{9.3}$$

$$\frac{d\mathbf{P}_i(t)}{dt} = \mathbf{F}(t, \hat{\boldsymbol{\eta}}_i(t))\mathbf{P}_i(t) + \mathbf{P}_i(t)\mathbf{F}(t, \hat{\boldsymbol{\eta}}_i(t))^\top + \mathbf{Q}. \tag{9.4}$$

Here $\hat{\boldsymbol{\eta}}_i(t)$ is the vector of predicted latent variables for subject i at time t; $\mathbf{P}_i(t)$ is the covariance matrix of prediction errors on the latent variables at time t, $Cov[\boldsymbol{\eta}_i(t) - \hat{\boldsymbol{\eta}}_i(t)]$. $\mathbf{F}(t, \hat{\boldsymbol{\eta}}_i(t))$ is the Jacobian matrix that consists of first-order partial derivatives of $\boldsymbol{f}$ with respect to each element of the latent variable vector, $\boldsymbol{\eta}_i(t)$, evaluated at $\boldsymbol{\eta}_i(t_{i,j-1}|t_{i,j-1})$, and $\mathbf{Q} = \mathbf{G}\mathbf{G}^\top$.

The correction step follows the equations below:

$$\boldsymbol{v}_i(t_{i,j}) = \boldsymbol{y}_i(t_{i,j}) - \left(\boldsymbol{\tau} + \boldsymbol{\Lambda}\boldsymbol{\eta}_i(t_{i,j}|t_{i,j-1}) + \mathbf{A}\mathbf{x}(t_{i,j})\right), \tag{9.5}$$

$$\mathbf{V}_i(t_{i,j}) = \boldsymbol{\Lambda}\mathbf{P}_i(t_{i,j}|t_{i,j-1})\boldsymbol{\Lambda}^\top + \mathbf{R}, \tag{9.6}$$

$$\boldsymbol{\eta}_i(t_{i,j}|t_{i,j}) = \boldsymbol{\eta}_i(t_{i,j}|t_{i,j-1}) + \mathbf{K}(t_{i,j})\boldsymbol{v}_i(t_{i,j}), \tag{9.7}$$

$$\mathbf{P}_i(t_{i,j}|t_{i,j}) = \mathbf{P}_i(t_{i,j}|t_{i,j-1}) - \mathbf{K}(t_{i,j})\boldsymbol{\Lambda}\mathbf{P}_i(t_{i,j}|t_{i,j-1}), \tag{9.8}$$

where $\boldsymbol{v}_i(t_{i,j})$ is a vector of prediction errors and $\mathbf{V}_i(t_{i,j})$ is the prediction error covariance matrix, and $\mathbf{K}(t_{i,j}) = \mathbf{P}_i(t_{i,j}|t_{i,j-1})\boldsymbol{\Lambda}^\top[\mathbf{V}_i(t_{i,j})]^{-1}$ is the Kalman gain function. The latent variable estimates are updated after the correction step, yielding $\boldsymbol{\eta}_i(t_{i,j}|t_{i,j})$ and $\mathbf{P}_i(t_{i,j}|t_{i,j})$. Note that the filtering step as described here involves using the CDEKF for latent variable estimation purposes. As such, the EKF is used in the correction step described in Eqs. (9.5)–(9.8). Other alternatives that do not require the computations of Jacobian matrices, such as the unscented Kalman filter, may also be used (Chow et al. 2007; Julier and Uhlmann 2004; Julier et al. 1995; Wan and Van der Merwe 2001). However, these alternative derivative-free approaches still involve appropriate selection of tuning parameters to optimize model estimation, which may or may not be intuitive to some researchers.

Step 2. Parameter Estimation Parameter estimation is performed by maximizing a log-likelihood function known as the prediction error decomposition (PED) function, computed using $\boldsymbol{v}_i(t_{i,j})$ and $\mathbf{V}_i(t_{i,j})$ as (Chow et al. 2007; Schweppe 1965):

$$\log[f(\mathbf{Y};\boldsymbol{\theta})] = -\frac{1}{2}\sum_{i=1}^{n}\sum_{j=1}^{T_i}\Big[p\log(2\pi) + \log|\mathbf{V}_i(t_{i,j})| + \boldsymbol{v}_i(t_{i,j})^\top \mathbf{V}_i(t_{i,j})^{-1}\boldsymbol{v}_i(t_{i,j})\Big] \tag{9.9}$$

with an optimization procedure of choice, where p is the number of observed variables. Standard errors associated with $\boldsymbol{\theta}$ are computed from the observed information matrix, obtained by computing the negative numerical Hessian matrix of $\log[f(\mathbf{Y}|\boldsymbol{\theta})]$, a matrix of numerical second derivatives of $\log[f(\mathbf{Y}|\boldsymbol{\theta})]$ with respective to the parameters. Wald-type confidence intervals for parameter estimates based on the standard errors an also be obtained.

Information criteria measures such as the AIC (Akaike 1973) and BIC (Schwarz 1978) can then be computed using $\log[f(\mathbf{Y}|\boldsymbol{\theta})]$ as (see Harvey 2001, p. 80):

$$\mathbf{AIC} = -2\log[f(\mathbf{Y}|\boldsymbol{\theta})] + 2q$$

$$\mathbf{BIC} = -2\log[f(\mathbf{Y}|\boldsymbol{\theta})] + q\log\left(\sum_{i}^{n} T_i\right),$$

where q is the number of time-invariant parameters in a model.

Step 3. Smoothing Kalman filtering occurs recursively forward in time by predicting and updating estimates. However, doing so only produces conditional latent variable estimates at a particular time point, $t_{i,j}$, and their corresponding covariance matrix, based on observed information up to time $t_{i,j}$. Since data are usually completely collected in social and behavioral sciences before analyses, these filtered estimates can be further refined using information from the entire observed time series, namely, $\{\boldsymbol{y}_i(t_{i,j});\ j = 1, \ldots, T_i\}$. The fixed interval smoother, in particular, occurs recursively backward in time by incorporating all available information from person i up to time T_i into the updated estimates. Using $\boldsymbol{\eta}_i(t_{i,j}|t_{i,j-1})$, $\mathbf{P}_i(t_{i,j}|t_{i,j-1})$, $\boldsymbol{\eta}_i(t_{i,j}|t_{i,j})$, and $\mathbf{P}_i(t_{i,j}|t_{i,j})$, the smoothing procedure can be implemented for $t_{i,j} = T_i - 1, \ldots 1$ and $i = 1, \ldots n$ as (Bar-Shalom et al. 2001; Chow et al. 2017; Harvey 2001; Shumway and Stoffer 2000):

$$\begin{aligned}\boldsymbol{\eta}_i(t_{i,j}|T_i) &= \boldsymbol{\eta}_i(t_{i,j}|t_{i,j}) + \widetilde{\mathbf{P}}_i(t_{i,j})(\boldsymbol{\eta}_i(t_{i,j+1}|T_i) - \boldsymbol{\eta}_i(t_{i,j+1}|t_{i,j})),\\ \mathbf{P}_i(t_{i,j}|T_i) &= \mathbf{P}_i(t_{i,j}|t_{i,j}) + \widetilde{\mathbf{P}}_i(t_{i,j})(\mathbf{P}_i(t_{i,j+1}|T_i) - \mathbf{P}_i(t_{i,j+1}|t_{i,j}))\widetilde{\mathbf{P}}_i(t_{i,j}),\end{aligned} \tag{9.10}$$

where $\widetilde{\mathbf{P}}_i(t_{i,j}) \triangleq \mathbf{P}_i(t_{i,j}|t_{i,j})\mathbf{F}(t, \hat{\boldsymbol{\eta}}_i(t))^\top[\mathbf{P}_i(t_{i,j+1}|t_{i,j})]^{-1}$. Examination of the smoothed latent variable estimates provides some indication of the dynamics of the latent variables. This step is particularly relevant to the explorations of TVPs, as the TVPs are now incorporated into the dynamic models as latent variables (as will be discussed in the next section), and researchers may not always have fully informed conceptualizations of how exactly the TVPs vary over time.

9.4 Motivating Model

One of the better-known and widely applied differential equation models in psychology is the damped linear oscillator model, a second-order ODE expressed as (Boker and Graham 1998):

$$\frac{d^2\eta_{1i}(t)}{dt^2} = \omega[\eta_{1i}(t) - \mu] + \zeta\frac{d\eta_{1i}(t)}{dt}, \tag{9.11}$$

where $\eta_{1i}(t)$ denotes a univariate, latent process of interest for unit i at any arbitrary time t, $\frac{d\eta_{1i}(t)}{dt}$ denotes the process's first derivative at time t, and $\frac{d^2\eta_{1i}(t)}{dt^2}$ denotes the second derivative at time t. The parameter ω governs the frequency of the oscillations, μ is the set point around which the system oscillates (usually set to 0), and ζ is a damping (when $\zeta < 0$) or amplification (when $\zeta > 0$) parameter that governs changes in the magnitude of the oscillations over time. The oscillatory properties of this model make it suitable for representing intraindividual variability, where the changes are described as "more or less reversible" (Nesselroade 1991, p.215). The damped oscillator model has been used to represent dynamic processes such as emotions (Chow et al. 2005; Deboeck et al. 2008), self-regulation in adolescence substance use (Boker and Graham 1998), and mood regulation in recent widows (Bisconti et al. 2004), among many other examples. In the illustrative examples in this chapter, we consider SDE variations of the damped linear oscillator model with TVPs. The TVPs considered include both the set point parameter, μ, and the damping parameter, ζ.

Every higher-order ODE (i.e., involving higher-order derivatives than just the first derivatives) can be expressed as a system of first-order ODEs. As a special case, the damped oscillator model can also be expressed as a system of two first-order ODEs. In particular, the SDE form of Eq. (9.11) is typically expressed as two first-order differential equations involving $\boldsymbol{\eta}_i(t) = \left[\eta_{1i}(t)\ \eta_{2i}(t)\right]'$ in Itô form as:

$$\begin{aligned} d\eta_{1i}(t) &= \eta_{2i}(t)dt \\ d\eta_{2i}(t) &= \left(\omega\left(\eta_{1i}(t) - \mu\right) + \zeta\eta_{2i}(t)\right)dt + \sigma_p dW(t) \end{aligned} \tag{9.12}$$

where $d\eta_{1i}$ is the first-order differential and $d\eta_{2i}$ is the second-order differential of the scalar latent process η_1 associated with unit i. $dW(t)$ is the differential of a univariate standard Wiener process (i.e., the process noise component in this model), and σ_p is the standard deviation of the process noise. This way of expressing higher-order ODEs as multiple first-order ODEs is a standard practice in ODE modeling to aid expression of higher-order ODEs in first-order vector form. This formulation also is in accordance with the model specification implemented in `dynr`.

The dynamic functions in (9.12) are assumed in the present context to be identified using a single manifest indicator measured at discrete but possibly irregularly spaced time points as:

$$\begin{aligned} y_i(t_{i,j}) &= \eta_{1i}(t_{i,j}) + \epsilon_i(t_{i,j}), \\ \epsilon_i(t_{i,j}) &\sim N\left(0, \sigma_\epsilon^2\right), \end{aligned} \tag{9.13}$$

namely, a univariate special case of Eq. (9.2) where $\boldsymbol{\tau}$ is zero and $\boldsymbol{\Lambda}$ contains only a single scalar loading fixed at unity on the latent process η_1 with the other loadings fixed at zero, and there are no covariates. Here, $y_i(t_{i,j})$ is the observed manifest indicator for unit i at discrete time, $t_{i,j}$; and $\epsilon_i(t_{i,j})$ is the corresponding measurement error with variance σ_ϵ^2.

To model the dynamics of the TVPs, we treat the TVPs as unknown latent variables to be inserted into $\boldsymbol{\eta}_i(t)$ and estimated with other latent variables in the system. Doing so would, in most cases, entail SDEs that are nonlinear because they involve the interaction between at least two latent variables. For instance, in Illustrative Example 3, we consider the scenario of time-varying damping/amplification parameter $\zeta(t)$. In such a scenario, even when the true model is known and is fitted directly to the data, the latent variable vector now includes a TVP as $\boldsymbol{\eta}_i(t) = \left[\eta_{1i}(t)\ \eta_{2i}(t)\ \zeta_i(t)\right]'$, and $\boldsymbol{\Lambda}$ becomes $\left[1\ 0\ 0\right]$. Consequently, the revised Eq. (9.12), when written in the form of Eq. (9.1), now involves the interaction between two latent variables, $\zeta_i(t)$ and $\eta_{2i}(t)$. Thus, the dynamic model of interest now becomes nonlinear in $\boldsymbol{\eta}_i(t)$, even though the damped oscillator function, conditional on $\zeta_i(t)$, is linear in form. The CDEKF algorithm as implemented in `dynr`, the `R` package used for model fitting purposes in the present study, can readily handle this form of nonlinearity.

The CDEKF algorithm can also handle any linear or nonlinear multivariate latent processes as alternatives to the damped oscillator model shown in Eqs. (9.11) and (9.12), provided that the functions are at least first-order differentiable with respect to the latent variables and are characterized by Gaussian-distributed process noises. These latent processes, in turn, are allowed to be apprehended through multiple observed indicators, but the measurement functions have to be linear, with Gaussian-distributed measurement errors. The fact that the measurement model is linear with Gaussian-distributed measurement noise also renders the one-step-ahead prediction errors, $\boldsymbol{v}_i(t_{i,j})$, Gaussian-distributed conditional on values of $\boldsymbol{\eta}_i(t_{i,j})$. Thus, estimation of the remaining time-invariant parameters in the extended damped linear oscillator model with TVPs can be performed by optimizing the prediction error decomposition function in (9.9).

9.5 Approximation Functions for TVPs

Across many scientific disciplines, it is often impossible to know the true underlying model for the phenomenon of interest (MacCallum 2003). Devising models for TVPs is even more challenging due to the scarcity of knowledge from the literature on plausible forms of their functional dynamics. Here we propose two flexible models as approximations for any TVP. The approximation functions considered include an Ornstein–Uhlenbeck (O-U) model and a stochastic noise model.

The O-U process, which is represented by a first-order SDE (Eq. (9.14)) maybe viewed as a continuous-time counterpart of a first-order autoregressive model in discrete-time modeling, and has been utilized to represent processes such as emotion regulation (Oravecz et al. 2009), which have the tendency to show exponential return to an equilibrium or "home base" after they are moved away from it by some random shocks or process disturbances. The O-U model for a TVP $\theta(t)$ is expressed as

$$d\theta_i(t) = \beta(\theta_i(t) - \theta_{0_{ou}})dt + \sigma_{ou} dW(t) \tag{9.14}$$

where $\theta_{0_{ou}}$ is the equilibrium or attractor of the O-U process; β, constrained to be greater than or equal to 0, is the velocity that the process returns to $\theta_{0_{ou}}$; and σ_{ou} is the diffusion parameter that governs the standard deviation of the random process noise.

Another approximation function considered in the present chapter is the stochastic noise model, written as:

$$d\theta_i(t) = \sigma_s dW(t). \tag{9.15}$$

In this model, no deterministic drift function is included. Instead, any changes in θ are posited to be driven completely by the random shocks captured in $dW(t)$. Thus, this function may be suited for capturing processes that show ongoing, noise-like shifts without the tendency to return to an equilibrium.

Both the O-U model and the stochastic noise model have specialized features that render them well-suited for representing change phenomena with particular characteristics. First, both functions are relatively simple. Second, both are built on the notion that the true value of the TVP, though unknown and possibly varying over time, is likely to vary much more gradually than the other latent processes in $\boldsymbol{\eta}_i(t)$. In particular, they both include as a special case the possibility of a completely stable, time-invariant parameter when there is no process noise in the system (namely, when $\sigma_{ou} = \sigma_s = 0$). In the O-U model, if $\theta_i(t)$ is a time-invariant parameter, its value may be set to equal its initial condition (IC) value, $\theta_i(t_{i,1})$, with $\theta_{0_{ou}}$, β, σ_{ou}, and the corresponding variance for $\theta_i(t_{i,1})$ in the IC covariance matrix, $\boldsymbol{\Sigma}_{\eta_1}$, all set to 0. Consequently, θ would just be a freely estimated (time-invariant) parameter, denoted herein as μ_{θ_1}. μ_{θ_1} is one of the entries in the IC latent variable vector, $\boldsymbol{\mu}_{\eta_1}$, and its value is estimated by optimizing the prediction error decomposition function. Other less restrictive special cases of the O-U process with $\beta > 0$ allow for transient deviations from an otherwise stable value as captured by $\theta 0_{ou}$. In a similar vein, a time-invariant parameter would be represented in the stochastic noise model by the value of μ_{θ_1} in $\boldsymbol{\mu}_{\eta_1}$, with σ_s and the corresponding IC variance of $\theta_i(t_{i,1})$ in $\boldsymbol{\Sigma}_{\eta_1}$ set to 0.

Despite the fact that the O-U model and the stochastic noise model are likely imperfect or misspecified functions of the true change mechanisms of the TVPs, the smoothing procedures summarized in Eq. (9.10) can still be used to provide smoothed estimates of the TVPs conditional on the observed data. Visual inspection of such smoothed estimates can, in turn, provide insights into the nature and over-time dynamics of the TVPs, as we show in the context of three illustrative examples.

9.6 Model Fitting via `dynr`

The model fitting was done in `R` (R Core Team 2016) through the package `dynr` (Ou et al. 2016). The `dynr` package provides an interface between `R` and the `C` language, where the models can be specified in `R` and the computations are carried out in `C`. The following components need to be provided for successful model fitting: the measurement model (with `prep.measurement()`), the measurement and dynamic noise components (with `prep.noise()`), the initial values of the latent state variables and their covariance matrix (with `prep.initial()`), and the dynamic model (with `prep.formulaDynamics()`). Once all these components are specified, they are combined into a `dynr` model object with `dynr.model()` for parameter estimation with this estimation done using the `dynr.cook()` function. There are also options to include transformation functions for parameters with restrictive ranges and upper/lower bounds of plausible estimated values. Once the estimation and smoothing procedures are complete, summary statistics can be extracted using the `summary()` command. It is also possible to extract parameter estimates using `coef()` and AIC or BIC measures using `AIC()` or `BIC()`. A sample script for one of the illustrative examples is included in the supplementary materials.

9.7 Illustrative Examples

9.7.1 Example 1: Stagewise Shifts in Set Point

For the first example, let us consider a variation of the damped oscillator model with distinct stagelike shifts in the set point parameter μ as the true data generation model. That is, μ in Eq. (9.11) or (9.12) is now expressed as $\mu(t)$, and its value is hypothesized to undergo two stagewise shifts. Denoting the first stage as the "reference stage," we consider a three-stage model for $\mu(t)$ as:

$$\mu_i(t) = \mu_1 + c_2(t)\mu_2 + c_3(t)\mu_3 \tag{9.16}$$

where $c_2(t)$ and $c_3(t)$ are time-dependent and binary indicators of stages 2 and 3 of the TVP values, while μ_2 and μ_3 are the differences in equilibria of stages 2 and 3 compared to that of stage 1 (μ_1). If the stage indicators, $c_2(t)$ and $c_3(t)$, are known and observed with perfect knowledge, then modeling of the time-varying set point, $\mu(t)$, is relatively trivial. However, such knowledge is not always available, and other approximation functions such as the O-U model and the stochastic noise model may have to be used.

Figure 9.1 illustrates the proposed continuous-time process with stagewise changes in set points around which the system of interest oscillates. One example of such a process would be mood swings in individuals with bipolar I or II disorder. Bipolar disorder (I or II) is a severe chronic disorder that is characterized by clear shifts in mood and energy levels between mood episodes, namely, alternating episodes of depression and (hypo)mania. During a depressive episode, an individual would feel sad and unmotivated and have low levels of energy. During a manic

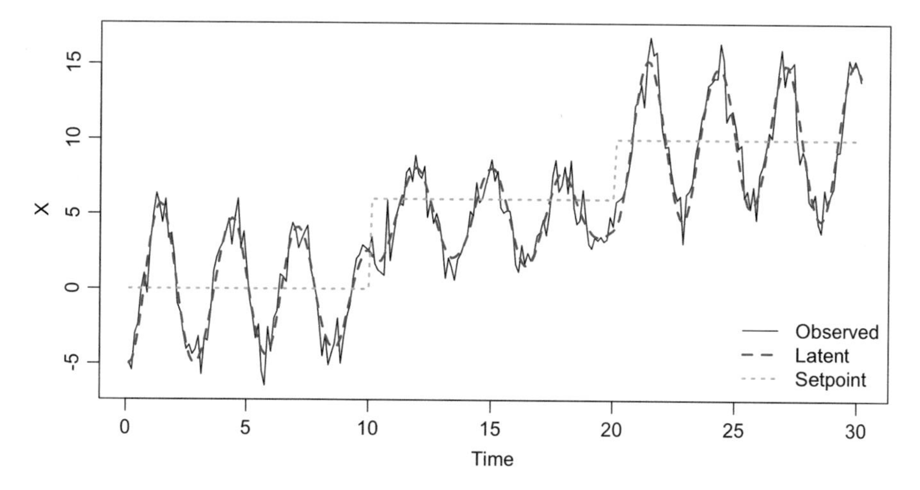

Fig. 9.1 Simulated observed and latent trajectory for one hypothetical subject generated using the damped oscillator model with stagewise shifts in set point considered in illustrative example 1

episode, the individual would feel elevated and excited and have high levels of energy (The National Institute of Mental Health 2016). Thus, on the episodic level, there is a sharp, stagewise change in mood (set point); within each episode, however, the individual's mood would fluctuate from moment to moment (or on an hourly or daily basis) around a set point as a result of environmental stimuli or daily events (Larson et al. 1990; Silk et al. 2011). This process, as well as other stagewise processes mentioned in this chapter, has some similarities to regime-switching models often used to capture psychological effect, in that they both allow the dynamics within each stage/regime to evolve in a continuous fashion, but the changes between stages/regimes are considered to be discrete (Chow et al. 2017; Hamaker and Grasman 2012). However, one key difference between the two approaches is that in modeling stagewise processes as a TVP, one does not need to model the probability of stage shifts explicitly.

Data were simulated for 10 subjects, each with 241 time points ($t_{i,0} = 0$, $t_{i,T} = 30$, $\Delta t = t_{i,j+1} - t_{i,j} = 0.125$ for all $i \in \{1, \ldots, 10\}$ and $j \in \{1, \ldots, 240\}$), using the parameter values specified in Table 9.2 under the assumption that the process of interest follows 3 stages. The sample size configuration was determined from a range of possible combinations of number of subjects and number of time points seen in literature of human behavioral dynamics. Of the studies we reviewed, the number of time points ranged from 50 time points per person in studies involving multiple participants' self-reports over time (e.g., Chow et al. 2004; Chow et al. 2005; Krone et al. 2016) to automated measures for assessing "micro-fluctuations"

Table 9.2 Parameter recovery in illustrative example 1 obtained from fitting the true data generation model and the two approximation models

		Estimate (SE)			
Parameter	True values	Stage-wise model (True)	O-U model	Stochastic noise model	Time invariant model
μ_1	0	−0.038 (0.044)	–	–	–
μ_2	6	5.979 (0.043)	–	–	–
μ_3	10	10.083 (0.044)	–	–	–
ω	−2	−2.000 (0.010)	−1.813 (0.030)	−1.823 (0.031)	−0.845 (0.038)
ζ	−0.1	−0.086 (0.007)	−0.084 (0.022)	−0.088 (0.023)	−0.276 (0.043)
σ_p^2	0.25	0.254 (0.046)	1.191 (0.431)	0.514 (0.416)	12.182 (0.706)
σ_ϵ^2	1	1.031 (0.031)	1.023 (0.031)	1.026 (0.032)	0.987 (0.031)
μ	–				5.509 (0.243)
β	–	–	0.036(0.016)	–	–
$\theta_{0_{ou}}$	–	–	14.707 (4.593)	–	–
σ_{ou}^2	–	–	1.166 (0.208)	–	–
σ_s^2	–	–	–	1.525 (0.233)	–
AIC		7195.03	7781.24	7809.23	8376.09
BIC		7235.54	7821.75	7838.17	8405.03
Computational time		49.74 s	53.75 s	23.25 s	24.31 s

that unfold over a few hundred (e.g., McCarthy et al. 2015) to over 10,000 time points per person (e.g., Kim et al. 2013). Table 9.2 summarizes the model fitting results with the true and approximation models. For illustrative purposes, Table 9.2 also includes results from fitting a stochastic damped oscillator model without TVPs (Eqs. (9.12)–(9.13)). As expected, fitting the true data generation model using the CDEKF and associated algorithms led to satisfactory recovery of all time-invariant parameters. The two approximation models performed reasonably well in recovering most of the parameters, except for an overestimation of the process noise variance, σ_p^2. The assumption of time invariance in parameters led to distorted parameter estimates. For example, the estimated frequency parameter ω and damping parameter ζ were estimated to be -0.360 and -0.761, which were far from the true value of -1 and -0.1, and indicated a very damped process with much slower oscillation compared to the true process. This again highlights the importance of considering TVPs.

The AIC and BIC both indicated that the true data generation model was preferred over the other approximation models. The smoothed estimates of the TVP trajectory for one arbitrary subject based on the three fitted models are plotted in Fig. 9.2a. It can be seen that the O-U and stochastic noise models, despite being misspecified compared to the true data generation model, still do a reasonable job in capturing the overall patterns of change in $\mu(t)$. For comparison purposes, the root-mean-square error (RMSE) obtained from each function for the plotted subject is also included in Fig. 9.2a. RMSE with respect to a TVP of interest $\theta(t)$ is calculated as follows for model m and subject i:

$$\text{RMSE}_{i,m} = \sqrt{\frac{\sum_{j=1}^{T_i} [\hat{\theta}_m(t_{i,j}) - \theta(t_{i,j})]^2}{T_i}} \tag{9.17}$$

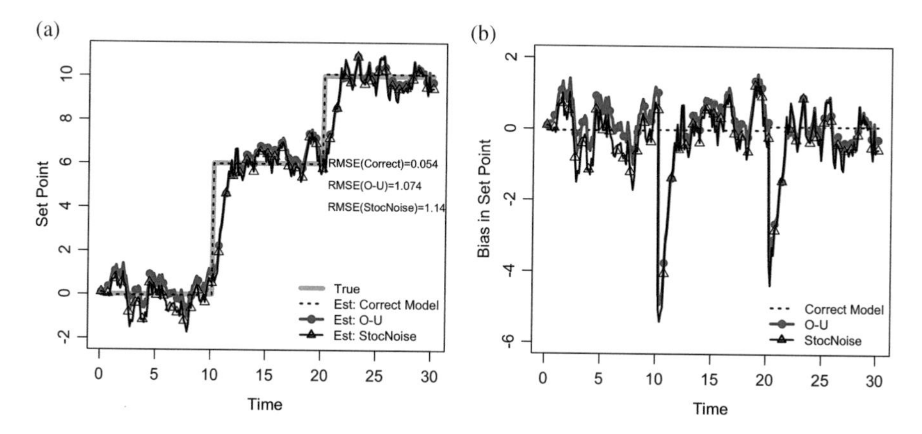

Fig. 9.2 Estimated TVP trajectories for one hypothetical subject in illustrative example 1: (**a**) estimated and true trajectories, with the corresponding RMSEs, (**b**) biases in the estimated TVP values in comparison with the true values

where $\hat{\theta}_m$ is the estimated process of θ by model m. Figure 9.2b shows the biases of the TVP estimates, $(\hat{\theta}(t_{i,j}) - \theta(t_{i,j}))$, for that particular subject at each available time point.

Interestingly, even though the stagewise changes in $\mu(t)$ posited in the data generation model were somewhat distinct from the dynamics postulated in the O-U model (e.g., dictating a return to the home base), the smoothed estimates from the O-U model still yielded reasonable approximations of the true trajectory of $\mu(t)$, with only a minor increase in the RMSE in the estimates for $\mu(t)$. Inspection of the parameter estimates from the O-U model in Table 9.2, particularly the high estimated value of $\theta_{0_{ou}}$, suggested the time-varying $\mu(t)$ was estimated as a process that was trying to approach its equilibrium but was not quite there yet by the end of the observations. Because both of the approximation models have stochastic components, the sudden jumps were captured as part of the influence of the process noise, thus resulting in overestimation of the process noise variance, σ_p^2 (see Table 9.2).

9.7.2 Example 2: Logistic Growth in Set Point

For the second example, we model the set point parameter, $\mu(t)$, as a logistic growth curve:

$$d\mu_i(t) = r\mu_i(t)\Big(1 - b\mu_i(t)\Big)dt. \tag{9.18}$$

A logistic growth curve is used to model population growth with limited resources. In Eq. (9.18), b is the inverse of the carrying capacity (i.e., maximum population size allowed by limited resources), and r defines the growth rate. When the population size is small (i.e., $\mu_i(t)$ close to 0), the per capita growth rate $r(1 - b\mu_i(t))$ will be close to r. On the other hand, when the population size approaches the carrying capacity (i.e., $\mu_i(t)$ gets close to $\frac{1}{b}$), the per capita growth slows down, and eventually the growth rate approaches 0. The logistic growth curve thus takes an elongated "S" shape and has the following two characteristics. First, the system develops in a monotonically increasing fashion. That is, the process does not regress. Second, the system has identifiable upper and lower bounds (Grimm and Ram 2009). Note that the logistic growth model is expressed here in ODE form. An alternative and arguably more familiar way of formulating Eq. (9.18) is to express it as the solution to the ODE in Eq. (9.18), as is used in many standard nonlinear growth curve models (Browne and Du Toit 1991).

An illustration of a damped oscillator whose set point is undergoing logistic growth is shown in Fig. 9.3. Given their characteristics, the logistic curve can thus serve as a reasonable model for many learning and life span developmental processes (Grimm and Ram 2009). Take the scenario of hormonal changes around puberty as an example. There is a shift of overall level of sex hormones from before

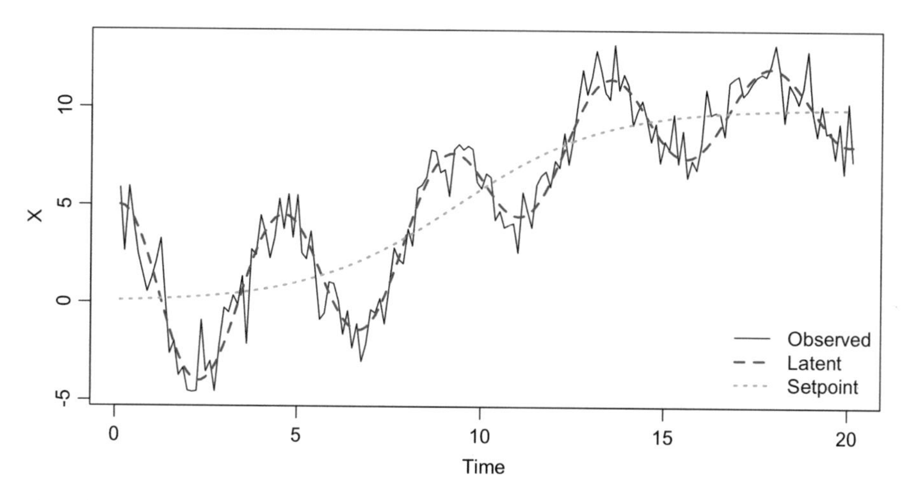

Fig. 9.3 A simulated set of observed and latent trajectories generated under the scenario of logistic growth as in example 2

to after the onset of puberty, though the change may unfold in a gradual way during puberty. On a faster time scale, however, the level of sex hormones has been shown to fluctuate from moment to moment, as subjected to the influence of factors such as time of day and day in the menstrual cycle (Liening et al. 2010; Marceau et al. 2011).

Similar to the previous example, data were simulated for 10 subjects, each with 241 time points ($t_{i,0} = 0$, $t_{i,T} = 30$, $\Delta t = 0.125$), using the parameter values specified in Table 9.3. As in illustrative example 1, both the AIC and BIC preferred the true model to the other approximation models (see Table 9.3). The true parameters were recovered very well when the true data generation model was fitted; the two approximation models also performed reasonably well in recovering all other parameters except for the process noise variance σ_p^2. The model with time-invariant parameters performed badly in parameter estimates. The smoothed estimates of the TVP for one arbitrary subject obtained using the three models are plotted in Fig. 9.4. Compared to the smoothed estimates obtained from fitting the true data generation model, the estimated set point trajectories from the two approximation models still captured the overall sigmoid-shaped changes in $\mu(t)$ reasonably accurately. Greater biases were observed in the TVP estimates obtained using the approximation models between time t of 0 and 15, during which the set point showed more pronounced rises in value as dictated by the logistic growth curve function (see Fig. 9.4). The overall patterns of changes in set point were well approximated, however, despite these biases in TVP estimation.

Table 9.3 Parameter recovery in illustrative example 2 obtained from fitting the true data generation model and the two approximation models

Parameter	True values	Estimate (SE) Logistic model (True)	O-U model	Stochastic noise model	Time invariant model
ω	−1	−1.000 (0.003)	−0.995(0.015)	−0.984 (0.020)	−0.360 (0.037)
ζ	−0.1	−0.090 (0.003)	−0.129 (0.017)	−0.153 (0.024)	−0.761 (0.097)
b	0.1	0.100(0.000)	–	–	–
r	0.5	0.500 (0.002)	–	–	–
σ_p^2	0.01	0.006 (0.003)	0.000 (0.000)	0.000 (0.000)	6.493 (0.826)
σ_ϵ^2	1	1.009 (0.030)	1.008 (0.030)	1.007 (0.030)	0.989 (0.031)
μ	–	–	–	–	7.833 (0.436)
β	–	–	0.053 (0.009)	–	–
$\theta_{0_{ou}}$	–	–	13.286 (1.282)	–	–
σ_{ou}^2	–	–	0.334 (0.043)	–	–
σ_s^2	–	–	–	0.647 (0.081)	–
AIC		6910.26	7331.19	7427.44	7729.81
BIC		6944.99	7371.70	7456.37	7758.75
Computational time		51.46 s	52.44 s	35.40 s	22.82 s

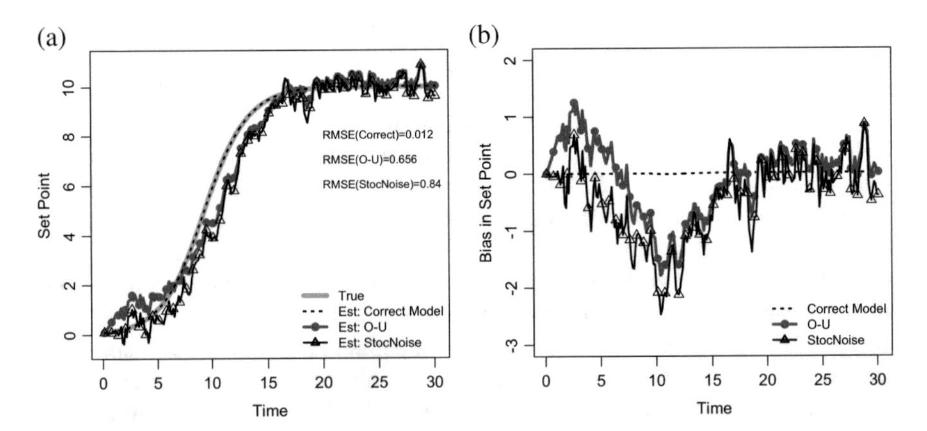

Fig. 9.4 Estimated TVP trajectories for one hypothetical subject in illustrative example 2: (**a**) estimated and true trajectories, with the corresponding RMSEs, (**b**) biases in the estimated TVP values in comparison with the true values

9.7.3 *Example 3: Stagewise Shifts in Damping*

In the last example, we allow the damping/amplification parameter, ζ, to undergo time-varying dynamics as:

$$\zeta_i(t) = \zeta_1 + c_2\zeta_2 + c_3\zeta_3 \tag{9.19}$$

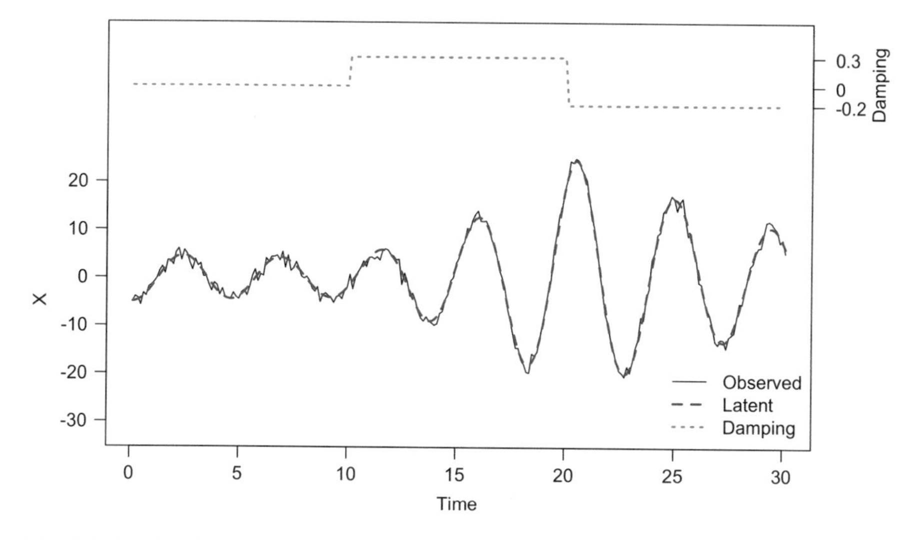

Fig. 9.5 Simulated observed and latent trajectory for one hypothetical subject generated using the damped oscillator model with stagewise shifts in ζ as considered in illustrative example 3. Damping = the damping/amplification parameter, $\zeta_i(t)$

This model assumes two stagewise shifts in the TVP as in example 1 but in the damping/amplification as opposed to the set point parameter. c_2 and c_3 are again indicators of stage 2 and stage 3. ζ_1, ζ_2, and ζ_3 represent, respectively, the value of the damping/amplification parameter in stage 1 and differences between the stage 1 value and the values in the two subsequent stages. Figure 9.5 shows a simulated example of such a case, in which the process goes through an oscillatory phase with no alterations in amplitude ($\zeta_i(t) = 0$), followed by a phase with amplification ($\zeta_i(t) > 0$), and ending with a phase with damping ($\zeta_i(t) < 0$).

Allowing the damping/amplification parameter to be time-varying provides a mathematical model for describing processes that undergo transient periods of increased/decreased fluctuations or instabilities. Consider again the bipolar disorder scenario. Some researchers have noted that the transitions between the depressive and manic episodes are often interspersed with a period of mood instability (Bonsall et al. 2011). Our proposed model with stagewise shifts in the damping/amplification parameter, as illustrated in Fig. 9.5, provides one way of representing such a transitional phase of increased mood instability. We did not allow the set point parameter to show concurrent stagewise shifts in the present illustration to ease presentation, but we should add that allowing for stagewise shifts in both the set point and the ζ parameter would be more consistent with the scenario of bipolar disorder described here.

For illustration purposes, data were simulated for 10 subjects, each with 241 time points ($t_{i,0} = 0$, $t_{i,T} = 30$, $\Delta t = 0.125$), using the parameter values specified in Table 9.4. The first stage has no damping effect ($\zeta_1 = 0$), followed by an amplification stage ($\zeta_2 = 0.3$) and a damping stage ($\zeta_3 = -0.2$). Based on the

Table 9.4 Parameter recovery in illustrative example 3 obtained from fitting the true data generation model and the two approximation models

		Estimate (SE)			
Parameter	True values	Stage-wise model (True)	O-U model	Stochastic noise model	Time invariant model
ζ_1	0	0.010 (0.009)	–	–	–
ζ_2	0.3	0.327 (0.004)	–	–	–
ζ_3	−0.2	−0.175 (0.003)	–	–	–
ω	−2	−2.004 (0.003)	−1.987 (0.005)	−1.984 (0.004)	−0.360 (0.037)
σ_p^2	0.25	0.167 (0.036)	0.094 (0.047)	0.104 (0.005)	6.493 (0.826)
σ_ϵ^2	1	1.030 (0.031)	1.019 (0.032)	1.020 (0.031)	0.989 (0.031)
ζ	–	–	–	–	−0.761 (0.097)
β	–	–	0.225 (0.004)	–	–
$\theta_{0_{ou}}$	–	–	0.020 (0.044)	–	–
σ_{ou}^2	–	–	0.026 (0.004)	–	–
σ_s^2	–	–	–	0.019 (0.003)	–
AIC		7193.59	7572.04	7596.46	7729.81
BIC		7228.31	7606.76	7619.61	7758.75
Computational time		33.43 s	31.99 s	17.13 s	24.37 s

model fitting results in Table 9.4, the true parameter values were recovered well when the correctly specified model was fitted to the data. The O-U model performed slightly better than the stochastic noise model in yielding smaller parameter biases, RMSE, AIC, and BIC compared to the stochastic noise model. However, both approximation models performed better compared to the model without TVPs, which resulted in very biased parameter estimates.

The smoothed estimates of the TVP trajectory for one arbitrary subject are shown in Fig. 9.6a. Fitting the correctly specified model recovers the three values of damping. By comparison, estimates from the two approximation models show less salient transitions across the stages. Rather, the TVP trajectories as estimated using the approximation functions convey the shifts from stage 1 to stage 2 more as a gradually increasing linear trend as opposed to a discrete shift. This set of results, compared to that from example 1, suggests that time-varying damping/amplification parameter may be more difficult to estimate than time-varying set point parameter and requires more replications of complete cycle to reliably distinguish discrete from other more gradual shifts in parameter value.

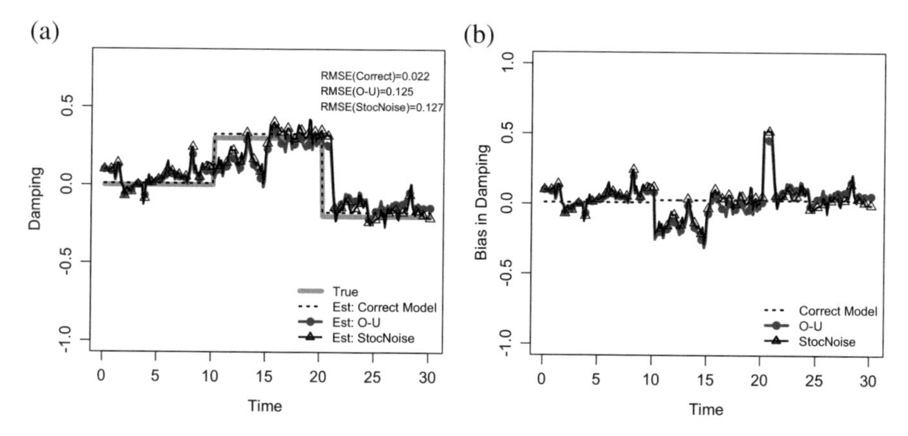

Fig. 9.6 Estimated TVP trajectories for one hypothetical subject in illustrative example 3: (**a**) estimated and true trajectories, with the corresponding RMSEs, (**b**) biases in the estimated TVP values in comparison with the true values

9.8 Simulation Study

When TVPs are included as additional latent variables in an ODE or SDE model, the resultant model, more often than not, would become a model that is nonlinear in the expanded set of latent variables. Even though the performance and features of the AIC and BIC as model comparison measures are well-known and well-investigated in the context of other models, their performance as model selection tools in the presence of such nonlinearity is unknown. In particular, the use of the CDEKF for handling estimation of the latent variable scores involves the use of Taylor series expansions in which higher-order terms are truncated. As such, the by-products used to compute the log-likelihood function—and by extension, the AIC and the BIC—may also contain truncation errors that may compromise the viability of using these measures as model comparison measures. A targeted Monte Carlo (MC) simulation study was performed to investigate the performance of the AIC and BIC in selecting the true TVP models. In addition, we also sought to evaluate the performance of the proposed estimation approach under conditions with correctly specified vs. incorrectly specified models.

We generated data using the oscillator model as shown in Eq. (9.11), with the corresponding set point ($\mu_i(t)$) following one of the three possible models: (1) a logistic growth process (Eq. (9.18)), (2) an O-U process (Eq. (9.14)), and (3) a stochastic noise process (Eq. (9.15)), each with 500 replications. Within each replication, data were fitted through each of the three processes to obtain parameter estimates as well as AIC and BIC. Starting values of parameter estimates were randomly sampled from distributions of possible parameter values. Boundaries were set for the parameters with variance parameters constrained within $(e^{-100}, e^{2.5})$ and other parameters constrained within $(-10, 10)$, with the exception of $\theta_{0_{ou}}$ having an upper bound of 20 when the data-generating model follows a logistic curve process based on inspection of data.

In the case of a model non-convergence, as indicated by nonpositive-definite Hessian matrix or parameter values at boundary values, another set of starting values were generated, and the model was reevaluated. If a model did not converge after five sets of random starting values, that particular replication was classified as a non-convergent trial. Convergence rates were tallied and reported in the result tables (Tables 9.5, 9.6, and 9.7). Also reported in the tables are (a) MC mean and medians of parameter estimates, (b) average SE estimates across MC replications (aSE), (c) standard deviations of parameter estimates within the MC sample (SD), (d) average confidence interval estimates across MC replications (aCI),[1] (e) MC sample percentile-based confidence intervals (MCCI), (f) average AIC and BIC from fitting each model (correctly specified and misspecified), and (g) percentage of the AIC/BIC successfully distinguishing the true data-generating models (i.e., AIC/BIC is the lowest for the true data generation models).

The computations were performed with `R` version 3.4.0, `dynr` package version 0.1.11-28, and `Sim.DiffProc` package version 3.7. Results show that all parameter estimates were recovered satisfactorily when the true data-generating models were fit. The MC means and medians were very close to the true parameter values. Both the aCIs and MCCIs included the true parameter values, and aCIs and MCCIs were similar. When models were misfitted, when the misfitted model was the O-U or stochastic noise model, the time-invariant dynamic parameters were still recovered well. However, when the misfitted model was the logistic growth model, the parameter estimates showed huge biases (Tables 9.6 and 9.7).

The AIC and BIC were mostly reliable in picking the true data-generating models (lowest AIC and BIC for the true model more than 98% of the time when the true mechanism was the logistic growth or O-U model). However, when the true data-generating model followed a stochastic noise process, the AIC (79.58% successfully distinguished) behaved less well compared to the BIC (99.1% successfully distinguished). Further inspection revealed that the AIC showed slightly higher instances of selecting the O-U model over the stochastic noise model even when the latter was the true data-generating model (AICs from the stochastic noise model were lower than those of the logistic growth model for 95.50% of times). These results were concordant with the AIC's general tendency to prefer overly complex models, especially since the stochastic noise process may also be obtained as a special case of the O-U model when $\beta = 0$ (see Eq. (9.14)). The simulation results from fitting the logistic growth model when the true data-generating model was the O-U model showed that it is inappropriate to use the logistic growth model as an approximation model, most likely due to the model's relatively restrictive

[1]The current version of `dynr` outputs Wald-type confidence intervals based on the standard errors of the (transformed) parameters. Wald-type confidence intervals are known to be inaccurate for variance parameters, particularly when the variance is near zero. The `dynr` team is working on options of other types of confidence interval estimates.

Table 9.5 Parameter recovery in simulation condition 1 where true time-varying set point follows a logistic growth process

Logistic growth model (true model)						
Parameter	*True*	*Estimate (mean/median)*	*aSE*	*SD*	*95% aCI*	*95% MCCI*
ω	−0.5	−0.501/−0.504	0.023	0.041	(−0.546, −0.456)	(−0.551, −0.451)
ζ	−0.2	−0.199/−0.192	0.037	0.051	(−0.273, −0.126)	(−0.292, −0.127)
σ_p^2	1	1.027/0.978	0.144	0.397	(0.746, 1.308)	(0.740, 1.298)
σ_ϵ^2	1	0.996/0.995	0.030	0.056	(0.936, 1.055)	(0.942, 1.058)
r	0.5	0.587/0.498	0.079	0.809	(0.432, 0.741)	(0.472, 0.531)
b	0.1	0.099/0.100	0.002	0.025	(0.095, 0.102)	(0.097, 0.103)
O-U Model						
Parameter	*True*	*Estimate (mean/median)*	*aSE*	*SD*	*95% aCI*	*95% MCCI*
ω	−0.5	−0.465/-0.466	0.030	0.067	(−0.524, −0.407)	(−0.562, −0.382)
ζ	−0.2	−0.262/−0.251	0.048	0.074	(−0.356, −0.169)	(−0.429, −0.160)
σ_p^2	1	1.339/1.380	0.144	0.607	(0.863, 1.815)	(0.547, 2.125)
σ_ϵ^2	1	0.994/0.992	0.030	0.030	(0.934, 1.053)	(0.940, 1.051)
β	–	0.140/0.063	0.243	0.632	(−0.095, 0.374)	(0.045, 0.083)
$\theta_{0_{ou}}$	–	12.861/12.824	1.499	1.713	(9.922, 15.799)	(10.012, 15.726)
σ_{ou}^2	–	0.422/0.003	0.174	0.507	(0.081, 0.762)	(0.000, 1.249)
Stochastic noise model						
Parameter	*True*	*Estimate (mean/median)*	*aSE*	*SD*	*95% aCI*	*95% MCCI*
ω	−0.5	−0.529/−0.525	0.033	0.033	(−0.594, −0.463)	(−0.603, −0.470)
ζ	−0.2	−0.266/−0.259	0.052	0.056	(−0.367, −0.165)	(−0.402, −0.174)
σ_p^2	1	0.673/0.673	0.187	0.213	(0.306, 1.039)	(0.252, 1.060)
σ_ϵ^2	1	1.002/1.001	0.031	0.030	(0.942, 1.062)	(0.946, 1.057)
σ_s^2	–	1.912/1.877	0.396	0.308	(1.136, 2.689)	(1.594, 2.428)
Model comparisons						
	Logistic growth (True)	*O-U*	*Stochastic noise*	*AIC/BIC selecting true model*		
AIC	7481.640	7505.013	7551.112	98.16% (426/434)		
BIC	7516.364	7545.524	7580.049	98.16% (426/434)		
% of convergence	87.4%	99.8%	99.6%			

Table 9.6 Parameter recovery in simulation condition 2 where true time-varying set point follows an O-U process

O-U model (true model)						
Parameter	*True*	*Estimate (mean/median)*	*aSE*	*SD*	*95% aCI*	*95% MCCI*
ω	−0.5	−0.504/-0.504	0.024	0.034	(−0.551, −0.457)	(−0.554, −0.464)
ζ	−0.2	−0.197/−0.193	0.024	0.042	(−0.273, −0.122)	(−0.279, −0.134)
σ_p^2	1	0.986/0.976	0.166	0.226	(0.660, 1.311)	(0.701, 1.301)
σ_ϵ^2	1	0.998/0.997	0.030	0.029	(0.939, 1.058)	(0.939, 1.059)
β	0.2	0.201/0.199	0.032	0.038	(0.138, 0.265)	(0.151, 0.269)
$\theta_{0_{ou}}$	−5	−5.001/−5.002	0.215	0.379	(−5.422, −4.580)	(−5.435, −4.634)
σ_{ou}^2	0.04	0.068/0.008	0.073	0.199	(−0.076, 0.211)	(0.000, 0.310)
Logistic growth model						
Parameter	*True*	*Estimate (mean/median)*	*aSE*	*SD*	*95% aCI*	*95% MCCI*
ω	−0.5	−0.165/−0.163	0.023	0.269	(−0.209, −0.120)	(−0.201, −0.121)
ζ	−0.2	−0.511/−0.477	0.079	0.371	(−0.665, −0.357)	(−0.634, −0.380)
σ_p^2	1	3.804/3.838	0.502	0.506	(2.821, 4.788)	(2.863, 4.546)
σ_ϵ^2	1	0.995/0.977	0.030	0.251	(0.936, 1.055)	(0.911, 1.047)
r	–	−3.225/-3.351	3.203	1.205	(−9.503, 3.054)	(−4.659, 0.053)
b	–	0.952/1.000	0.307	1.185	(0.350, 1.555)	(−2.442, 1.000)
Stochastic noise model						
Parameter	*True*	*Estimate (mean/median)*	*aSE*	*SD*	*95% aCI*	*95% MCCI*
ω	−0.5	−0.534/−0.534	0.029	0.040	(−0.592, −0.477)	(−0.601, −0.485)
ζ	−0.2	−0.249/−0.244	0.045	0.053	(−0.337, −0.160)	(−0.370, −0.169)
σ_p^2	1	0.777/0.759	0.172	0.283	(0.439, 1.115)	(0.428, 1.176)
σ_ϵ^2	1	1.001/1.001	0.031	0.029	(0.941, 1.061)	(0.941, 1.064)
σ_s^2	–	1.069/1.042	0.240	0.233	(0.598, 1.540)	(0.881, 1.334)
Model comparisons						
	O-U (True)	*Logistic growth*	*Stochastic noise*	*AIC/BIC selecting true model*		
AIC	7482.371	7826.754	7521.707	99.79% (466/467)		
BIC	7517.095	7861.478	7550.644	99.79% (466/467)		
% of convergence	99%	96.2%	98.2%			

Table 9.7 Parameter recovery in simulation condition 3 where true time-varying set point follows a stochastic noise process

Stochastic noise model (true model)						
Parameter	*True*	*Estimate (mean/median)*	*aSE*	*SD*	*95% aCI*	*95% MCCI*
ω	−0.5	−0.502/-0.501	0.025	0.028	(−0.551, −0.453)	(−0.558, −0.454)
ζ	−0.2	−0.199/−0.196	0.039	0.037	(−0.275, −0.122)	(−0.279, −0.136)
σ_p^2	1	0.987/0.980	0.150	0.173	(0.692, 1.282)	(0.717, 1.319)
σ_ϵ^2	1	1.000/1.000	0.030	0.032	(0.940, 1.060)	(0.940, 1.060)
σ_s^2	0.04	0.057/0.039	0.025	0.189	(0.008, 0.106)	(0.003, 0.100)
O-U model						
Parameter	*True*	*Estimate (mean/median)*	*aSE*	*SD*	*95% aCI*	*95% MCCI*
ω	−0.5	−0.503/−0.504	0.026	0.052	(−0.554, −0.451)	(−0.566, −0.453)
ζ	−0.2	−0.190/−0.195	0.039	0.176	(−0.267, −0.113)	(−0.286, −0.133)
σ_p^2	1	0.937/0.939	0.168	0.195	(0.606, 1.267)	(0.545, 1.291)
σ_ϵ^2	1	1.021/1.001	0.031	0.405	(0.960, 1.082)	(0.940, 1.058)
β	–	0.106/0.018	0.167	0.416	(−0.222, 0.433)	(−0.085, 1.061)
θ_{0ou}	–	−0.047/0.099	2.522	2.277	(−4.991, 4.896)	(−6.964, 5.233)
σ_{ou}^2	–	0.291/0.050	0.391	1.209	(−0.475, 1.057)	(0.000, 1.785)
Logistic growth model						
Parameter	*True*	*Estimate (mean/median)*	*aSE*	*SD*	*95% aCI*	*95% MCCI*
ω	−0.5	−0.475/-0.476	0.028	0.098	(−0.530, −0.421)	(−0.533, −0.418)
ζ	−0.2	−0.241/−0.223	0.049	0.272	(−0.337, −0.145)	(−0.325, −0.151)
σ_p^2	1	1.280/1.239	0.177	0.566	(0.933, 1.627)	(0.942, 1.673)
σ_ϵ^2	1	1.001/0.994	0.032	0.224	(0.937, 1.064)	(0.934, 1.059)
r	–	−0.511/−0.214	2.708	1.854	(−5.818, 4.796)	(−5.079, 3.897)
b	–	4.650/3.102	2.569	4.791	(−0.387, 9.686)	(−4.895, 9.988)
Model comparisons						
	Stochastic noise (True)	*O-U*	*Logistic growth*	*AIC/BIC selecting true model*		
AIC	7449.796	7475.394	7586.414	79.58% (265/333)		
BIC	7478.733	7515.906	7621.138	99.10% (330/333)		
% of convergence	100%	82.8%	79.8%			

constraints (i.e., in imposing a very specific form of change and does not allow for stochasticity via the inclusion of a process noise component). That is, estimates for dynamic parameters such as ω and ζ were far from their true values. This highlights some of the detrimental consequences of using inappropriate models to represent TVPs.

9.9 Discussion

In this chapter, we proposed an SDE modeling framework with TVPs as a way to capture multi-time scale and self-organizing processes in continuous time. We considered several variations of the damped oscillator model in the illustrative examples: two with time-varying set points and one with time-varying damping/amplification. We also performed an MC study that included several variations of time-varying set point processes. Furthermore, we presented and evaluated the use of two functions—the O-U model and the stochastic noise model—as plausible functions for approximating changes in the TVPs in the absence of further knowledge concerning their true change mechanisms.

We showed in all three of the illustrative examples considered and in the MC simulation study that the TVPs can be recovered very well and the smoothed estimates provided accurate reflections of the underlying dynamic processes when the true models were fitted to the simulated data. Approximating the true models of TVP with the O-U model and the stochastic noise model yielded satisfactory estimates of other time-invariant parameters, but assuming time invariance in the parameters when they did in fact vary over time resulted in substantial biases in the parameter estimates in the process model. The smoothed estimates of the TVP trajectories from these two approximation models also revealed the general change patterns of the true TVP processes reasonably well, though not as accurately as the smoothed estimates obtained from fitting the true models. Overall, our results confirmed the plausibility of using these functions as approximations for model and data exploration purposes. We further showed that the smoothed latent parameter estimates can be inspected to facilitate later confirmatory modeling of the trajectories of the TVP.

Our simulation study offered some initial results validating the utility of the AIC and BIC as model selection tools for continuous-time models with TVPs. As noted earlier, because models with TVPs incorporated as latent variables become nonlinear in nature, the log-likelihood function used for parameter estimation purposes involves approximations stemming from the use of the EKF to linearize nonlinearities in the model. By extension, computations of the AIC and the BIC, which capitalize on the same log-likelihood function, now also involve approximations due to the linearizations. No previous work has systematically investigated whether the AIC and BIC computed under such approximations are still viable as model selection tools. Our simulation results confirmed their utility, indicating that both the AIC and BIC are still reliable in selecting the true data-generating models in most instances, except in one special case in which the AIC tended to select an overly complex model (the O-U model) over the simpler true data-generating model (the stochastic noise model).

The stochastic noise model, as briefly mentioned in simulation results, can be obtained as a special case of the O-U model if the β parameter is 0. However, apart from that, the stochastic noise model is also a special case of any other model wherein the deterministic portion of the model (the drift function) is equal to 0. This model imposes the notion that there is no regularity or predictability in a process over time. Hence, if the stochastic noise model is preferred over other models that do impose some form of regularity in the process, this has implications on the nature of whether and how the TVPs vary over time.

Although not considered in one of the illustrative examples in this chapter, it is also possible to allow the frequency parameter, ω, to be modeled as a TVP. Our preliminary results (not shown here) suggested that similar results and modeling conclusions would still hold in scenarios involving a single, time-varying frequency parameter. Relatedly, even though we only considered examples that are extensions of the well-known damped oscillator process, the procedures we proposed are, in principle, readily generalizable to other dynamic processes of similar characteristics to those considered in the chapter. The relative flexibility of the CDEKF algorithms and the modeling interface provided as part of the `dynr` package make such generalizations viable and potentially very doable.

Our illustrations and explorations of SDEs with TVPs are far from comprehensive. More MC studies should be conducted to evaluate the performance of the CDEKF algorithms, the AIC, and the BIC in fitting and facilitating the selection of SDEs with TVPs under different SDE models and parameters of different natures, over various effect size and sample size settings. Also worth the investigation is the signal-to-noise ratio this proposed data exploration scheme can tolerate. In fitting an approximation model such as the O-U model or stochastic noise model, the "true" noise in the data can manifest in more than one place through one or several of the process noise parameters in the approximation models. Hence, it is important to know to what extent the approximation models could still capture the underlying changes in the TVPs with increased process and measurement noise. Such MC studies can also help us determine the extent of impact if TVPs are modeled as fixed and approaches researchers could take in data exploration phase when it is not obvious if or how many parameters are time-varying.

When it is not apparent of which parameters might be reasonably allowed to be time-varying, one plausible strategy would be to allow all parameters to be TVPs but only to the extent that the resultant model still satisfies the observability and controllability conditions needed to identify the latent variables and recover the parameters accurately (Bar-Shalom et al. 2001). In brief, the TVPs are not linked directly to observed indicators but rather, are identified indirectly through other latent variables with direct indicators. Thus, one may run into identifiability issues if more latent variables (specifically, TVPs) are defined than there is sufficient information from the data to identify them.

In this chapter, our examples of TVPs are all proposed to have more gradual changes than other latent system (endogenous) variables. In order to achieve a TVP process on a slower time scale, the process variances of the TVPs in all our data generation conditions were set to be much smaller than those of other latent variables. This constraint—or rather, assumption—in the data generation process is similar to the assumption imposed in Molenaar et al. (2009), in which the authors used a random walk model to represent the TVPs. This assumption of slow-varying TVPs is necessary to ensure that the data are of sufficient density to identify the dynamics of the TVPs, the failure of which would also result in failure to identify the dynamics of the system as a whole. However, the relative differences in time scales between the TVPs and other system variables are still very much an active area of research and likely depend on characteristics and nature of the model and TVPs of interest. For instance, if the TVP is a "trend" or time-varying intercept, then this parameter may change at similar time scales as other system variables. In contrast, if the TVP of interest is the frequency or damping parameter in the damped oscillator model, a variance parameter, or any other parameters whose effects on the dynamics of the system can only be reliably observed over multiple replications of the process, then sufficient number of replications and data density are needed to allow these parameters to be recovered accurately. Thus, even though it is not always necessary to dictate that the time scales of the TVPs be much slower than the time scales of the latent variables, we find it noteworthy to state this at least as a cautionary constraint for the reader to consider.

Throughout, we assumed the process noises of the TVPs to be Gaussian distributed. This assumption may not be tenable for certain parameters, for instance, variance parameters whose permissable values are always positive. One way to circumvent this issue is to apply transformations to the TVPs and specify the unconstrained TVPs (e.g., the log of a variance parameter), as opposed to the constrained parameters (e.g., the exponent of the log variance parameter), to be characterized by Gaussian-distributed process noises. In `dynr`, there are multiple ways in which such transformations can be specified. One way is to use the function "prep.trans" to specify transformation functions to be applied to any of the parameters in the model. An alternative route is to capitalize on `dynr`'s flexibility in handling nonlinear functions of the latent variables to specify the required constraints explicitly as part of the dynamic functions of the TVPs. We did not apply any of these parameter transformation approaches in the illustrative or MC examples, but it is worthwhile to examine the relative performance of the proposed estimation approach in the presence of such transformations in future studies.

Our work in this chapter is closely related to that of Bringmann et al. (2017)'s where they implemented discrete-time autoregressive models with TVPs under the generalized additive modeling framework with smooth regression splines. The smoothing procedure inherent to the approach by Bringmann et al. (2017) has added utility in "smoothing" out noisy fluctuations in the TVPs as well as system variables and can be readily used to represent changes in modeling parameters over other specified observed variables (e.g., spatial variables). However, the implemented approach is currently restricted to one special case of a discrete-time model and only handles missingness via case deletion, as opposed to using raw data (approximate) maximum likelihood as in the present approach. It would be interesting to consider extensions of generalized additive modeling framework to continuous-time models, or incorporation of smoothing features into the proposed approach.

Several limitations of the proposed estimation procedures should be noted. First, due to the use of a Jacobian matrix $\mathbf{F}$ in the prediction step in (9.4), only drift functions that are at least first-order differentiable with respect to elements in $\boldsymbol{\eta}_i(t)$ may be used in the dynamic model. This requirement, however, is not met for some dynamical systems, for instance, examples that show sudden shifts in the TVP, similar to the stagewise true model used in our illustrative examples. When preexisting indicators of the exact change points are available (e.g., as reflected in how we fit the true model to the simulated data), estimation can be handled in a straightforward manner. When direct indicators of these change points are not available, we resorted to using other differentiable functions, such as the O-U model, as an approximation for such discrete shifts. For better approximation results, we may need to utilize some other filtering methods, such as iterated particle filtering (Ionides et al. 2015), the unscented Kalman filter (Chow et al. 2007), or other regime-switching extensions (Chow et al. 2017) as alternatives for systems that are not first-order differentiable.

A fourth-order Runge–Kutta procedure is used to solve the two ODEs in Eqs. (9.3) and (9.4) in the prediction step of the CDEKF—namely, to obtain conditional expectations of the latent variables for time $t_{i,j}$ given the estimates from up to time $t_{i,j-1}$, that is, $E\big(\boldsymbol{\eta}_i(t_{i,j}|\mathbf{Y}_i(t_{i,j-1})\big)$ and $Cov\big(\boldsymbol{\eta}_i(t_{i,j}|\mathbf{Y}_i(t_{i,j-1})\big)$. Using numerical solvers such as the Runge–Kutta procedure opens up more possibilities of fitting ODE/SDE models that do not have closed-form solutions. In exchange for this added flexibility, however, is the need to select appropriate Δt or time step for the numerical integration as the truncation errors embedded in the numerical solutions are direct functions of Δt. Δt that is too large would yield poor estimation results; in contrast, Δt that is too small can be computationally inefficient and may also lead to other numerical problems. For the examples given in this chapter, Δt was fixed at 0.125 throughout all the examples. The exact same observations with a Δt of 1 would return inaccurate inferred model dynamics and associated parameter estimates, even with the same model fitting procedures and specifications. In some cases, rescaling of the time units—for example, recoding time from seconds to minutes—may be necessary. Other alternatives, such as adaptive ODE solvers (Kulikov and Kulikova 2014; Kulikova and Kulikov 2014) and multiple shooting methods (Gander and Vandewalle 2007; Kiehl 1994; Stoer and Bulirsch 2013), may also be used in place of the fourth-order Runge–Kutta procedure to improve the robustness of the estimation procedures to choices of Δt. In addition, the CDEKF algorithms as currently implemented in `dynr` rely on the use of first-order Taylor series expansion to approximate the uncertainties implicated in $\mathbf{P}(t_{i,j})$. When the system is highly nonlinear, retaining only the first-order terms in the Taylor series expansion may be inadequate, and higher-order terms may have to be incorporated instead (Bar-Shalom et al. 2001).

The smoothing algorithm used in the present study can also be improved in a number of ways. First, in the illustrative examples considered, the smoothed estimates of the TVP obtained from the approximation models generally recovered the true patterns of the TVPs reasonably well. However, there were noticeable delays in recovering the true values of the TVPs around all the transition points in the stagewise examples. One way to circumvent this estimation delay is to reverse the time ordering of the data (so that the last time of observation T now becomes the first time point t_1 and so on), repeat the model fitting procedure, and generate a set of refined smoothed estimates as a weighted average of the smoothed estimates obtained from forward and reverse smoothing. This revised smoothing procedure is similar in principle to the zero-phase noncausal filtering utilized in the field of engineering (Kormylo and Jain 1974; Powell and Chau 1991). Finally, the prediction error decomposition function used for parameter optimization purposes in (9.9) uses only the one-step-ahead prediction errors and associated covariance matrix from the prediction step. One alternative, which replaces the PED function with a penalized log-likelihood function (Fahrmeir and Wagenpfeil 1997) constructed using an iterated extended Kalman filter (e.g., Bar-Shalom et al. 2001, p. 404), may be used to improve the quality of the parameter estimates and, consequently, the quality of the smoothed estimates.

In conclusion, modeling in continuous time using SDEs with TVPs provides renewed possibilities for capturing self-organization across different levels of analysis and time scales. In cases where the true underlying mechanisms of interest are unknown, we have illustrated the viability of several approximation models for representing TVP processes that unfold on much slower time scales compared to the changes manifested by other system variables. We hope our subset of examples provides a starting point for researchers to pursue other modeling ideas and applications involving self-organizing phenomena in continuous time.

References

Akaike, H. (1973). Information theory and an extension of the maximum likelihood principle. In B. N. Petrov & F. Csaki (Eds.), *Second international symposium on information theory,* (pp. 267–281). Budapest: Akademiai Kiado. https://doi.org/10.1007/978-1-4612-1694-015

Baltes, P. B., & Nesselroade, J. R. (1979). History and rationale of longitudinal research. In J. R. Nesselroade & P. B. Baltes (Eds.), *Longitudinal research in the study of behavior and development* (pp. 1–39). New York: Academic.

Bar-Shalom, Y., Li, X. R., & Kirubarajan, T. (2001). *Estimation with applications to tracking and navigation: Theory algorithms and software.* New York: Wiley. https://doi.org/10.1002/0471221279

Bertenthal, B. (2007). Dynamical systems: It is about time! In S. Boker & M. Wenger (Eds.), *Data analytic techniques for dynamical systems.* Mahwah, NJ: Lawrence Erlbaum Associates, Inc.

Bisconti, T. L., Bergeman, C. S., & Boker, S. M. (2004). Emotional well-being in recently bereaved widows: A dynamical systems approach. *The Journals of Gerontology Series B: Psychological Sciences and Social Sciences, 59*(4), 158–167. https://doi.org/10.1093/geronb/59.4.P158

Boker, S. M., & Graham, J. (1998). A dynamical systems analysis of adolescent substance abuse. *Multivariate Behavioral Research, 33*(4), 479–507. https://doi.org/10.1207/s15327906mbr3304_3

Bonsall, M. B., Wallace-Hadrill, S. M. A., Geddes, J. R., Goodwin, G. M., & Holmes, E. A. (2011). Nonlinear time-series approaches in characterizing mood stability and mood instability in bipolar disorder. In *Proceedings of the Royal Society of London B: Biological Sciences*. http://rspb.royalsocietypublishing.org/content/early/2011/08/12/rspb.2011.1246. https://doi.org/10.1098/rspb.2011.1246

Bosma, H. A. & Kunnen, E. S. (Eds.). (2011). *Identity and emotion: Development through self-organization*. Cambridge: Cambridge University Press.

Bringmann, L., Hamaker, E. L., Vigo, D. E., Aubert, A., Borsboom, D., & Tuerlinckx, F. (2017). Changing dynamics: Time-varying autoregressive models using generalized additive modeling. *Psychological Methods, 22*(3), 409–425. https://doi.org/10.1037/met0000085

Browne, M. W., & Du Toit, S. H. (1991). Models for learning data. In L. M. Collins & J. L. Horn (Eds.), *Best methods for the analysis of change: Recent advances, unanswered questions, future directions* (pp. 47–68). Washington, DC: American Psychological Association. https://doi.org/10.1037/10099-004

Chow, S.-M., Ferrer, E., & Nesselroade, J. R. (2007). An unscented Kalman filter approach to the estimation of nonlinear dynamical systems models. *Multivariate Behavioral Research, 42*(2), 283–321. https://doi.org/10.1080/00273170701360423

Chow, S.-M., Hamaker, E. J., Fujita, F., & Boker, S. M. (2009). Representing time-varying cyclic dynamics using multiple-subject state-space models. *British Journal of Mathematical and Statistical Psychology, 62*, 683–716. https://doi.org/10.1348/000711008X384080

Chow, S.-M., Lu, Z., Sherwood, A., & Zhu, H. (2016). Fitting nonlinear ordinary differential equation models with random effects and unknown initial conditions using the Stochastic Approximation Expectation Maximization (SAEM) algorithm. *Psychometrika, 81*, 102–134. https://doi.org/10.1007/s11336-014-9431-z

Chow, S.-M., Nesselroade, J. R., Shifren, K., & McArdle, J. J. (2004). Dynamic structure of emotions among individuals with Parkinson's disease. *Structural Equation Modeling, 11*, 560–582. https://doi.org/10.1207/s15328007sem1104_4

Chow, S. M., Ou, L., Ciptadi, A., Prince, E. B., You, D., Hunter, M. D., & Messinger, D. S. (2018). Representing Sudden Shifts in Intensive Dyadic Interaction Data Using Differential Equation Models with Regime Switching. *Psychometrika, 83*(2), 476–510.

Chow, S.-M., Ram, N., Boker, S. M., Fujita, F., & Clore, G. (2005). Emotion as a thermostat: Representing emotion regulation using a damped oscillator model. *Emotion, 5*(2), 208–225. https://doi.org/10.1037/1528-3542.5.2.208

Chow, S.-M., Zu, J., Shifren, K., & Zhang, G. (2011). Dynamic factor analysis models with time-varying parameters. *Multivariate Behavioral Research, 46*, 303–339. https://doi.org/10.1080/00273171.2011.563697

Deboeck, P. R., Boker, S. M., & Bergeman, C. S. (2008). Modeling individual damped linear oscillator processes with differential equations: Using surrogate data analysis to estimate the smoothing parameter. *Multivariate Behavioral Research, 43*, 497–523. https://doi.org/10.1080/00273170802490616

Del Negro, M., & Otrok, C. (2008). *Dynamic Factor Models with Time-Varying Parameters: Measuring Changes in International Business Cycles* (Staff Reports No. 326). New York: Federal Reserve Bank of New York.

Fahrmeir, L., & Wagenpfeil, S. (1997). Penalized likelihood estimation and iterative Kalman smoothing for non-Gaussian dynamic regression models. *Computational Statistics & Data Analysis, 24*(3), 295–320. https://doi.org/10.1016/S0167-9473(96)00064-3

Gander, M. J., & Vandewalle, S. (2007). Analysis of the parareal time-parallel time-integration method. *SIAM Journal on Scientific Computing, 29*(2), 556–578. https://doi.org/10.1137/05064607X

Gottlieb, G., Cairns, R., Elder, J., & Costello, E. (1996). Developmental psychobiological theory. *Developmental Science*, 63–77. https://doi.org/10.1017/CBO9780511571114.005

Grimm, K. J., & Ram, N. (2009). Nonlinear growth models in M*plus* and SAS. *Structural Equation Modeling, 16*(4), 676–701. https://doi.org/10.1080/10705510903206055

Hamaker, E. L., & Grasman, R. P. P. P. (2012). Regime switching state-space models applied to psychological processes: Handling missing data and making inferences. *Psychometrika, 77*(2), 400–422. https://doi.org/10.1007/s11336-012-9254-8

Harvey, A. C. (2001). *Forecasting, structural time series models and the Kalman filter*. Cambridge: Cambridge University Press.

Ionides, E. L., Nguyen, D., Atchadé, Y., Stoev, S., & King, A. A. (2015). Inference for dynamic and latent variable models via iterated, perturbed Bayes maps. *Proceedings of the National Academy of Sciences, 112* (3), 719–724. https://doi.org/10.1073/pnas.1410597112

Julier, S. J., & Uhlmann, J. K. (2004). Unscented filtering and nonlinear estimation. *Proceedings of the IEEE, 92*(3), 401–422. https://doi.org/10.1109/JPROC.2003.823141

Julier, S. J., Uhlmann, J. K., & Durrant-Whyte, H. F. (1995). A new approach for filtering nonlinear systems. In *Proceedings of the American control conference* (pp. 1628–1632). Seattle, WA. https://doi.org/10.1109/ACC.1995.529783

Kelso, S. (1995a). *Dynamic patterns: The self-organization of brain and behavior.* Cambridge, MA: The MIT Press.

Kelso, S. (1995b). How nature handles complexity. In *Dynamic patterns: The self-organization of brain and behavior* (Chap. 1). Cambridge, MA: The MIT Press.

Kiehl, M. (1994). Parallel multiple shooting for the solution of initial value problems. *Parallel Computing, 20*(3), 275–295. https://doi.org/10.1016/S0167-8191(06)80013-X

Kim, J., Nakamura, T., Kikuchi, H., Sasaki, T., & Yamamoto., Y. (2013). Covariation of depressive mood and locomotor dynamics evaluated by ecological momentary assessment in healthy humans. *PLoS ONE, 8*(9), e74979. https://doi.org/10.1371/journal.pone.0074979

Kitagawa, G. (1998). A self-organizing state-space model. *Journal of the American Statistical Association, 93*(443), 1203–1215. https://doi.org/10.2307/2669862

Kormylo, J., & Jain, V. (1974). Two-pass recursive digital filter with zero phase shift. *IEEE Transactions on Acoustics, Speech, and Signal Processing, 22*(5), 384–387. https://doi.org/10.1109/TASSP.1974.1162602

Krone, T., Albers, C. J., & Timmerman, M. E. (2016). Bayesian dynamic modelling to assess differential treatment effects on panic attack frequencies. *Statistical Modelling, 16*(5), 343–359. https://doi.org/10.1177/1471082X16650777

Kulikov, G., & Kulikov, M. (2014). Accurate numerical implementation of the continuous-discrete extended Kalman filter. *IEEE Transactions on Automatic Control, 59*(1), 273–279. https://doi.org/10.1109/TAC.2013.2272136

Kulikova, M. V., & Kulikov, G. Y. (2014). Adaptive ODE solvers in extended Kalman filtering algorithms. *Journal of Computational and Applied Mathematics*, *262*, 205–216. https://doi.org/10.1016/j.cam.2013.09.064

Larson, R., Raffaelli, M., Richards, M. H., Ham, M., & Jewell, L. (1990). Ecology of depression in late childhood and early adolescence: A profile of daily states and activities. *Journal of Abnormal Psychology*, *99*(1), 92–102. https://doi.org/10.1037/0021-843X.99.1.92

Lewis, M. D., & Ferrari, M. (2001). Cognitive-emotional self-organization in personality development and personal identity. In H. A. Bosma & E. S. Kunnen (Eds.), *Identity and emotion: Development through self-organization* (pp. 177–198). Cambridge: Cambridge University Press. https://doi.org/10.1017/CBO9780511598425.015

Liening, S. H., Stanton, S. J., Saini, E. K., & Schultheiss, O. C. (2010). Salivary testosterone, cortisol, and progesterone: two-week stability, interhormone correlations, and effects of time of day, menstrual cycle, and oral contraceptive use on steroid hormone levels. *Physiology & Behavior, 99*(1), 8–16. https://doi.org/10.1016/j.physbeh.2009.10.001

MacCallum, R. C. (2003). Working with imperfect models. *Multivariate Behavioral Research, 38*(1), 113–139. Retrieved from http://dx.doi.org/10:1207=S15327906MBR38015 (PMID: 26771126) https://doi.org/10.1207/S15327906MBR3801_5

Magnusson, D., & Cairns, R. B. (1996). Developmental science: Toward a unified framework. In R. B. Cairns, J. Elder Glen H., & E. J. Costello (Eds.), *Developmental science.* (pp. 7–30). New York, NY: Cambridge University Press.

Marceau, K., Ram, N., Houts, R. M., Grimm, K. J., & Susman, E. J. (2011). Individual differences in boys' and girls' timing and tempo of puberty: Modeling development with nonlinear growth models. *Developmental Psychology, 47*(5), 1389–1409. https://doi.org/10.1037/a0023838

McCarthy, D. E., Minami, H., Yeh, V. M., & Bold, K. W. (2015). An experimental investigation of reactivity to ecological momentary assessment frequency among adults trying to quit smoking. *Addiction, 110*(10), 1549–1560. Retrieved from http://dx.doi.org/10.1111/add.12996 (ADD-14-0754.R2) https://doi.org/10.1111/add.12996

Molenaar, P. C. M. (1994). Dynamic latent variable models in developmental psychology. In A. von Eye & C. Clogg (Eds.), *Latent variables analysis: Applications for developmental research* (pp. 155–180). Thousand Oaks, CA: Sage Publications.

Molenaar, P. C. M., & Newell, K. M. (2003). Direct fit of a theoretical model of phase transition in oscillatory finger motions. *British Journal of Mathematical and Statistical Psychology, 56*, 199–214. https://doi.org/10.1348/000711003770480002

Molenaar, P. C. M., Sinclair, K. O., Rovine, M. J., Ram, N., & Corneal, S. E. (2009). Analyzing developmental processes on an individual level using nonstationary time series modeling. *Developmental Psychology, 45*(1), 260–271. https://doi.org/10.1037/a0014170

Nesselroade, J. R. (1991). The warp and woof of the developmental fabric. In R. Downs, L. Liben, & D. Palermo (Eds.), *Visions of development, the environment, and aesthetics: The legacy of Joachim F. Wohlwill* (pp. 213–240). Hillsdale, NJ: Lawrence Erlbaum Associates.

Newell, A. (1990). *Unified theories of cognition.* Cambridge, MA: Harvard University Press.

Newtson, D. (1993). The dynamics of action and interaction. In L. B. Smith & E. Thelen (Eds.), *A dynamic systems approach to development: Applications* (pp. 241–264). Cambridge, MA: MIT Press.

Newtson, D., Hairfield, J., Bloomingdale, J., & Cutino, S. (1987). The structure of action and interaction. *Social Cognition, 5*(3), 191–237. https://doi.org/10.1521/soco.1987.5.3.191

Nowak, A., & Lewenstein, M. (1994). Dynamical systems: A tool for social psychology? In R. Vallacher & A. Nowak (Eds.), *Dynamical systems in social psychology* (Chap. 2). San Diego, CA: Academic.

Oravecz, Z., Tuerlinckx, F., & Vandekerckhove, J. (2009). A hierarchical Ornstein–Uhlenbeck model for continuous repeated measurement data. *Psychometrika, 74*(3), 395–418. https://doi.org/10.1007/s11336-008-9106-8

Ou, L., Hunter, M. D., & Chow, S.-M. (2016). dynr: Dynamic modeling in R. [Computer software manual].

Ou, L., Hunter, M. D., & Chow, S.-M. (2017). *What's for dynr: A package for linear and nonlinear dynamic modeling in R* (manuscript submitted for publication).

Oud, J. H. L. (2007a). Comparison of four procedures to estimate the damped linear differential oscillator for panel data. In K. van Montfort, J. H. L. Oud, & A. Satorra (Eds.), *Longitudinal models in the behavioral and related sciences* (pp. 19–39). Mahwah, NJ: Lawrence Erlbaum Associates.

Oud, J. H. L. (2007b). Continuous time modeling of reciprocal relationships in the cross-lagged panel design. In S. Boker & M. Wenger (Eds.), *Data analytic techniques for dynamical systems* (pp. 87–129). Mahwah, NJ: Lawrence Erlbaum Associates, Inc.

Oud, J. H. L., & Jansen, R. A. (2000). Continuous time state space modeling of panel data by means of SEM. *Psychometrika, 65*(2), 199–215. https://doi.org/10.1007/BF02294374

Oud, J. H. L., & Singer, H. (2008). Continuous time modeling of panel data. *Statistica Neerlandica, 62*, 4–28. https://doi.org/10.1111/j.1467-9574.2007.00376.x

Pagan, A. (1980). Some identification and estimation results for regression models with stochastically varying coefficients. *Journal of Econometrics, 13*, 341–363. https://doi.org/10.1016/0304-4076(80)90084-6

Powell, S. R., & Chau, P. M. (1991). A technique for realizing linear phase IIR filters. *IEEE Transactions on Signal Processing, 39*(11), 2425–2435.

R Core Team. (2016). R: A language and environment for statistical computing [Computer software manual]. Vienna, Austria. https://www.R-project.org/
Ram, N., & Gerstorf, D. (2009). Time structured and net intraindividual variability: Tools for examining the development of dynamic characteristics and processes. *Psychology and Aging, 24*, 778–791. https://doi.org/10.1037/a0017915
Schwarz, G. (1978). Estimating the dimension of a model. *The Annals of Statistics, 6*(2), 461–464. https://doi.org/10.1214/aos/1176344136
Schweppe, F. C. (1965). Evaluation of likelihood functions for Gaussian signals. *IEEE Transactions on Information Theory, IT-11*, 61–70. https://doi.org/10.1109/TIT.1965.1053737
Shiffman, S., Stone, A. A., & Hufford, M. R. (2008). Ecological momentary assessment. *Annual Review of Clinical Psychology, 4*, 1–32. https://doi.org/10.1146/annurev.clinpsy.3.022806.091415
Shumway, R. H., & Stoffer, D. S. (2000). *Time series analysis and its applications*. New York: Springer. https://doi.org/10.1007/978-1-4757-3261-0
Silk, J. S., Forbes, E. E., Whalen, D. J., Jakubcak, J. L., Thompson, W. K., Ryan, N. D., et al. (2011). Daily emotional dynamics in depressed youth: A cell phone ecological momentary assessment study. *Journal of Experimental Child Psychology, 110*(2), 241–25. https://doi.org/10.1016/j.jecp.2010.10.007
Stock, J., & Watson, M. (2008). Forecasting in dynamic factor models subject to structural instability. In J. Castle & N. Shephard (Eds.), *The methodology and practice of econometrics, a Festschrift in honour of Professor David F. Hendry.* Oxford: Oxford University Press.
Stoer, J., & Bulirsch, R. (2013). *Introduction to numerical analysis* (Vol. 12). : Springer Science & Business Media.
Tarvainen, M. P., Georgiadis, S. D., Ranta–aho, P. O., & Karjalainen, P. A. (2006). Time–varying analysis of heart rate variability signals with Kalman smoother algorithm. *Physiological Measurement, 27*, 225–239. https://doi.org/10.1088/0967-3334/27/3/002
The National Institute of Mental Health. (2016). *Bipolar disorder.* Retrieved April 22, 2017, from https://www.nimh.nih.gov/health/topics/bipolar-disorder/index.shtml
Turvey, M. (1990). Coordination. *American Psychologist, 45*(8), 938–953. doi: 10.1037/0003-066X.45.8.938
Vallacher, R. R., & Nowak, A. (Eds.). (1994). *Dynamical systems in social psychology*. San Diego, CA: Academic.
Voelkle, M. C., & Oud, J. H. L. (2013). Continuous time modelling with individually varying time intervals for oscillating and non-oscillating processes. *British Journal of Mathematical and Statistical Psychology 66*, 103–126. https://doi.org/10.1111/j.2044-8317.2012.02043.x
Wan, E., & Van der Merwe, R. (2001). The unscented Kalman filter. In S. Haykins (Ed.), *Kalman filtering and neural networks* (pp. 221–280). New York: Wiley. https://doi.org/10.1002/0471221546.ch7
Weiss, A. A. (1985). The stability of the AR(1) process with an AR(1) coefficient. *Journal of Time Series Analysis, 6*, 181–186. https://doi.org/10.1111/j.1467-9892.1985.tb00408.x

Chapter 10
Robustness of Time Delay Embedding to Sampling Interval Misspecification

Steven M. Boker, Stacey S. Tiberio, and Robert G. Moulder

10.1 Introduction

Real-world applications of intensively sampled longitudinal data often have unequal intervals between observed data points. For instance, experience sampling methods may have data collection methods that randomize the time of day when prompts are given to participants on their mobile devices (e.g., Dufau et al. 2011; Hektner et al. 2006). Daily diary studies may collect data in the morning and the evening, but these intervals may differ as much as 3 or 4 h from 1 day to the next (e.g., Erbacher et al. 2012). Long-term cognitive aging studies may schedule to gather yearly observations, but practicalities of contacting and scheduling may result in observation intervals that differ by 3 or 4 months from 1 year to the next (e.g., Salthouse et al. 2004; Salthouse and Tucker-Drob 2008). The problem of sampling interval misspecification has been addressed in a variety of contexts including latent growth curve analysis, autoregressive modeling, and continuous-time autoregressive modeling. This chapter addresses the question of misspecification of sampling interval in the context of time delay embedding.

Electronic Supplementary Material The online version of this article (https://doi.org/10.1007/978-3-319-77219-6_10) contains supplementary material, which is available to authorized users.

S. M. Boker (✉) · R. G. Moulder
Department of Psychology, The University of Virginia, Charlottesville, VA, USA
e-mail: boker@virginia.edu; rgm4fd@virginia.edu

S. S. Tiberio
Oregon Social Learning Center, Eugene, OR, USA
e-mail: StaceyT@oslc.org

K. van Montfort et al. (eds.), *Continuous Time Modeling in the Behavioral and Related Sciences*, https://doi.org/10.1007/978-3-319-77219-6_10

This problem can be particularly acute when using discrete-time autoregressive models including cross-lag panel models. Discrete-time forward prediction models such as auto- and cross-regressive models are susceptible to time interval misspecification since the auto- and cross-regressive parameters are a function of the time interval between observations. This is often ignored in discrete-time modeling by assuming that the lag (i.e., time interval between observations) is equal to one. This essentially ignores the time interval problem, but does not solve it.

Continuous-time models remove the dependence of model parameters on the sampling interval. Analysts can be tempted to use the mean time interval between observations when estimating the continuous-time parameters. Or, it may be that the original data collection did not save individual intervals between observations, and so a mean or planned interval is all that is available. Others have addressed the problems that can occur when the sampling interval is ignored in continuous-time autoregressive models (e.g., Delsing and Oud 2008; Oud 2007).

Some authors addressing time interval misspecification in latent growth curve and continuous-time autoregressive models have used a technique in which the actual time interval is retained as a variable at each observation for each participant and used to modify the model for each data row (e.g., McArdle and Woodcock 1997). Then a full information maximum likelihood optimization can be used to correct for the model sampling interval misspecification. This *row-wise loading correction* can be accomplished reasonably easily using software such as OpenMx using its *definition variable* functionality (Neale et al. 2016).

A second method that has been used when estimating models from time interval misspecified data: *interpolation splines with resampling* (e.g., Klump et al. 2013). This method fits a spline through the data, and then the values of the spline at equal time intervals are used as input to the model estimation software. While this method can be effective in some cases, there are pathological cases where the method can go seriously awry, producing effects that are artifacts of what boils down being a longitudinal imputation technique. As in any imputation technique if the imputation model is misspecified, effects can be produced by imputation that do not actually exist in the data.

Another commonly used technique for correcting for sampling interval misspecification is the *method of planned missingness* (McArdle 1994; McArdle and Woodcock 1997). The argument is that the data could have been sampled at equal intervals, and so if one chooses a planned sampling interval that is sufficiently small, one can restructure existing data to have approximately equal sampling intervals by inserting missing observations in between existing observations.

We decided to use a Monte Carlo simulation to compare degree of combined bias and increased parameter variability produced when using uncorrected sampling intervals in a time delay embedding method for estimation of parameters of differential equations from interval misspecified data. We used *latent differential equation* (LDE) models (Boker 2007; Boker et al. 2010, 2004) to compare uncorrected time delayed data to the three methods of sampling interval correction described above when estimating parameters for two common differential equation models.

The results of the Monte Carlo simulation were counterintuitive. Time delay embedding is surprisingly robust to misspecifications of sampling intervals in commonly used sampling designs. The current chapter explores these conditions and points out why this robustness occurs. While we did not explore other parameter estimation techniques for continuous-time modeling (e.g., Oud 2007; Oud and Jansen 2000; Voelkle and Oud 2013), we believe that the time delay embedding approach could be usefully applied by those estimation approaches as something like a hybrid of the state space and SEM versions of the *exact discrete* method.

At this point, it should be noted that neither time delay embedding nor any of the other methods described above protect a model from bias induced by missingness conditioned on the outcome variable (Little and Rubin 1989, i.e., when data are missing not at random). If the reason that unequal intervals exist between observations is that a participant chose not to respond conditioned on their response, then parameter bias due to missingness is still likely to exit. One must be mindful of this possibility, which is outside the boundary of the intention of the current chapter. Here, we will consider the case where the time interval misspecification is independent of the values of the variables for which the model is estimated. One important and common example of independent time interval misspecification is when an irregular sampling interval is part of the experimental design.

10.1.1 Time Delay Embedding

Time delay embedding is a technique by which a time series is reorganized so that short time-ordered sections of the time series appear as rows of a matrix. When the *embedding dimension*, D, is chosen appropriately, Takens' embedding theorem (Takens 1985) guarantees that the dynamics of the time series will be represented in the rows of this matrix. While this sounds complicated, in practice it is straightforward. Suppose we have a time series $[x_1, x_2, \ldots, x_{p-1}, x_p]$, and then we can construct a time delay embedded matrix with an embedding dimension $D = 5$ as

$$\mathbf{X}^{(5)} = \begin{bmatrix} x_1 & x_2 & x_3 & x_4 & x_5 \\ x_2 & x_3 & x_4 & x_5 & x_6 \\ x_3 & x_4 & x_5 & x_6 & x_7 \\ \vdots & \vdots & \vdots & \vdots & \vdots \\ x_{p-4} & x_{p-3} & x_{p-2} & x_{p-1} & x_p \end{bmatrix}. \tag{10.1}$$

Each row of this matrix contains five columns of time-ordered values. Takens' theorem guarantees that if the embedding dimension D has at least the number of degrees of freedom in the generating equation plus one, the time dynamics of the generating equation will be captured. For the purposes of this chapter, we will be considering first- and second-order differential equations with error. For the second-order equation, we would need, at a minimum, four columns in our time delay embedded matrix. What is the maximum number of columns we could use? Suppose our second-order system describes an oscillation with a period of 1 week. Once the total elapsed time between the first and last column equals the period of the oscillation, then we exceed what is commonly known as the *Nyquist limit* (e.g., Jagerman and Fogel 1956), and the time delay embedding no longer captures the dynamics of interest. If our experiment sampled once per day and we wanted to capture a weekly cycle with time delay embedding, we could not have more than six columns in the matrix. However, if we had sampled once per hour, we could use many more columns without running into the Nyquist limit. While rules of thumb for choosing D have been in use for decades (Sauer et al. 1991), recent work has suggested that one may estimate an optimal embedding dimension for a particular model applied to a particular data set (Hu et al. 2014). For the purposes of our simulation, we will choose an embedding dimension of 5 and parameters for the equations that will simulate data that would reasonably correspond to a time delay embedding of 5.

10.1.2 Latent Differential Equations

One way that a time delay embedded matrix is useful is that it preprocesses the time series in such a way that *convolution filtering* can be quickly applied to it. Convolution filtering is an operation in which a *kernel* is multiplied with subselections of a data set (Savitzky and Golay 1964). The kernel is a matrix that is chosen in order to perform some useful operation. In computer graphics, kernels exist for operations such as smoothing, sharpening, edge detection, and blending (Goldman 1983). In the case of LDE, we choose a kernel such that its convolution with the data will produce estimates of derivatives of the time series. If we time delay embed the time series, a simple matrix multiplication of the time delay embedded matrix and the kernel matrix corresponds to a convolution operation. This allows the LDE method to be computationally efficient and to be estimated using standard structural equation modeling (SEM) software. The LDE models used in the simulations are presented in path diagram form and described later in the chapter.

10.2 Methods

Monte Carlo simulations were performed in R (R Core Team 2016) using both univariate and multivariate first- and second-order continuous-time differential equations with parameters selected to be useful in studies of behavioral and psychological regulation. Each parameter was allowed to take on two values: one at the slower end of the typical range and one at the faster end of the range. Thus, there was a slow and a fast exponential decay system for the first-order equation. Similarly there were slow and fast oscillating systems each with slow and fast damping. Thus data for six model and parameter conditions were simulated.

"Slow" and "fast" are relative to the chosen time scale, so these parameters were selected to exhibit behavior that was appropriate for the time interval that would represent the average elapsed time from the first to the last column of the time delay embedding. As an example, one might be interested in weekly cycles in affect and were using a time delay embedding dimension of 5. One might have measured once a day, and so one would have seven observations per cycle. This oscillation would be fast relative to the time delay embedding given that the total elapsed time from the first column to the last column of the time delay embedding comprises $5/7 = 0.7$ of a cycle. However, suppose one measures once per hour and still is interested in the 7-day cycle. In that case, the oscillation would be relatively slow given that a time delay embedding dimension of 5 would only comprise $5/(24 \times 7) = 0.03$ of a cycle.

Given these parameter choices (two cases for the first-order system and four cases for the second-order system), numerical integration was used to create full data sets. 1000 data sets were simulated for the first-order system (500 for each parameter condition) and 1000 data sets for the second-order system (250 for each parameter condition). Then, each resulting data set was degraded by removing all but a target number of observations according to three degradation strategies. For each Monte Carlo iteration, this created three data sets that were identical other than their time interval misspecification. For each Monte Carlo iteration, each data set was then fit by one of the four time interval correction strategies: (1) time delay embedding with no correction, (2) time delay embedding with row-wise loading correction, (3) insertion of latent variables to create a "planned missingness" data set which was then time delay embedded and no further correction, and (4) interpolation splining with fixed time interval resampling which was then time delay embedded with no further correction.

The 1000 simulated data sets by 2 order conditions by 2 uni-/multivariate conditions by 2 error conditions by 3 degradation conditions by 4 time-correction conditions (in all, 96,000 models) were fit in OpenMx (Neale et al. 2016) using the appropriate LDE model with a time delay embedding dimension of 5. The three time interval degradation methods and four misspecification correction methods are described in detail in the next two sections followed by a description of the two models.

10.2.1 Simulation Methods

For each of 1000 Monte Carlo iterations, we first chose parameters for the target differential equation model: either a first-order or second-order linear system with process (time-dependent) error or measurement (time-independent) error. Next, we numerically integrated the selected system at a high time resolution using the `ode()` function in R, producing a densely time-sampled simulated data matrix with $T = 1000$ observations. There were two data simulation model types: first-order univariate linear and second-order linear differential equations. Each of these had a univariate versus multivariate condition as well as measurement error versus a measurement error plus process noise condition.

The first-order univariate linear system with measurement error (time-independent white noise) was specified as

$$\begin{aligned} \dot{F}_i &= \zeta F_i \\ x_i &= F_i + e_i \end{aligned} \tag{10.2}$$

where $\dot{F}_i$ is the first derivative of F with respect to time at time index i and e_i is a residual term drawn from a normal distribution with mean $\mu = 0$ and standard deviation $\sigma = 0.25$. The scores for x_i were saved, in a $T \times 1$ matrix, $\mathbf{X}$, resulting in a densely sampled univariate time series. For each Monte Carlo iteration, the parameter ζ was drawn from the set $\{-0.02, -0.04\}$.

The first-order multivariate simulation with measurement error consisted of a simulated error-free latent variable that was indicated by three manifest variables with independent error terms

$$\begin{aligned} \dot{F}_i &= \zeta F_i \\ x_i &= b_1 F_i + e_{ix} \\ y_i &= b_2 F_i + e_{iy} \\ z_i &= b_3 F_i + e_{iy} \end{aligned} \tag{10.3}$$

where e_i is a residual term drawn from a normal distribution with mean $\mu = 0$ and standard deviation $\sigma = 0.25$. The scores x_1, x_2, and x_3 were stored into a $T \times 3$ matrix, $\mathbf{X}$, resulting in a densely time-sampled multivariate time series. For each Monte Carlo iteration, the parameter ζ was drawn from the set $\{-0.02, -0.04\}$.

The first-order systems with process noise were simulated in the same way except that process error ϵ was added to the latent variable F during the numerical integration, resulting in

$$\begin{aligned} \dot{F}_i &= \zeta F_i + \epsilon_i \\ x_i &= F_i + e_i \end{aligned} \tag{10.4}$$

where ϵ_i was drawn from a normal distribution with mean $\mu = 0$ and standard deviation $\sigma = 0.25$. The first-order multivariate system was modified in the same way, adding a process error ϵ_i to the latent variable F and then proceeding as in Eq. (10.3).

The second-order differential equations were simulated in a similar manner so that, for instance, the second-order linear system with measurement error was specified as

$$\begin{aligned} \ddot{F}_i &= \zeta \dot{F}_i + \eta F_i \\ x_i &= F_i + e_i \end{aligned} \tag{10.5}$$

where $\ddot{F}_i$ is the second derivative of F_i with respect to time at time index i. For each Monte Carlo iteration, the parameter ζ was drawn from the set $\{-0.01, -0.03\}$, and η was drawn from the set $\{-0.0632, -0.2527\}$. The multivariate and process error conditions were specified by modifying Eq. (10.5) in the same way that the equivalent first-order equation (10.2) was modified to produce Eqs. (10.3) and (10.4). Thus, a data set was simulated for each of the $2 \times 2 \times 2$ simulation conditions. This process was repeated 1000 times, and each of these densely time-sampled data sets was degraded in three ways to produce the time sampling misspecifications described in the next section.

10.2.2 Normally Distributed Time Interval Misspecification

There are several reasons why time intervals are irregular in real data. Most common is that intended target observation intervals are not achieved due to scheduling problems. In order to simulate this condition, we first simulated a full data set with a very small time interval between samples and with process error and measurement error. We then selected a fixed long target time interval to approximate the target sampling interval of the experiment. For each target observation time, we drew a normally distributed random number with mean 5.5 and standard deviation $\sigma = 6.75$ as an offset from the target time and then selected the appropriate sample from the full data set while keeping track of the actual time of observation. This results in a data set with time interval misspecification that is normally distributed about a fixed interval similar to a daily diary study where samples were not measured at exactly the same time of day or a yearly longitudinal study where samples were not measured at the same day of the year.

10.2.3 Missing Sleep Intervals

In longitudinal methods variously called *experience sampling* (Csikszentmihalyi and Larson 1987), *ecological momentary assessment* (Shiffman et al. 2008; Stone and Shiffman 1994), and *ambulatory assessment* (Trull and Ebner-Priemer 2013), participants are commonly beeped at randomized intervals throughout the day, but are not disturbed during normal sleeping hours. To simulate this type of sampling misspecification, we created a degraded data set with normally distributed time interval misspecification as in the previous method, and then we further removed blocks of one third of the target observations for each simulated day. Thus if the target data sampling interval was 1 h, a block of eight target observations per day were designated as missing values (NAs) in the same time block every simulated day.

10.2.4 Rectangularly Distributed Time Interval Misspecification

A worst case scenario for time interval misspecification is when the time of each occasion of measurement is drawn from a rectangular distribution of the numbers between zero and the time of the end of the study. In this case, there is no planned target interval between samples. At each moment during the study, there is a probability that a measurement would occur, and thus time intervals between samples can range very widely. To simulate this, we chose a target number, N, of samples and then drew a vector of integers from a rectangular distribution on the interval from 1 to the total number of numbers in the full simulated data set. Duplicated integers were removed, and then this vector was sorted and used to select a degraded data set.

10.2.5 Time Interval Correction Methods

For each Monte Carlo iteration and for each time interval degradation method, we fit a latent differential equation (Boker 2007; Boker et al. 2004) model to the time interval misspecified data. In order to isolate the effects of time interval misspecification, we fit the same model using full-information maximum likelihood and the same time delay embedding dimension, $D = 5$, for every correction method described below.

10.2.5.1 No Correction

In this condition, we ignored the time interval misspecification and proceeded with the time delay embedding as if there were a fixed interval between samples that was equal to the mean interval between all samples in the target degraded data set. Note that the no correction method retains missing values due to a sleep cycle. Thus, if a 1-h sampling interval was planned but eight observations were missing from each day, eight NAs were retained in the data set as placeholders for the missing observations.

10.2.5.2 Row-Wise Loading Correction

Row-wise loading correction produces a fixed factor loading matrix for each row of the time delay embedded data. The loadings are corrected to account for the intervals actually existing on each row of the data. If row i in a time delay embedded matrix with five columns has observation times $\{t_1, t_2, t_3, t_4, t_5\}$, then the 5×3 loading matrix $\mathbf{L}_i$ can be constructed by subtracting half the interval between t_1 and t_5 from each value so that the filter is centered around the full interval and the elapsed basis for the linear part reflects the actual elapsed time between each observation. Thus, $\mathbf{L}_i$ becomes

$$\mathbf{L}_i = \begin{bmatrix} 1 & t_1 - ((t_5 + t_1)/2) & (t_1 - ((t_5 + t_1)/2))^2 \\ 1 & t_2 - ((t_5 + t_1)/2) & (t_2 - ((t_5 + t_1)/2))^2 \\ 1 & t_3 - ((t_5 + t_1)/2) & (t_3 - ((t_5 + t_1)/2))^2 \\ 1 & t_4 - ((t_5 + t_1)/2) & (t_4 - ((t_5 + t_1)/2))^2 \\ 1 & t_5 - ((t_5 + t_1)/2) & (t_5 - ((t_5 + t_1)/2))^2 \end{bmatrix}. \tag{10.6}$$

By using the *definition variable* method of OpenMx, one may insert a loading matrix into the model that is correct for each row. R code to accomplish this is included in supplementary material.

10.2.5.3 Planned Missingness

One popular method for accounting for unequal intervals in longitudinal data is the method of *planned missingness* (Graham et al. 2006; McArdle 1994). This method inserts missing values into the data set when the interval between observations is longer than some specified interval. In order to apply this method, we found the smallest interval s between observations in the target degraded data set, then divided the interval between first and last observations into p equal intervals of length s, and created a vector Y of p missing values (NAs). For each observation in the degraded data set x_i with elapsed time of observation t_i, we found the index, j, into Y such that $s(j-1) \le t_i \le s(j)$ and the set $y_j = x_i$. In this way we created a data vector with approximately fixed interval sampling. The new vector was time delay embedded and fit with the model in the same manner as the other correction methods.

10.2.5.4 Interpolation Spline

Sometimes the data are appropriate for splining. Interpolation splines with resampling from the predicted values of the spline have been used as a method for producing equal interval time series (e.g. Hu et al. 2014; Klump et al. 2013). To implement this method, we used the `na.spline()` function from the R zoo package. We submitted the degraded vector to the interpolation spline function and then sampled predicted values of the interpolation spline at the median interval as found in the degraded data. In this way, we created a time series with true fixed intervals, but with estimated values at the fixed intervals. In a way, the interpolation spline method is the obverse of the planned missingness method. Interpolation splines give exactly equal intervals but with an estimated value of the variable, whereas planned missingness gives exact values of the variables but at approximately equal intervals. The interpolated data were then time delay embedded and fit with the model in the same manner as above.

10.2.6 Structural Equation Models

Each data set was fit using an LDE structural equation model appropriate to the simulated data. Thus first-order simulated data was fit with a first-order LDE. Path diagrams of first- and second-order univariate models are shown in Fig. 10.1. The models for both measurement error and process error simulations were the same, where measurement error is estimated by residual variances of the manifest variables and process error is estimated by the residual variance labeled $Ve\dot{F}$ in the first-order model in Fig. 10.1a and $Ve\ddot{F}$ in the second-order model in Fig. 10.1b.

Path diagrams of first- and second-order multivariate models are shown in Fig. 10.2. Note that the embedding dimension is the same as it was in the univariate models, but since there are now three variables being embedded, each line of the time delay embedded matrix has 15 columns.

Boker et al. (2016) present an example of OpenMx code and an extensive discussion of each of these models. Supplementary material has example code that creates simulated data conforming to this chapter and implements the row-wise time correction. Note that supplementary material relies on the GLLAfunctions.R source file from Boker et al. (2016).

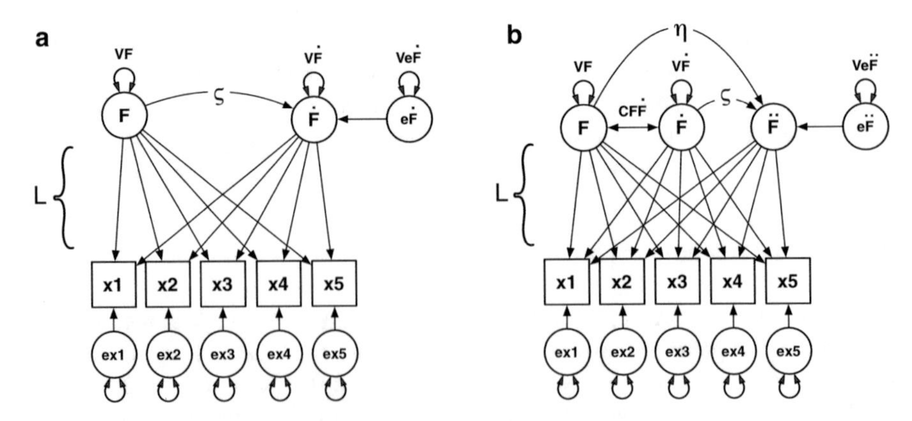

Fig. 10.1 Univariate latent differential equation (LDE) path models used to fit the time misspecified data. (**a**) First-order linear univariate LDE. (**b**) Second-order univariate linear LDE

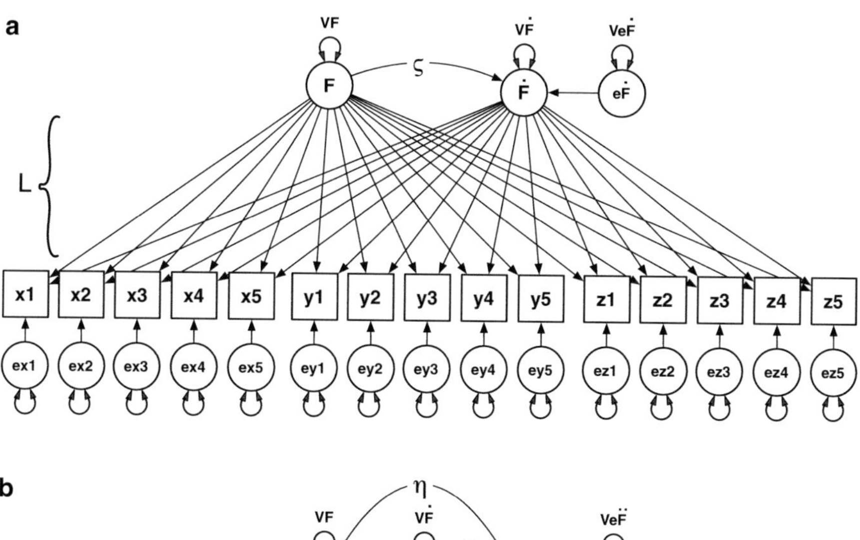

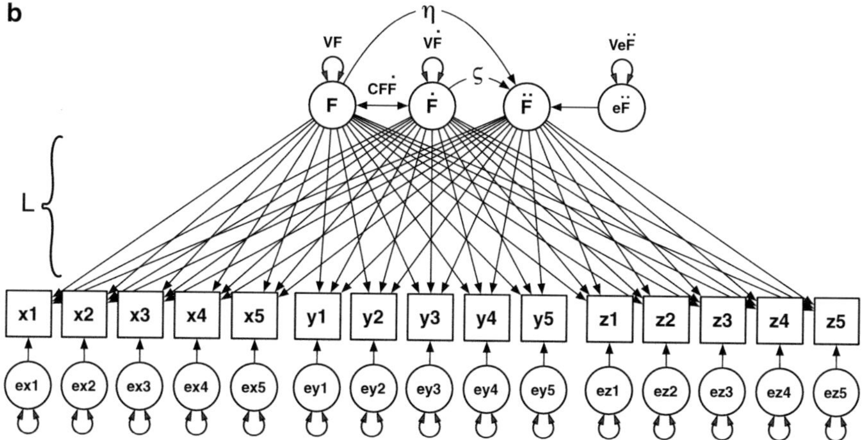

Fig. 10.2 Multivariate latent differential equation (LDE) path models used to fit the time misspecified data. (**a**) First-order linear multivariate LDE. (**b**) Second-order multivariate linear LDE

10.3 Results

For each parameter in each simulation, misspecification, and correction condition, the root-mean-square error (RMSE) in parameter estimation was calculated as

$$\text{RMSE} = \sqrt{1/N \sum_{i=0}^{N} (p_i - p_{sim})^2} \tag{10.7}$$

where N is the number of iterations in the Monte Carlo simulation, p_i is an estimated parameter for iteration i, and p_{sim} is the parameter value used to create

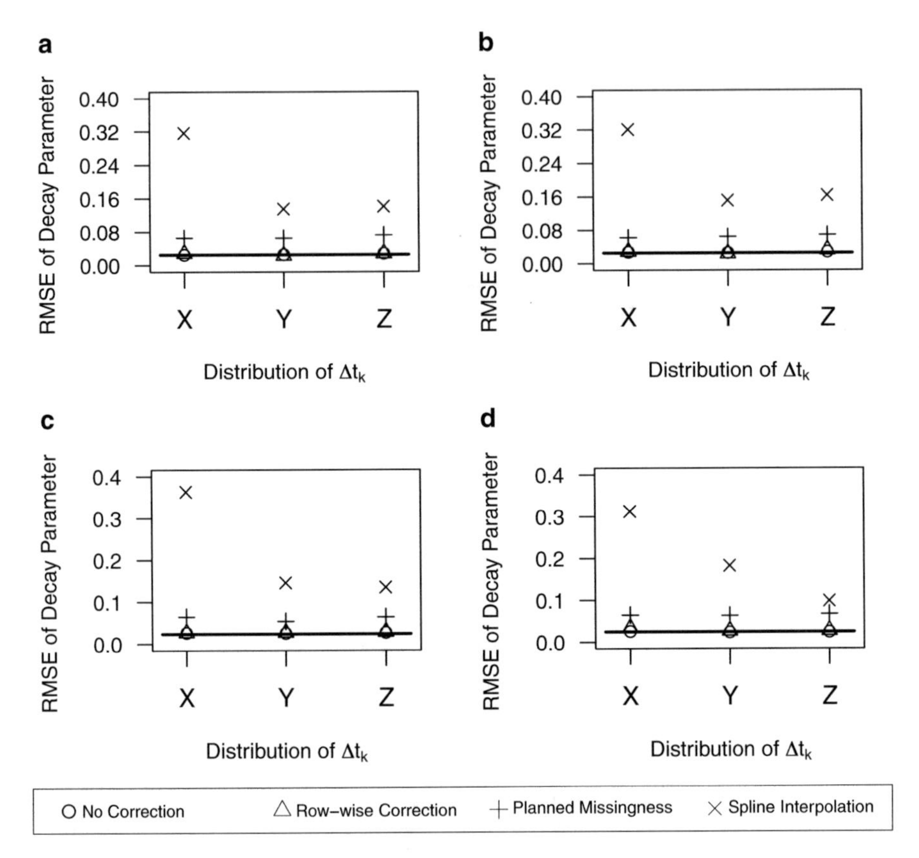

Fig. 10.3 Root-mean-square error (RMSE) of the decay parameter, ζ, for 100 simulated first-order trajectories for each misspecification method and each correction method. (**a**) Univariate indicator and no process error. (**b**) Univariate indicator with process error. (**c**) Multivariate indicators and no process error. (**d**) Multivariate indicators with process error. In each graph, the black bar plots RMSE values from a baseline model with equal interval data. On the horizontal axis, X is uniform distribution misspecification, $U \sim (1, 1000)$; Y is normally distributed misspecification, $N \sim (\mu = 5.5, \sigma = 6.75)$; and Z is sleep misspecification

the simulated data. Figure 10.3 plots the RMSE of the estimated ζ parameter for each first-order differential equation condition. Note that both the row-wise correction and no-correction conditions produce approximately the same RMSE as the estimates using the full data set for every misspecification condition, both error types and both uni- and multivariate data. The planned missingness correction produced a small increase in the RMSE in every condition. The interpolation spline produced the worst performance, with RMSE much larger than any of the other methods of correction. For first-order processes where the spline interpolation correction was used, the degree of bias in the estimated decay parameters was greatest for uniform time interval degradation relative to the normal and sleep time interval degradation methods, whereas the degree of bias in the estimated decay

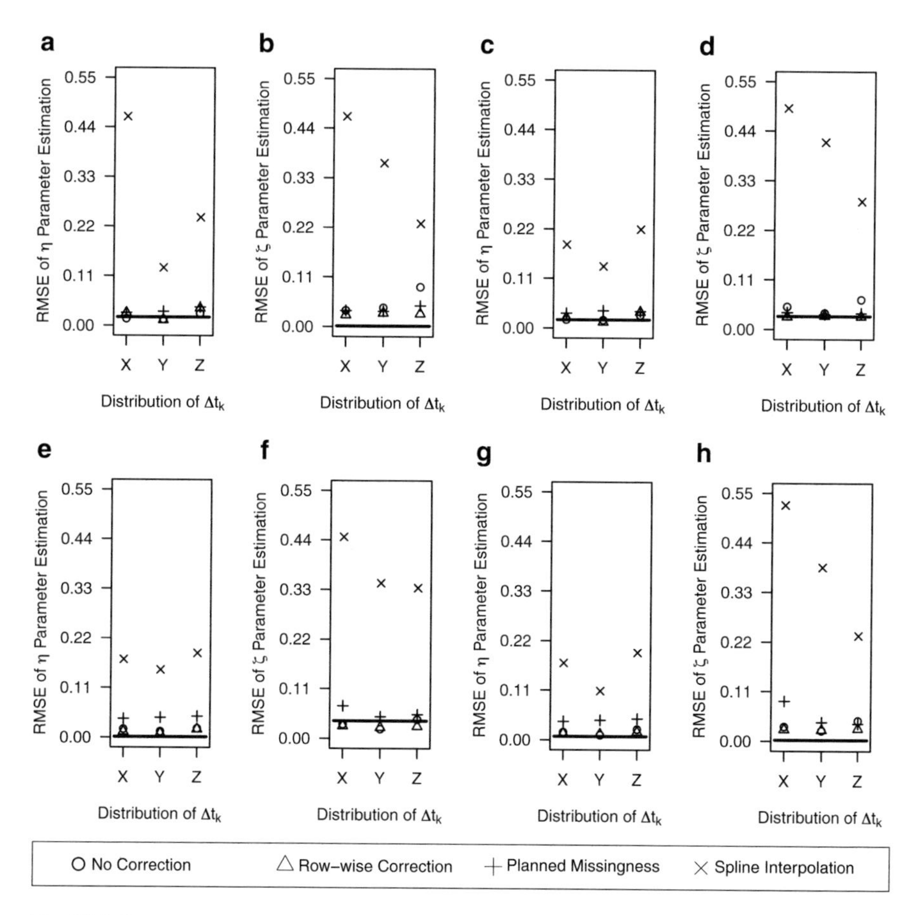

Fig. 10.4 Root-mean-square error (RMSE) of the (**a**, **c**, **e**, **g**) η and (**b**, **d**, **f**, **h**) ζ parameters for 100 simulated first-order trajectories for each misspecification method and each correction method. (**a**, **b**) Univariate indicator and no process error. (**c**, **d**) Univariate indicator with process error. (**e**, **f**) Multivariate indicators and no process error. (**g**, **h**) Multivariate indicators with process error. In each graph, the black bar plots RMSE values from a baseline model with equal interval data. On the horizontal axis, X is uniform distribution misspecification, $U \sim (1, 1000)$; Y is normally distributed misspecification, $N \sim (\mu = 5.5, \sigma = 6.75)$; and Z is sleep misspecification

parameters was nearly equivalent for normal versus sleep time interval degradation methods.

Figure 10.4 plots the RMSE of the estimated ζ and η parameters for each second-order differential equation condition. Again, the uncorrected and row-wise correction performed the best, and RMSE of the parameter estimates was nearly identical to each other in almost all conditions. However, when estimating ζ in the univariate case and sleep misspecification, the uncorrected version performed worse than the row-wise corrected version. The planned missingness correction performed better for the second-order models than it did for the first-order models, although when estimating η in the multivariate condition, it performed somewhat worse than either the row-wise corrected or uncorrected versions. When the spline interpolation correction was used to estimate the ζ parameter in the second-order processes, the greatest to least amount of bias in the RMSE was observed for the uniform, normal, and sleep time interval degradation methods, respectively, across all conditions (i.e., uni- and multivariate data, with or without process error). On the contrary, no discernible pattern emerged in RMSE estimates for the estimation of η in the first-order processes using the spline interpolation correction across conditions.

In order to understand if the RMSE was significantly worse than baseline for each condition, we performed a regression analysis, and the results are shown in Tables 10.1 and 10.2. From these tables, it is clear that the interpolation method is always significantly worse than baseline. On the other hand, no correction and row-

Table 10.1 Significance levels from regression results predicting RMSE difference from baseline for first-order models combining the conditions with and without process error

	RMSE estimating ζ		
Correction	Normal	Uniform	Sleep
Univariate			
None	ns	ns	ns
Row-wise	0.04	0.001	ns
Interpolation	<0.001	<0.001	<0.001
Planned missing	0.02	<0.001	0.02
Multivariate			
None	ns	ns	ns
Row-wise	ns	ns	ns
Interpolation	<0.001	<0.001	<0.001
Planned missing	ns	ns	ns

Table 10.2 Regression results predicting RMSE difference from baseline for second-order models combining the conditions with and without process error

	RMSE estimating η			RMSE estimating ζ		
Correction	Normal	Uniform	Sleep	Normal	Uniform	Sleep
Univariate						
None	ns	ns	ns	ns	ns	0.03
Row-wise	ns	ns	ns	ns	ns	ns
Interpolation	<0.001	<0.001	<0.001	<0.001	<0.001	<0.001
Planned missing	0.01	ns	ns	ns	ns	ns
Multivariate						
None	ns	0.009	0.003	ns	ns	ns
Row-wise	ns	ns	0.02	ns	ns	ns
Interpolation	<0.001	<0.001	<0.001	<0.001	<0.001	<0.001
Planned missing	0.03	<0.001	<0.001	ns	ns	ns

wise correction are almost always comparable to the performance of the baseline model fit to equal interval data.

10.4 Discussion

The results of the simulation are surprising but convergent. When using time delay embedding, ignoring time misspecification when fitting LDE first- and second-order models almost always performed as well as the best correction method: row-wise loading correction using full-information maximum likelihood. When we started this work, we had planned to warn users not to ignore a violation of the assumption of equal interval sampling and show the consequences of that mistake. But after the simulation, we found we needed to answer a different question: Why is time delay embedding so robust to violations of equal interval sampling? Our explanation involves two characteristics; one due to time delay embedding and one due to the ipsative nature of time interval misspecification.

10.4.1 Effect of Time Delay Embedding

Previous work has shown that time delay embedding improves precision of estimates in first- and second-order latent differential equations due to induced canceling effects (von Oertzen and Boker 2010). A graphical illustration of why this happens is shown in Fig. 10.5a. When a perturbation is added to a single value of the true scores for a sine curve, estimates of the frequency are biased in one direction when the perturbation is to the right of center of the time delay embedded matrix, $\mathbf{X}^{(5)}$, and are biased by an equivalent amount when the perturbation is to the left of center of $\mathbf{X}^{(5)}$. The reason that this cancelation of bias occurs is because the convolution filter values, i.e., the loadings of $\mathbf{L}$ in Eq. (10.6), are symmetric about the center column of $\mathbf{X}^{(5)}$. Thus equal bias is applied with a positive and negative sign as the convolution filter is applied to the data.

Similarly, consider the time misspecification illustrated in Fig. 10.5b. In this example, a perturbation of the time interval occurs such that two observations, x_5 and x_6, were not recorded so that the time series skips directly from x_4 to x_7, thereby violating the assumption of equal time intervals between samples. This longer time interval first is at the rightmost column of the time delay embedded matrix, $\mathbf{X}^{(5)}$, and then in each subsequent row, the perturbation moves one column to the left. When the time perturbation is at the right of the center in $\mathbf{X}^{(5)}$, the estimated wavelength of the sine is biased to be longer, but when the time perturbation is to the right of the center, the bias is toward shorter wavelengths. In this way, the bias induced by the time interval misspecification tends to cancel.

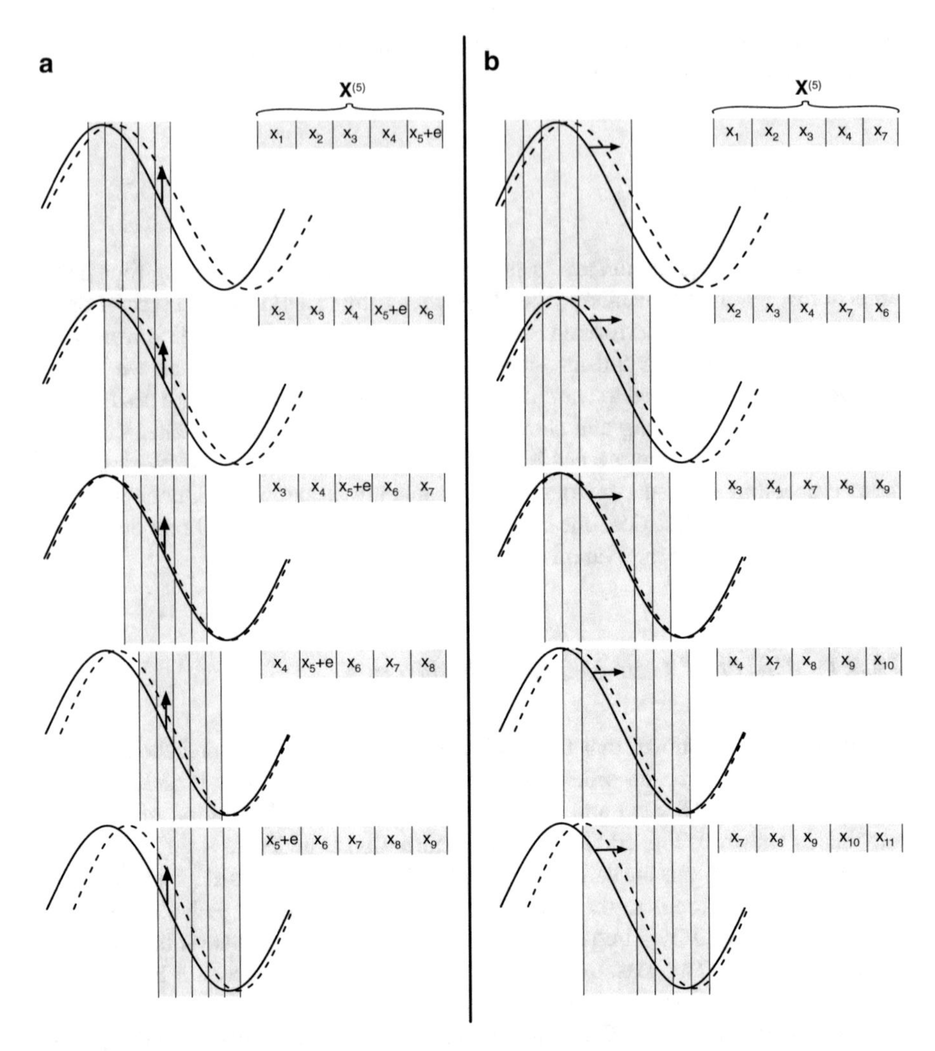

Fig. 10.5 Graphical depiction of error cancelation induced by time delay embedding. (**a**) Example of a single perturbation, e, canceling its own bias. The true signal sine curve (solid line) and the biased estimate (dotted line) resulting at each row of a five-dimensional time delay embedding $\mathbf{X}^{(5)}$. When the perturbation is to the right of the center column, estimates are biased toward longer wavelength, and when the perturbation is to the left of the center, estimates are biased toward shorter wavelengths. (**b**) Example of time interval misspecification where observations x_5 and x_6 are skipped. When the longer interval is to the right of the center, estimates are biased toward longer wavelengths, and when the longer interval is to the left of the center, estimates are biased toward shorter wavelengths

10.4.2 Effect of Ipsative Time Interval Misspecification

The previous section provides an explanation that relies on bias cancelation. This explanation would fail if, for instance, time intervals were ever increasing and one took the reference time interval to be the first interval encountered. In this case, one would always have asymmetric effects of time perturbation, leading to a positive bias and therefore a large RMSE. In the method presented here, the reference time interval is taken to be the total elapsed time between the first and last observation in the time series divided by the total number of observations in the time series minus one.

Using the mean time interval as the reference time interval creates an interesting balance. If the source of the time misspecification is missing data, then the mean time interval becomes longer. Suppose 20 observations are missing from a full time series consisting of 101 observations sampled at regular interval t. This results in a time series of 80 intervals in which dynamics are happening that really occurred over an elapsed time of $100t$. As long as the first and last observations are still present, the total elapsed time in the degraded data set is still $100t$. So when we choose a reference time interval of $100t/80 = 1.25t$, we are automatically rescaling time so that the total elapsed time evolution of the dynamics is not affected by the missingness. Thus, one can maintain a constant scale for time when we are estimating our parameters.

Suppose now that one does not have a missing data problem. We have 101 observations that occur over a time interval of $100t$. But now, suppose just one of the observations, x_{50}, is sampled too early by an amount of time ϵ. Thus, the interval between x_{49} and x_{50} is $t - \epsilon$, but it follows that the interval between x_{50} and x_{51} must be $t + \epsilon$. This is what is meant by the ipsative nature of time misspecification. Holding the total elapsed time and number of observations constant, every interval that is smaller than the average produces a "time debt" that must be paid by a longer interval or intervals somewhere else in the time series. Conversely, longer intervals produce a "time surplus" that must be balanced by the "time debt" from other intervals. It has become apparent to us that the combination of the ipsative nature of time misspecification along with the bias canceling properties of time delay embedding has produced the observed surprising robustness.

10.4.3 Problems and Limitations

While this chapter explored a large number of conditions ($2 \times 2 \times 2 \times 3 \times 4$) that could reasonably be expected to be encountered in intensive longitudinal studies, not every possible condition was explored. For instance, we did not consider short time series. Given a close examination of Fig. 10.5, one may see that when the full time series is embedded, the first four and last four rows of $\mathbf{X}^{(5)}$ can contain bias that is not canceled. This problem is known as a *boundary effect* or *edge effect* in the convolution literature. The shorter the time series and the larger the embedding dimension, the greater is this boundary effect. Further work needs to be done to understand how to compensate for these boundary effects. It is likely that the row-wise correction method will turn out to be a significant improvement over no correction in time series with few observations.

Another potential problem that was not addressed by this chapter is the effect of so-called Nyquist limit violations. When there is time interval misspecification and the mean interval is greater than half the wavelength of the smallest feature that one wishes to recover, does the sampling theorem hold? There are hints from computer vision that the noisy intervals can do better than what would be expected by the sampling theorem (Cook 1986; Hennig and Wörgötter 2004). Our intuition is that randomized intervals may allow one to recover features that equal interval sampling may miss (see Voelkle and Oud 2013).

This chapter does not address questions of multi-time scale analyses. The simulations were not mixtures of source signals with differing time scales. This is an active area of research, and it will be important to address problems in non-equal time interval data when such mixtures are suspected.

10.5 Conclusions

Time delay embedding combined with convolution filtering is a surprisingly robust way of estimating dynamics from time series. This chapter simulated common time misspecifications and fits common models using one of the popular convolution filtering techniques, latent differential equations. For almost all of the 96 conditions explored in the simulation, row-wise correction of time misspecification and just simply ignoring the time misspecification resulted in root-mean-squared error that was not significantly different from estimates made using the full time series.

Interpolation splining should be used with caution. In almost every condition, interpolation splines performed significantly worse: RMSE effect sizes were mostly larger by a factor of 10 or more than those of row-wise correction and no correction. It is apparent that interpolation splines can induce bias into the estimation procedure. The method of planned missingness uniformly performed slightly worse than row-wise correction as evidenced by a slightly larger RMSE in almost every condition. In most conditions, other than when fitting univariate first-order models, the increased RMSE due to the method of planned missingness was not statistically significant.

Overall, in only 3 out of 18 time misspecification and modeling conditions did the RMSE of row-wise correction differ from the RMSE of the same model fit to equal time interval data. Likewise, in only 3 out of 18 conditions was the RMSE of no correction significantly different from the RMSE of the same model fit to equal time interval data. We can comfortably recommend using row-wise corrections for time misspecification. However, sometimes actual sample intervals are not available, but the overall elapsed time of the experiment is available. In this case, we recommend using a time delay embedding approach with continuous-time modeling due to its robustness to time interval misspecification. However, uncorrected time delay embedding may not be appropriate for short time series due to boundary effects.

We would like to conclude by reminding the readers that only by using some form of continuous-time modeling can one obtain parameter estimates that are invariant with respect to the sampling interval. Equal interval data does not remove the problem of time interval dependency in discrete-time models.

Acknowledgements Funding for this work was provided in part by NIH Grant 1R21DA024304–01. Any opinions, findings, and conclusions or recommendations expressed in this material are those of the authors and do not necessarily reflect the views of the National Institutes of Health.

References

Boker, S. M. (2007). Specifying latent differential equations models. In S. M. Boker & M. J. Wenger (Eds.), *Data analytic techniques for dynamical systems in the social and behavioral sciences* (pp. 131–159). Mahwah, NJ: Lawrence Erlbaum Associates.

Boker, S. M., Deboeck, P. R., Edler, C., & Keel, P. K. (2010). Generalized local linear approximation of derivatives from time series. In S.-M. Chow & E. Ferrar (Eds.), *Statistical methods for modeling human dynamics: An interdisciplinary dialogue* (pp. 161–178). Boca Raton, FL: Taylor & Francis.

Boker, S. M., Neale, M. C., & Rausch, J. (2004). Latent differential equation modeling with multivariate multi-occasion indicators. In K. van Montfort, J. H. L. Oud, & A. Satorra (Eds.), *Recent developments on structural equation models: Theory and applications* (pp. 151–174). Dordrecht: Kluwer Academic Publishers. https://doi.org/10.1007/978-1-4020-1958-6_9

Boker, S. M., Staples, A., & Hu, Y. (2016). Dynamics of change and changes in dynamics. *Journal of Person-Oriented Research, 2*(1–2), 34–55. https://doi.org/10.17505/jpor.2016.05

Cook, R. L. (1986). Stochastic sampling in computer graphics. *ACM Transactions on Graphics, 5*(1), 51–72. https://doi.org/10.1145/7529.8927

Csikszentmihalyi, M., & Larson, R. (1987). Validity and reliability of the experience-sampling method. *The Journal of Nervous and Mental Disease, 175*(9), 526–536. https://doi.org/10.1097/00005053-198709000-00004

Delsing, M. J. M. H., & Oud, J. H. L. (2008). Analyzing reciprocal relationships by means of the continuous-time autoregressive latent trajectory model. *Statistica Neerlandica, 62*(1), 58–82. https://doi.org/10.1111/j.1467-9574.2007.00386.x

Dufau, S., Duñabeitia, J. A., Moret-Tatay, C., McGonigal, A., Peeters, D., Alario, F.-X., et al. (2011). Smart phone, smart science: How the use of smartphones can revolutionize research in cognitive science. *PLoS ONE, 6*(9), 1–3. https://doi.org/10.1371/journal.pone.0024974

Erbacher, M. K., Schmidt, K. M., Boker, S. M., & Bergeman, C. S. (2012). Measuring positive and negative affect in older adults over 56 days: Comparing trait level scoring methods using the partial credit model. *Applied Psychological Measurement, 13*(2), 146–164.

Goldman, R. N. (1983). An urnful of blending functions. *IEEE Computer Graphics and Application, 3*, 49–54. https://doi.org/10.1109/MCG.1983.263276

Graham, J. W., Taylor, B. J., Olchowski, A. E., & Cumsille, P. E. (2006). Planned missing data designs in psychological research. *Psychological Methods, 11*(4), 323–343. https://doi.org/10.1037/1082-989X.11.4.323

Hektner, J. M., Schmidt, J. A., & Csikszentmihalyi, M. (2006). Experience sampling method: Measuring the quality of everyday life. Thousand Oaks, CA: SAGE Publications, Incorporated.

Hennig, M. H., & Wörgötter, F. (2004). Eye micro-movements improve stimulus detection beyond the Nyquist limit in the peripheral retina. In *Advances in neural information processing systems* (pp. 1475–1482).

Hu, Y., Boker, S. M., Neale, M. C., & Klump, K. (2014). Latent differential equations with moderators: Simulation and application. *Psychological Methods, 19*(1), 56–71. https://doi.org/10.1037/a0032476

Jagerman, D., & Fogel, L. (1956). Some general aspects of the sampling theorem. *IRE Transactions on Information Theory, 2*(4), 139–146. https://doi.org/10.1109/TIT.1956.1056821

Klump, K. L., Keel, P. K., Kashy, D. A., Racine, S., Burt, S. A., Neale, M., et al. (2013). The interactive effects of estrogen and progesterone on changes in binge eating across the menstrual cycle. *Journal of Abnormal Psychology, 122*(1), 131–137. https://doi.org/10.1037/a0029524

Little, R. J. A., & Rubin, D. B. (1989). The analysis of social science data with missing values. *Socialogical Methods and Research, 18*, 292–326. https://doi.org/10.1177/0049124189018002004

McArdle, J. J. (1994). Structural factor analysis experiments with incomplete data. *Multivariate Behavioral Research, 29*(4), 409–454. https://doi.org/10.1207/s15327906mbr2904_5

McArdle, J. J., & Woodcock, J. R. (1997). Expanding test rest designs to include developmental time lag components. *Psychological Methods, 2*(4), 403–435. https://doi.org/10.1037/1082-989X.2.4.403

Neale, M. C., Hunter, M. D., Pritikin, J. N., Zahery, M., Brick, T. R., Kirkpatrick, R., et al. (2016). Openmx 2.0: Extended structural equation and statistical modeling. *Psychometrika, 81*(2), 535–549. PMCID: 25622929. https://doi.org/10.1007/s11336-014-9435-8

Oud, J. H. L. (2007). Continuous time modeling of reciprocal relationships in the cross-lagged panel design. In S. M. Boker & M. J. Wenger (Eds.), *Data analytic techniques for dynamical systems* (pp. 87–130). Mahwah, NJ: Lawrence Erlbaum Associates.

Oud, J. H. L., & Jansen, R. A. R. G. (2000). Continuous time state space modeling of panel data by means of SEM. *Psychometrika, 65*(2), 199–215. https://doi.org/10.1007/BF02294374

R Core Team. (2016). *R: A language and environment for statistical computing [Computer software manual]*. Vienna, Austria. Retrieved from https://www.R-project.org/

Salthouse, T. A., Schroeder, D. H., & Ferrer, E. (2004). Estimating retest effects in longitudinal assessments of cognitive functioning in adults between 18 and 60 years of age. *Developmental Psychology, 40*(5), 813–822. https://doi.org/10.1037/0012-1649.40.5.813

Salthouse, T. A., & Tucker-Drob, E. M. (2008). Implications of short-term retest effects for the interpretation of longitudinal change. *Neuropsychology, 22*(6), 800–811. https://doi.org/10.1037/a0013091

Sauer, T., Yorke, J. A., & Casdagli, M. (1991). Embedology. *Journal of Statistical Physics, 65*(3,4), 95–116. https://doi.org/10.1007/BF01053745

Savitzky, A., & Golay, M. J. E. (1964). Smoothing and differentiation of data by simplified least squares. *Analytical Chemistry, 36*, 1627–1639. https://doi.org/10.1021/ac60214a047

Shiffman, S., Stone, A. A., & Hufford, M. R. (2008). Ecological momentary assessment. *Annual Review of Clinical Psychology, 4*, 1–32. https://doi.org/10.1146/annurev.clinpsy.3.022806.091415

Stone, A. A., & Shiffman, S. (1994). Ecological momentary assessment (EMA) in behavioral medicine. *Annals of Behavioral Medicine, 16*, 199–202.

Takens, F. (1985). Detecting strange attractors in turbulence. In A. Dold & B. Eckman (Eds.), *Lecture notes in mathematics 1125: Dynamical systems and bifurcations* (pp. 99–106). Berlin: Springer.

Trull, T. J., & Ebner-Priemer, U. (2013). Ambulatory assessment. *Annual Review of Clinical Psychology, 9*, 151–176. https://doi.org/10.1146/annurev-clinpsy-050212-185510

Voelkle, M. C., & Oud, J. H. (2013). Continuous time modelling with individually varying time intervals for oscillating and non-oscillating processes. *British Journal of Mathematical and Statistical Psychology, 66*(1), 103–126. https://doi.org/10.1111/j.2044-8317.2012.02043.x

von Oertzen, T., & Boker, S. M. (2010). Time delay embedding increases estimation precision of models of intraindividual variability. *Psychometrika, 75*(1), 158–175. https://doi.org/10.1007/s11336-009-9137-9

Chapter 11
Recursive Partitioning in Continuous Time Analysis

Andreas M. Brandmaier, Charles C. Driver, and Manuel C. Voelkle

11.1 Introduction

In psychological research, the analysis of large-scale data sets has been gaining increasing importance and bears novel challenges for empirical researchers. Such "big" data sets arise, for example, through affordable and diverse internet-based sampling, integrative data analysis, or intensive collection methods (e.g., day reconstruction, Kahneman et al. 2004, or momentary experience sampling, Mehl and Conner 2012). In the context of longitudinal assessment, "big" data with many measurement occasions have also been called intensive longitudinal data (ILD; Walls and Schafer 2006). ILD reflects quantitative data, usually from multiple individuals at a large number of time points, often irregularly sampled at various intervals and different numbers per person. Longitudinal data analysis, whether considered "big" or not, plays an increasing role in empirical psychological research, especially within developmental and life span psychology. Longitudinal data sets of human development and aging are specifically characterized by their inherent heterogeneity of between- and within-person changes, which challenge the very interpretability of averaged data. Walls and Schafer (2006) noted the complexity and variety of individual trajectories in intensive longitudinal data sets

Electronic Supplementary Material The online version of this article (https://doi.org/10.1007/978-3-319-77219-6_11) contains supplementary material, which is available to authorized users.

A. M. Brandmaier (✉) · C. C. Driver
Center for Lifespan Psychology, Max Planck Institute for Human Development, Berlin, Germany
e-mail: brandmaier@mpib-berlin.mpg.de; driver@mpib-berlin.mpg.de

M. C. Voelkle
Department of Psychology, Humboldt-Universität zu Berlin, Berlin, Germany
e-mail: manuel.voelkle@hu-berlin.de

K. van Montfort et al. (eds.), *Continuous Time Modeling in the Behavioral and Related Sciences*, https://doi.org/10.1007/978-3-319-77219-6_11

and emphasized the need to move beyond simple time-graded population average effects. Only rarely does it seem plausible that a single model should hold for everybody, and we often expect large differences between persons. For example, while one person may appear quite stable in a measure of emotion regulation, another may reveal highly erratic behavior, and a third person may show a rather quick return to equilibrium after disturbances through exogenous shocks to their emotion regulation system. The discrepancy or even inconsistency between person-specific and group-average models has long been criticized and dates back to the observation that average learning curves must not represent the learning curve of any single individual (Tucker 1966). Although it has been argued repeatedly that the privileged unit of analysis in psychology should be the individual (Nesselroade et al. 2007), many analyses rely solely on average effects. Thus, there is a dire need for formal treatments of heterogeneity in longitudinal data analysis. Continuous time models are a useful device to address the issue of heterogeneity of measurement occasions in complex research designs, where the exact timing of measurement occasions may vary both within and between persons. In addition, heterogeneity may also appear at the level of parameter estimates, that is, some persons or groups of persons may be better represented by different dynamics than others. If the reasons for such heterogeneity are unobserved, a hierarchical modeling approach may be useful, in which between-person differences in parameters are explicitly modeled using distributional assumptions, or a mixture modeling approach may be useful to retrieve unobserved group membership. If heterogeneity is observed, that is, if there are variables that may explain differences, for example, mean differences in latent or observed variables, their effects can be explicitly tested by either including them in the model or by adding them to a multiple group model as group indicators. However, how to proceed if there is a large number of potentially relevant variables, whose roles (e.g., moderators, predictors of mean differences, higher-order interaction terms) in the model are unknown, and it is difficult to hypothesize about them? Here, SEM trees (Brandmaier et al. 2013b) and forests (Brandmaier et al. 2016), which are a combination of decision trees and structural equation models (SEM), offer a viable approach to deciding which variables to select by choosing those variables that add the largest predictive power to a multivariate model.

11.2 Decision Trees

The history of decision trees can be traced back to the late 1950s, but their popularity only started growing with the automatic interaction detector (AID) by Morgan and Sonquist (1963). The method's fame culminated with the seminal work of Breiman et al. (1984) and Quinlan (1986). The idea behind a decision tree is simple. A tree is a set of decision nodes that are connected by branches, with a root node at which the decision process starts and a set of end nodes ("leaves"), at which the decision process ends and a prediction is made. The decision nodes along the way determine which branch to take next. This allows one to formalize the association of different patterns of decisions (or predictors) and a prediction (or outcome). Learning a decision tree from data is achieved by growing the tree in a stepwise fashion. The sample is subdivided according to one out of potentially many predictors such that it best explains differences between the resulting subdivisions with respect to an outcome of interest. One proceeds by independently examining the subdivisions again for a best predictor and, again, continues in the resulting subdivisions. Stepwise procedures that divide data and "reapply" themselves to their results are referred to as "recursive" in computer science, and, thus, the decision tree paradigm is also known as recursive partitioning. The fact that the approach of recursively splitting data can be represented in a tree structure gave rise to the name decision trees. Figure 11.1 shows a simple decision tree with two binary predictors and a binary outcome.

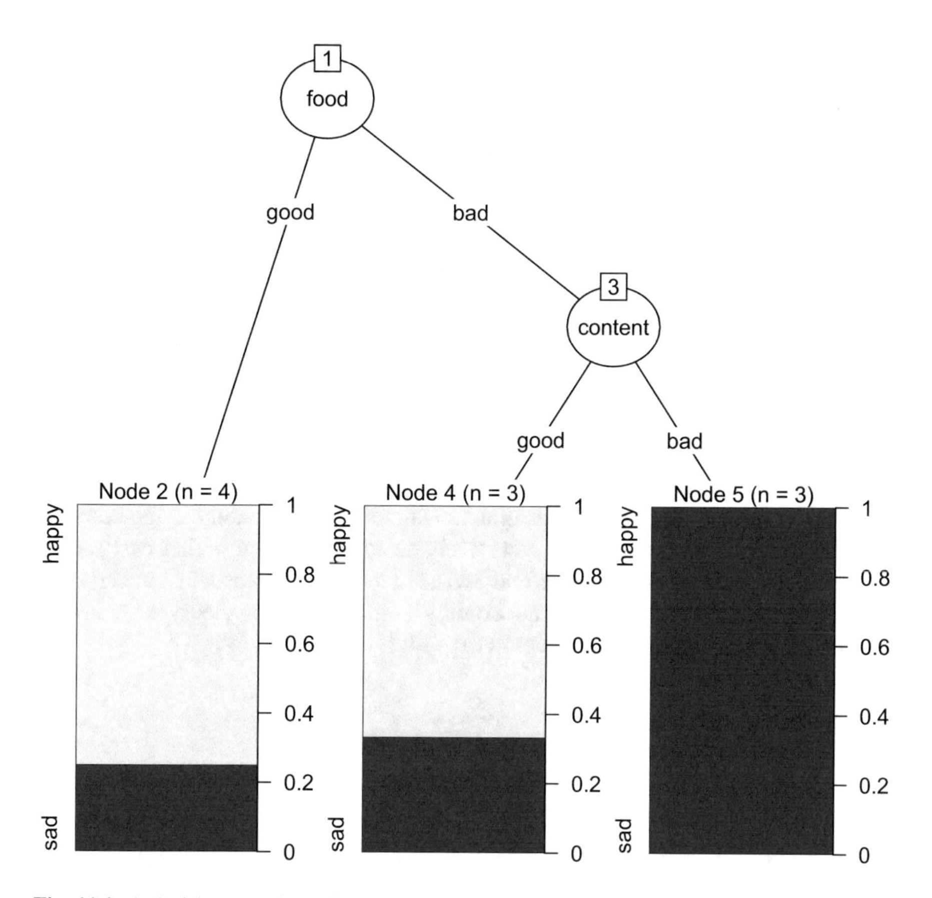

Fig. 11.1 A decision tree that selected predictors according to maximal reduction in entropy in self-reported mood according to the data presented in Table 11.1 created with R packages rpart (Therneau et al. 2015) and partykit (Hothorn and Zeileis 2015)

Table 11.1 A hypothetical data set of a researcher's self-reported mood (happy/sad) after having visited a scientific conference

Food	Content	Fees	Mood
Bad	Good	Low	Happy
Good	Bad	Low	Happy
Good	Good	Low	Sad
Bad	Bad	Low	Sad
Good	Good	High	Happy
Good	Bad	High	Happy
Bad	Good	High	Happy
Bad	Good	High	Sad
Bad	Bad	High	Sad
Bad	Bad	High	Sad

Predictors of the researcher's dichotomous mood rating ("mood") are binary ratings of three conference properties: quality of the scientific content ("content"), food quality at the preconference dinner ("food"), and level of conference fees ("fees")

11.2.1 Cross-Sectional Decision Trees

In the following, we illustrate how decision trees, such as the one in Fig. 11.1, are learnt from data. Table 11.1 shows a hypothetical data set of a researcher's self-reported mood (with either positive or negative valence) after having visited scientific conferences around the globe. Here, conferences are distinguished by the researcher's dichotomous ratings of their scientific quality ("content"), the quality of the restaurant in which the preconference dinner took place ("food"), and the conference fees ("fees"). What conference features are most important to maximize the researcher's happiness? A decision tree provides a systematic approach to creating a predictive model to answer this question. The decision tree algorithm proceeds as follows. First, we determine the amount of uncertainty in the outcome (mood), and then we proceed by recursively selecting predictors that best reduce uncertainty in the outcome. To this end, we need to select a measure of uncertainty. A generic measure for quantifying uncertainty from information theory is Shannon's entropy. Entropy is defined as shown in Eq. (11.1):

$$I = -\sum_{i=1}^{M} p_i \log p_i \tag{11.1}$$

with p_i being the probability of an outcome i (e.g., being happy) among all possible outcomes M (i.e., happy vs. sad, then $M = 2$). Entropy is larger with more unpredictable outcomes and zero for perfectly predictable outcomes. To start, all pair-wise associations of mood and any of the potential predictors (content, food, and fees) are examined, and we select the one that best reduces uncertainty in the outcome, that is, the one that maximally decreases entropy. Unconditional entropy in mood is $-\frac{5}{10}\log\frac{5}{10} - \frac{5}{10}\log\frac{5}{10} = 0.69$.[1] If we split the data set into two subgroups according to the predictor "food," we obtain an entropy for mood conditional on "bad" food of $-\frac{2}{6}\log\frac{2}{6} - \frac{4}{6}\log\frac{4}{6} = 0.64$ and an entropy for mood conditional on "good" food of $-\frac{1}{4}\log\frac{1}{4} - \frac{3}{4}\log\frac{3}{4} = 0.56$. In the first condition, entropy was higher, and thus prediction was less reliable than in the second condition. However, in both conditions, uncertainty in the prediction was lower than in the unconditional model. To obtain the total reduction of uncertainty, or put differently, the *information gain,* we compute the difference between the unconditional entropy and the conditional entropies weighted by sample size:

$$0.69 - \left(\frac{6}{10} \cdot 0.64 + \frac{4}{10} \cdot 0.56\right) = 0.082$$

We find that conditioning on a conference's food quality does indeed reduce entropy in the outcome by 0.082 nats.[2] Now, when we evaluate the other predictors, we find that both have a lower reduction of uncertainty. The reduction of uncertainty in mood conditional on scientific content is:

$$0.69 - \left(\frac{5}{10} \cdot 0.67 + \frac{5}{10} \cdot 0.67\right) = 0.02$$

and the reduction of uncertainty in mood conditional on fees is zero, because the entropy (i.e., uncertainty about the outcome) is identical in both resulting leaves:

$$0.69 - \left(\frac{4}{10} \cdot 0.69 + \frac{6}{10} \cdot 0.69\right) = 0$$

Hence, "food" will eventually be selected as the first split (see Fig. 11.1). This approach to selecting predictors is now applied in each of the resulting branches until a stopping criterion is reached. This criterion may either be a minimum number of cases, a maximum height (respectively, depth) of the tree, thresholds such as a certain entropy or a test statistic exceeding a critical value, or simply the fact that no predictors are left to predict with.

The simplified case of binary predictors and binary outcomes can easily be extended to other variable types. For example, if predictors are ordinal or continuous, decision trees select a dichotomous split for each variable, such that we can adhere to the binary tree scheme. For categorical variables, the decision tree may also select the best partition of all categories into two sets; however, with categorical variables, the number of tests to perform for selection of the best split within a

[1]Often, entropy is normalized by dividing by $\log M$ such that maximal entropy is 1. Here, we report raw entropy values.

[2]Natural unit of information (nat) is a unit of entropy based on natural logarithms, similar to bits that are based on base 2 logarithms.

categorical variable grows exponentially with the number of categories and is thus only advisable for categorical variables with few categories.

When *outcome* variables are continuous, we need to change the criterion for selecting best fit. For example, in CART (Classification and Regression Tree), a criterion of variance reduction is typically applied, that is, the predictor that maximally decreases variance in the outcome is selected. This is consistent with the notion that when selecting between models, those with the highest explained variance (typically measured as R^2) are preferred. Furthermore, variance reduction conforms with the idea of uncertainty reduction because the entropy of a normally distributed variable is proportional to the logarithm of its variance.

11.2.2 SEM Trees

To allow for multivariate outcomes in decision trees, univariate decision trees can be extended to multivariate trees. This can be achieved by requiring the a priori specification of a multivariate model of the multiple outcomes of interest. To this end, we divide our variables into two distinct sets, one set for which we have strong theory-driven assumptions (the parametric outcome model) and another set for which we have no or only weak assumptions (the potential predictors for which we learn a nonparametric association with the outcome model). Using the same logic as in classic decision trees, recursive partitioning then proceeds such that predictors are selected that best describe differences with respect to the multivariate outcomes model. One particularly useful instance of this idea is based on SEM as a formal language for specifying multivariate outcomes. SEM is a general modeling technique that encompasses many statistical models of normally distributed variables. The resulting SEM trees (Brandmaier et al. 2013b) can be conceived as trees in which each leaf is a SEM with a different set of parameter estimates. The decision nodes of such a tree represent a hierarchical structure describing properties of cases (in the social sciences typically persons) that are best described by the respective leaf of the tree. With these model-based trees, we achieve a great flexibility in modeling, because we can examine predictors and interactions of predictors in a variety of latent variable models such as latent factor models, latent growth curve models, latent difference score models, or any other type of model representable via SEM.

SEM trees evaluate split candidates, that is, the set of all potential predictors, using a likelihood-ratio criterion. For each potential split variable, a likelihood-ratio test is performed that compares the likelihood of the presplit model, that is, the model for the complete sample in a particular node of the tree, with that of the postsplit model, that is, a multi-group model in which the two resulting groups are independently modeled. This corresponds to a null hypothesis that there are no differences between the groups resulting from a given split. The corresponding test statistic has a χ^2-distribution with degrees of freedom corresponding to the number of free parameters in the outcome model. This criterion can also be considered from an information-theoretic perspective. One can show that the expected log-likelihood ratio between the presplit and postsplit model corresponds to the Kullback-Leibler divergence, which is a measure of relative entropy or information gain achieved if the postsplit model is used over the presplit model (also see Brandmaier et al. 2013a).

Trees are especially useful for finding structures in high-dimensional data because they are relatively simple to understand and interpret. Decision trees visualize a hierarchy of decisions that can either be read as a hierarchy of conditional effects describing a sample or as a hierarchical set of rules that can simply be traced along all decision nodes from the root to a leaf yielding specific predictions. Generally, trees are regarded as a robust technique in the sense that they require less data preprocessing (e.g., trees are invariant to standardization of variables and work in principle for all types of scales) and fewer assumptions about the predictors than many other techniques. Trees are nonparametric estimators of predictors' associations with the outcome and make no specific assumptions about the predictors' distribution.

SEM trees have been gaining increasing momentum in psychological research. Brandmaier et al. (2013b) described an application of SEM trees to modeling learning curves of a verbal and performance composite score in the Wechsler Intelligence Scale for Children and in a factor model of verbal intelligence from the Wechsler Adult Intelligence Scale. The method was also used to explore correlates of individual differences in trajectories of declining cognitive functioning in old age (Brandmaier et al. 2013a). Kuroki (2014) explored different pathways to decline in episodic memory using diverse predictors, including genetic variants, based on data from the Wisconsin Longitudinal Study. Brandmaier et al. (2017) demonstrated an application of SEM trees in combination with quadratic latent growth curve models to explore constellations of correlates of terminal decline in well-being using data from the German Socio-Economic Panel. Ammerman et al. (2017) used SEM trees to develop an empirical diagnostic criteria for non-suicidal self-injury (NSSI) disorder. They used trees to find optimal cutoffs in NSSI behavior frequency such that participants were maximally different with respect to a common factor model of various symptomatology and cognitive-affective deficits. Jacobucci et al. (2016) presented a theoretical and empirical comparison of SEM trees and finite mixture models as complementary approaches to uncovering heterogeneity in latent variable models. A recent simulation study by Usami et al. (2017) examined the method's sensitivity to misspecification in longitudinal models.

11.2.3 Latent Growth Curve SEM Trees

To illustrate SEM trees and how they may help us select important variables from a set of variables for which we may have little or no prior knowledge on how they relate to the outcome of interest, we apply them to a latent growth curve model (LGCM). Within SEM, LGCMs are widely used to capture changes in longitudinal data, for example, on human development (Meredith and Tisak 1990; Brandmaier et al. 2015; Muthén and Curran 1997; Ferrer and McArdle 2010). The latent growth curve model assumes individual parametric trajectories of change for each individual and allows individuals to vary in the parameters of the trajectory. In contrast to continuous time models, which will be introduced in the next paragraph, LGCMs can be said to model the overall shape of change rather than the dynamics of change. Generally, LGCMs are an important statistical approach to analyze longitudinal data that allows us to capture interindividual variability in intraindividual change over time or, in other words, between-person differences in within-person changes. LGCMs can be regarded as factor models of repeated indicators over time. The fact that the factors represent evolving temporal processes is why they are also referred to as chronometric factor models (following the term psychometric factor models in which the indicators are psychological measurements; McArdle and Epstein 1987). Fixed factor loadings encode different shapes of change over time; by default a latent intercept factor with constant loadings and a latent slope factor with loadings represent a linear increase over time. However, other types of change are possible including polynomial, exponential, or piecewise definitions of factors; and finally, in fully latent curve models, the loadings can also be estimated from the data (McArdle 1988; Meredith and Tisak 1990). By conditioning on hypothesized time-invariant predictors, we can test to what extent predictors may serve as explanations of individual differences in level or change. For example, we can explicitly test whether a hypothesized vascular risk factor explains variance in change of perceptual speed by regressing the latent slope factor on the risk factor in a LGCM of repeated assessments of perceptual speed.

Here, we use a simple model of linear change to illustrate the utility of SEM trees to explore predictors of interindividual differences in change. SEM trees and forests are implemented in the R package semtree by Brandmaier and Prindle (2017), which can be freely downloaded from the package website (http://brandmaier.de/semtree) or from CRAN (https://cran.r-project.org/). The package implements all functions used in this chapter. First, we load some simulated demonstration data shipped with the semtree package. This data set contains five repeated measurements of 400 persons, of which half were younger (agegroup = 0) and half were older adults (agegroup = 1). One half of each age group participated in a training program, whereas the other half served as no-contact control group. The following R code demonstrates how the semtree package and the example data set can be loaded:

```
require(semtree)
data(lgcm)
head(lgcm)
```

The first five rows of the complete data set are shown in Table 11.2.

Table 11.2 First five rows of simulated data provided in the semtree package

o1	o2	o3	o4	o5	Agegroup	Training	Noise
4.49	3.32	2.71	1.31	0.38	0	1	1
4.91	5.19	4.76	4.85	4.93	0	1	0
4.51	3.49	2.41	0.24	−1.01	0	1	1
4.91	5.19	4.95	4.68	4.72	0	1	0
4.96	5.12	5.39	4.56	4.44	0	1	0

The following code shows how a simple linear LGCM with equidistant measurements (see Fig. 11.2 for an illustration) can be specified using OpenMx:

```
require(OpenMx)

manifests<-c("o1","o2","o3","o4","o5")
latents<-c("icept","slope")

model <- mxModel("Unnamed_Model",
type="RAM",manifestVars = manifests, latentVars = latents,
mxPath(from="icept",to=manifests,free=FALSE, value=1.0, arrows=1,
        connect="single"),
mxPath(from="slope",to=manifests, FALSE,arrows=1,
        connect="single", values=0:4 ),
mxPath(from="one",to=latents,TRUE, value=1,      arrows=1, label=c
    ("mu_icept","mu_slope"), connect="single" ),
mxPath(from=latents,connect = "unique.pairs", free=TRUE, value
    =1.0,      arrows=2, label=c("var_icept" ,"cov_icept_slope","
    var_slope") ),
mxPath(from=manifests,to=manifests, connect = "single",
        free = TRUE, label="var_residual", arrow=2, value=.1),
mxPath(from="one",to=manifests, free=FALSE, value=0, arrows=1,
connect="single") ,
mxData(observed=lgcm, type="raw") )
```

In the above model, free parameters are the intercept mean (`mu_icept`) and variance (`var_icept`) and linear slope mean (`mu_slope`) and variance (`var_icept`) and the covariance of intercept and slope (`cov_icept_slope`) and the residual error variance (`var_residual`). Tree construction is simply done by passing the specified SEM and a `data.frame` object containing the data to the `semtree()`-function:

```
tree <- semtree(model=model, data=lgcm)
```

A more fine-grained control over meta-parameters guiding the tree creation process can be gained by using a control object:

```
control <- semtree.control(min.N = 50, bonferroni = TRUE, max.
    depth = 3, exclude.heywood = TRUE)
tree2 <- semtree(model=model, data=lgcm, control=control)
```

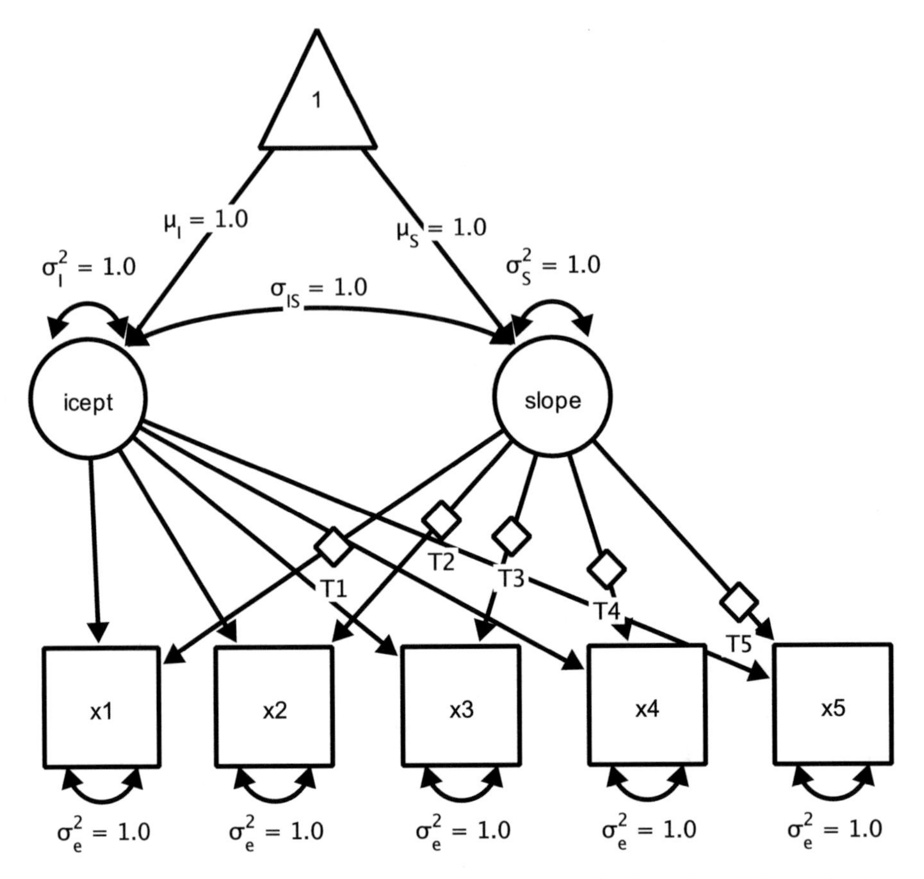

Fig. 11.2 Path diagram of a linear LGCM corresponding to the OpenMx model specification. The model has six free parameters, the intercept mean (μ_I in the path diagram or `mu_icept` in code) and variance (σ_S^2 or `var_icept`), the slope mean (μ_S or `mu_slope`) and variance (σ_S^2 or `var_slope`), the intercept-slope-covariance (σ_{IS} or `cov_icept_slope`), and the residual error variance (σ_e^2 or `var_residual`). The slope loadings are indicated as definition variables (labelled T1 to T5), allowing for person-specific intervals between measurement occasions

If no control object is given, the default constructor is invoked. In the above example, some defaults are overridden: a minimum number of 50 cases is required to be in a leaf to continue splitting (`min.N`), Bonferroni correction is applied at each level when split variables are evaluated for further splitting (`bonferroni`), the tree is grown to a depth[3] of 3 (`max.depth`), and Heywood cases, that is, those in which variances were estimated to be negative, are not considered as valid splits (`exclude.heywood`). The semtree package offers various facilities to plot or print a tree. Useful commands are:

[3]Computer scientists like to grow trees from top to bottom, so height as a property of a tree is simply replaced by depth.

```
plot(tree)
plot(prune(tree,max.depth=1))
print(tree)
summary(tree)
```

For large trees, the `prune()` command is particularly useful as it prunes a tree back to a desired height given by parameter `max.depth`. A tabular representation of the hierarchical decisions encoded in a tree can be obtained with `toTable()` (e.g., see Table 5 in Brandmaier et al. 2017), and for LaTeXusers, a script-based representation of a tree based on the pstricks package can be obtained with command `toLatex()`. The resulting tree is illustrated in Fig. 11.3, where we have replaced the numerical

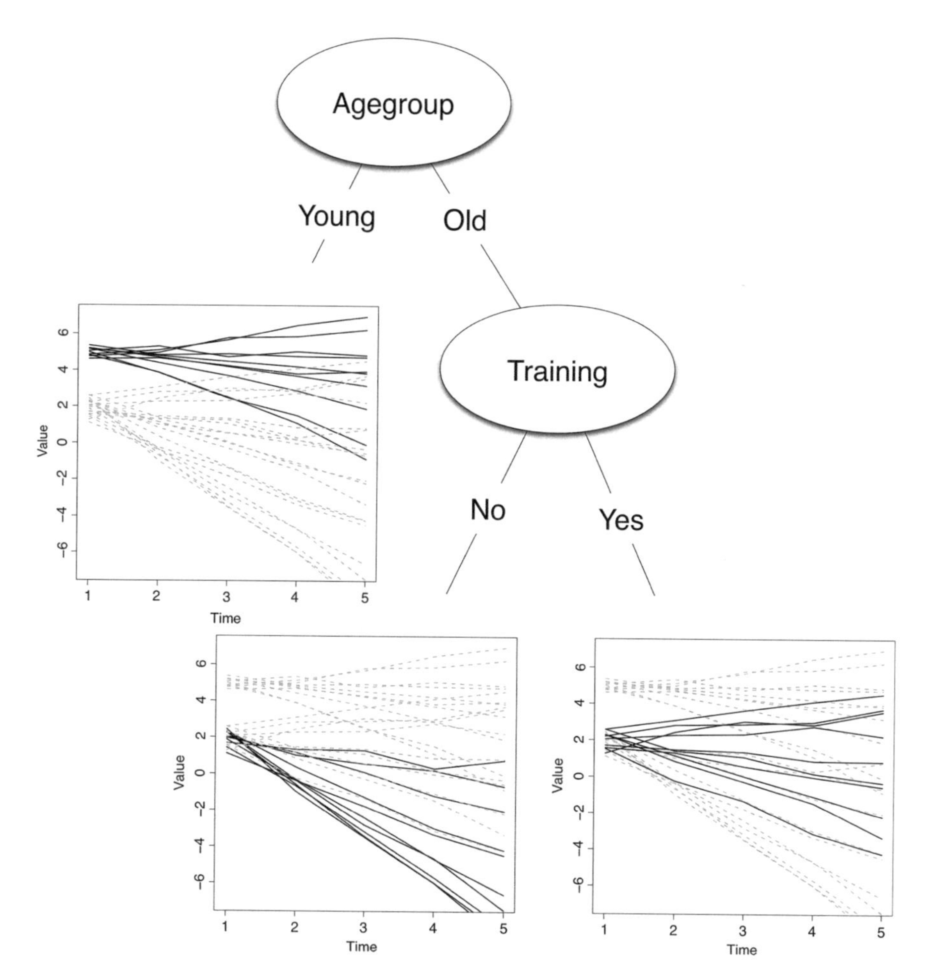

Fig. 11.3 LGCM SEM Tree using sample data from the semtree package. Simulated persons differ in whether they are young or old and whether they participated in a training intervention. Training is only predictive of mean differences for old people and shows up as a difference in mean slope, whereas the age group predicts differences in both intercept and slope

parameter estimates by randomly drawn trajectories from the leaf models. We can see from the tree that simulated participants differ in whether they are young or older and whether they participated in a training intervention. Training is only predictive of mean differences for older persons and shows up as a difference in mean slope, whereas the age group predicts differences in both intercept and slope.

Further ways to represent results of SEM trees were shown by Brandmaier et al. (2017), who presented an empirical application of SEM trees with growth curve models with both a linear and a quadratic trend to examine terminal decline in self-reported well-being. Using data from the German Socio-Economic Panel, they found that health-related and psychosocial factors are most strongly predictive of individual differences in terminal decline of well-being.

11.3 Continuous Time Modeling

Continuous time models are dynamic models for the analysis of longitudinal data that adequately account for unequal assessment intervals within as well as across individuals. As such, they can be considered a generalization of discrete time models, such as the popular autoregressive and cross-lagged model presented in Eq. (11.2), in which time is assumed to proceed in discrete steps of $\Delta t = 1$ for all measurement occasions $u = 1, \ldots, U$.

$$\boldsymbol{\eta}_u = \mathbf{A}^* \cdot \boldsymbol{\eta}_{u-1} + \boldsymbol{\zeta}_u \tag{11.2}$$

The vector $\boldsymbol{\eta}_u \in \mathbb{R}^v$ represents v latent variables at measurement occasion u. $\mathbf{A}^* \in \mathbb{R}^{v \times v}$ denotes the autoregressive and cross-lagged matrix with autoregressive effects on the main diagonal and cross-lagged effects in the off-diagonals, while $\boldsymbol{\zeta}_u \in \mathbb{R}^v$ represents the discrete time dynamic error term at measurement occasion u.

In contrast to discrete time equation (11.2), time is considered a continuous variable that ranges over the entire range of real numbers $t = 1, \ldots, T \in \mathbb{R}$ in continuous time models. The corresponding continuous time equation is obtained by taking the derivative of vector $\boldsymbol{\eta}(t)$ with respect to time $\frac{\mathrm{d}\boldsymbol{\eta}(t)}{\mathrm{d}t}$, so that after multiplication with $\mathrm{d}t$:

$$\mathrm{d}\boldsymbol{\eta}(t) = (\mathbf{A} \cdot \boldsymbol{\eta}(t))\,\mathrm{d}t + \mathbf{G} \cdot \mathrm{d}\mathbf{W}(t). \tag{11.3}$$

Vector $\boldsymbol{\eta}(t) \in \mathbb{R}^v$ represents the state of the v latent variables at time point t and $\mathbf{A}$ constitutes the so-called drift matrix, which is the continuous time analog to the autoregressive and cross-lagged matrix $\mathbf{A}^*$. $\mathbf{W}(t)$ denotes the famous Wiener process, the limiting form of the discrete time random walk, with a continuous time covariance matrix $\mathbf{Q} = \mathbf{G}\mathbf{G}^{\mathrm{T}}$. Solving Eq. (11.3) for any arbitrary time interval $\Delta t = t - t_0$ and starting point t_0 yields:

$$\boldsymbol{\eta}(t) = \exp(\mathbf{A} \cdot \Delta t) \cdot \boldsymbol{\eta}(t_0) + \int_{t_0}^{t} \exp(\mathbf{A} \cdot (t - s)) \cdot \mathbf{G} \cdot \mathrm{d}\mathbf{W}(s) \tag{11.4}$$

As described elsewhere (Driver et al. 2017; Voelkle et al. 2012; Oud and Jansen 2000), Eqs. (11.3) and (11.4) can be extended easily in various ways, such as the inclusion of intercepts, time-invariant predictors, or time-varying predictors. Furthermore, $\boldsymbol{\eta}(t)$ may be directly observed or may be latent with standard (LISREL) measurement model:

$$\mathbf{y}(t) = \boldsymbol{\Lambda}\boldsymbol{\eta}(t) + \boldsymbol{\tau} + \boldsymbol{\varepsilon}(t) \tag{11.5}$$

where $\boldsymbol{y}(t) \in \mathbb{R}^c$ denotes c manifest variables, $\boldsymbol{\Lambda} \in \mathbb{R}^{c \times v}$ the factor loading matrix, $\boldsymbol{\tau} \in \mathbb{R}^v$ the vector of manifest intercepts, and $\boldsymbol{\varepsilon}(t) \in \mathbb{R}^v$ the manifest error terms. As illustrated above and discussed in more detail in Voelkle et al. (2012) and Oud and Delsing (2010), discrete and continuous time models can be formulated easily as structural equation models. Although in principle any SEM software with basic programming functionality, such as the matrix exponential function, may be used to fit continuous time models, the R package ctsem (Driver et al. 2017) provides a particularly user-friendly way for model specification and estimation, along with additional features to ease the process. The ctsem package interfaces either to OpenMx (Neale et al. 2016), a powerful package for extended SEM, or to the Stan software (Stan Development Team 2016) for Bayesian estimation. For the purpose of the present paper, we will focus on the SEM implementation via OpenMx. As will be demonstrated in the next section, this makes it possible to apply the SEM Tree algorithms implemented in the package semtree to continuous time models specified via ctsem.

11.4 Continuous Time SEM Trees

Combining the benefits of continuous time models and SEM trees allows us to explore predictors of differences in dynamics. In practice, the combination of continuous time modeling and SEM trees is straightforward, because both approaches rely on model specifications in OpenMx (among available alternatives, such as lavaan specification in semtree or Stan in ctsem). Using the strengths of both R packages, we can relatively easily set up CT SEM trees in the following way. First, we use the ctsem package as frontend for model specification. Then, we unwrap the internal OpenMx model specification and pass it to the semtree package as a template model. In the following, we illustrate how this can be done using a simple example.

First, we create a model for an univariate Gaussian process with a single indicator measured repeatedly at 100 time points.

```
model <- ctModel(n.latent = 1, n.manifest = 1, Tpoints=100,
+ LAMBDA = diag(1));
```

With `n.latent=1`, we specify a single process, measured with a single manifest variable (`n.manifest`) and loading fixed to one (`LAMBDA=diag(1)`). Freely estimated parameters in this model are the residual error variance, the drift parameter (corresponding to the autoregression in discrete models), the process mean and variance at first occasion of measurement, and the diffusion parameter (variance of the innovation term in discrete models).

Note that naming conventions in ctsem require wide-format data columns to adhere to a format, in which the repeatedly measured variable name is followed by an underscore and "T," followed by an integer denoting the index of the measurement occasion, where the first measurement occasion has the index 0.

We estimate parameters for the complete sample from some data (assumed to be in a `data.frame` called `data`) by running:

```
result <- ctFit(data, model, transformedParams=FALSE)
```

Setting `transformedParams=TRUE` adds internal parameter transformations such as logarithmic transformations of variances so that variances are always larger than zero. When parameters are transformed, SEM Tree results will constrain the transformed parameters in the leaves, that is, parameters must be retransformed manually. Second, it is important that the data contain only columns corresponding to the observed time points and time lags (see Driver et al. 2017, for a tutorial on preparing data for ctsem). Next, we unwrap the OpenMx model specification, the data as processed by `ctFit`, and merge the data with the potential predictors (here assumed to be in a `data.frame` called `covariates`):

```
semtree.model <- result$mxobj
semtree.data <- semtree.model$data@observed
semtree.data$id <- rownames(semtree.data)
merge(semtree.data, covariates, by="id")
```

Finally, we run the semtree as described before:

```
tree <- semtree(model = semtree.model, data = semtree.data)
```

11.5 CT SEM Trees to Explore Dynamics in Perceptual Speed

To illustrate the utility of CT SEM trees, we applied CT SEM trees to data from the COGITO study (Schmiedek et al. 2010). In this study, the authors examined whether intensive cognitive training in one hundred daily sessions would enhance broad cognitive abilities. In the study, 101 younger (age, 20–31 years; $M = 25$ years) and 103 older participants (age, 64–80; $M = 71$ years), practiced multiple tests of perceptual speed, working memory, and episodic memory in 100 daily sessions,

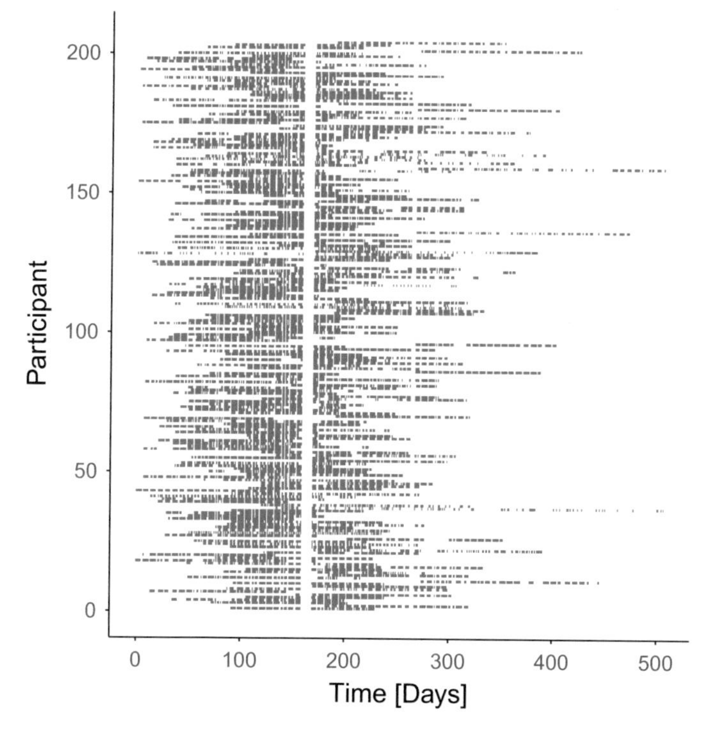

Fig. 11.4 Measurement density in the COGITO study. The x-axis displays time in days relative to the start of the study; the y-axis displays the individual participants. Time intervals considerably differ within and between persons. Also note the large gap common to all participants caused by the Christmas break. Continuous time models readily account for such unequal time intervals in the statistical analysis

each lasting 1 h. Participants performed 12 tasks in each session assessing working memory, episodic memory, and perceptual speed. To balance cognitive challenge and motivation, difficulty levels were adapted based on pretest performance. The COGITO data illustrate the exigent need for appropriate modeling of individual time points of measurement. Figure 11.4 shows all 204,000 individual measurements distributed over time. We can observe large differences between and within persons, with some individuals' 100 measurements stretched over up to 500 days. By the same token, there are stretches of time with no or few measurements, for example, the Christmas break, which appears as a clearly visible gap in Fig. 11.4 along with other regular gaps corresponding to weekends or holidays.

For this chapter, we focus on the perceptual speed comparison tasks, which were presented in three different perceptual domains: figural, spatial, and numerical. Participants had to indicate as quickly as possible whether two presented items were identical. Figure 11.5 (left panel) shows three exemplary reaction time trajectories aligned to the first day of measurement for three different participants. Here, we examine the question to what extent individual differences in response time

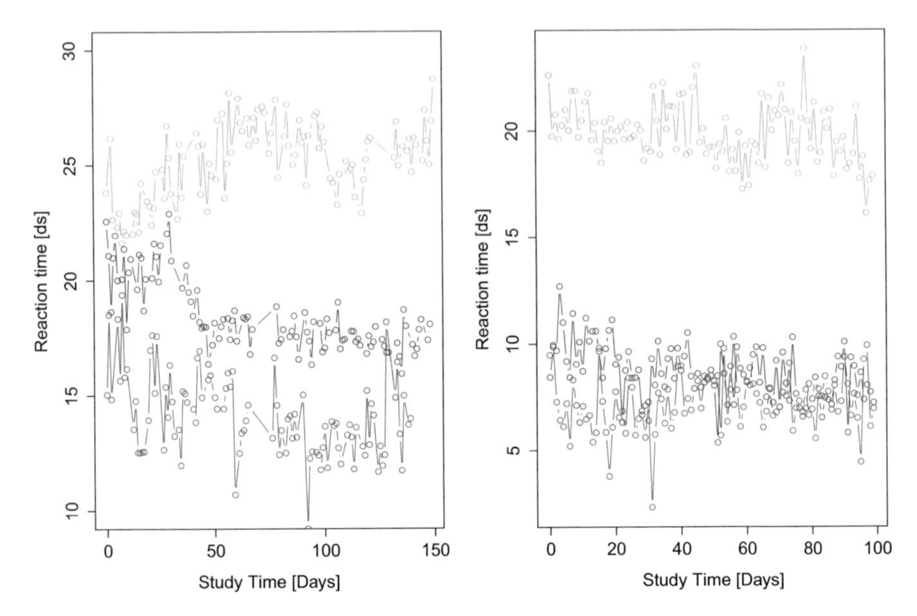

Fig. 11.5 Left: response time dynamics of three different participants in the COGITO study in deciseconds. Trajectories were derived as mean response times across all trials from a single day. Note the pronounced differences in time lags between measurement occasions and across persons. Right: three randomly sampled trajectories from the dynamic continuous time model we used to fit the COGITO data. The model includes three latent processes, which capture the within-person dynamics, stable between-person differences, and a saturating trend

dynamics over the course of 100 days of cognitive training can be predicted by sex (0 = men), age group, personality (Big Five), positive or negative affect (PANAS; see Watson et al. 1988), or stress (see Cohen et al. 1983). We aggregated response times as median response time over all trials of each day. From this response time trajectory, we performed a person-specific outlier removal procedure that removed days for which the median response time over trials was more than three standard deviations below the mean response time across all days. We also removed all days on which the median response time over trials was lower than a fixed threshold of 250 ms. Before model fitting, we rescaled response times by a factor of 1/100 such that reported parameter estimates are in units of deciseconds (i.e., to obtain estimates in milliseconds, one needs to multiply estimates by 100). Individual sum scores were computed for all ten positive affect items, all ten negative affect items, and all stress items.

Computation time can be reduced by requiring ctsem to use OpenMx's state space representation, which allows parameter estimation with the more efficient Kalman filter (see Driver et al. 2017). To speed up processing, we discretized predictors based on quartiles of their distributions, which reduced computation time by an approximate factor of 30 but, of course, comes at the risk of missing more informative split points in each variable. The longitudinal dynamics are modeled as a superposition of three latent processes: the first process captures the within-person dynamics, the second process stable between-person differences, and the third process models a saturating trend. The latter is necessary to account for the general decline in the overall response times across the entire study period. In the model specification provided below, the first process, daily fluctuations, is represented by a diffusion parameter (`diffusion1`), a drift parameter (`drift1`), and an initial variance parameter (`stationary`). The initial variance parameter is not estimated, but deterministically set to ensure stationary variance for this process, avoiding potential overfitting issues. The second process has no drift and no diffusion,[4] but encodes stable interindividual differences, that is, the process is equivalent to a random intercept term (`traitvar`). The third process is represented by a drift coefficient (`trenddrift`), no diffusion, an average trend (`trend_cint`), and individual differences in this trend (`t0trendvar`). Furthermore, the trend and trait process are allowed to covary (`traittrendcov`). The right-hand panel in Fig. 11.5 illustrates the dynamics of this model with three selected individual trajectories.

```
model<-ctModel(type='omx',Tpoints=100,n.latent=3,n.manifest=1,
    manifestNames="rt",
        DIFFUSION=matrix(c('diffusion1',0,0,
                0,.0001,0,
                0,0,.0001),3,3),
        DRIFT=matrix(c('drift1',0,0,
                0,-.00001,0,
                0,0,'trenddrift'),3,3),
        T0MEANS=matrix(c(0,0,0),ncol=1),
        CINT=matrix(c(0,0,'trend_cint'),ncol=1),
        T0VAR=matrix(c('stationary',0,0,
                0,'traitvar','traittrendcov',
                0,0,'t0trendvar'),3,3),
        LAMBDA=matrix(c(1,1,1),ncol=3))
```

Figure 11.6 shows the resulting SEM Tree. We find that age group is at the root and is thus the most informative unconditional split. Conditional on the first split, we find extraversion for the younger and positive affect for the older participants to be the most informative next splits. Further conditioning on low and moderate extraversion (leftmost branch in Fig. 11.6), sex is the next best predictor for young adults, whereas conditional on high extraversion, positive affect is the next best predictor for young adults. For older adults with low positive affect, conscientiousness is the next best predictor, and for those with high positive affect, extraversion is the best predictor. It can be seen that four of the eight leafs (node ids #7, #8, #14, #15) are determined by participants' age, positive affect, and extraversion, which seem to contribute largely to predictions of individual differences in dynamics. For the remaining four leafs (node ids #4, #5, #11, #12), it is either extraversion or positive affect and sex or conscientiousness. As further

[4]For numerical reasons, zero diffusion and drift parameters are fixed to very small non-zero values.

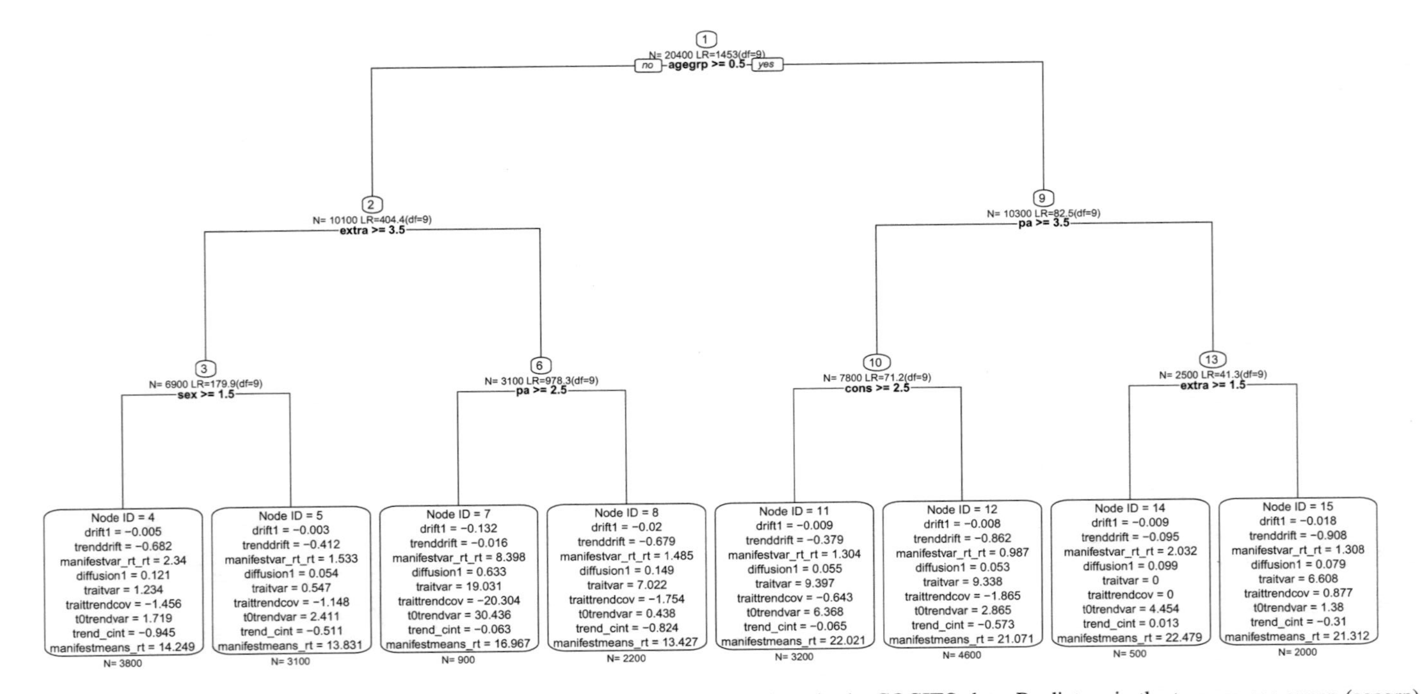

Fig. 11.6 The first three levels of a continuous time SEM Tree modeling response times in the COGITO data. Predictors in the tree are age group (agegrp), extraversion (extra), conscientiousness (cons), and sex

Table 11.3 Parameter estimates across the leaves of the CT SEM Tree

	4	5	7	8	11	12	14	15
drift1	−0.005	−0.003	−0.132	−0.020	−0.009	−0.008	−0.009	−0.018
trenddrift	−0.682	−0.412	−0.016	−0.679	−0.379	−0.862	−0.095	−0.908
manifestvar_rt_rt	2.340	1.533	8.398	1.485	1.304	0.987	2.032	1.308
diffusion1	0.121	0.054	0.633	0.149	0.055	0.053	0.099	0.079
traitvar	1.234	0.547	19.031	7.022	9.397	9.338	0.000	6.608
traittrendcov	−1.456	−1.148	−20.304	−1.754	−0.643	−1.865	0.000	0.877
t0trendvar	1.719	2.411	30.436	0.438	6.368	2.865	4.454	1.380
trend_cint	−0.945	−0.511	−0.063	−0.824	−0.065	−0.573	0.013	−0.310
manifestmeans_rt	14.249	13.831	16.967	13.427	22.021	21.071	22.479	21.312

Rows correspond to parameters and columns correspond to numbered leafs

apparent from the parameter estimates given in Table 11.3, we obtain subgroups with differences in all model parameters. Maybe most striking are the rather high within-person temporal dependencies in reaction times across study days as evidenced by `drift1` coefficients close to zero. So, when subjects deviate from their general trend, they tend to remain that way. In this regard, only young adults with high levels of extraversion but low levels of positive affect are a slight exception. As compared to the others, this group exhibits a substantial amount of variability, not only at the within-person level (`manifestvar_rt_rt` and `diffusion1`) but also at the between-person level (`traitvar`). Given this specific combination (young adults, high levels of extraversion and low levels of positive affect), a greater variability in reaction times seems reasonable. However, we hasten to add that the present analyses are primarily for illustrative purposes and that further research is necessary before the resulting pattern of parameter estimates can be interpreted substantially.

11.6 From Trees to Forests

A particular challenge with decision trees that was identified early on is their potential instability. Instability refers to the observation that decision trees may vary slightly and are sometimes considerably different if the sample at hand is only marginally changed. We saw this effect, when we (Brandmaier et al. 2013b) looked at longitudinal data obtained with the Wechsler Scale of Intelligence (McArdle and Epstein 1987). We created a SEM tree to discover predictors of developmental differences in children's verbal performance. After we created a SEM tree using a linear LGCM, we bootstrapped the original data and created a tree on each bootstrap iteration. The individual bootstrapped trees looked slightly different each time. Whereas mother's education was the strongest predictor in the original tree using the complete sample, father's education would sometimes show up as first split, and mother's education would either be a subsequent predictor or not show up at all. Generally, when highly correlated predictors compete for a split in the tree, only one of them can be chosen, which may lead to situations where the other predictor does not even show up in any of the subsequent splits. Second, by using a "greedy" (i.e., locally optimal) search strategy to retrieve optimal predictors, a tree finds one particular conditional structure although alternative tree structures may explain the outcome equally well or even better (the locally optimal strategy may not represent the global optimum). How can we handle this problem in a sensible way? Recently, Brandmaier et al. (2016) extended SEM trees to SEM forests to address the issue of tree instability. The idea behind SEM forests is again simple and straightforward, drawing upon the seminal work on random forests by Leo Breiman (2001). In random forests, instead of relying on a single tree, multiple trees from random samples of the original data set are created. In machine learning, this is referred to as an ensemble technique. The random samples create diversity among the trees. To further increase diversity, one further subsamples the available predictors each time the tree is branched. Therefore, all variables (independent of their total impact on the outcome or on their interrelation with other predictors) have a chance to enter some of the trees, at least occasionally. Brandmaier et al. (2017) demonstrated the utility of model-based forests for modeling terminal decline in well-being. They used data from the nationwide German Socio-Economic Panel (SOEP) to explore the association of constellations of diverse survey variables with changes of well-being in the last years of life. The authors used a latent growth curve model, similar to the one in our simulation, with an additional latent factor representing acceleration in change by setting factor loadings such that they represent quadratic change and thus curvature of well-being trajectories. In a single tree, they found objective health-related indicators interacting with social participation (frequency of involvement in social and cultural activities). The single tree allowed plotting of group-average trajectories based on health and social participation status, and the forest corroborated these results when the importance of predictors was rank-ordered. Growing SEM forests can be done with the `semforest()`-function of the semtree package:

```
forest <- semforest(model = semtree.model, data = semtree.data)
```

SEM forests usually consist of hundreds or thousands of individual trees. This makes their visual inspection and interpretation more difficult, if not impossible. Instead, one typically relies on aggregate measures such as variable importance (Brandmaier et al. 2016). Variable importance is based on random permutations of the out-of-bag samples of the individual trees (i.e., the part of the data not used to build each particular tree) and quantifies, for each variable, the average reduction of uncertainty about the model-predicted distribution after permutation. If a certain predictor is not important for predicting the multivariate outcome, permutations of that predictor should only marginally increase the uncertainty about the outcome, whereas the permutation of important predictors will show a large increase of uncertainty (expressed as model misfit in a χ^2-metric). For a given forest, variable importance is computed and plotted as follows:

```
vim <- varimp(forest)
plot(vim)
```

11.7 Conclusions

SEM trees and forests address the problem of variable selection (or "feature selection" in machine learning parlance). The central premise in feature selection approaches is that the full data set contains irrelevant variables that can be ignored because they add no information to answer a particular prediction problem. A typical goal in feature selection is to select a small subset of variables from a large set while granting optimal (up to some threshold) predictive accuracy of the model. The overarching, more general theme is model selection that is the general problem of selecting between different models (that may or may not differ in the number of variables they rely on, but may be further distinguished by fit functions or other statistical properties of the method). Trees and even more so (random) forests have proven a useful data-analytic tool in practice. When comparing a large number of data-driven statistical learning approaches (including support vector machines, linear discriminant analysis, regularized regression), Fernández-Delgado et al. (2014) found that, on average, random forests had the best performance in predicting outcomes when no prior information was available. Like any other exploratory method, however, trees make certain assumptions that limit the approach. For example, the recursive approach to growing a tree makes locally optimal selections (i.e., splits are selected that are optimal given the current structure of conditional effects), but a set of locally optimal selections does not necessarily yield the globally optimal result. In other words, there may be trees that perform better in some chosen metric (e.g., R^2 or Kullback-Leibler divergence to the population distribution), but they may feature variables along the way that are not locally optimal. However, searching all possible trees is a combinatorially difficult problem that in most practical cases can only be approximated but never solved exactly. Second, with their focus on interactions, trees are an inefficient structure if there are only weak interactions and/or many independent main effects. Relatedly, if there is only a conditional effect of two variables but no unconditional effect (in ANOVA language, an interaction but neither of the two main effects), a single tree will not be able to recover the structure since it will select neither of the two variables in the first place. Third, trees can, in theory, approximate arbitrary, nonlinear, and even noncontinuous associations between predictors and outcomes; however, if the association is linear, the hierarchical approximation with a set of dichotomous splits is inefficient, and the underlying linearity is hard, if not impossible, to guess from a given tree structure. All of the above problems are mitigated in random forests and SEM forests due to the resampling of trees, which allows one to search a much larger space of alternative hypotheses.

When embarking upon data-driven, that is, exploratory data analysis, researchers need to make sure that they do not overfit the data and thus jeopardize the generalizability of their findings. When we carefully adjust for the level of overfitting to the data (e.g., Brandmaier et al. 2013a,b), SEM trees provide a formal venue to search large sets of heterogeneous predictors from different domains for the most influential effects on a hypothesized model while maintaining generalizable results. Statistical control for overfitting is possible in various ways, including cross-validation (i.e., using independent subsets of the original data for hypothesis generation with a tree and hypothesis confirmation with the resulting SEM), conservative statistical tests with correction for multiple testing, or pruning of overfitted trees using a holdout set. Generally, we advise researchers to report results of a SEM Tree as exploratory, and we recommend validating exploratory results on an independent data set, preferably custom-tailored to the hypothesized effect.

In the process of scientific inquiry, trees and forests support researchers in generating novel hypotheses and challenging established theory-driven models (see Brandmaier et al. 2016). The flexibility of SEM as a general modeling framework for various research designs and hypotheses, and the flexibility of trees and forests to account for diversity of predictors, makes the method suitable as an "off-the-shelf" tool for exploratory data analysis after a first step of theory-driven, confirmatory modeling.

In the present chapter, we have shown how to combine recursive partitioning techniques with dynamic continuous time models. This combination became possible because of two recent developments: first, by the extension of recursive partitioning techniques to the broad class of structural equation models (i.e., SEM trees) and the possibility to formulate and estimate continuous time models as SEM and, second, through recent software developments to fit SEM trees/forests and continuous time models (i.e., semtree and ctsem). Since both approaches operate in the same SEM framework and can interface to a common software package (OpenMx), the combination is theoretically straightforward and easy to implement. While the present chapter is the first to introduce continuous time SEM trees to the literature, additional testing of the method will be necessary in future work. Both SEM trees and SEM-based continuous time models can be computationally challenging, and their combination is likely to exacerbate this problem. Nevertheless, we believe continuous time SEM trees hold a lot of potential for applied research targeting better understanding of mechanisms of change in heterogeneous samples. The door is open to explore this potential.

Acknowledgments We thank John J. Prindle, John J. McArdle, Timo von Oertzen, and Ulman Lindenberger who have each played essential roles in the development of SEM trees and forests. We would also like to thank Florian Schmiedek, Annette Brose, and Ulman Lindenberger for sharing the COGITO data with us. We are grateful to Julia Delius for her helpful assistance in language and style editing.

References

Ammerman, B. A., Jacobucci, R., Kleiman, E. M., Muehlenkamp, J. J., & McCloskey, M. S. (2017). Development and validation of empirically derived frequency criteria for NSSI disorder using exploratory data mining. *Psychological Assessment, 29*(2), 221–231. https://doi.org/10.1037/pas0000334

Brandmaier, A. M., & Prindle, J. J. (2017). *semtree: Recursive partitioning for structural equation models [Computer software manual]*. Retrieved from http://www.brandmaier.de/semtree (R package version 0.9.11)

Brandmaier, A. M., Prindle, J. J., McArdle, J. J., & Lindenberger, U. (2016). Theory-guided exploration with structural equation model forests. *Psychological Methods, 21*(4), 566–582. https://doi.org/10.1037/met0000090

Brandmaier, A. M., Ram, N., Wagner, G. G., & Gerstorf, D. (2017). Terminal decline in well-being: The role of multi-indicator constellations of physical health and psychosocial correlates. *Developmental Psychology, 53*(5), 996–1012. https://doi.org/10.1037/dev0000274

Brandmaier, A. M., von Oertzen, T., Ghisletta, P., Hertzog, C., & Lindenberger, U. (2015). Lifespan: A tool for the computer-aided design of longitudinal studies. *Frontiers in Psychology, 6*, 272. https://doi.org/10.3389/fpsyg.2015.00272

Brandmaier, A. M., von Oertzen, T., McArdle, J. J., & Lindenberger, U. (2013a). Exploratory data mining with structural equation model trees. In J. J. McArdle & G. Ritschard (Eds.), *Contemporary issues in exploratory data mining in the behavioral sciences* (pp. 96–127). New York: Routledge.

Brandmaier, A. M., von Oertzen, T., McArdle, J. J., & Lindenberger, U. (2013b). Structural equation model trees. *Psychological Methods, 18*(1), 71–86. https://doi.org/10.1037/a0030001

Breiman, L. (2001). Random forests. *Machine Learning, 45*(1), 5–32. https://doi.org/10.1023/A:1010933404324

Breiman, L., Friedman, J. H., Olshen, R. A., & Stone, C. J. (1984). *Classification and regression trees*. Belmont: Wadsworth International.

Cohen, S., Kamarck, T., & Mermelstein, R. (1983). A global measure of perceived stress. *Journal of Health and Social Behavior, 24*(4), 385–396. https://doi.org/10.2307/2136404

Driver, C., Oud, J. H. L., & Voelkle, M. C. (2017). Continuous time structural equation modeling with r package ctsem. *Journal of Statistical Software, 77*(5), 1–35. https://doi.org/10.18637/jss.v077.i05

Fernández-Delgado, M., Cernadas, E., Barro, S., & Amorim, D. (2014). Do we need hundreds of classifiers to solve real world classification problems? *Journal of Machine Learning Research, 15*, 3133–3181.

Ferrer, E., & McArdle, J. J. (2010). Longitudinal modeling of developmental changes in psychological research. *Current Directions in Psychological Science, 19*(3), 149–154. https://doi.org/10.1177/0963721410370300

Hothorn, T., & Zeileis, A. (2015). partykit: A modular toolkit for recursive partytioning in R. *Journal of Machine Learning Research, 16*, 3905–3909.

Jacobucci, R., Grimm, K. J., & McArdle, J. J. (2016). A comparison of methods for uncovering sample heterogeneity: Structural equation model trees and finite mixture models. *Structural Equation Modeling: A Multidisciplinary Journal, 24*(2), 270–282. https://doi.org/10.1080/10705511.2016.1250637

Kahneman, D., Krueger, A. B., Schkade, D. A., Schwarz, N., & Stone, A. A. (2004). A survey method for characterizing daily life experience: The day reconstruction method. *Science, 306*(5702), 1776–1780. https://doi.org/10.1126/science.1103572

Kuroki, Y. (2014). *Identifying diverse pathways to cognitive decline in later life using genetic and environmental factors*. Unpublished doctoral dissertation, University of Southern California.

McArdle, J. J. (1988). Dynamic but structural equation modeling of repeated measures data. In J. Nesselroade & R. Cattell (Eds.), *Handbook of multivariate experimental psychology* (Vol. 2, pp. 561–614). New York: Plenum Press.

McArdle, J. J., & Epstein, D. (1987). Latent growth curves within developmental structural equation models. *Child Development, 58*(1), 110–133. https://doi.org/10.2307/1130295

Mehl, M. R., & Conner, T. S. (2012). *Handbook of research methods for studying daily life*. New York: Guilford Press.

Meredith, W., & Tisak, J. (1990). Latent curve analysis. *Psychometrika, 55*(1), 107–122. https://doi.org/10.1007/BF02294746

Morgan, J. N., & Sonquist, J. A. (1963). Problems in the analysis of survey data, and a proposal. *Journal of the American Statistical Association, 58*(302), 415–434. https://doi.org/10.1080/01621459.1963.10500855

Muthén, B., & Curran, P. (1997). General longitudinal modeling of individual differences in experimental designs: A latent variable framework for analysis and power estimation. *Psychological Methods, 2*(4), 371–402. https://doi.org/10.1037/1082-989X.2.4.371

Neale, M. C., Hunter, M. D., Pritikin, J. N., Zahery, M., Brick, T. R., Kirkpatrick, R. M., et al. (2016). OpenMx 2.0: Extended structural equation and statistical modeling. *Psychometrika, 81*(2), 535–549. https://doi.org/10.1007/s11336-014-9435-8

Nesselroade, J. R., Gerstorf, D., Hardy, S. A., & Ram, N. (2007). Focus article: Idiographic filters for psychological constructs. *Measurement: Interdisciplinary Research & Perspectives, 5*(4), 217–235. https://doi.org/10.1080/15366360701741807

Oud, J. H. L., & Delsing, M. J. M. H. (2010). Continuous time modeling of panel data by means of SEM. In K. van Montfort, J. H. L. Oud, & A. Satorra (Eds.), *Longitudinal research with latent variables* (pp. 201–244). New York: Springer.

Oud, J. H. L., & Jansen, R. A. R. G. (2000). Continuous time state space modeling of panel data by means of SEM. *Psychometrika, 65*(2), 199–215. https://doi.org/10.1007/BF02294374

Quinlan, J. (1986). Induction of decision trees. *Machine Learning, 1*(1), 81–106. https://doi.org/10.1007/BF00116251

Schmiedek, F., Lövdén, M., & Lindenberger, U. (2010). Hundred days of cognitive training enhance broad cognitive abilities in adulthood: Findings from the COGITO study. *Frontiers in Aging Neuroscience, 2*, 27. https://doi.org/10.3389/fnagi.2010.00027

Stan Development Team. (2016). *Stan modeling language users guide and reference manual [Computer software manual]* (version 2.9.0)

Therneau, T., Atkinson, B., & Ripley, B. (2015). *rpart: Recursive partitioning and regression trees [Computer software manual]*. Retrieved from https://CRAN.R-project.org/package=rpart (R package version 4.1-10)

Tucker, L. (1966). Learning theory and multivariate experiment: Illustration by determination of generalized learning curves. In R. B. Cattell (Ed.), *Handbook of multivariate experimental psychology* (pp. 476–501). Chicago: Rand McNally.

Usami, S., Hayes, T., & Mcardle, J. J. (2017). Fitting structural equation model trees and latent growth curve mixture models in longitudinal designs: The influence of model misspecification. *Structural Equation Modeling: A Multidisciplinary Journal, 24*(4), 585–598. https://doi.org/10.1080/10705511.2016.1266267

Voelkle, M. C., Oud, J. H. L., Davidov, E., & Schmidt, P. (2012). An SEM approach to continuous time modeling of panel data: Relating authoritarianism and anomia. *Psychological Methods, 17*(2), 176–192. https://doi.org/10.1037/a0027543

Walls, T. A., & Schafer, J. L. (2006). *Models for intensive longitudinal data*. Oxford: Oxford University Press.

Watson, D., Clark, L. A., & Tellegen, A. (1988). Development and validation of brief measures of positive and negative affect: The PANAS scales. *Journal of Personality and Social Psychology, 54*(6), 1063–1070. https://doi.org/10.1037/0022-3514.54.6.1063

Chapter 12
Continuous versus Discrete Time Modeling in Growth and Business Cycle Theory

Omar Licandro, Luis A. Puch, and Jesús Ruiz

12.1 Introduction

The time dimension is of fundamental importance for macroeconomic theory, since most macroeconomic problems deal with intertemporal trade-offs. In modeling time, economists move from discrete to continuous time according to their methodological needs, as if both ways of representing time were equivalent. For example, growth theory is mainly written in continuous time, but business cycle theory is to a large extent written in discrete time. However, they refer to each other as being two pieces of the same framework.

The view that continuous and discrete time representations are equivalent is mainly supported by limit properties: the discrete time version of the standard dynamic general equilibrium model does converge to its continuous time representation when the period length tends to zero. However, this view hides a fundamental problem of timing. In continuous time, investment at time t becomes capital at time $t+dt$. The discrete time equivalent is that period t investment transforms into capital at period $t+1$. Thus, the speed at which investment becomes capital depends directly on the length of the period, and this is of fundamental importance as far as one deals with intertemporal trade-offs. Also, there is the issue of self-fulfilling prophecies in dynamic equilibrium models for which business cycle fluctuations may be driven by beliefs or *animal spirits*. A *sunspot* shock can be defined over the parametric space for local indeterminacy of equilibria, which in turn may critically depend on the time dimension as far as we set empirically plausible parameterizations for the model.

O. Licandro
School of Economics, University of Nottingham, Nottingham, UK
e-mail: Omar.Licandro@nottingham.ac.uk

L. A. Puch (✉) · J. Ruiz
Department of Economics and ICAE, Universidad Complutense de Madrid, Madrid, Spain
e-mail: lpuch@ccee.ucm.es; jruizand@ccee.ucm.es

K. van Montfort et al. (eds.), *Continuous Time Modeling in the Behavioral and Related Sciences*, https://doi.org/10.1007/978-3-319-77219-6_12

Our discussion in this paper is mostly related to these two issues in macroeconomic dynamics.

Several authors have exploited these fundamental differences to study the properties of discrete versus continuous time models. The classical references discuss differences between discrete and continuous time representations arising from uncertainty (cf. Burmeister and Turnovsky 1977) or adjustment costs (cf. Jovanovic 1982). Later on, Carlstrom and Fuerst (2005) focus on determinacy in monetary models, or Hintermaier (2003) and Bambi and Licandro (2005) on the dependency of the conditions for indeterminacy on the frequency of the discrete time representation of the model of Benhabib and Farmer (1994). Key ingredients for local indeterminacy typically relate to some form of market friction such as either imperfect competition (increasing returns to scale in production or market power in trade) or externalities (own production or consumption depends on other agents' in the same or the other side of the market).

Finally, there is a literature that looks to models in continuous time with discrete elements. Benhabib (2004) builds upon distributed lag structures to make the pure continuous time and discrete time frameworks emerge as special cases of a system of differential equations with delays. Anagnostopoulos and Giannitsarou (2013) propose a general continuous time model where certain events take place discretely, whereas Licandro and Puch (2006) use optimal control theory with delays to characterize the gap between discrete and continuous time models. As these authors, we discuss next a general framework where the pure continuous time and discrete time representations emerge as special cases. Different from them, we stress the unified framework provided by optimal control with delays. See Kolmanovskii and Myshkis (1998) and recent applications by Boucekkine et al. (2005), Licandro et al. (2008), and others.

Before using optimal control with delays, we introduce continuous and discrete time representations in a standard macroeconomic framework. We start with a baseline description of continuous versus discrete time modeling in growth and business cycle theory. To this purpose, we introduce the Solow growth model and the Ramsey model of the business cycle together with a description of the economic equilibrium. This description follows selected sections in Farmer (1999) or Novales et al. (2008). Then we turn to the issue of local indeterminacy in a canonical growth model of the environment which is subject to a pollution externality building upon Fernández et al. (2012). We show with a simple illustration that the time period of the model critically modifies its parameterization and thus the empirically plausible space for local indeterminacy. This leads to differences in transitional dynamics that are quantitatively meaningful. Finally, with a time-to-build example, we show that the discrete time representation of the standard optimal growth model implicitly imposes a particular form of time-to-build to the continuous time representation. Time-to-build in discrete time is analyzed by Kydland and Prescott (1982). An alternative version with this assumption in continuous time is in Asea and Zak (1999) and Collard et al. (2008). Here we show that the discrete time version is a true representation of the continuous time problem under some sufficient conditions.

The organization of the paper is as follows. We start by introducing a simple dynamic system that has proven useful to discuss questions in economic growth theory. Thus, Sect. 12.2 describes the Solow growth model in continuous and discrete time and their uses. Then we move forward to the use of optimal control theory and the concept of competitive equilibrium. Doing so, we describe the Ramsey model for the business cycle in Sect. 12.3, and the description is both in continuous and discrete time. Section 12.4 discusses some of the consequences of the differences in the time dimension for indeterminacy, and we focus on steady states and transitional dynamics. In Sect. 12.5, we propose a time-to-build extension of the theory by using optimal control theory with delays. Section 12.6 concludes.

12.2 The Solow Growth Model

The workhorse of economic growth theory is the *Solow model* in discrete time (cf. Solow 1956). The Solow growth model is characterized by the following set of equations:

$$\begin{aligned} Y_t &= F(K_t, A_t L_t), \\ K_{t+1} &= (1-\delta)K_t + I_t, \quad K_0 = \bar{K}_0, \\ I_t &= s\,Y_t, \quad 0 < s < 1, \\ A_t &= \gamma^{\,t} A_0, \quad \gamma \equiv 1 + g > 1, \quad A_0 = \bar{A}_0, \\ L_t &= 1, \quad (\text{might be} \quad L_t = \gamma_L^{\,t}\, L_0). \end{aligned} \tag{12.1}$$

The key assumptions are (1) savings (equal to investment, I_t) are a constant fraction, s, of output, Y_t, and (2) $F(\bullet)$ is a *neoclassical production function* in capital, K_t, and labor, L_t, where A_t represents the state of the production technology. In particular, let $k_t = \frac{K_t}{A_t L_t}$ and $F(\bullet)$ homogenous of degree one; then the equilibrium of the model is described by

$$\gamma k_{t+1} = (1-\delta)k_t + s\, f(k_t), \tag{12.2}$$

a first-order nonlinear difference equation. Note $F(\bullet)$ neoclassical implies $f'(k) > 0,\ f''(k) < 0,\ \forall k > 0$, and we further assume $\lim_{k\to\infty} f'(k) = 0,\ \lim_{k\to 0} f'(k) = \infty$. Figure 12.1 summarizes the equilibrium of the Solow growth model written in efficiency units, k, with $A_0 = 1$ so that $y_t = k_t^{\alpha}$ corresponds to $Y_t = K_t^{\alpha}\,(A_t L_t)^{1-\alpha}$.

Even in this simple representation of an economic model with dynamics induced by the accumulation of a stock (of physical capital K_t in this case), there is already an important approximation. Such an approximation comes from the fact that the definition of the capital stock is primarily established from cumulative investment in continuous time and not in discrete time. Therefore

$$K(t) = \int_{-\infty}^{t} I(s)\, e^{-\delta(t-s)} ds, \tag{12.3}$$

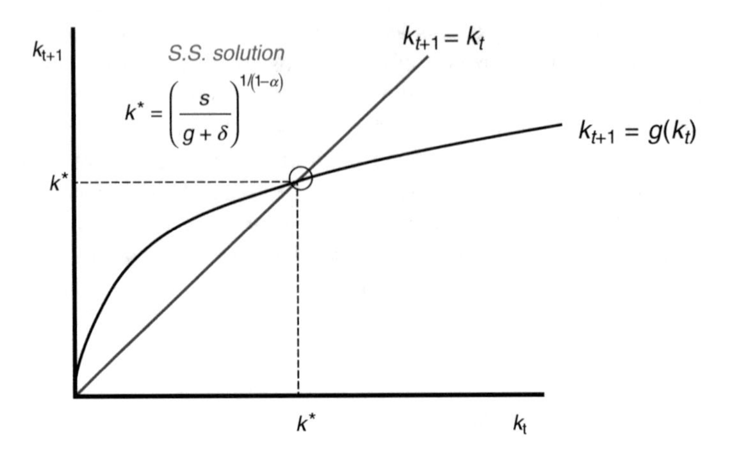

Fig. 12.1 The steady-state solution and the dynamics of the Solow growth model in discrete time, with $k_{t+1} = \eta\left[(1-\delta)\,k_t + s\,k_t^{\alpha}\right] \equiv g(k_t)$, where $\eta = 1/(1+g)$ and $k^* = k_{ss}$

with the additional simplifying assumption here that investment in different capital vintages can be added up at an exponential price $q(s) \equiv \mathrm{e}^{-\delta(t-s)}$, a simplification that is labeled the *law of permanent inventory* and corresponds to the assumption of exponential depreciation at the rate δ.

If this is the case, the accumulation of physical capital in continuous time is described by

$$K'(t) = I(t) - \delta K(t), \tag{12.4}$$

a differential equation whose solution is the integral equation (12.3) above, as it can be obtained from direct differentiation of the integral equation. With (12.4) rather than $K_{t+1} = (1-\delta)K_t + I_t$, one arrives at the continuous time version of the Solow model, with all the rest of the system (12.1) remaining the same as above, provided

$$\begin{aligned} &A(t) = A(0)\,\mathrm{e}^{g\,t}, \quad g > 0, \quad \text{note above } \gamma = (1+g), \\ &L(t) = 1 \quad \text{for instance, and with the assumptions above,} \end{aligned}$$

are defined correspondingly. Consider further again $k(t) = \frac{K(t)}{A(t)\,L}$. Then

$$k'(t) = s\,f(k(t)) - (\delta+g)\,k(t) \quad (\text{or } \dot{k}_t = s\,f(k_t) - (\delta+g)\,k_t, \text{ with } \dot{k}_t \equiv k'(t)), \tag{12.5}$$

describes now the equilibrium of the model in continuous time: a first-order nonlinear differential equation. Notice the use of the two alternative notations in continuous time (more on this below).

Regarding the solution of the equilibrium representation in (12.5), we know that under $f(k(t)) = k(t)^{\alpha}$, and with defined $z(t) \equiv k(t)^{1-\alpha}$, it is obtained that

$z(t) = (z(0) - \bar{z})\, e^{\phi t}$, $\phi = -(1-\alpha)(\delta + g)$, by guessing $z(t) = c\, e^{\phi t} + \bar{z}$, with $\bar{z} = s/(\delta + g)$. Unfortunately, there is no exact solution for the discrete time case. Thus, the discrete time assumption involves not only a key approximation in the specification of the model but also an approximation in obtaining the solution of (12.2) either recursively from an initial condition or by linearizing about the steady state.

All in all, dynamic economic models in continuous time form the basis for economic growth theory, as far as those models are more analytically tractable that their discrete version counterparts. However, the discrete time version of the models has a clear advantage when it comes to the issue of bringing the model to the data. The full characterization of the dynamics in discrete time often involves linear approximation about the steady-state solution though. Precisely, the steady-state k_{ss} is found from solving Eq. (12.2) irrespective of time:

$$(\gamma + \delta - 1)k_{ss} = s\, f(k_{ss}).$$

Notice that if k_t converges to k_{ss}, then K_t converges to a trend. One can then (log-)linearize around k_{ss} to obtain (one may further consider linearization versus log-linearization):

$$\hat{k}_{t+1} \simeq \frac{(1-\delta) + s\, f_k(k_{ss})}{\gamma} \hat{k}_t,$$

where $\hat{k}_t \equiv \left(\frac{k_t - k_{ss}}{k_{ss}}\right) \simeq \log k_t - \log k_{ss}$, and therefore, backward substitution implies (from an approximated convergence result as before)

$$\log k_{t+1} - \log k_{ss} \simeq a^t (\log k_0 - \log k_{ss}),$$

where $a = \frac{(1-\delta)+s\, f_k(k_{ss})}{\gamma}$ ($a < 1$ since $f(\bullet)$ is concave). Figure 12.2 illustrates again the equilibrium relation, $g(k)$, as it corresponds to representation (12.2) above, together with a linear approximation around its steady state. Richer approximations will allow us to characterize equilibrium dynamics of growth models also in a stochastic environment and with arbitrary precision. Indeed, one can consider a version of system (12.1) above, but now stochastic with,

$$Y_t = \upsilon_t\, F(K_t, A_t L_t), \quad \upsilon_t \sim D[A, B],$$

where $D[A, B]$ represents the probability distribution of υ_t. The equilibrium of the stochastic growth economy is then

$$\gamma k_{t+1} = (1-\delta)k_t + \upsilon_t\, s\, f(k_t),$$

and $(\gamma + \delta - 1)k_{ss} = \bar{\upsilon}\, s\, f(k_{ss})$ is the steady state, provided $\bar{\upsilon} \equiv E[\upsilon_t]$. Further parameterization of the shock process υ_t may be required for business

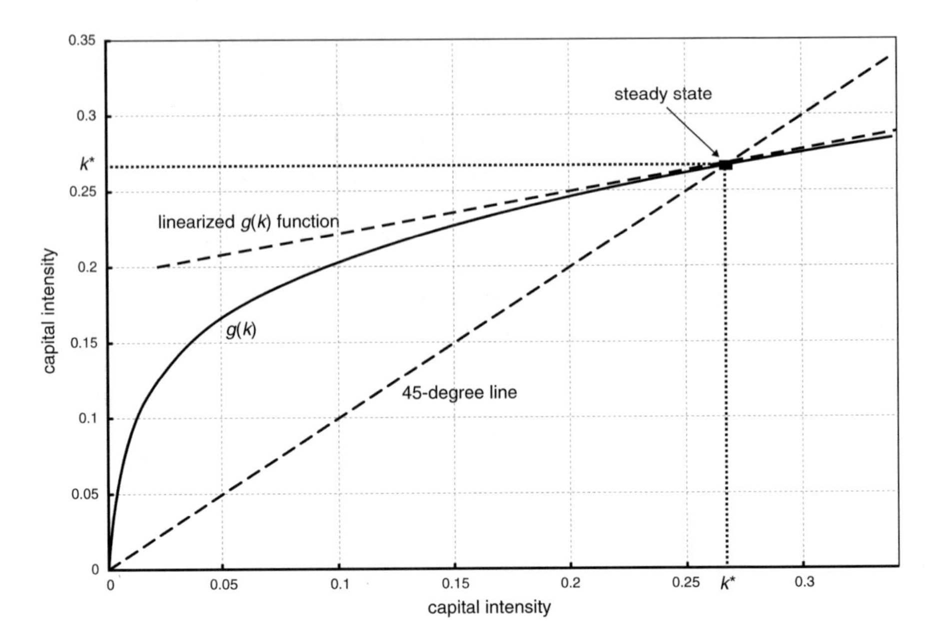

Fig. 12.2 The policy function $g(k)$, with $k_{t+1} = \eta\left[(1-\delta)\,k_t + s\,k_t^{\alpha}\right] \equiv g(k_t)$, where $\eta = 1/(1+g)$ and $k^* = k_{ss}$, and a linear approximation

cycle purposes. For instance, Fig. 12.3 depicts simulated output under $\theta_t\,k_t^{\alpha}$ with $\theta_t = \bar{\theta}^{1-\rho}\,\theta_{t-1}^{\rho}\,\varepsilon_t$, so that $\log\varepsilon_t \sim_{\text{iid}} N(0, \sigma_{\varepsilon}^2)$ and ρ is the persistence parameter of the process. The example suggests that it makes sense to think of fluctuations as caused by shocks to productivity of a neoclassical growth economy, in a richer environment though.

Summarizing, we have presented here the basic framework of the Solow growth model to show that the issue of continuous versus discrete time representation arises. We have also revised the basic methodology to deal with the model and its solution, introducing the issue of linear approximation to nonlinear models in discrete time around stable steady states. The interested readers may refer to Farmer (1999) and Novales et al. (2008) for further details. Next we introduce optimization and economic equilibrium in the framework of the *Ramsey growth model* as the building blocks of modern business cycle theory while focusing on the consequences for policy functions and equilibrium determination of changes in the frequency of decisions in the model.

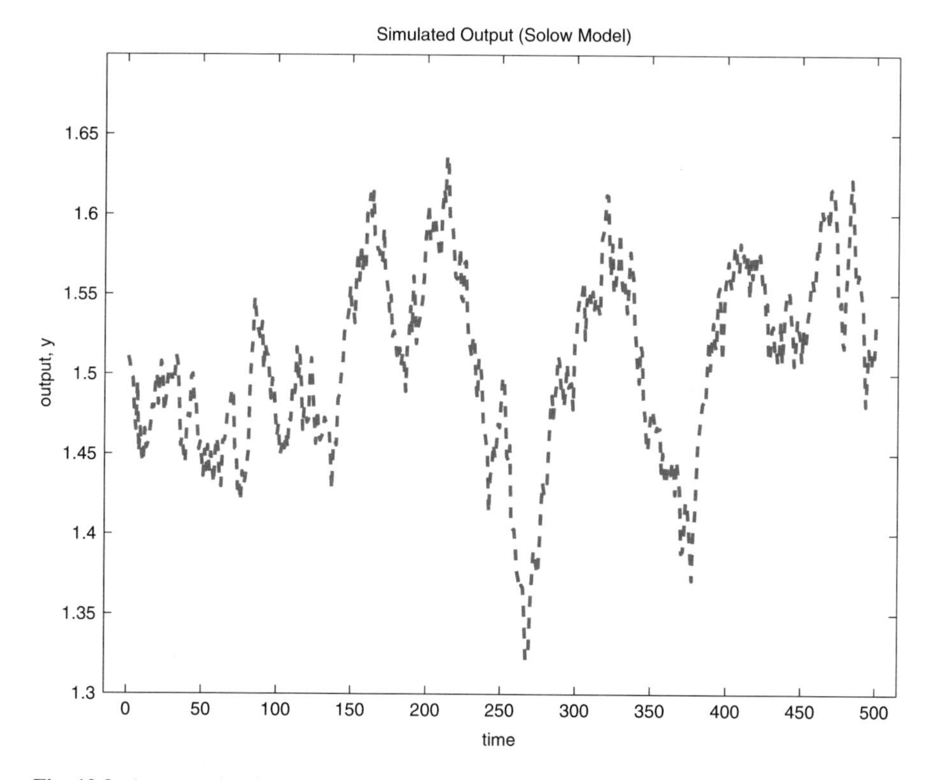

Fig. 12.3 An example of output fluctuations in the stochastic Solow model: output values around the deterministic steady state over 500 periods. Parameters: $\alpha = 0.36,\ g = 0.02,\ \delta = 0.1,\ s = 0.25,\ \rho = 0.95,\ \sigma_\varepsilon = 0.01$

12.3 The Ramsey Model and the Business Cycle

Rather than assuming that savings are a constant fraction of income as in the Solow model, now it is assumed that there exists a representative household that confronts a consumption-saving trade-off in discrete time. Moreover, it is assumed that such a representative household orders infinite sequences of consumption streams by using a well-behaved felicity function given by $U(c_t)$, such that $U'(c_t) > 0,\ U''(c_t) < 0,\ \forall c_t > 0$, according to

$$\sum_{t=0}^{\infty} \beta^t U(c_t), \quad \beta \in (0, 1),$$

where β is the subjective discount factor of the household and captures its degree of impatience in discrete time. Again we abstract from population growth, and we consider $L = 1$. Also, we abstract from exogenous technical progress A_t, and we assume that technology frontier is given by some $f(k_t)$, under the regularity conditions above.

Without loss of generality, let us specify the problem of a representative household that takes the decision to invest in physical capital in a centralized framework. Thus,

$$k_{t+1} - k_t = f(k_t) - c_t - \delta k_t, \tag{12.6}$$

and therefore, the decision problem of the *social planner* in discrete time can be written as

$$\begin{aligned} &\max_{\{c_t\,,\,k_{t+1}\}} \sum_{t=0}^{\infty} \beta^t U(c_t) \\ &\text{subject to (12.6), and given } k_0. \end{aligned} \tag{12.7}$$

The assumption of a representative household is more general than it first appears, and there are conditions under which the market allocation with many agents can be achieved by a social planner. This result can be established by using the two fundamental theorems of welfare economics (cf. Debreu 1959). At this point, however, it just allows us to focus on quantities and abstract from market prices. Later on, when necessary, we will be more precise on the type of market economy that supports the centralized problem at hand. This implies, in the case of the economy we are describing here, that a social planner maximizes the preferences of the representative household subject to the feasibility constraint of the economy.

The key issue here is that the dynamic optimization approach breaks the tight connection between output and savings in the short run, the one that we had in the Solow growth model. The discrete time framework allows us to bring the model to the data. These two elements have made some elaborated extensions of the model in (12.7) to form the basis for the theory of business cycles. Thus, according to modern business cycle theory, if productivity shocks are persistent and of the right magnitude, business cycle fluctuations are what growth theory predicts. It is the case though that, for a given volatility of the exogenous state and then of the endogenous state, the volatility of the control variable differs with the frequency of decisions in the model. Next, we illustrate this issue in the basic Ramsey model in continuous versus discrete time.

Let us characterize further the problem (12.7) above. The first-order conditions of this problem are given by

$$\begin{aligned} &U'(c_t) = \lambda_t, \\ &\lambda_t = \lambda_{t+1}\beta\left[f'(k_{t+1}) + 1 - \delta\right], \\ &\lim_{T\to\infty} \lambda_{t+T}k_{t+T+1} = 0, \end{aligned}$$

all $t \geq 0$, where λ_t is the shadow price in utility units of a unit of capital at time t. From the first-order conditions, the standard Euler equation condition is obtained:

$$\frac{U'(c_t)}{\beta U'(c_{t+1})} = 1 + r(k_{t+1}) \tag{12.8}$$

where we are assuming that the return to capital is just the extra output that the economy gets from an extra unit of capital, that is, $r(k) = \frac{\partial[f(k)-\delta k]}{\partial k} = f'(k) - \delta$.

The continuous time version of the problem basically involves the discount factor, $\beta = 1/(1+\rho)$, where now ρ is the subjective discount rate, and the approximation of first differences by derivatives, that is,

$$U'(c_{t+1}) - U'(c_t) \simeq U''(c_t)\,\dot{c}_t, \text{ since } \lim_{dt\to 0} \frac{U'(c_{t+dt})-U'(c_t)}{dt} = U''(c_t)\,\dot{c}_t,$$
$$k_{t+1} - k_t \simeq \dot{k}_t \ .$$

Here we are introducing as in (12.5) the notation $\dot{x}_t \equiv x'(t)$ that we will use for the rest of the paper and that correspondingly substitutes the notation $x(t)$ by the notation x_t to be used both in continuous and discrete time, except for Sect. 12.5 below where we combine both $\dot{x}_t$ and $x(t-d)$.

Consequently, the Euler equation (12.8) and the aggregate resource constraint (12.6) are, respectively, transformed into

$$\frac{1}{\beta(1+r)} = 1 + \frac{U''(c_t)}{U'(c_t)}\dot{c}_t \quad \underbrace{\rightarrow}_{\frac{1}{\beta(1+r)}=\frac{1+\rho}{1+r}\simeq 1+\rho-r} \quad \frac{U''(c_t)}{U'(c_t)}\dot{c}_t = \rho - r \tag{12.9}$$

$$\dot{k}_t = f(k_t) - c_t - \delta k_t, \tag{12.10}$$

where we assume that ρ and r are "small." Clearly though, the smaller the time period, the worse the approximation. However, given the approximations above, Eqs. (12.9) and (12.10) are exactly the optimality conditions of the continuous time problem

$$\begin{array}{l} \max\limits_{\{c_t\}} \int_0^\infty e^{-\rho t}\, U(c_t)dt \\ \text{subject to (12.10), and given } k_0, \end{array} \tag{12.11}$$

and obtained from direct application of optimal control theory in problem (12.11).

Beyond the precision of the approximation, the policy function of the discrete time version of the problem can be quite different from the policy function of the continuous time version. We can consider as an example the analytical case, that is, the case of full depreciation ($\delta = 1$), and logarithmic utility, $U(c) = \log c$. We further retain the assumption of a quasi-Cobb-Douglas production function in k, that is, $f(k_t) = A\, k_t^{\alpha}$. Under such a parameterization, in the discrete time version of the model, it is immediate to obtain the policy function, which is of the form

$$c_t = (1 - \alpha\beta) A k_t^{\alpha}. \tag{12.12}$$

On the other hand, by implementing a linear approximation around the steady-state solution to the optimality conditions of the discrete time problem, the policy function would be

$$c_t = c_{ss} + \left(\frac{1}{\beta} - \mu_2\right)(k_t - k_{ss}),$$
$$\text{where } k_{ss} = (A\alpha\beta)^{\frac{1}{1-\alpha}},\ c_{ss} = (1-\alpha\beta)Ak_{ss}^{\alpha},\ \mu_2 = \frac{\left(\frac{1+\alpha^2\beta}{\alpha\beta}\right) - \left[\left(\frac{1+\alpha^2\beta}{\alpha\beta}\right)^2 - 4\frac{1}{\beta}\right]^{1/2}}{2}. \tag{12.13}$$

Such a policy function is obtained by imposing the stability condition that cancels the eigenvector associated to the unstable eigenvalue of the dynamical system formed by the linear approximation around the steady state of the aggregate resource constraint and the Euler equation. Also, notice that the solution to the Ramsey problem for the feasible parameter space is of the saddle form, that is, it is a determinate solution.

Finally, by implementing a linear approximation around the steady-state solution to the optimality conditions of the continuous time problem, the policy function obtained from eliminating the unstable manifold is

$$\begin{aligned} &c_t = c_{ss} - \frac{h}{\mu_2}(k_t - k_{ss}) \\ &\text{where } k_{ss} = (A\alpha\beta)^{\frac{1}{1-\alpha}},\ c_{ss} = (1-\alpha\beta)A\,k_{ss}^{\alpha},\ h = \frac{1+\rho-\alpha}{\alpha}(1-\alpha)(1+\rho), \\ &\mu_2 = \frac{\rho - \left[\rho^2 + 4h\right]^{1/2}}{2}. \end{aligned} \tag{12.14}$$

Figure 12.4 depicts all three representations of the policy function in a phase diagram. The differences between the solutions in discrete versus continuous time are apparent. In particular, it is shown that at a given volatility in the state variable, the control variable fluctuates more in continuous time than in discrete time.

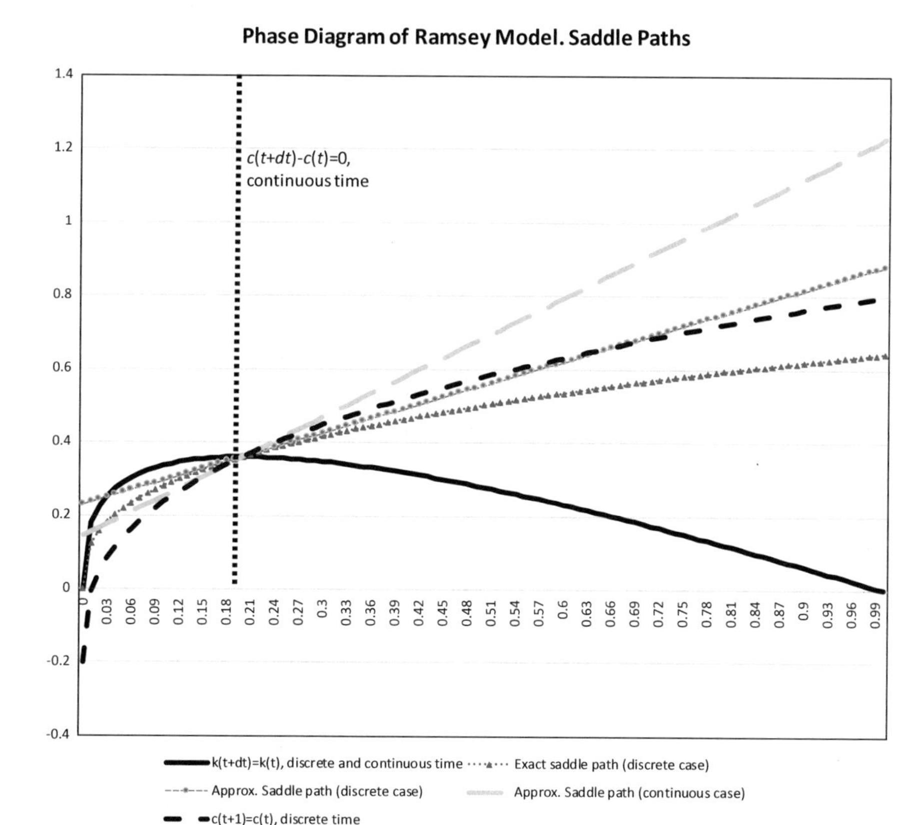

Fig. 12.4 Saddle paths of the Ramsey model

12.4 Indeterminacy of Equilibria in Continuous and Discrete Time

The dynamic properties in discrete time versions of continuous time models can fundamentally differ, particularly when the time domain at which agents take decisions differs. Convergence speeds to long-run equilibrium significantly differ between continuous and discrete time versions, the shorter the model period. This section builds upon Fernández et al. (2012) to show that transitional dynamics of pollution differ significantly, the model being written in either continuous or discrete time and in the latter case with the amplitude of the model period.

12.4.1 A Model of the Environmental Kuznets Curve

Fernández et al. (2012) study the existence of an environmental Kuznets curve (EKC), the hypothesis being that with development pollution goes up first and then down, associated to a neoclassical growth model with a pollution externality. The pollution externality goes into the utility function of the households, and it is shown that the non-separability in household preferences for consumption and pollution is crucial for the indeterminacy result to arise. Moreover, it is necessary for indeterminacy that the concern for the environment is large enough. Thus, both non-separability and enough environmentalism are needed for an environmental Kuznets curve pattern to be present.

The EKC result implies that economic growth could be compatible with environmental improvement if appropriate policies are taken. But before adopting a policy, it is important to understand the nature and causal relationship between economic growth and environmental quality, where the key question is whether economic growth can be part of the solution rather than the cause of environmental problem. In this section, we sketch the environment in Fernández et al. (2012) and their main result, with a focus on the fact that the predictions of the theoretical model vary substantially with the length of the model period which is typically large, particularly when analyzing climate issues in economic models.

In the model economy, there is a continuum of identical competitive firms that maximize profits from operating a neoclassical technology. Thus,

$$\max_{\{n_t,k_t\}} \quad y_t - \omega_t n_t - r_t k_t - \tau_P k_t$$
$$\text{subject to} \quad y_t = A k_t^{\alpha} n_t^{1-\alpha}, \quad \alpha \in (0,1),$$

with A being the state of technology, y_t the aggregate output, and k_t, n_t the two production factors: capital and labor. Firms rent capital from households at the interest rate r_t, pay wages w_t on labor, and pay a constant pollution tax τ_P on the level of the capital stock. Moreover, notice that with the presence of the pollution externality, it turns out that equilibria of the representative agent economy are no longer Pareto optimal (as the solution of the social planner is). Therefore, we have to specify the market economy and, as a consequence, have to deal with the determination of market prices in an environment here with continuum of identical firms and identical households. In addition, the existence of the externality may preclude the property that equilibria are (at least locally) uniquely determined by preferences and technology, and therefore determinate, in the sense of being isolated from their neighbors.

Environmental pollution, P_t, is a side product of the capital stock used by the firms but can be reduced by means of abatement activities made by the government, z_t:

$$P_t = \frac{k_t^{\chi_1}}{z_t^{\chi_2}}, \quad \chi_1,\ \chi_2 > 0. \tag{12.15}$$

The households solve

$$\max_{\{c_t, n_t\}} \int_0^\infty e^{-\rho t} \left[\frac{\left(c_t P_t^{-\eta}\right)^{1-\sigma} - 1}{1-\sigma} - \gamma h_t \right] dt \ , \quad \gamma, \ \sigma > 0,$$

$$\text{subject to} \quad c_t + \dot{k}_t + \delta k_t = (1-\tau)(\omega_t h_t + r_t k_t) + T_t, \quad \text{given } k_0,$$

where c_t is private consumption, P_t aggregate pollution, and h_t the labor supply.[1] σ is the inverse of the intertemporal elasticity of substitution of consumption, η is the weight of pollution in utility and γ the marginal disutility of work. Households receive income from labor and capital that can be used to consume, save, and pay taxes at a constant rate $\tau \in (0, 1)$ on the two sources of income. Finally, households receive lump-sum transfers from the government, T_t.

The problem faced by the households in the discrete time version of the economy is

$$\max_{\{c_t, n_t, k_{t+1}\}} \sum_{t=0}^{\infty} \beta^t \left[\frac{\left(c_t P_t^{-\eta}\right)^{1-\sigma} - 1}{1-\sigma} - \gamma n_t \right]$$

$$\text{subject to} \quad c_t + k_{t+1} - (1-\delta)k_t = (1-\tau)(\omega_t n_t + r_t k_t) + T_t, \quad \text{given } k_0.$$

The government sector in turn chooses an income tax rate τ and an environmental tax τ_P, and it uses these revenues to finance abatement activities, z_t, and transfers to households, T_t, balancing its budget every period. Thus, the instantaneous government budget constraint is

$$T_t + z_t = \tau\,(\omega_t n_t + r_t k_t) + \tau_P k_t \ , \tag{12.16}$$

where

$$z_t = \phi\,[\tau\,(\omega_t n_t + r_t k_t) + \tau_P k_t\ ] \ , \quad \phi \in [0, 1], \tag{12.17}$$

ϕ being the ratio that defines the allocation of government spending to abatement activities.

12.4.2 Transitional Dynamics

It can be shown that the dynamics of the economy in continuous time is described by

$$\begin{bmatrix} \dot{k}_t \\ \dot{\lambda}_t \end{bmatrix} = \begin{bmatrix} \mu_1 & \Omega \\ 0 & \mu_2 \end{bmatrix} \begin{bmatrix} k_t - k_{ss} \\ \lambda_t - \lambda_{ss} \end{bmatrix}, \tag{12.18}$$

[1] We assume indivisible labor as in Hansen (1985). In equilibrium, $h_t = n_t$.

whereas in discrete time it is

$$\begin{bmatrix} k_{t+1} - k_{ss} \\ \lambda_{t+1} - \lambda_{ss} \end{bmatrix} = \begin{bmatrix} \widetilde{\mu}_1 & \widetilde{\Omega} \\ 0 & \widetilde{\mu}_2 \end{bmatrix} \begin{bmatrix} k_t - k_{ss} \\ \lambda_t - \lambda_{ss} \end{bmatrix}, \tag{12.19}$$

where $\lambda(t)/\lambda_t$ is the co-state/multiplier associated to the household's budget constraint. Both dynamic systems are a linear approximation around the steady state, and therefore, $\{\mu_1, \mu_2, \Omega\}$ and $\{\widetilde{\mu}_1, \widetilde{\mu}_2, \widetilde{\Omega}\}$ are nonlinear functions of structural parameters of the model.

As far as the transition matrices are triangular, the elements in the main diagonal are the eigenvalues of the dynamical systems. Since one of the variables in the system is predetermined (k_t) and the other is free (λ_t), indeterminacy of equilibria arises only when the two roots $\{\mu_1, \mu_2\}$ have negative real parts or $\{\widetilde{\mu}_1, \widetilde{\mu}_2\} \in (0, 1)$. Fernández et al. (2012) show that indeterminacy will arise, if and only if $\sigma + (\xi_1 - \xi_2)\eta(1 - \sigma) < 0$. Under indeterminacy, they show that when the economy is initially placed on the steady state and the agents eventually coordinate to choose a level of labor above its steady-state value, the capital stock begins to increase in the following period and continues rising for several periods, but at an ever-decreasing rate; after reaching the turning point, the capital stock begins to decrease toward the steady state. The same behavior will be exhibited by labor, output, and abatement activities. The pollution exhibits an overshooting in the first periods of the transition, leading to an inverted U-shaped pattern, the same pattern followed by the CO2 emissions by some developed economies.

To compute the dynamic response of pollution, we do the following:

Step 1 Given structural parameters, we solve for the systems (12.18) and (12.19) above

$$\text{Continuous time:} \begin{cases} \lambda_t = \lambda_{ss} + e^{\mu_2 t}(\lambda_0 - \lambda_{ss}), & \text{where “}ss\text{” denotes steady state} \\ k_t = k_{ss} + e^{\mu_1 t}(k_0 - k_{ss}) + \frac{\Omega}{\mu_2 - \mu_1}(\lambda_0 - \lambda_{ss})\left(e^{\mu_2 t} - e^{\mu_1 t}\right), \end{cases}$$

$$\text{Discrete time:} \begin{cases} \lambda_t = \lambda_{ss} + \widetilde{\mu}_2^t(\lambda_0 - \lambda_{ss}), \\ k_t = k_{ss} + \widetilde{\mu}_1^t(k_0 - k_{ss}) + \frac{\widetilde{\Omega}}{\widetilde{\mu}_2 - \widetilde{\mu}_1}(\lambda_0 - \lambda_{ss})\left(\widetilde{\mu}_2^t - \widetilde{\mu}_1^t\right). \end{cases}$$

Step 2 With the expressions above, together with (12.15)–(12.17) above, we obtain a path for pollution for $k_0 = k_{ss}$ and λ_0 such that initial employment is $h_0 = n_{ss}$. Indeed, transition paths to the long-run indeterminate equilibrium can be indexed by the initial conditions in the control variable employment.

Figure 12.5a–c shows the transitional dynamics for pollution in the theoretical economy with parameters chosen for a model period of a quarter, a year, and 5 years. The figures illustrate that overshooting is bigger the smaller the model period, and the speed of convergence is slower in the discrete versus the continuous version of the model.

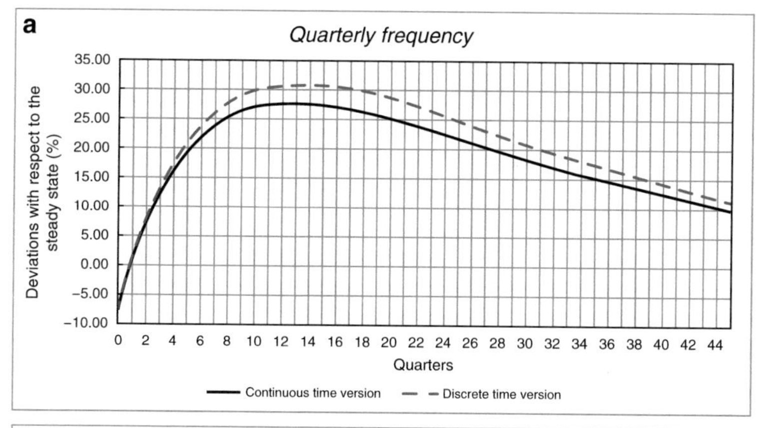

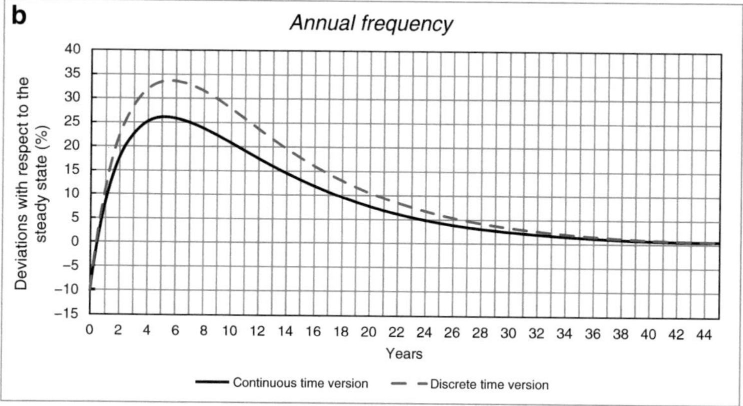

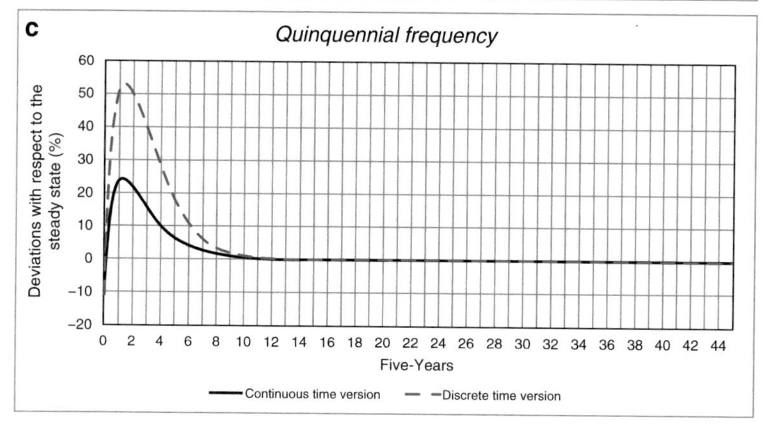

Fig. 12.5 Indeterminacy case with $n_0 > n_{ss}$. (**a**) Parameters quarterly: $\alpha = 0.33$, $\chi_1 = 1.3$, $\chi_2 = 0.6$, $\tau_P = 4\%$, $\tau = 20\%$, $\sigma = 2.5$, $\eta = 3.5$, $A = 1$, $\phi = 0.1$, $\delta = 0.025$ (10% per year), $\rho = 0.01$ (4% per year); γ is chosen to match n_{ss}, and $n_0 = 0.4$. (**b**) Parameters yearly: parameters above, but for $\delta = 0.1$ (10% per year), $\rho = 0.04$ (4% per year). (**c**) Parameters 5-year: parameters above, but for $\delta = 0.5$ (10% per year), $\rho = 0.2$ (4% per year)

12.5 A Continuous Time Model with Time-to-Build

Continuous time methods in economic dynamics have proven useful to give rise to substantial progress in modern growth theory and business cycle theory. The advantages of continuous time modeling are mainly technical, as far as continuous time systems turn out to be more tractable from the point of view of mathematical convenience. However, there is a crucial limitation of continuous time representations which is related to bringing the model to the data and the estimation of the structural parameters of dynamic models. An important subset of those parameters is key for economic policy design and evaluation.

Several approaches have been taken in the literature to approximate continuous time systems by discrete time systems. Also, inference in continuous time has developed substantially in recent years. The goal of this section is to illustrate the potential of using optimal control theory with delays to bridge the gap between continuous and discrete time representations in economic dynamics. We characterize the link between the two representations with a simple example of a growth economy with time-to-build.

12.5.1 The Environment

Let us primarily recover the centralized description of the economy as in Sect. 12.3, when we are under the conditions that the market allocation can be achieved by a social planner. Let us assume that time is continuous and introduce a simple time-to-build technology in an otherwise standard one-sector growth model. For simplicity, all variables are in per capita terms. Let $d > 0$ be the planned horizon of an investment project, i.e., the time-to-build delay. The technology to produce one unit of the investment good available at time $t + d$ requires a flow of $\frac{1}{d}$ units of the final good in the time interval $[t, t + d]$. Consequently, the only relevant decision at time t is the amount of *planned investment* $i(t)$, which will become operative at time $t + d$.

The stock of *planned capital* at time $t \geq -d$ is given by

$$k(t) = k(-d) + \int_{-d}^{t} i(s)ds. \tag{12.20}$$

The implicit assumption of zero depreciation makes $i(t)$ to be net investment. By definition of $i(t)$, $k(t)$ becomes operative at time $t + d$. Initial conditions need to be specified: $k(-d) = \bar{k} > 0$ and $i(t) = i_0(t) \geq 0$ for all $t \in [-d, 0]$. Consequently, $k(t) = k_0(t)$ for all $t \in [-d, 0]$ is computed using (12.20).

Final output is produced using a standard neoclassical technology $f(k)$, assumed to be C^2, increasing and concave for $k > 0$ and verifying Inada conditions (cf. Inada 1963; Uzawa 1963). Operative capital at time t was already planned at time $t - d$, implying that production at time t is $f(k(t-d))$.

The production of the final good is allocated to consumption $c(t)$ and to net investment expenditures $x(t)$. At time $t \geq 0$, the amount of the final good employed in the production of investment goods is given by

$$x(t) = \frac{1}{d} \int_{t-d}^{t} i(s)\, \mathrm{d}s. \tag{12.21}$$

It corresponds to investment expenditures associated to all active investment projects. Under these assumptions, the feasibility constraint for $t \geq 0$ takes the following form:

$$f(k(t-d)) = c(t) + x(t). \tag{12.22}$$

12.5.2 The Planner's Problem

As in Sect. 12.3 above, in the described environment, the solution of a benevolent social planner is the competitive equilibrium allocation. Let such a planner maximize the utility of the representative household

$$\max \int_{0}^{\infty} u(c(t))\, \mathrm{e}^{-\rho t} \tag{P}$$

subject to (12.22) and

$$\dot{x}(t) = \frac{1}{d}\left(i(t) - i(t-d)\right), \tag{12.23}$$

$$\dot{k}(t) = i(t). \tag{12.24}$$

Constraints (12.23) and (12.24) result from time differentiation of (12.21) and (12.20), respectively. The initial conditions are $x(0) = x_0 = \frac{1}{d}\int_{-d}^{0} i_0(s)\, ds$, $k(t) = k_0(t)$, and $i(t) = i_0(t)$ for all $t \in [-d, 0]$, as specified previously. The instantaneous utility function $u(t)$ is C^2, increasing and concave for $c > 0$, and verifies Inada conditions.

Using optimal control theory with delays (cf. Bambi et al. 2014; Boucekkine et al. 2005; Kolmanovskii and Myshkis 1998), necessary first-order conditions for this problem are

$$u'(c(t)) = \phi(t), \tag{12.25}$$

$$\lambda(t) + \frac{1}{d}\mu(t) = \frac{1}{d}\mu(t+d)\ \mathrm{e}^{-\rho d}, \tag{12.26}$$

$$-\phi(t+d)\, f'(k(t))\ \mathrm{e}^{-\rho d} = \dot{\lambda}(t) - \rho\lambda(t), \tag{12.27}$$

$$\phi(t) = \dot{\mu}(t) - \rho\mu(t), \tag{12.28}$$

and the transversality conditions

$$\lim_{t\to\infty} k(t)\,\lambda(t)\ \mathrm{e}^{-\rho t} = 0, \tag{12.29}$$

$$\lim_{t\to\infty} x(t)\,\mu(t)\ \mathrm{e}^{-\rho t} = 0. \tag{12.30}$$

The Lagrange multiplier $\phi(t)$ is associated to constraint (12.22), and the co-states $\lambda(t)$ and $\mu(t)$ are associated to the states $k(t)$ and $x(t)$, respectively. Advanced terms appearing in (12.26) and (12.27), related to the delays in (12.23) and (12.24), make explicit the trade-offs. Marginal investment at time t has three different effects on utility. Firstly, it increases planed capital, which marginal value is $\lambda(t)$. Secondly, it rises investment expenditures, with marginal costs $\frac{\mu(t)}{d}$. Finally, when the project will be finished at $t+d$, investment expenditures will end.

12.5.3 Discrete Time as a Representation of Continuous Time

In this section, we establish the correspondence between the proposed continuous time model with a time-to-build delay in Sect. 12.5.2 and the discrete time representation of the neoclassical growth model in Sect. 12.3. Let us assume that the initial function $i_0(t)$ is piecewise continuous and that feasible trajectories $i(t)$, for $t \geq 0$, belong to the family of piecewise continuous functions.

Proposition 12.1 *Under $d = 1$, the optimal conditions (12.25)–(12.30) of problem (P) become*

$$k(t) - k(t-1) = f(k(t-1)) - c(t) \tag{12.31}$$

$$\frac{u'(c(t))}{u'(c(t+1))} = \beta\left(1 + f'(k(t))\right), \tag{12.32}$$

where $\beta \equiv e^{-\rho}$.

Proof From (12.20) and (12.21), under $d = 1$, we get $x(t) = k(t) - k(t-1)$. The feasibility constraint (12.31) results from substituting the relation between x and k on Eq. (12.22). Differentiating (12.26) and substituting $\dot{\lambda}$ and $\dot{\mu}$ by (12.27) and (12.28), after some rearrangements, we get (12.32).

The equilibrium path of the neoclassical growth model is indeed represented by (12.31) and (12.32) for given initial conditions. Notice the correspondence with system (12.6) and (12.8) in Sect. 12.3 under the assumption here of zero depreciation, that is, $\delta = 0$.

Corollary 12.1 *The steady-state solution of (12.31) and (12.32) is saddle-path stable for* $t \geq 0$.

Corollary 12.1 implies that for every $s \in [0, 1)$, the optimal sequence $\{c_{s+i}, k_{s+i}\}$ for $i = \{0, 1, 2, 3 \ldots\}$ is the solution of the discrete time neoclassical Ramsey growth model of Sect. 12.3, given $k(-1) = k_0(-1)$. However, in continuous time with delay $d = 1$, the steady-state solution involves the solution path for all $s \in [0, 1)$, which depends on the boundary function $k_0(t)$, for $t \in [-1, 0)$, defining initial conditions (cf. Collard et al. 2008).

Corollary 12.2 *Under* $d = 1$ *and* $k(t) = k_0 > 0$ *for* $t \in [-1, 0)$, *the optimal solution* $k(t), c(t)$ *of problem (P) is constant in the interval* $[i-1, i)$ *for* $i = \{1, 2, 3 \ldots\}$, *and it corresponds to the stable brand of the discrete problem in Proposition 12.1.*

Corollary 12.2 states the dynamic properties of the neoclassical growth model in its correspondence with the continuous time representation with delays. All in all, the example illustrates the particular form of a delay in the continuous time model that the discrete time model imposes. The recent literature on continuous time with delays should help to take advantage of the analytical tractability of models in continuous time while providing precise quantitative statements about the issues of interest.

12.6 Concluding Remarks

In this paper, we have explored the differences arising from modeling time as discrete or continuous. This has been done in the basic framework of dynamic macroeconomic models and focusing on appropriate approximation, dynamic indeterminacy, and delays. We have shown that the differences between continuous and discrete representations arise from investment decisions at time t that become productive at a time that depends on the model period. Agents are then committed to their decisions until the period when the return of their investments is realized. This modifies not only the structure and the solution of the models but also the economic interpretation of the results. The recent literature on continuous time models with delays should help to bridge the gap between continuous and discrete time representations in economic dynamics.

Acknowledgements We thank Mauro Bambi, Raouf Boucekkine, Fabrice Collard, Esther Fernández, Gustavo Marrero, and Alfonso Novales for insightful discussions. We also thank a reviewer and the editors for their suggestions. Finally, we thank the financial support from the Spanish Ministerio de Economía y Competitividad (grant ECO2014-56676) and the Bank of Spain (grant Excelencia 2016–17).

References

Anagnostopoulos, A., & Giannitsarou, C. (2013). Indeterminacy and period length under balanced budget rules. *Macroeconomic Dynamics, 17*, 898–919. https://doi.org/10.1017/S1365100511000745

Asea, P., & Zak, P. (1999). Time-to-build and cycles. *Journal of Economic Dynamics and Control, 23*, 1155–1175. https://doi.org/10.1016/S0165-1889(98)00052-9

Bambi, M., Gozzi, F., & Licandro, O. (2014). Endogenous growth and wave-like business fluctuations. *Journal of Economic Theory, 154*, 68–111. https://doi.org/10.1016/j.jet.2014.08.004

Bambi, M., & Licandro, O. (2005). *(In)determinacy and Time-to-Build* (Economics Working Papers ECO2004/17). European University Institute.

Benhabib, J. (2004). Interest rate policy in continuous time with discrete delays. *Journal of Money, Credit and Banking, 36*, 1–15. https://doi.org/10.1353/mcb.2004.0001

Benhabib, J., & Farmer, R. (1994). (In)determinacy and increasing returns. *Journal of Economic Theory, 63*, 19–41. https://doi.org/10.1006/jeth.1994.1031

Boucekkine, R., Licandro, O., Puch, L. A., & Río, F. (2005). Vintage capital and the dynamics of the AK model. *Journal of Economic Theory, 120*, 39–72. https://doi.org/10.1016/j.jet.2004.02.006

Burmeister, E., & Turnovsky, S. J. (1977). Price expectations and stability in a short-run multi-asset macro model. *American Economic Review, 67*, 213–218.

Carlstrom, C. T., & Fuerst, T. S. (2005). Investment and interest rate policy: A discrete time analysis. *Journal of Economic Theory, 123*, 4–20. https://doi.org/10.1016/j.jet.2004.05.002

Collard, F., Licandro, O., & Puch, L. A. (2008). The short-run dynamics of optimal growth model with delays. *Annals of Economics and Statistics, 90*, 127–143. https://doi.org/10.2307/27739821

Debreu, G. (1959). *Theory of value: An axiomatic analysis of economic equilibrium*. New Haven: Yale University Press.

Farmer, R. (1999). *Macroeconomics of self-fulfilling prophecies* (2nd ed.). Cambridge: The MIT Press.

Fernández, E., Pérez, R., & Ruiz, J. (2012). The environmental Kuznets curve and equilibrium indeterminacy. *Journal of Economic Dynamics and Control, 36*(11), 1700–1717. https://doi.org/10.1016/j.jedc.2012.05.004

Hansen, G. (1985). Indivisible labor and the business cycle. *Journal of Monetary Economics, 16*, 309–327. https://doi.org/10.1016/0304-3932(85)90039-X

Hintermaier, T. (2003). On the minimum degree of returns to scale in sunspot models of the business cycle. *Journal of Economic Theory, 110*, 400–409. https://doi.org/10.1016/S0022-0531(03)00034-6

Inada, K. (1963). On a two-sector model of economic growth: Comments and a generalization. *Review of Economic Studies, 30*(2), 119–127. https://doi.org/10.2307/2295809

Jovanovic, B. (1982). Selection and the evolution of industry. *Econometrica, 50*, 649–670. https://doi.org/10.2307/1912606

Kolmanovskii, V., & Myshkis, A. (1998). *Introduction to the theory and applications of functional differential equations*. Boston: Kluwer Academic Publishers.

Kydland, F., & Prescott, E. C. (1982). Time-to-build and aggregate fluctuations. *Econometrica, 50*, 1345–1370. https://doi.org/10.2307/1913386

Licandro, O., & Puch, L. A. (2006). *Is Discrete Time a Good Representation of Continuous Time?* (Economics Working Papers ECO2006/28). European University Institute.

Licandro, O., Puch, L. A., & Sampayo, A. R. (2008). A vintage model of trade in secondhand markets and the lifetime of durable goods. *Mathematical Population Studies, 15*, 249–266. https://doi.org/10.1080/08898480802440828

Novales, A., Fernández, E., & Ruiz, J. (2008). *Economic growth: Theory and numerical solution methods*. Berlin: Springer Science.

Solow, R. (1956). A contribution to the theory of economic growth. *Quarterly Journal of Economics, 70*, 65–94. https://doi.org/10.2307/1884513

Uzawa, H. (1963). On a two-sector model of economic growth II. *Review of Economic Studies, 30*(2), 105–118. https://doi.org/10.2307/2295808

Chapter 13
Continuous Time State Space Modelling with an Application to High-Frequency Road Traffic Data

Siem Jan Koopman, Jacques J. F. Commandeur, Frits D. Bijleveld, and Sunčica Vujić

13.1 Introduction

We present a practical treatment of continuous time state space modelling. The main features of the analysis are highlighted and explored in some generality. We further present and discuss the main results of an empirical study related to road safety analysis. This application of the continuous time methodology in time series analysis shows how it can be used in practice.

A time series is a set of observations which are sequentially ordered over time. In a discrete time state space analysis, the time series observations are assumed to be equally spaced in time. Although missing data may give rise to different time gaps between available observations in discrete time series also, these time gaps are then always multiples of the time unit specific to the time series at hand (e.g. a year for annual data, a month for monthly data, etc.). In this chapter, on the other hand, we concentrate on the continuous time state space model and some of its special cases. In continuous state space models, the time gaps between

Electronic Supplementary Material The online version of this article (https://doi.org/10.1007/978-3-319-77219-6_13) contains supplementary material, which is available to authorized users.

S. J. Koopman (✉) · J. J. F. Commandeur · F. D. Bijleveld
Department of Econometrics, Vrije Universiteit Amsterdam, Amsterdam, The Netherlands
e-mail: s.j.koopman@vu.nl; j.j.f.commandeur@vu.nl; f.d.bijleveld@vu.nl

S. Vujić
Department of Economics, University of Antwerp, Antwerp, Belgium
e-mail: suncica.vujic@uantwerpen.be

K. van Montfort et al. (eds.), *Continuous Time Modeling in the Behavioral and Related Sciences*, https://doi.org/10.1007/978-3-319-77219-6_13

consecutive observations of a series are typically allowed to vary freely from one pair of consecutive observations to the next. The expositions in this chapter rely mostly on the textbooks by Harvey (1989) and by Durbin and Koopman (2012). For an introduction to state space time series analysis, we refer to Commandeur and Koopman (2007).

The literature on continuous time modelling in statistics and econometrics is extensive. It is beyond the scope of this paper to present a full review of this literature. A major and key reference to continuous time models in econometrics is the review of Bergstrom (1984) where various results on parameter estimation for dynamic structural models in continuous time are provided. The many benefits of continuous time modelling are also illustrated. In the statistics literature, there is a considerable focus on smoothing methods that are formulated in continuous time. For example, a standard treatment using the continuous time approach is developed by Green and Silverman (1994). But also the earlier contributions of Wahba (1978) and Silverman (1985) have been of key importance in the development of signal extraction and spline smoothing in continuous time. The connections between spline smoothing and continuous time state space analysis are first established in the work of Wecker and Ansley (1983).

In this review chapter, we provide a detailed account of a continuous time state space approach to time series analysis. The outline of this chapter is as follows. In Sect. 13.2 we formulate the general continuous time state space model and discuss two well-known special cases. Section 13.3 discusses the estimation of the unobserved states together with the unknown model parameters. Finally, in Sect. 13.4 we apply the methodology to an empirical example consisting of road traffic speed data.

13.2 A Continuous Time Modelling Framework

Let t_τ denote the time point at which observation τ in the series was measured, $\tau = 1, 2, \ldots, T$. Note that τ is an integer denoting the *number* of the observation in the time series, while t_τ is the *time* at which this observation was made. Thus, unlike τ, t_τ can be any non-negative number, for example, 10 years, 200 days, 300.405 ms, etc. The only requirement is that $t_1 < t_2 < t_3 < \cdots < t_T$. The general linear Gaussian state space model for the T-dimensional observation sequence $y_1, \ldots, y_T$ is given by

$$y_\tau = Z_\tau \alpha_\tau + \varepsilon_\tau, \qquad \varepsilon_\tau \sim \text{NID}(0, H_\tau), \tag{13.1}$$

$$\alpha_{\tau+1} = T_\tau \alpha_\tau + R_\tau \eta_\tau, \qquad \eta_\tau \sim \text{NID}(0, Q_\tau), \qquad \tau = 1, \ldots, T, \tag{13.2}$$

where α_τ is the state vector, ε_τ and η_τ are disturbance vectors and the system matrices Z_τ, T_τ, R_τ, H_τ and Q_τ are fixed and known. A selection of the elements of the system matrices may depend on an unknown parameter vector. Equation (13.1) is referred as the *observation* or *measurement equation*, while Eq. (13.2) is called the *state* or *transition equation*. The $p \times 1$ observation vector y_τ contains the p observations at time point t_τ, and the $m \times 1$ state vector α_τ is unobserved. The $p \times 1$ irregular vector ε_τ has zero mean and $p \times p$ variance matrix H_τ.

The $p \times m$ matrix Z_τ links the observation vector y_τ with the unobservable state vector α_τ and may consist of regression variables. The $m \times m$ transition matrix T_τ in (13.2) determines the dynamic evolution of the state vector. The $r \times 1$ disturbance vector η_τ for the state vector update has zero mean and $r \times r$ variance matrix Q_τ. The observation and state disturbances ε_τ and η_τ are assumed to be serially independent and independent of each other at all time points. In many standard cases, $r = m$ and matrix R_τ is the identity matrix I_m. In other cases, matrix R_τ is a $m \times r$ selection matrix with $r < m$. Although matrix R_τ can be specified freely, it is often composed of a selection from the first r columns of the identity matrix I_m. It further implies that often we can treat the matrix R_τ as a constant matrix that does not vary with τ. Similarly, all system matrices are assumed to be (deterministically) varying with τ, but in many cases of practical interest, most system matrices are fixed for all τ.

The initial state vector α_1 is assumed to be generated as

$$\alpha_1 \sim \text{NID}(a_1, P_1),$$

independently of the observation and state disturbances ε_τ and η_τ, where initial mean a_1 and initial variance P_1 can be treated as given and known in almost all stationary processes for the state vector. For nonstationary processes and regression effects in the state vector, the associated elements in the initial mean a_1 can be treated as unknown and need to be estimated. For an extensive discussion of *initialisation* in state space analysis, we refer to Durbin and Koopman (2012, Chapter 5).

13.2.1 Local Level and Local Linear Trend Models

By appropriate choices of the vectors α_τ, ϵ_τ and η_τ, and of the matrices Z_τ, T_τ, H_τ, R_τ and Q_τ, a wide range of different continuous time state space models can be derived from (13.1) and (13.2). Here we focus on the continuous time equivalents of the discrete local level and local linear trend models. Other model formulations can be considered as well since our state space framework allows for many different linear dynamic specifications that are commonly used in time series analysis. However, the arguments for continuous time formulations are similar, and therefore our treatment below remains relatively general.

Let $\delta_\tau = t_\tau - t_{\tau-1}$ denote the amount of time elapsed between two consecutive observations τ and $\tau - 1$. Also defining

$$\alpha_\tau = \mu_\tau, \quad \eta_\tau = \xi_\tau, \quad Z_\tau = T_\tau = R_\tau = 1, \quad H_\tau = \sigma_\varepsilon^2, \quad Q_\tau = \delta_\tau \sigma_\xi^2,$$

(all variables are scalars) for $\tau = 1, \ldots, T$, model (13.1) and (13.2) reduces to the univariate continuous local level model as given by

$$\begin{aligned} y_\tau &= \mu_\tau + \varepsilon_\tau, & \varepsilon_\tau &\sim \text{NID}(0, \sigma_\varepsilon^2), \\ \mu_{\tau+1} &= \mu_\tau + \xi_\tau, & \xi_\tau &\sim \text{NID}(0, \delta_\tau \sigma_\xi^2), \end{aligned} \tag{13.3}$$

for $\tau = 1, \ldots, T$. Note that (13.3) reduces to the discrete local level model when the observations are equally spaced, i.e. when $\delta_\tau = t_\tau - t_{\tau-1} = 1$, say, for all $\tau = 1, \ldots, T$.

The local level model can be regarded as the most basic version of a state space model. It is intuitive as it can be interpreted as a model representation for y_τ that is generated by the normal distribution with a time-varying mean μ_τ and a fixed variance σ_ε^2. The continuous time formulation only applies to the dynamic process of the time-varying mean. The local level model also provides a statistical specification for the exponentially weighted moving average (EWMA) forecasting method that is very popular amongst professional practitioners. The forecast function of the local level model is equivalent to the EWMA, but the state space treatment also provides statistical standard errors to the point forecasts; see the discussion below. A full discussion and treatment of the local level model is provided by Harvey (1989) and Durbin and Koopman (2012, Chapter 2).

By defining

$$\alpha_\tau = \begin{pmatrix} \mu_\tau \\ \nu_\tau \end{pmatrix}, \quad \eta_\tau = \begin{pmatrix} \xi_\tau \\ \zeta_\tau \end{pmatrix}, \quad T_\tau = \begin{bmatrix} 1 & \delta_\tau \\ 0 & 1 \end{bmatrix}, \quad Z_\tau = \begin{pmatrix} 1 & 0 \end{pmatrix},$$

$$H_\tau = \sigma_\varepsilon^2, \quad \text{Var}(\eta_\tau) = Q_\tau = \delta_\tau \begin{bmatrix} \sigma_\xi^2 + \frac{1}{3}\delta_\tau^2 \sigma_\zeta^2 & \frac{1}{2}\delta_\tau \sigma_\zeta^2 \\ \frac{1}{2}\delta_\tau \sigma_\zeta^2 & \sigma_\zeta^2 \end{bmatrix}, \quad \text{and} \quad R_\tau = \begin{bmatrix} 1 & 0 \\ 0 & 1 \end{bmatrix},$$

the scalar notation of (13.1) and (13.2) leads to

$$\begin{aligned} y_\tau &= \mu_\tau + \varepsilon_\tau, & \varepsilon_\tau &\sim \text{NID}(0, \sigma_\varepsilon^2), \\ \mu_{\tau+1} &= \mu_\tau + \delta_\tau \nu_\tau + \xi_\tau, \\ \nu_{\tau+1} &= \nu_\tau + \zeta_\tau, \end{aligned} \tag{13.4}$$

for $\tau = 1, \ldots, T$, and we obtain the univariate continuous local linear trend model. Unlike in the discrete local linear trend model, we see that the disturbances of the level and the slope component are correlated through the off-diagonal elements $\frac{1}{2}\delta_\tau^2 \sigma_\zeta^2$ in matrix Q_τ in the continuous local linear trend model. However, as mentioned by Harvey (1989, p. 487), "this difference is unlikely to be of any great importance".

The treatment above for the local linear trend model has many connections with the statistical literature on spline smoothing. Reviews of methods related to spline smoothing are given in Silverman (1985), Wahba (1990) and Green and Silverman (1994, Chapter 2). Some of these connections with the approach given above and the more traditional methods are given by Wahba (1990) but are also discussed in Wecker and Ansley (1983). These connections are also highlighted in Durbin and Koopman (2012, Chapter 3).

13.2.2 *Multivariate Continuous Time State Space Models*

The treatments as set out for univariate time series above can be easily extended to multivariate time series. This is one of the advantages of the state space approach since multivariate spline smoothing methods are not widespread.

In case we let y_τ denote a $p \times 1$ vector of observations, a multivariate local linear trend model can be applied to the p time series simultaneously:

$$\begin{aligned} y_\tau &= \mu_\tau + \varepsilon_\tau, & \varepsilon_\tau &\sim \text{NID}(0, \Sigma_\varepsilon), \\ \mu_{\tau+1} &= \mu_\tau + \eta_\tau, & \eta_\tau &\sim \text{NID}(0, \Sigma_\eta), \end{aligned} \tag{13.5}$$

for $\tau = 1, \ldots, T$, where μ_τ, ε_τ, and η_τ are $p \times 1$ vectors, Σ_ε is a $p \times p$ variance matrix, and

$$\Sigma_\eta = \delta_\tau \begin{bmatrix} \Sigma_\xi + \frac{1}{3}\delta_\tau^2 \Sigma_\zeta & \frac{1}{2}\delta_\tau \Sigma_\zeta \\ \frac{1}{2}\delta_\tau \Sigma_\zeta & \Sigma_\zeta \end{bmatrix}$$

is a $2p \times 2p$ matrix, Σ_ξ and Σ_ζ being the $p \times p$ variance matrices of the level and the slope disturbances, respectively.

13.3 State Space Methods for Continuous Time Models

The model formulations as discussed above are all special cases of the general linear Gaussian state space model. We can therefore rely on the associated methods for signal extraction, parameter estimation and forecasting. The most important and well-known method for this class of state space models is the Kalman filter that allows the (predictive and filtered) estimation of the unobserved state vector α_τ when the system matrices have given values. It also enables the computation of the log-likelihood function of the model, for a given parameter vector, via the prediction error decomposition. It allows the maximisation of the log-likelihood function with respect to the parameter vector, in order to obtain its maximum likelihood estimate. On the basis of these parameter estimates, signal extraction and forecasting can take place. We next provide more details of this central part of the state space methodology.

In a similar way as in discrete state space models, in continuous state space models for given values of all system matrices—and for known initial conditions a_1 and P_1—the state vector can be estimated in three different ways, yielding what are known as the *filtered*, the *predicted* and the *smoothed* state vector. Depending on the types of state estimates required in the analysis, the estimates of the state vector can be obtained by performing one or two passes through the observed time series:

1. A *forward* pass, from $\tau = 1, \ldots, T$, using a recursive algorithm known as the *Kalman filter* enables the computation of filtered and predicted states and prediction errors, including their variances; from the prediction error and their variances, we can compute the log-likelihood function of the given continuous state space model;
2. A *backward* pass, from $\tau = T, \ldots, 1$, using all filtered and associated variables from the Kalman filter and using recursive algorithms known as *state and disturbance smoothers* enables the computation of smoothed estimates of states and disturbances; it requires the storage of the Kalman filter variables.

In continuous time state space models, the standard Kalman, state and disturbance smoothing filters can be used; see Durbin and Koopman (2012, Chapter 4) for technical details. A specific difference of substance between discrete and continuous time models is that the variance matrix Q_τ in Eq. (13.2) of the state space formulation of the model (containing the variances of the state disturbances) is typically time-invariant in the discrete case while it becomes a time-varying matrix for continuous time state space models.

We have discussed these continuous time state space methods above as if the disturbance variances are given and known. In practice, of course, these parameters are unknown, and they have to be estimated. Just as in the discrete time series situation, the parameter estimates are obtained via maximum likelihood methods which are discussed in Durbin and Koopman (2012, Chapter 7). It requires an optimisation algorithm, and for this purpose quasi-Newton methods are typically used. Each time new parameter values are proposed by the search-for-the-maximum algorithm, the Kalman filter is used to compute the log-likelihood function. In many applications, it is found that the maximum is found quickly and the estimation process does not take much computing time.

13.4 An Application in Road Safety and Traffic Control

We consider our continuous time modelling approach to a full day of measurements of the speed of passing motor vehicles on a fixed location in the right lane of a Dutch motorway, starting at midnight and ending at midnight of the following day. For our analysis of this interesting and important time series for road safety studies, we have considered the continuous time models and methods as set out in the previous sections. All computations are implemented in the `OxMetrics` object-oriented programming environment of Doornik (2013) together with the `SsfPack` library of state space routines of Koopman et al. (2008). Initial analyses are carried out by means of the discrete time versions of our models using the `STAMP` software of Koopman et al. (2007).[1]

There is a total of 25,539 passages in this series meaning that we also have 25,539 observations. The time of each passage is measured as the number of milliseconds elapsed since the start of the measurements and the difference between the time of the last and the first observation of the series, i.e. $t_T - t_1$, is 86,396 ms which indeed corresponds to a full $86{,}396/60^2 = 24$ h. The average time lapse between consecutive observations in the series is 3.383 ms with a minimum of 0.038 ms and a maximum of 1417.4 ms. The variance of the time lapses δ_τ is 230.170.

From the perspective of road safety, it is of interest to analyse passages of cars at different speed levels. In our analyses, we consider two groups of speed levels: slow passages with a speed of less than 100 km/h (but faster than 75 km/h as we discard very slow passages which may be due to measurement failings) and fast passages with a speed of higher than 120 km/h. These two different groups constitute a total of 14,435 passages (9010 slow and 5425 fast passages).

The analyses of the two series are based on the continuous time local linear trend model (13.4). To enforce a smoother evolving signal in this highly noisy time series of speed passages, we restrict the variance of the level component to be zero. The remaining variances are estimated by the method of maximum likelihood (ML). We yield the following estimation results. At convergence of the ML process, the parameter estimates for the variance of the slope disturbances and the measurement errors are, respectively, given by $\sigma_\zeta^2 = 1.061 \times 10^{-6}$ and $\sigma_\varepsilon^2 = 31.966$ for the slow passages and $\sigma_\zeta^2 = 5.434 \times 10^{-6}$ and $\sigma_\varepsilon^2 = 43.295$ for the fast passages.

The recorded speed levels of the passages for the slow and the fast groups are presented in Figs. 13.1 and 13.3, respectively, together with their estimated trend components which are also presented separately in Figs. 13.2 and 13.4, respectively. We learn from these graphs that the number of passages of motor vehicles on the motorway diminishes during the night. It is especially observable for the fast-speed passages, between roughly 8000 ms after midnight (i.e. around half past three in the

[1]The programming code of our analyses is available as supplementary material at the book website http://www.springer.com/us/book/9783319772189.

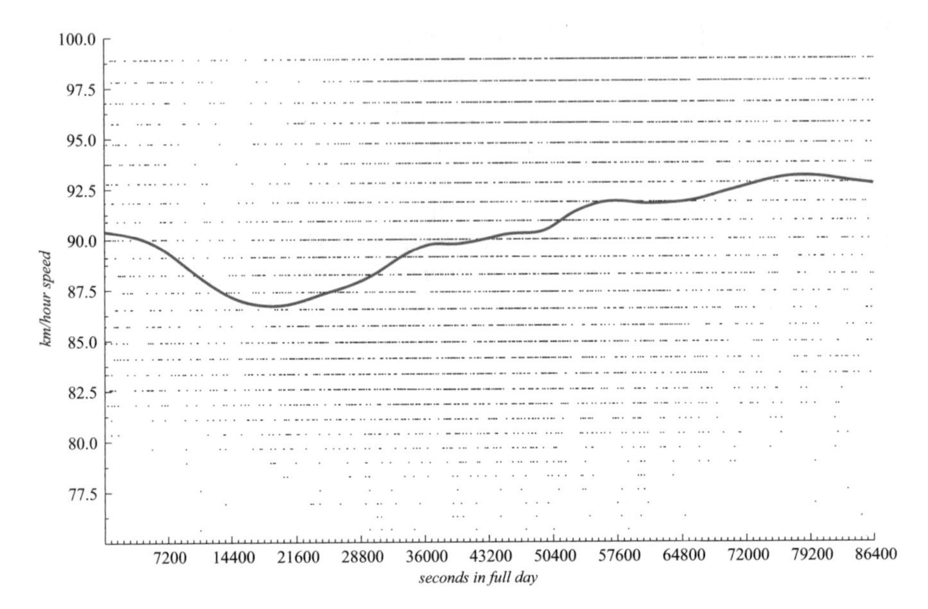

Fig. 13.1 Slow-speed passages: speed measures (in km/hour) of slow passages of motor vehicles during a full day on a fixed location in the right lane of a Dutch motorway (in tiny dots) together with the smoothed estimated trend component (solid line) from the continuous time local linear trend model. The horizontal x-axis represents the time index measured in seconds of a full day starting at midnight 0:00 h

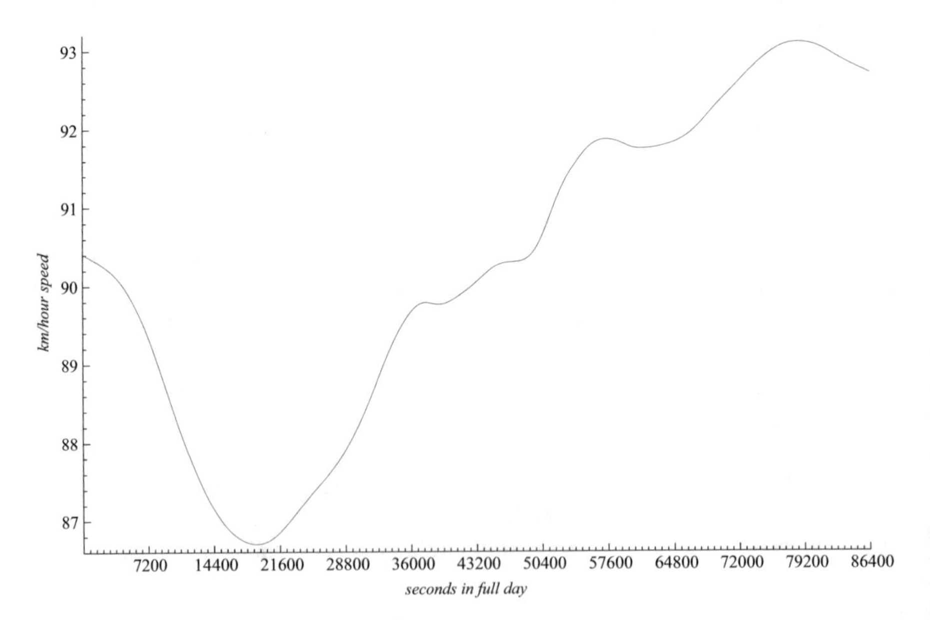

Fig. 13.2 Slow-speed trend: the smoothed estimated trend component from the continuous time local linear trend model. The horizontal x-axis represents the time index measured in seconds of a full day starting at midnight 0:00 h

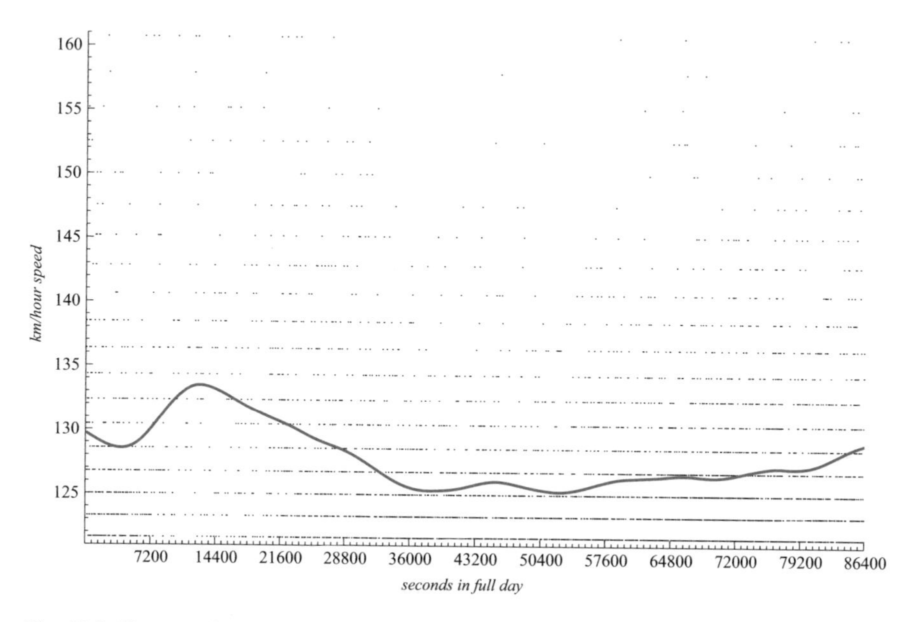

Fig. 13.3 Fast-speed passages: speed measures (in km/hour) of fast passages of motor vehicles during a full day on a fixed location in the right lane of a Dutch motorway (in tiny dots) together with the smoothed estimated trend component (solid line) from the continuous time local linear trend model. The horizontal x-axis represents the time index measured in seconds of a full day starting at midnight 0:00 h

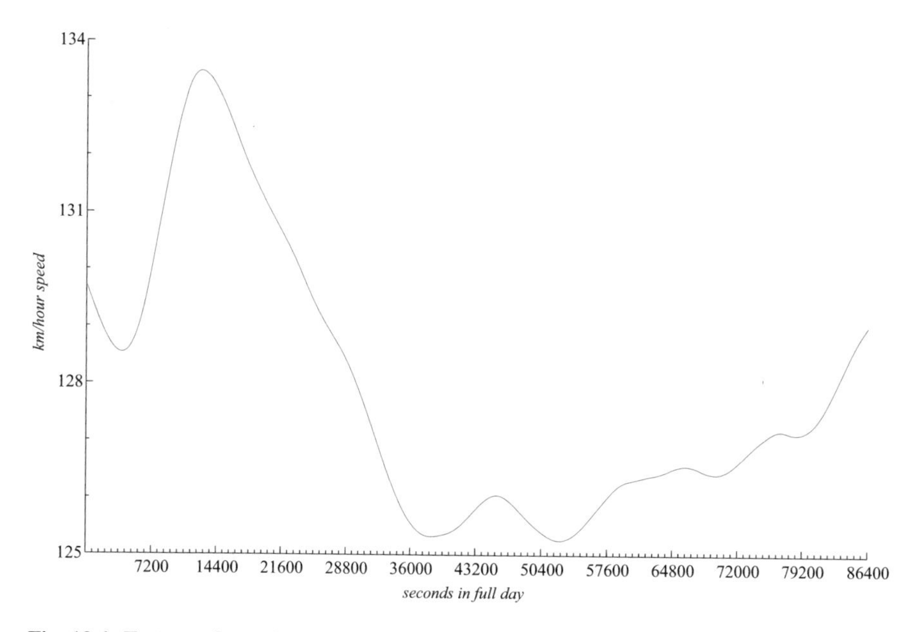

Fig. 13.4 Fast-speed trend: the smoothed estimated trend component from the continuous time local linear trend model. The horizontal x-axis represents the time index measured in seconds of a full day starting at midnight 0:00 h

morning) and 20,000 ms after midnight (i.e. around six o'clock in the morning), that the number of passages is clearly much smaller. At the same time, we can conclude that the speed of the fast group increases somewhat as it is quiet on the motorway during these night hours. In contrast, somewhat later in the night and up to the early morning hours, the speed of the slow group diminishes to clearly lower speed levels which is possibly due to a relatively intensified presence of more heavy trucks that generally drive slower and on the right lane of the road. This possible explanation can be investigated in more detail since our data set has information on *following distances* between two passing vehicles. In future research we plan to formally test such hypotheses by using statistical procedures based on the continuous time modelling framework developed in this chapter (Figs. 13.1 and 13.3).

We have shown in our current analysis that the unequal time lapses between consecutive vehicles can be handled effectively using our continuous time trend model. In this particular application that is highly relevant for road safety studies, there are several directions in which the results of our analysis can be improved. Diagnostic tests on the one-step ahead prediction errors indicate that neither the assumption of independence nor the assumption of normality of the residuals is satisfied: the Box-Ljung statistic for independence has values for $Q(10)$ that are too high; their values should be smaller than 16.95 in order to be non-significant at the usual 5% level. Also, the values for the Bowman-Shenton test for normality are too high; their values should be smaller than 5.99 to be non-significant.

Although our reported initial findings are highly interesting, the continuous time trend model appears to be somewhat away from a correct model specification for the analysis of traffic speed data. Further research can be conducted in order to obtain a more satisfactory model that is capable of capturing the remaining autocorrelation and non-normality of the data. However, this research falls outside the scope of our current review on continuous time state space modelling.

13.5 Conclusions

We have discussed the basic principles of a model-based continuous time approach using the state space methodology. The methodology is especially designed for the analysis of irregularly spaced data. We have highlighted the potential of this approach in an illustration of high-frequency intra-daily time series of speed measures from vehicles that pass a certain point at a motorway.

Acknowledgements We thank Rijkswaterstaat, The Netherlands (WVL), for providing us with the data set.

References

Bergstrom, A. R. (1984). Gaussian estimation of structural parameters in higher order continuous time dynamic models. In Z. Griliches & M. Intriligator (Eds.), *The handbook of econometrics* (Vol. 2, pp. 1145–1212). Amsterdam: North-Holland. https://doi.org/10.1016/s1573-4412(84)02012-2.

Commandeur, J. J. F., & Koopman, S. J. (2007). *An introduction to state space time series analysis*. Oxford: Oxford University Press.

Doornik, J. A. (2013). *Object-oriented matrix programming using Ox 7.00*. London: Timberlake Consultants Press.

Durbin, J., & Koopman, S. J. (2012). *Time series analysis by state space methods* (2nd ed.). Oxford: Oxford University Press. https://doi.org/10.1093/acprof:oso/9780199641178.001.0001.

Green, P., & Silverman, B. W. (1994). *Nonparametric regression and generalized linear models: A roughness penalty approach*. London: Chapman & Hall. https://doi.org/10.1007/978-1-4899-4473-3.

Harvey, A. C. (1989). *Forecasting, structural time series models and the Kalman filter*. Cambridge: Cambridge University Press.

Koopman, S. J., Harvey, A. C., Doornik, J. A., & Shephard, N. (2007). *STAMP 8.0: Structural time series analyser, modeller and predictor*. London: Timberlake Consultants.

Koopman, S. J., Shephard, N., & Doornik, J. A. (2008). *Statistical algorithms for models in state space form: Ssfpack 3.0*. London: Timberlake Consultants.

Silverman, B. W. (1985). Some aspects of the spline smoothing approach to non-parametric regression curve fitting. *Journal of the Royal Statistical Society, Series B, 47*, 1–52.

Wahba, G. (1978). Improper priors, spline smoothing, and the problems of guarding against model errors in regression. *Journal of the Royal Statistical Society, Series B, 40*, 364–372.

Wahba, G. (1990). *Spline models for observational data*. Philadelphia: SIAM. https://doi.org/10.1137/1.9781611970128.

Wecker, W. E., & Ansley, C. F. (1983). The signal extraction approach to nonlinear regression and spline smoothing. *Journal of the American Statistical Association, 78*, 81–89. https://doi.org/10.1080/01621459.1983.10477935.

Chapter 14
Continuous Time Modelling Based on an Exact Discrete Time Representation

Marcus J. Chambers, J. Roderick McCrorie, and Michael A. Thornton

14.1 Introduction

Time series modelling in the social sciences often involves data that are generated in finer time intervals than the sampling interval pertaining to the available data. In economics, for example, macroeconomic time series represent the aggregation of a large number of decisions made by microeconomic agents within the chosen sampling interval. Even today, however, the issues in macroeconomics addressed using aggregate time series data almost never tie models or conclusions to parameters governing the pre-aggregated behaviour of economic agents. Instead, at best, agents' preferences are modelled through a so-called representative agent.

On the other hand, when estimating and making inferences about parameters of interest, econometrics has tended to embody and adapt developments in the statistical analysis of time series. Notably, its response to Box-Jenkins models,

Electronic Supplementary Material The online version of this article (https://doi.org/10.1007/978-3-319-77219-6_14) contains supplementary material, which is available to authorized users.

M. J. Chambers (✉)
Department of Economics, University of Essex, Essex, UK
e-mail: mchamb@essex.ac.uk

J. R. McCrorie
School of Economics and Finance, University of St. Andrews, St. Andrews, UK
e-mail: mccrorie@st-andrews.ac.uk

M. A. Thornton
Department of Economics and Related Studies, University of York, York, UK
e-mail: michael.thornton@york.ac.uk

K. van Montfort et al. (eds.), *Continuous Time Modeling in the Behavioral and Related Sciences*, https://doi.org/10.1007/978-3-319-77219-6_14

which as 'black-box' forecasting models in the 1960s and 1970s outperformed structural econometric models that incorporated restrictions based on economic theory, was to create a unit root/cointegration paradigm that embodied the best features of both approaches. Even so, the one pervasive characteristic of time series econometrics has been its use of linear-in-variables discrete time series models, such as autoregressive (AR) or autoregressive moving average (ARMA) models and their vector counterparts, as the basis of model specification.

One aim of this chapter is to draw attention to a modelling issue that still perhaps does not take on the importance it deserves: the potential incompatibility of using such linear time series models, if naïvely specified, in a context where the data are generated in finer time intervals than the interval pertaining to the available data. This is because linear discrete time models are not time-invariant, meaning that, on a strict interpretation, parameter estimates are tied only to a particular sampling frequency. Such discrete time models therefore do not readily admit an economic interpretation in the absence of a treatment of temporal aggregation bias. One potential remedy to this problem is to formulate a structural model in continuous time with the property that equidistant data generated from its solution satisfy a linear discrete time model. The essence of the method relies on the derivation of a system of stochastic difference equations that satisfy exactly a linear stochastic difference equation system with constant coefficients. Such a discrete time model is called an *exact discrete time model*, and, through it, the structural AR or ARMA specification can be embodied in statistical inference independently of the sampling interval.[1]

The approach based on an exact discrete time model has been historically associated with A.R. (Rex) Bergstrom[2] who, perhaps more than any other econometrician, presaged the advent of continuous time models in econometrics and finance; see, for example, Bergstrom (1966, 1983), although it was Phillips (1972) who provided the first implementation of the methods discussed in this chapter.[3] There are, however, some costs in following this approach, notably that in multivariate models, identifying the parameters of the structural continuous time model on the basis of discrete time data is considerably more challenging than identifying the parameters of an (albeit time-varying) discrete time model using the same data (see Sect. 14.2.2 below). This chapter discusses the development of, and issues arising

[1] McCrorie (2009) lists a number of contributions that use an exact discrete time model.

[2] Rex Bergstrom spent over 20 years of his academic career at the University of Essex and had both direct and indirect influences on the current authors. He taught both Marcus Chambers and Roderick McCrorie at the Masters level and supervised the PhD thesis of Chambers (1990). Chambers, in turn, was the PhD supervisor of McCrorie (1996) and Thornton (2009).

[3] This paper was based on Phillips's M.A. dissertation supervised by Bergstrom at the University of Auckland in 1969. It represented the first of many contributions by Phillips on continuous time econometrics; Yu (2014) provides a survey of this work.

in, the formulation of structural continuous time models and the estimation of their parameters using an exact discrete time model, in a way that we hope will facilitate future applications in interdisciplinary areas.

Throughout this chapter, we focus mainly on continuous time models specified as systems of linear stochastic differential equations, although in Sect. 14.4 we also briefly discuss non-linear systems in the macroeconomic modelling literature that have antecedents in linear-in-variables approaches. Recent developments have enabled non-linear systems to be estimated directly; see Wymer (1997, 2012) for details. Discussion of non-linear diffusion-type models in finance, which appear in the survey by Aït-Sahalia (2007), is outside the scope of this chapter.[4] See Aït-Sahalia and Jacod (2014) for a comprehensive treatment of this topic.

The advantages of formulating econometric models in continuous time, over and above the issue of embodying an ARMA-type specification independently of the sampling frequency, were discussed by Bergstrom (1990, 1996), inter alios. Specifically, continuous time models can take account of the interaction among variables during the observation interval; they permit a more accurate representation of the partial adjustment processes in dynamic disequilibrium models, as discussed in Sect. 14.4.3 below; they allow a proper distinction to be made in estimation between stock variables (measured at points in time) and flow variables (measured as integrals of a rate of flow over the observation period); and they can be used to generate forecasts of the (unobservable) continuous time paths of the variables.

In view of the backgrounds and expertise of the authors, this chapter is written from the viewpoint of economics and, more specifically, econometrics. It therefore mostly neglects the treatment of the estimation of continuous time models in other areas of the social sciences and science more generally, such as engineering. Material relevant to other disciplines can be found in other contributions to this volume. The plan of this chapter is as follows. Section 14.2 is broadly concerned with continuous time methods in econometrics and contains seven subsections. The first lays the groundwork for subsequent subsections and explains how an exact discrete time model corresponding to a linear continuous time system can be obtained and provides a worked example for a second-order differential equation system. Section 14.2.2 deals with the fundamental problem of identification of the parameters of a continuous time system from discrete time data, and Sect. 14.2.3 discusses how the process of temporal aggregation can distort inferences relating to Granger causality. Section 14.2.4 explores various issues of nonstationarity that are important when analysing economic and financial time series, while Sect. 14.2.5 summarises recent work that enables the information contained in observations made at different sampling frequencies to be used in the estimation of a continuous

[4]Most non-linear models are not directly amenable to the derivation of exact discrete time representations and typically result in transition densities that have no closed-form solution. See, however, Phillips and Yu (2009), Fergusson and Platen (2015) and Thornton and Chambers (2016), for examples where a closed-form density is apposite.

time system. The remaining two subsections deal with Gaussian estimation as well as alternative (frequency domain) methods.

The final sections of the chapter have a more practical aim. Section 14.3 is devoted to computational issues and reports the results of a small simulation exercise (the code for which is contained in Appendix), while Sect. 14.4 is concerned with empirical applications. Sects. 14.4.1 and 14.4.2 contain new applications to consumer prices and inflation and to oil prices and the macroeconomy, respectively, while Sect. 14.4.3 discusses applications of the continuous time methodology in the arena of macroeconometric modelling. Section 14.5 contains some concluding comments.

14.2 Continuous Time Models in Econometrics

14.2.1 Linear Continuous Time Systems and Exact Discrete Time Models

We will be concerned with an $n \times 1$ vector of variables, denoted $x(t)$, whose dynamic evolution is determined by a stochastic differential equation system in continuous time. Bergstrom (1983, 1984) provided a rigorous foundation for the specification of such systems and pioneered the development of the exact discrete time approach for first- and second-order systems, subsequently extended by Chambers (1999) to systems of order greater than two. A higher-order system is specified as

$$d[D^{p-1}x(t)] = [A_{p-1}D^{p-1}x(t) + \ldots + A_1Dx(t) + A_0x(t)]dt + \zeta(dt), \quad t > 0, \tag{14.1}$$

where $A_{p-1}, \ldots, A_0$ are $n \times n$ parameter matrices, D denotes the mean square differential operator satisfying

$$\lim_{\delta \to 0} E\left|\frac{x_i(t+\delta) - x_i(t)}{\delta} - Dx_i(t)\right|^2 = 0, \quad i = 1, \ldots, n,$$

$x(0), \ldots, D^{p-1}x(0)$ are a set of initial conditions,[5] and $\zeta(dt)$ is an $n \times 1$ vector of random measures with $E[\zeta(dt)] = 0$, $E[\zeta(dt)\zeta(dt)'] = \Sigma dt$ (Σ being an $n \times n$ symmetric positive definite matrix), and $E[\zeta(\Delta_1)\zeta(\Delta_2)'] = 0$ for any disjoint

[5]The initial conditions are usually assumed to be fixed which imparts a type of nonstationarity on an otherwise stable system. This is a different type of nonstationarity to that which has dominated the econometrics literature in recent years and which we discuss in Sect. 14.2.4.

intervals, Δ_1 and Δ_2, on the real line $-\infty < t < \infty$.[6] Under these assumptions, the random measure vector $\zeta(dt)$ is similar to vector white noise, and the system (14.1) can be regarded as a continuous time autoregressive system of order p, which we shall denote CAR(p). The system could be extended to include a deterministic linear trend function with the addition of a term of the form $[\gamma_0 + \gamma_1 t]dt$ on the right-hand side of (14.1), where γ_0 and γ_1 are $n \times 1$ vectors of unknown parameters, or to include exogenous variables, but to do so would result in additional complexity that we wish to avoid here. The system (14.1) is interpreted as meaning that $x(t)$ satisfies the stochastic integral equation

$$D^{p-1}x(t) - D^{p-1}x(0) = \int_0^t [A_{p-1}D^{p-1}x(r) + \ldots + A_1 Dx(r) + A_0 x(r)]dr + \int_0^t \zeta(dr)$$

for all $t > 0$; see Bergstrom (1983) for further details.

The objective is to estimate the elements of the matrices $A_{p-1}, \ldots, A_0$ and Σ from a sample of data observed at discrete points in time, i.e. not observed continuously. The elements of these matrices will often be known functions of an underlying vector of structural parameters although we avoid emphasising such dependencies here for reasons of notational simplicity.[7] The exact representation approach derives the law of motion for the observations that is consistent with their having been generated by the stochastic differential equation system (14.1). The nature of the observations themselves depends on the form of variables that comprise the vector $x(t)$. In the most general (mixed sample) case, the vector $x(t)$ can be partitioned into an $n^s \times 1$ subvector of stock variables (x^s) and an $n^f \times 1$ subvector of flow variables (x^f), where $n^s + n^f = n$, so that

$$x(t) = \begin{pmatrix} x^s(t) \\ x^f(t) \end{pmatrix}.$$

Stock variables are assumed to be observable at equally spaced discrete points in time of length h, resulting in the sequence

$$\left\{x^s_{th} = x^s(th)\right\}_{t=0}^{T} = \{x^s_0, x^s_h, \ldots, x^s_{Th}\},$$

[6]The use of a vector of random measures to specify the disturbance vector in a continuous time model in the econometrics literature is due to Bergstrom (1983) who built on the work of Rozanov (1967). A common alternative is to replace $\zeta(dt)$ with $\Sigma^{1/2}dW(t)$ where $dW(t)$ denotes the increment in a vector of Wiener processes and $\Sigma^{1/2}(\Sigma^{1/2})' = \Sigma$. Note, though, that the latter specification imposes Gaussianity on the system, whereas the distribution of $\zeta(dt)$ is unspecified beyond its first two moments.

[7]Such dependencies are, however, emphasised in Sect. 14.2.2 where we discuss issues of identification.

while flow variables are observable as an integral of the underlying rate of flow over the sampling interval of length h, yielding the sequence

$$\left\{x_{th}^f = \frac{1}{h}\int_{th-h}^{th} x^f(r)dr\right\}_{t=1}^T = \left\{\frac{1}{h}\int_0^h x^f(r)dr, \ldots, \frac{1}{h}\int_{Th-h}^{Th} x^f(r)dr\right\}.$$

Examples of stock variables in economics include the money stock, exchange rates, interest rates and other asset prices, all of which are observable (at least in principle) at points in time. Examples of flow variables include consumers' expenditure, income, exports, imports and cumulative rainfall in Brazil, each of which is measured as the accumulation of a rate of flow over a time interval (corresponding with the sampling interval). Although we assume that the observations are equally spaced, it is possible to extend the setup to allow for irregularly spaced observations. This can be achieved by introducing an index $i = 1, \ldots, N$, where N denotes sample size, and denoting the sampling intervals by $h_i = t_i - t_{i-1}$. For notational convenience, however, we shall assume that the observations are equally spaced. Also, for the purposes of clarity, we will, for the time being, assume that $x(t) = x^s(t)$ so that all n variables are of the stock variety. The consequences of relaxing this assumption will be discussed in due course.

The first step in deriving an exact discrete time representation is to write the model in a suitable state space form. In order to do this, we can define the $np \times 1$ state vector

$$y(t) = [x(t)', Dx(t)', \ldots, D^{p-1}x(t)']',$$

which satisfies the first-order stochastic differential equation system

$$dy(t) = Ay(t)dt + \phi(dt), \quad t > 0, \tag{14.2}$$

where

$$A = \begin{pmatrix} 0 & I & 0 & \ldots & 0 & 0 \\ 0 & 0 & I & \ldots & 0 & 0 \\ \vdots & \vdots & \vdots & & \vdots & \vdots \\ 0 & 0 & 0 & \ldots & 0 & I \\ A_0 & A_1 & A_2 & \ldots & A_{p-2} & A_{p-1} \end{pmatrix}, \quad \phi(dt) = \begin{pmatrix} 0 \\ 0 \\ \vdots \\ 0 \\ \zeta(dt) \end{pmatrix}.$$

The solution to (14.2) is given by

$$y(t) = e^{At}y(0) + \int_0^t e^{A(t-r)}\phi(dr), \quad t > 0, \tag{14.3}$$

where $y(0)$ denotes the vector of initial conditions and the matrix exponential is defined by its series expansion

$$e^{tA} = I + tA + \frac{1}{2!}(tA)^2 + \ldots = \sum_{j=0}^{\infty} \frac{(tA)^j}{j!}.$$

Noting that $y(th)$ contains the observable vector $x(th)$, the solution (14.3) can be manipulated to relate $y(th)$ to $y(th-h)$ and thereby $x(th)$ to $x(th-h)$. This is achieved by rewriting the system at the observation points as

$$y(th) = e^{Ath}y(0) + \int_0^{th-h} e^{A(th-r)}\phi(dr) + \int_{th-h}^{th} e^{A(th-r)}\phi(dr)$$

$$= e^{Ah}\left[e^{A(th-h)}y(0) + \int_0^{th-h} e^{A(th-h-r)}\phi(dr)\right] + \int_{th-h}^{th} e^{A(th-r)}\phi(dr).$$

The term in square brackets is simply $y(th-h)$ which results in the following first-order stochastic difference equation for $y(th)$:

$$y(th) = Fy(th-h) + \epsilon_{th}, \quad t = 1, \ldots, T, \tag{14.4}$$

where $F = e^{Ah}$ and

$$\epsilon_{th} = \int_{th-h}^{th} e^{A(th-r)}\phi(dr)$$

is an i.i.d. random vector with mean vector zero and covariance matrix

$$\Sigma_\epsilon = \int_0^h e^{As}\Sigma_\phi e^{A's}ds,$$

$\Sigma_\phi dt$ being the covariance matrix of $\phi(dt)$.[8]

Although the system (14.4) implicitly embodies the dynamics of the observable vector $x_{th} = x(th)$, the remaining elements of $y(th)$ are unobservable. The Bergstrom approach derives the exact discrete time model by eliminating the unobservable elements from this system using appropriate substitutions.[9] This process results in the ARMA$(p, p-1)$ representation

$$x_{th} = F_1 x_{th-h} + \ldots + F_p x_{th-ph} + \eta_{th}, \quad t = p, \ldots, T, \tag{14.5}$$

[8]In fact, Σ_ϕ is an $np \times np$ matrix of zeros except for the $n \times n$ bottom right-hand corner block which is equal to Σ.

[9]Wymer (1972) provided the first treatment of higher-order systems in the econometrics literature using the framework (14.1)–(14.4) but subsequently derived an approximate discrete model.

where η_{th} is an MA$(p-1)$ process. Note that this equation holds only for period p onwards owing to the first available observation being $x_0 = x(0)$ (recall that we are assuming that x comprises purely stock variables at this point). It is, however, possible to derive an additional $p-1$ equations that relate $x_h, \ldots, x_{ph-h}$ to the lagged values and to x_0; see, for example, Theorem 2.2 of Bergstrom (1986) for the mixed sample case when $p = 2$ and Theorem 2 of Chambers (1999) also for the mixed sample case but for $p \geq 2$.

To see how this approach works in practice, consider the case where $p = 2$. The observable vector is $x(th)$, and the unobservable vector in this case is $Dx(th)$, the equations for which from (14.4) are

$$x(th) = F_{11}x(th-h) + F_{12}Dx(th-h) + \epsilon_{1,th}, \tag{14.6}$$

$$Dx(th) = F_{21}x(th-h) + F_{22}Dx(th-h) + \epsilon_{2,th}, \tag{14.7}$$

where the F_{ij} $(i, j = 1, 2)$ are the $n \times n$ submatrices of F and $\epsilon_{th} = (\epsilon_{1,th}', \epsilon_{2,th}')'$. The objective is to eliminate $Dx(th-h)$ from (14.6) using the information in (14.7), and for this purpose, Bergstrom (1983, Assumption 4) assumes that the matrix F_{12} is nonsingular. From (14.6), we obtain, using this assumption,

$$Dx(th-h) = F_{12}^{-1}\left[x(th) - F_{11}x(th-h) - \epsilon_{1,th}\right], \tag{14.8}$$

while lagging (14.7) by one period yields

$$Dx(th-h) = F_{21}x(th-h) + F_{22}Dx(th-2h) + \epsilon_{2,th-h}. \tag{14.9}$$

Substituting the right-hand side of (14.8) for $Dx(th-h)$ in (14.9) and the one-period lag of (14.8) for $Dx(th-2h)$ in (14.9) results in

$$x_{th} = F_1 x_{th-h} + F_2 x_{th-2h} + \eta_{th}, \quad t = 2, \ldots, T, \tag{14.10}$$

where $F_1 = F_{11} + F_{12}F_{22}F_{12}^{-1}$, $F_2 = F_{12}[F_{21} - F_{22}F_{12}^{-1}F_{11}]$, and the disturbance vector is given by $\eta_{th} = \epsilon_{1,th} - F_{12}F_{22}F_{12}^{-1}\epsilon_{1,th-h} + F_{12}\epsilon_{2,th-h}$ which is clearly seen to be MA(1) due to ϵ_{th} being an i.i.d. process.

Although the ARMA(2,1) representation in (14.10) holds for $t = 2, \ldots, T$, it is possible to supplement it, for purposes of computing the unconditional likelihood function, with an equation that relates x_h to x_0. In the case of the second-order system considered here, the relevant equation is given by (14.6) evaluated at $t = 1$, giving

$$x_h = F_{11}x_0 + F_{12}Dx(0) + \epsilon_{10}. \tag{14.11}$$

Note that this equation also includes the unobservable component $Dx(0)$, and there are two main ways of treating it. The first is to make an assumption about its value, an example being $Dx(0) = 0$, which implies that, at time $t = 0$, the system was in equilibrium. Alternatively the $n \times 1$ vector $Dx(0)$ can be treated as part of the unknown parameter vector whose value is estimated by maximisation of the likelihood function, although in this case it is not possible to obtain a consistent estimator of its value.[10]

As mentioned earlier, not all variables are observed as stocks, and so the above techniques have to be modified in the presence of flow variables or mixtures of stocks and flows. This is particularly important in macroeconometric modelling where many variables, such as consumers' expenditures and national income, are measured as flows. Early contributions dealing with the problems associated with flow variables can be found in Phillips (1974) and Wymer (1976). In subsequent work, Bergstrom (1984, Theorem 8) presented an exact discrete time model for a first-order system, while an exact discrete model for flow variables when $p = 2$ was derived by Bergstrom (1983, Theorem 3) and extended to the mixed sample case by Bergstrom (1986, Theorems 2.1 and 2.2).[11] In these cases, the exact discrete time model can be shown to be an ARMA(p, p) system, the presence of flows increasing the order of the moving average disturbance by one. These results were subsequently extended to the general $p \geq 2$ case by Chambers (1999).

A feature of the results cited above is that all require an assumption of invertibility of certain matrices; for example, Bergstrom (1983) requires A_0 to be nonsingular in addition to F_{12}. The nonsingularity of A_0 rules out important cases such as unit roots and cointegration (see Sect. 14.2.4), but can be relaxed as follows. Our demonstration applies to the case $p = 2$ but can be generalised to larger values of p. Recalling the definition of the observed flow variables, x^f_{th}, we can integrate (14.6) and (14.7) over the interval $(th - h, th]$ to obtain

$$x^f_{th} = F_{11}x^f_{th-h} + F_{12}z_{th-h} + v_{1,th}, \quad (14.12)$$

$$z_{th} = F_{21}x^f_{th-h} + F_{22}z_{th-h} + v_{2,th}, \quad (14.13)$$

where we have defined

$$z_{th} = \int_{th-h}^{th} Dx^f(r)dr = x^f(th) - x^f(th - h),$$

$$v_{th} = \begin{pmatrix} v_{1,th} \\ v_{2,th} \end{pmatrix} = \int_{th-h}^{th} \int_{s-h}^{s} e^{A(s-r)}\phi(dr)ds.$$

[10]Note that this inconsistency arises owing to no new information on $Dx(0)$ becoming available as $T \to \infty$.

[11]Bergstrom (1986) also includes results for a system that contains exogenous stock and flow variables.

The vector z_{th} is unobservable and can be eliminated from the system using the same steps that led to (14.10), the result being

$$x^f_{th} = F_1 x^f_{th-h} + F_2 x^f_{th-2h} + \eta^f_{th}, \quad t = 2, \ldots, T, \tag{14.14}$$

where F_1 and F_2 are defined following (14.10) and $\eta^f_{th} = v_{1,th} - F_{12}F_{22}F_{12}^{-1} v_{1,th-h} + F_{12}v_{2,th-h}$ is now an MA(2) process which follows by noting that v_{th} can be written under the white noise assumption as the sum of a pair of single intervals with respect to $\zeta(dr)$ over the intervals $(th-2h, th-h]$ and $(th-h, th]$; details can be found in McCrorie (2000). Although the autoregressive matrices remain the same functions of the underlying parameters as in the case of stock variables, the presence of flows affects the serial correlation properties of the disturbance vector, increasing the moving average order by one, a feature which needs to be incorporated in any estimation algorithm.

Although autoregressive models, in both discrete and continuous time, dominate the time series econometrics literature, there has been considerable interest in continuous time ARMA (CARMA) processes in the statistics literature, where the focus has been on state space approaches rather than exact discrete time representations. Results on maximum likelihood estimation based on an appropriate state space model are contained in Zadrozny (1988), while a survey of recent results on CARMA processes can be found in Brockwell (2014). It is, however, possible to derive an exact discrete time model corresponding to a CARMA system. Chambers and Thornton (2012) extend (14.1) to the CARMA(p, q) system

$$D^p x(t) = A_{p-1} D^{p-1} x(t) + \ldots + A_0 x(t) + u(t) + \Theta_1 Du(t) + \ldots + \Theta_q D^q u(t), \tag{14.15}$$

for $t > 0$, where $u(t)$ is an $n \times 1$ continuous time white noise process and $A_0, \ldots, A_{p-1}$ and $\Theta_1, \ldots, \Theta_q$ are $n \times n$ matrices of coefficients.[12] The interpretation of a white noise process in continuous time can be problematic (see, e.g., the discussion and results in Bergstrom 1984), but the interpretation of $u(t)$ in (14.15) is that it satisfies $E[u(t)] = 0$ and, for $t_2 > t_1$, has autocovariance properties

$$E\left[\int_{t_1}^{t_2} u(r)dr \int_{t_1}^{t_2} u(s)'ds\right] = \Sigma\,(t_2 - t_1),$$

$$E\left[\int_{t_1}^{t_2} u(r)dr \int_{t_1}^{t_2} u(\tau + s)'ds\right] = 0, \quad |\tau| > t_2 - t_1,$$

where Σ is an $n \times n$ positive definite symmetric matrix.

[12]The coefficient matrix multiplying $u(t)$ is set to an identity in order to identify the parameters of the model in view of $u(t)$ having covariance matrix Σ.

The presence of the MA component in (14.15) means that a different state space form is more useful in deriving the exact discrete model than the one defined in (14.2). Chambers and Thornton (2012) employed the state space representation used by Zadrozny (1988) in which the $np \times 1$ state vector is defined as $w(t) = [w_1(t)', \ldots, w_p(t)']'$ and with $w_1(t) = x(t)$. The state space form is based on the following set of p equations in the derivatives of the components of $w(t)$, given by

$$Dw_1(t) = A_{p-1}w_1(t) + w_2(t) + \Theta_{p-1}u(t), \tag{14.16}$$

$$Dw_2(t) = A_{p-2}w_1(t) + w_3(t) + \Theta_{p-2}u(t), \tag{14.17}$$

$$\vdots \quad \vdots$$

$$Dw_{p-1}(t) = A_1w_1(t) + w_p(t) + \Theta_1u(t), \tag{14.18}$$

$$Dw_p(t) = A_0w_1(t) + u(t), \tag{14.19}$$

in which we define $\Theta_j = 0$ for $j > q$. Combining the expressions for $Dw_1(t), \ldots, Dw_p(t)$ above, the relevant state space form can be written as

$$Dw(t) = Cw(t) + \Theta u(t), \tag{14.20}$$

where

$$C = \begin{bmatrix} A_{p-1} & I & 0 & \ldots & 0 \\ A_{p-2} & 0 & I & \ldots & 0 \\ \vdots & & & & \vdots \\ A_1 & 0 & 0 & \ldots & I \\ A_0 & 0 & 0 & \ldots & 0 \end{bmatrix}, \quad \Theta = \begin{bmatrix} \Theta_{p-1} \\ \Theta_{p-2} \\ \vdots \\ \Theta_1 \\ I \end{bmatrix}.$$

Utilising this state space model, Chambers and Thornton (2012) show that the exact discrete time model for a vector of stock variables is of ARMA$(p, p-1)$ form, while, for a vector of flow variables or mixed sample data, it is of ARMA(p, p) form. The presence of the continuous time MA disturbance therefore does not affect the MA order of the exact discrete time model. This means, in effect, that there are additional parameters in the CARMA model that can be used to pick up the dynamics in the discrete time model that are not present in a CAR representation, a feature that has been shown to have empirical content by Chambers and Thornton (2012).

More recently, Thornton and Chambers (2017) have shown that exact discrete time representations corresponding to CARMA systems are not unique.[13] The discrete time representations for CAR(p) systems with mixed sample data,

[13]Hence the presence of the phrase '*an* exact discrete time representation' rather than '*the* exact discrete time representation' in the title of this chapter.

developed in Bergstrom (1983) and in Chambers (1999), rely on differencing the stock variables and are of ARMA(p, p) form. Once the stock variables are reintegrated (or 'undifferenced'), these representations correspond to a discrete time ARMA$(p + 1, p)$ process. Thornton and Chambers (2017), however, work with an augmented state space form[14] that more naturally incorporates both stock and flow variables and show that the differencing of the stock variables identifies the representation among a wider class of ARMA$(p+1, p)$ processes and that the more parsimonious ARMA(p, p) is also among this class.

14.2.2 Identification

To a large extent, we motivated the formulation of continuous time models as linear stochastic differential systems because equispaced data generated by such systems satisfy ARMA specifications that are typical in time series analysis but whose parameters, unlike those in naïvely specified discrete time models, are not tied to the sampling interval. The principal counterpoint to this advantage of estimating the parameters of structural continuous time models on the basis of discrete data is that one can 'join up the dots', as Robinson (1992) described it, in an uncountably infinite number of ways. The problem is multivariate in character and can be illustrated using the following simple example for a stock variable.[15] Suppose that the $n \times 1$ finite-variance vector $x(t)$ satisfies the stochastic differential equation system

$$dx(t) = A(\theta)x(t)dt + \zeta(dt), \quad t > 0, \tag{14.21}$$

subject to the initial condition $x(0) = y_0$, where A is an $n \times n$ matrix whose elements are now explicitly assumed to be known functions of a $p \times 1$ vector θ of unknown parameters ($p \leq n^2$), y_0 is a non-random $n \times 1$ vector, and $\zeta(dt)$ is an uncorrelated vector random measure of the type described in Sect. 14.2.1 with covariance matrix $\Sigma(\mu)dt$, the elements of Σ being known functions of a $q \times 1$ vector μ of unknown parameters ($q \leq \frac{1}{2}n(n + 1)$). The exact discrete time model is obtained from the solution of (14.21) subject to the initial condition, giving a sequence of equispaced discrete time data $x(0), x(h), \ldots, x(Th)$ that satisfies the

[14]The state space form in (14.20) is augmented by an additional n^f elements in a vector $y_0(t)$ that corresponds to the aggregated or observed flow variables.

[15]This identification problem is therefore different in nature and on top of the classical identification problem which seeks to avoid observational equivalence through model and estimator choice; see, for example, Chambers and McCrorie (2006). In open systems, namely, systems involving exogenous variables, the solution of the stochastic differential equation depends on a continuous time record of the exogenous variables and so some sort of approximation of the time paths is necessary to achieve identification; see, in particular, Bergstrom (1986), Hamerle et al. (1991, 1993) and McCrorie (2001) for explicit discussion of this issue.

stochastic difference equation system

$$x(th) = F(\theta)x(th - h) + \epsilon_{th}, \quad t = 1, \ldots, T, \tag{14.22}$$

where $F(\theta) = e^{A(\theta)h}$ and ϵ_{th} is white noise with covariance matrix

$$\Omega_\epsilon(\theta, \mu) = E(\epsilon_{th}\epsilon_{th}') = \int_0^h e^{A(\theta)r}\Sigma(\mu)e^{A(\theta)'r}dr;$$

see Bergstrom (1984, Theorem 3).

In the context of (14.21), the identification problem relates directly to the fact that there are, in principle, many different matrices that share the same exponential F in (14.22); see, for example, Phillips (1973), Hansen and Sargent (1983) and Hamerle et al. (1993). These matrices are *aliases* of A in the sense that, through taking the place of A in (14.21), they generate the same equidistant discrete time data. The aliasing problem of identifying structural continuous time parameters on the basis of discrete time data is clearly more severe than simply identifying the parameters of discrete time models (but, to reiterate, there is a trade-off in that naïvely specified discrete time models suffer from a lack of time invariance). If Gaussianity is assumed, the problem in the context of (14.21) is to find a necessary and sufficient condition such that the pair $[A(\theta), \Sigma(\mu)]$ is identifiable in $[F(\theta), \Omega_\epsilon(\theta, \mu)]$. In any particular application, the forms of A and Σ are heavily governed by the role of the parameter vectors θ and μ, although for the purpose of simplifying the discussion that follows, the dependence of A and Σ on θ and μ will be suppressed.

McCrorie (2003) offered a framework for the identification problem by considering the following Hamiltonian matrix M that allows the pair $[A, \Sigma]$ to be treated together: if

$$M = \begin{pmatrix} -A & \Sigma \\ 0 & A' \end{pmatrix},$$

then, as an application of Van Loan (1978, Theorem 1),

$$e^{Mh} = \begin{pmatrix} F^{-1} & F^{-1}\Omega_\epsilon \\ 0 & F' \end{pmatrix}.$$

The following theorem (McCrorie 2003), which is a consequence of Theorem 2 of Culver (1966), contains the basic result on identification in terms of when the matrix exponential mapping is bijective in general.

Theorem 14.1 *For the prototypical model* (14.21), $[A, \Sigma]$ *is identifiable in* $[F, \Omega_\epsilon]$ *if the eigenvalues of M are strictly real and no Jordan block of M belonging to any eigenvalue appears more than once.*

Note that the eigenvalues of M are simply the eigenvalues and reverse eigenvalues of A, and so if A has no complex eigenvalues and there is no confluence in its eigenvalues, the aliasing problem reduces essentially to a univariate problem involving the exponential function which, when viewed as real-valued, is bijective. Unfortunately, both restrictions are not generally appropriate for economic time series: they rule out plausible cyclical behaviour resulting from complex eigenvalues and plausible trend behaviour resulting from multiple unit roots (multiple zero eigenvalues of A). In the complex eigenvalue case, several authors achieve identification through additional restrictions: Phillips (1973) uses Cowles Commission-type restrictions (see also Blevins 2017), and Hansen and Sargent (1983) show there are restrictions inherent in the requirement that Ω_ϵ be positive semidefinite. Hansen and Sargent (1991) use cross-equation restrictions implied by the rational expectations hypothesis. Bergstrom et al. (1992) and Bergstrom and Nowman (2007) use prior bounds on the parameters as a means of achieving identification, in the way researchers do for large-scale structural VAR models today. The results of Hansen and Sargent (1983) show that without importing a priori restrictions beyond the problem in hand, identification can only be local; see Appendix 1 of McCrorie (2009) for some examples. In practice, one has *jointly* to solve the aliasing identification problem and the classical identification problem of avoiding observational equivalence through model and estimator choice. In the context given here, the general problem relates not to the matrices A and Σ but to the underlying parameter vectors θ and μ. The incorporation of exogenous variables in open systems can be useful (e.g. Hamerle et al. 1993; Bergstrom et al. 1992), as can the aspect that the matrices in terms of the underlying parameter vectors are often heavily restricted. Nevertheless, finding necessary and sufficient conditions to solve the identification problem for estimating continuous time models on the basis of discrete data remains open even for the most basic of models.

14.2.3 Granger Causality

Formulating a structural model in continuous time offers a means of resolving the problem that discrete time models, whose estimated parameters are tied to the sampling frequency, do not readily lend themselves to economic interpretation. A parallel problem that has also been downplayed in the econometrics literature is the tendency for naïvely specified discrete time models to generate spurious Granger causality relationships when the time intervals in which the data are generated are finer than the sampling interval.[16] To define (global) Granger non-causality between two variables $x_1(t)$ and $x_2(t)$, let $I_j(t)$ $(j = 1, 2)$ denote the sigma algebra generated by $x_j(t)$ up to time t (this is interpreted as an information set), let $\bar{I}(t)$

[16] McCrorie and Chambers (2006, Section 3.1) outline and discuss the concept of Granger causality in the context of continuous and discrete time models.

denote all other information up to time t, and let $E(A|B)$ denote the conditional expectation of A given B. Then x_2 *does not Granger cause* x_1 if

$$E(x_1(t+k)|I_1(t), I_2(t), \bar{I}(t)) = E(x_1(t+k)|I_1(t), \bar{I}(t)) \tag{14.23}$$

for all t and $k > 0$; see Florens and Fougère (1996) and Comte and Renault (1996). If the above condition does not hold, then x_2 is said to Granger cause x_1. A brief survey and discussion of the literature of Granger causality in continuous time can be found in McCrorie and Chambers (2006, Section 3).

In the context of temporal aggregation, a coarsely sampled process, omitting information useful for predicting an economic time series, will exhibit bidirectional Granger causality with another coarsely sampled process provided that they are correlated, even if there is only unidirectional causality in the finer time interval. Inferences made about the underlying behaviour of economic agents from observed time series can, therefore, be distorted. For example, Christiano and Eichenbaum (1987) find evidence for the money stock Granger-causing output with quarterly US data that seems to be overturned when moving to a finer sampling interval. Some authors, for example, Marcellino (1999) and Breitung and Swanson (2002), have tried to approach the temporal aggregation problem through the lens of fixed-interval time aggregation; however, this approach relies on constructing corrections to estimates through knowing the time unit in which the data are generated. Otherwise, a distortional effect owing to temporal aggregation will remain.[17]

Specifying a structural continuous time model allows a priori restrictions to be imposed on the observed discrete data independently of the sampling interval, enabling Granger causality relationships to be preserved, and thereby facilitates obtaining efficient estimates of the structural parameters that are devoid of temporal aggregation bias. Such considerations matter materially in empirical work. For example, Harvey and Stock (1989) find evidence, using US data, of the money stock not Granger-causing output on the basis of a continuous time model but obtain a strong reversal of this conclusion when temporal aggregation is ignored in discrete time VARs. McCrorie and Chambers (2006) also consider the issue of money-income causality in discrete time models where the temporal aggregation restrictions were imposed exactly, approximately and not at all. They find that accounting for temporal aggregation restrictions can have an important bearing on Granger causality tests, even when the restrictions are only approximately imposed. In an application to exchange rates, Renault et al. (1998) used a continuous time model, and the methods of this chapter to distinguish between 'true' and 'spurious' causality and on the basis of their data suggested that there was a 'discrete time illusion' of spurious causality observed between the German mark and the Swiss franc at certain sampling frequencies.

[17] Thornton and Chambers (2013) provide a recent discussion of temporal aggregation in macroeconomics with continuous time models in view.

The above discussion motivates formulating continuous time models as a way of countering the problem that some observed Granger causality relationships in naïvely specified discrete time models are spurious. In practice, there exists a *trade-off* between preserving a priori information on Granger causality relationships in estimation with solving the problem of identifying the parameters of a structural continuous time model on the basis of discrete time data as discussed in Sect. 14.2.2. Both issues are in the background regardless of the model formulated. For example, naïvely specifying a discrete time model on its own, common throughout econometric time series analysis, is insufficient as it gives no reference point to assess whether the magnitude of temporal aggregation is important.

14.2.4 Nonstationarity

Economic time series data are inherently nonstationary, and the nonstationarity can manifest itself in a variety of forms. A second-order stationary time series is one for which the mean, variance and autocovariances are time-independent. Examination of the solution to the state space representation of the continuous time system given in (14.3) shows immediately that the mean depends on time because $E[y(t)] = e^{At}y(0)$, assuming $y(0)$ is fixed. This is the form of nonstationarity referred to in the title of Bergstrom (1985). However, in recent years, nonstationarity has come to be associated with a different concept, namely, that of unit roots and stochastic trends, which are consistent with the earlier observation of Granger (1966) concerning the shape of the spectral density function at the origin.

A discrete time process, x_{th} ($t = 1, \ldots, T$), is said to have a unit root if it has the representation

$$\Delta_h x_{th} = u_{th}, \quad t = 1, \ldots, T, \tag{14.24}$$

where $\Delta_h x_{th} = x_{th} - x_{th-h}$ and u_{th} is a second-order stationary random process. Solving the difference equation from an initial value x_0 yields the representation for the level process in the form

$$x_{th} = x_0 + \sum_{j=1}^{t} u_{jh}, \quad t = 1, \ldots, T, \tag{14.25}$$

where the partial sum of the stationary process u_{th} represents the stochastic trend. One way of thinking about a unit root is that the process requires differencing once to become stationary, as in (14.24), while the stochastic trend representation (14.25) leads to the levels process being described as integrated of order one, often denoted I(1).

In continuous time, the equivalent representation to (14.24) is

$$Dx(t) = u(t), \quad t > 0, \tag{14.26}$$

where $u(t)$ is a second-order stationary continuous time process and $x(0)$ will be taken to be fixed. In this case, the process $x(t)$ requires differentiating once to become stationary, and the stochastic trend representation for the level is given by

$$x(t) = x(0) + \int_0^t u(r)dr, \quad t > 0, \tag{14.27}$$

assuming the integral (which represents the continuous time stochastic trend) exists. If $x(t)$ is observed as a discrete time process at integer values of t at intervals of length h, then integrating (14.26) once over the interval $(th - h, th]$ reveals that

$$x(th) = x(th - h) + \int_{th-h}^{th} u(r)dr$$

and hence the discrete time process has a unit root. This is also true of an observed flow variable obtained by a further integration of the model above.

Following Phillips (1987), a large literature has emerged on unit root processes in discrete time, and much effort has been expended in the search for tests for a unit root that have good properties. Many economic time series have been found to display unit root-type properties, but one of the challenges facing economics in the mid-1980s was how to reconcile economic theory with these apparent features. In particular, individual series with unit roots can wander freely over time, driven by the stochastic trends, whereas much of economics implies the existence of stable relationships among variables (an example being consumers' expenditure and income). The solution to this apparent dichotomy, proposed by Engle and Granger (1987), was the concept of cointegration. An $n \times 1$ vector, x_{th}, of I(1) series is said to be cointegrated if there exist a set of $1 \leq r < n$ linear combinations of the form $\beta' x_{th}$ that are stationary, where β is an $n \times r$ matrix of cointegrating parameters whose columns are the r cointegrating vectors. Cointegration has subsequently become an essential concept in the analysis of multivariate nonstationary economic time series.

In terms of continuous time processes, Phillips (1991) showed that a vector process that is cointegrated in continuous time is also cointegrated in terms of the discrete time observations.[18] This is an important result because it implies that discrete time methods can be used to test for cointegration even if the researcher is interested in formulating a model in continuous time. If evidence of r cointegrating vectors is found, let $m = n - r$ and partition $x(t) = [x_1(t)', x_2(t)']'$, where $x_1(t)$ is $r \times 1$ and $x_2(t)$ is $m \times 1$. Then there exists an $r \times m$ matrix, B, of cointegrating vector such that $x_1(t) - Bx_2(t)$ is a stationary continuous time process. Note that these cointegrating relationships have been normalised on $x_1(t)$, which is an identification condition. The continuous time model can then be represented in terms of an error

[18] Stock (1987) had earlier provided an example that cointegration as a property was invariant to temporal aggregation.

correction model (ECM) of the form

$$Dx(t) = -JAx(t) + u(t), \quad t > 0, \tag{14.28}$$

where $J = [I_r, 0_{r\times m}]'$, $A = [I_r, -B]$ and $u(t)$ is a stationary process. The ECM representation (14.28) embodies two key features of the cointegrated system. The first r equations are of the form

$$Dx_1(t) = -[x_1(t) - Bx_2(t)] + u_1(t), \quad t > 0,$$

in which x_1 is responding to the disequilibrium (or error) depicted by $x_1(t) - Bx_2(t)$. Such systems are often motivated by $x_1(t) = Bx_2(t)$ representing an equilibrium or optimal level of x_1 given the level of x_2. The remaining m equations in (14.28) are the stochastic trends driving the system; they are given by

$$Dx_2(t) = u_2(t), \quad t > 0,$$

subject to an initial value $x_2(0)$.

In continuous time cointegrated systems of the form (14.28) interest centres on estimation of the matrix B. Equispaced discrete time observations generated by this system satisfy

$$x(th) = e^{-JAh}x(th-h) + v(th), \; v(th) = \int_{th-h}^{th} e^{-JA(th-r)}u(r)dr, \; t = 1, \ldots, T.$$

Using the fact that $AJ = I_r$, the infinite series representation for the matrix exponential can be used to show that $e^{-JAh} = I_n - fJA$ where $f = 1 - e^{-h}$. It then follows that $x(th)$ satisfies the discrete time ECM

$$\Delta_h x(th) = -JAx(th-h) + w(th), \; w(th) = v(th) + e^{-h}JAx(th-h), \; t = 1, \ldots, T,$$

where $w(th)$ is a stationary disturbance vector in view of $Ax(th) = x_1(th) - Bx_2(th)$ being stationary. Phillips (1991) recommended the use of spectral regression estimators that treat the dynamics nonparametrically. Such methods exploit the stationary nature of $u(t)$ to the full without requiring any particular parametric form for the dynamics and were shown to have good finite sample properties in the simulation study of Chambers (2001). Frequency domain methods can also be used in cointegrated systems in which the dynamics are modelled parametrically, for example, in CAR(p) models such as (14.1) that embed cointegration by setting $A_0 = CA$, where A is defined following (14.28) and C is an $n \times r$ matrix of rank r. Chambers and McCrorie (2007) show that maximisation of a frequency domain likelihood function leads to estimates of B that are asymptotically mixed normal and to estimates of the autoregressive parameters that govern the dynamics that are asymptotically normal. As is the case with cointegrated systems in discrete time, the estimates of B converge to the limiting distribution at rate T, while those of the autoregressive parameters converge at rate $\sqrt{T}$. The exact discrete model corresponding to a first-order cointegrated system with mixed sample data was derived by Chambers (2009), and such models can be estimated based on the time domain Gaussian likelihood outlined earlier.[19] The effects of sampling frequency in the context of cointegrated continuous time CAR systems were also analysed by Chambers (2011).

14.2.5 Mixed Frequency Data

Time series data in economics are available at a variety of frequencies. Observations on macroeconomic aggregates, such as consumers' expenditure, investment and national income, are typically available quarterly; variables such as the money supply and price indices used to compute measures of inflation are usually observed monthly, while financial variables, such as asset prices (interest rates, exchange rates, stock prices, etc.), can be observed almost continuously, but daily closing prices are often used. The extant approach to dealing with observations at different frequencies is to aggregate all variables to the lowest frequency, thereby potentially throwing away information contained in the high-frequency observations that could be exploited for gains in modelling. For example, it might be possible to use high-frequency financial variables to predict fluctuations in real economic activity before the low-frequency observations are available. In recent years, a number of advances in the analysis of mixed frequency data have been made, and the topic has assumed added significance following the financial crisis of 2008.

In the context of continuous time models, an often overlooked but nevertheless important contribution that incorporates observations at different frequencies was made by Zadrozny (1988). He considered the general problem of estimating a CARMA(p, q) system with mixed sample data available at mixed frequencies and recommended the use of state space forms and the Kalman filter for constructing the Gaussian likelihood function. More recently, and in keeping with the exact discrete time modelling approach, Chambers (2016) derived the exact discrete model corresponding to a CAR(1) system with mixed sample data observed at mixed frequencies. Suppose, for simplicity, that there are two vectors of stock variables, a low-frequency one, x_2 $(n_2 \times 1)$, observed at unit intervals of time, and a high-frequency vector, x_1 $(n_1 \times 1)$, observed at time intervals of length $0 < h < 1$ where it is convenient to assume that $k = h^{-1}$ is an integer. For example, if x_2 is observed quarterly and x_1 monthly, then $h = 1/3$ and x_1 is observed $k = 3$ times more frequently than x_2. Then, for each integer t, the $(kn_1 + n_2) \times 1$ vector

$$X_t = [x'_{1t}, x'_{1,t-h}, x'_{1,t-2h}, \ldots, x'_{1,t-(k-1)h}, x'_{2t}]', \quad t = 1, \ldots, T,$$

[19]Other time domain approaches to cointegrated models in continuous time can be found in Comte (1999) and Corradi (1997).

can be defined. The underlying continuous time model is assumed to be a CAR(1) system in the $n \times 1$ vector $x(t) = [x_1(t)', x_2(t)']'$ of the form

$$dx(t) = Ax(t)dt + \zeta(dt), \quad t > 0,$$

where $\zeta(dt)$ is defined following (14.1). The objective is to use the mixed frequency data to estimate the $n \times n$ matrix A and the $\frac{1}{2}n(n+1)$ elements of the covariance matrix, Σ, of $\zeta(dt)$. Theorem 1 of Chambers (2016)[20] shows that the discrete time observations satisfy the exact discrete time model

$$\begin{aligned}
x_{1t} &= B_{11,1}x_{1,t-h} + \ldots + B_{11,k}x_{1,t-1} + B_{12,0}x_{2,t-1} + \eta_{1t}, \\
x_{1,t-h} &= B_{11,1}x_{1,t-2h} + \ldots + B_{11,k-1}x_{1,t-1} + B_{12,1}x_{2,t-1} + \eta_{1,t-h}, \\
&\vdots \qquad\qquad\qquad \vdots \\
x_{1,t-(k-1)h} &= B_{11,1}x_{1,t-1} + B_{12,k-1}x_{2,t-1} + \eta_{1,t-(k-1)h}, \\
x_{2t} &= \sum_{j=1}^{k} B_{21,j}x_{1,t-jh} + B_{22}x_{2,t-1} + \eta_{2t},
\end{aligned}$$

where the $B_{ij,k}$ matrices are of dimension $n_i \times n_j$ $(i, j = 1, 2)$ and the $(kn_1+n_2)\times 1$ vector

$$\eta_t = [\eta_{1t}', \eta_{1,t-h}', \eta_{1,t-2h}', \ldots, \eta_{1,t-(k-1)h}', \eta_{2t}']'$$

is a vector white noise process. It is important to stress that all of the autoregressive matrices in the mixed frequency discrete time representation are only functions of the elements of the matrix A, while the covariance matrix of η_t depends only on A and Σ. By way of comparison, a discrete time vector autoregression in the vector X_t would be significantly over-parameterised.

Similar exact discrete time models can be derived for the cases where both the high- and low-frequency observations are on flow variables and where they are mixtures of stocks and flows. The main difference when flow variables are present is that the disturbance vector becomes an MA(1) process but the parsimony over unrestricted VAR and VARMA systems remains. Simulation results in Chambers (2016) for stationary and cointegrated systems show that utilising the mixed frequency data reduces bias and mean squared error of Gaussian estimates compared with the situation where high-frequency variables are aggregated to the low frequency. Furthermore, in an empirical application testing long-run purchasing power parity restrictions between the UK and the USA, inferences are found to be unfavourable to the restrictions when using the information in daily frequency exchange rates, but

[20]The model considered by Chambers (2016) also includes a vector of intercepts and deterministic trends.

the restrictions are not rejected when the exchange rates are aggregated to weekly and monthly frequencies. A possible explanation for this finding is that the estimates of the two key parameters of interest have large standard errors using the aggregated series but are more precisely determined when using the high-frequency data.

14.2.6 Gaussian Estimation Using an Exact Discrete Model

The exact discrete time model in the form of (14.10), allied with an additional set of $p-1$ conditions relating the initial observations to the initial state vector in the case of stock variables or p such conditions in the case of flows or a mixed sample, provides a basis for the construction of the likelihood function. It is usually assumed that the $nT \times 1$ vector $\eta = (\eta_h', \ldots, \eta_{Th}')'$ is Gaussian with mean vector zero and covariance matrix $\Omega_\eta = E(\eta\eta')$, which is equivalent to specifying $\zeta(dt)$ in (14.1) to be the increment in a Brownian motion process. Under such an assumption the log-likelihood is of the form

$$\log L(\theta) = -\frac{nT}{2}\log 2\pi - \frac{1}{2}\log|\Omega_\eta| - \frac{1}{2}\eta'\Omega_\eta^{-1}\eta, \qquad (14.29)$$

where θ denotes the parameter vector of interest (i.e. the elements of the autoregressive matrices $A_0, \ldots, A_{p-1}$ and the covariance matrix Σ). Section 14.3 discusses computational aspects associated with (14.29).

The asymptotic properties of estimates obtained by maximising (14.29) depend, of course, on the set of assumptions made concerning the model (14.1). Bergstrom (1983) provided a set of conditions that ensures that the vector, $\hat{\theta}$, that maximises (14.29) is almost surely consistent and, furthermore, that $\sqrt{T}(\hat{\theta} - \theta)$ is asymptotically normal and efficient in the Cramer sense. These conditions include such things as identification of θ in a closed bounded set Θ over which maximisation takes place, stationarity and ergodicity of $x(t)$ and continuity and differentiability of the autoregressive matrices and covariance matrix of (14.1) in cases where the elements of these matrices may depend, possibly non-linearly, on an underlying parameter vector of smaller dimension. The issue of identification has an added dimension in continuous time models owing to the phenomenon of aliasing which was discussed in Sect. 14.2.2.

In finite samples, the problem of estimation bias has the potential to beset *all* Gaussian/maximum likelihood (ML) methods including those based on the exact discrete time model. It is particularly relevant when estimating mean reversion parameters, as demonstrated in Phillips and Yu (2005) and Yu (2012). In a sampling experiment using a common interest rate model, Phillips and Yu (2009) showed that the estimation bias can be more important than the bias arising from using an approximate rather than an exact solution of the continuous time model. Wang et al. (2011) decompose the overall bias into separate terms arising from estimation and from discretisation, finding that when using Euler and trapezoidal approximations to the exact discrete model, both approximate methods dominate the exact method for empirically realistic cases. They also show that the sign of the discretisation bias is opposite to that of the estimation bias in such cases, meaning that the bias in the approximate methods is less than for estimation based on the exact discrete model. In addition the asymptotic variance of the estimator based on the Euler approximation is smaller than for the ML estimator of the mean reversion parameter in the exact discrete model, supporting a conclusion that the Euler approximation would be preferred to ML estimation of the exact discrete model in certain circumstances, such as when mean reversion in a univariate linear diffusion is slow.

It should be borne in mind that the exact discrete model is the only model that exactly incorporates restrictions implied by economic theory and other a priori information on the observed discrete data, and methods have been proposed to reduce finite sample estimation bias. Phillips and Yu (2005, 2009) propose jackknife techniques and a simulation-based indirect inference method and show they are successful in reducing finite sample bias in univariate diffusion models. Jackknife methods can also be expected to work successfully in higher-order and multivariate continuous time models, as indicated by the results of Chambers (2013) for stationary autoregressions and Chambers and Kyriacou (2013) in unit root models. The application of such techniques to more general continuous time systems is worthy of further investigation.

14.2.7 Alternative Approaches

Although we have emphasised the exact discrete time modelling approach to the estimation of continuous time systems, it is not the only suitable method. As mentioned above, Zadrozny (1988) has shown how Kalman filtering techniques can be used to compute the Gaussian likelihood function in CARMA systems with mixed sample and mixed frequency data that can also include exogenous variables. State space forms and the Kalman filter were also used in a sequence of contributions by Harvey and Stock (1985, 1988, 1989) that built on earlier work by Jones (1981) and focused on CAR systems that may contain integrated and cointegrated variables. Singer (1995) also proposed a filtering method and used analytic derivatives to facilitate computing the likelihood function. The evaluation of the likelihood function using the exact discrete model approach treats the entire observation vector simultaneously, whereas the Kalman filter is a recursive method that is usually defined stepwise from observation to observation. However, both methods should produce the same value for the likelihood function. Bergstrom (1985) offers some comparison between the methods, as do Singer (2007) and Oud and Singer (2008) though for methods extended to deal with panel data.

The main advantage of these approaches is that it is not necessary to derive the full exact discrete time model, merely the first-order difference equation satisfied by the state vector that includes unobservable components as well as the observed variables. Another advantage of this approach is that the Kalman filter produces optimal estimates of the unobservable components of the state vector which may be of interest in certain applications. A disadvantage is that it is less readily comparable to alternative discrete time models, a shortcoming that is clearly not shared by the exact discrete time representation. Furthermore, Bergstrom (1985) provided some arguments as to why the exact discrete time approach has computational advantages over the Kalman filter approach, although no formal testing of these claims appears to have been conducted and will, no doubt, depend on a whole variety of factors.[21]

Alternative frequency domain methods can also be used to estimate stationary CARMA systems. The spectral density matrix of the continuous time process $x(t)$ in (14.15) is given by

$$F(\lambda) = \frac{1}{2\pi} A(-i\lambda)^{-1} \Theta(-i\lambda) \Sigma \Theta(i\lambda)' [A(i\lambda)']^{-1}, \quad -\infty < \lambda < \infty, \tag{14.30}$$

where $i^2 = -1$,

$$A(z) = z^p I_n - A_{p-1} z^{p-1} - \ldots - A_1 z - A_0,$$
$$\Theta(z) = I_n + \Theta_1 z + \ldots + \Theta_{q-1} z^{q-1} + \Theta_q z^q.$$

Assuming that $x(t)$ is comprised of stock variables, the spectral density matrix of the discretely observed vector x_{th} is given by

$$F^d(\lambda) = \frac{1}{h} \sum_{j=-\infty}^{\infty} F\left(\frac{\lambda + 2\pi j}{h}\right), \quad -\pi < \lambda \leq \pi,$$

the so-called folding formula. Robinson (1993) provides formulae that enable $F^d(\lambda)$ to be computed exactly so that a frequency domain version of the Gaussian likelihood function (or Whittle likelihood) can be constructed. Flow variables are also easily handled within this framework, as are mixed samples. Suppose $x(t) = [x^s(t)', x^f(t)']'$, and we partition $F(\lambda)$ as

$$F(\lambda) = \begin{pmatrix} F^{ss}(\lambda) & F^{sf}(\lambda) \\ F^{fs}(\lambda) & F^{ff}(\lambda) \end{pmatrix}.$$

[21] Such factors include, but are not restricted to, the order of the continuous time system, the dimension of the vector $x(t)$, the sample size, the way in which the likelihoods are programmed and the optimisation algorithm used.

Then the spectral density matrix of the continuous time process

$$X(t) = \begin{pmatrix} x^s(t) \\ \frac{1}{h}\int_{t-h}^{t} x^f(r)dr \end{pmatrix}$$

is given by Robinson (1993) as

$$F_X(\lambda) = \begin{pmatrix} F^{ss}(\lambda) & \frac{1-e^{-ih\lambda}}{ih\lambda}F^{sf}(\lambda) \\ \frac{e^{ih\lambda}-1}{ih\lambda}F^{fs}(\lambda) & \frac{4\sin^2 h\lambda/2}{h^2\lambda^2}F^{ff}(\lambda) \end{pmatrix}, \quad -\infty < \lambda < \infty.$$

The terms multiplying components of the spectral density matrix involving flow variables arise through the frequency response function of the integral determining the observed process (and the squared frequency response function for $F^{ff}(\lambda)$). The spectral density of the process observed at discrete points in time, i.e. for $X(th)$ $(t = 1, 2, \ldots)$, is then subject to the folding formula yielding

$$F_X^d(\lambda) = \frac{1}{h}\sum_{j=-\infty}^{\infty} F_X\left(\frac{\lambda + 2\pi j}{h}\right), \quad -\pi < \lambda \leq \pi.$$

Fourier methods for the estimation of even more general continuous time systems were earlier proposed by Robinson (1976).

14.3 Computational Issues

Except in the simplest cases, estimates of the parameters of continuous time models do not have closed-form solutions and typically require optimisation using programmable statistical software such as R, Matlab or Gauss. Fortunately, the growth in computing power has expanded the scope and the dimension of feasible models, provided the sparse nature of many of the matrices involved in computing the likelihood and the possibility of in-sample convergence is exploited.

Firstly, the translation of an autoregressive model in continuous time to a discrete time model free of dependence on any sampling frequency involves the calculation of a matrix exponential, as in Eq. (14.3), and functions thereof. Owing to results by Van Loan (1978), the functions of the exponential can be computed as products of submatrices of a single, larger dimensional matrix exponential. Chambers (1999), McCrorie (2000) and Thornton and Chambers (2016) provide expressions pertaining to the exact discrete time model, while Harvey and Stock

(1985) and Zadrozny (1988) provide similar expressions for application of the Kalman filter.

Moler and Van Loan (1978), in a celebrated article in the numerical analysis literature,[22] showed that computation of the matrix exponential is a notoriously ill-conditioned problem, to the extent that of 19 methods considered, only 3 or 4 were potentially suitable in general, including a scaling and squaring method that employs Padé approximation to the scalar exponential (see Higham 2009). Jewitt and McCrorie (2005) discuss the computational issues behind computing matrix exponentials and their functions with continuous time econometrics in view. Standard methods are not always robust. For example, taking the partial sums of the Taylor series following Eq. (14.3) can be ill-conditioned because round-off error can propagate in computing higher and higher powers in a way that eventually dominates analytical convergence. A popular alternative is to exploit an eigenvalue decomposition when A is diagonalisable, i.e. when A is similar to a diagonal matrix Λ containing the eigenvalues of A. If $A = Q\Lambda Q^{-1}$, then $e^A = Qe^{\Lambda}Q^{-1}$, where e^{Λ} is, conveniently, a diagonal matrix whose elements are exponentials of the corresponding elements of Λ. The method relies, however, on an a priori assumption that the matrix A is diagonalisable, which is inconsistent with the property of cointegration that economic data plausibly satisfy. It is also possible that Q itself is ill-conditioned; see, for example, Higham and Al-Mohy (2010, Section 4). The main recommendation of Jewitt and McCrorie (2005) is that, for the type of matrices liable to be seen in econometric modelling, the problem is not likely to be ill-conditioned should any of the three standard methods discussed therein, including the scaling and squaring method also recommended by Zadrozny (1988), be used and supported by calculations made to at least standard IEEE double precision.

Hereafter, computation of the likelihood using the exact discrete representation diverges from calculation using the Kalman filter. The Kalman filter may be applied to Eq. (14.4) in association with an observation equation that synthesises the observed series, x_{th}, from the state vector, $y(th)$. Well-known methods have been developed to cope with irregularly spaced data and with observation noise; see, for example, Harvey (1989). The likelihood is often evaluated using the T prediction error vectors for the observed series, which, being optimal linear predictions, are uncorrelated. Such routines have the advantage/incur the expense, depending on requirements, of estimating the full state vector at the observation time points.

The evaluation of the likelihood in Eq. (14.29), in contrast, does not attempt an optimal prediction of x_{th} but rather models the time dependence between the η_{th} vectors parametrically. The computation of (14.29) is potentially troublesome as it involves the calculation of the determinant and inverse of the $nT \times nT$ covariance matrix Ω_{η}. However, the MA nature of η_{th} ensures that Ω_{η} is a sparse block-Toeplitz matrix with no more than $n(2p-1)$ non-zero elements in any row or column in the case of stocks and no more than $n(2p+1)$ non-zero elements in any row or column when flows are present. This sparsity can be exploited for computational advantages

[22]This paper was reprinted 25 years later with an update as Moler and Van Loan (2003).

including speed and accuracy. Let M denote the $nT \times nT$ lower triangular Cholesky matrix with typical elements m_{ij} satisfying $MM' = \Omega_\eta$, and let $\xi = M^{-1}\eta$ with typical element ξ_i. Then $\eta'\Omega_\eta^{-1}\eta = \xi'\xi$ and $|\Omega_\eta| = |MM'| = |M|^2$ so that the log-likelihood can be written as

$$\log L(\theta) = -\frac{nT}{2}\log 2\pi - \frac{1}{2}\log|M|^2 - \frac{1}{2}\xi'\xi$$

$$= -\frac{nT}{2}\log 2\pi - \sum_{i=1}^{nT}\log m_{ii} - \frac{1}{2}\sum_{i=1}^{nT}\xi_i^2, \tag{14.31}$$

which follows because $|M| = \prod_{i=1}^{nT} m_{ii}$. Bergstrom (1983) showed that the elements of ξ can be computed recursively from the system $M\xi = \eta$, while Bergstrom (1990, chapter 7) showed that the elements of M converge rapidly to fixed limits as computations proceed within the matrix, resulting in savings in computational storage requirements.

One of the important features of a continuous time model is that the form of an exact discrete time representation is invariant to the sampling frequency of the observations. We are able to illustrate this aspect in the context of a small simulation exercise using a simple first-order stochastic differential equation in a scalar random variable $x(t)$, given by

$$dx(t) = ax(t)dt + \zeta(dt), \quad t > 0, \tag{14.32}$$

where we take $x(0) = 0$ for convenience and $E[\zeta(dt)^2] = \sigma^2 dt$. Assuming $x(t)$ to be a stock variable, suppose that the sequence $x_0, x_h, x_{2h}, \ldots, x_{Th}$ is observed, where h denotes the sampling interval and T is the number of discrete time observations. Then the exact discrete time model satisfied by the sequence of observations is a discrete time AR(1), regardless of the sampling interval; it is given by

$$x_{th} = f_h x_{th-h} + \eta_{th}, \quad t = 1, \ldots, N, \tag{14.33}$$

where $f_h = e^{ah}$ and η_{th} is white noise with variance $\sigma_\eta^2 = \sigma^2(e^{2ah} - 1)/(2a)$. Acknowledging that x_{th} is subject to temporal aggregation means that we focus on estimating a and σ^2 regardless of the sampling interval. However, ignoring this feature means that f_h would be estimated directly and estimates would suggest differing degrees of serial correlation depending on the value of h. Associated patterns of variation would also be observed in estimates of σ_η^2 owing to its dependence on h.

In order to assess these features, we consider values of $h \in \{\frac{1}{12}, \frac{1}{6}, \frac{1}{4}, \frac{1}{3}, \frac{1}{2}, 1\}$ and $a \in \{-2, -1, -0.5, -0.1\}$ with $\sigma^2 = 1$. A total of 100,000 replications of each parameter combination were conducted, and we set the data span equal to $N = Th = 100$; this is the number of observations when $h = 1$. As the sampling interval falls, the number of observations, $T = N/h$, rises to a maximum of 1200 when $h = 1/12$. The data are generated at this highest frequency ($h = 1/12$), and then the lower-frequency observations are selected, so that, for example, the observations for $h = 1/6$ correspond to every second observation in the $h = 1/12$ sequence, while those for $h = 1$ correspond to every twelfth observation. The maximum likelihood estimator of a can be shown to be equal to

$$\hat{a}_{ML} = \frac{1}{h} \log \hat{f}_h,$$

where $\hat{f}_h$ denotes the ordinary least squares (OLS) estimator of f_h in the autoregression (14.33). Clearly this is only feasible if $\hat{f}_h > 0$, and it is only for smaller values of f_h and T that it becomes a problem. In fact, the only cases where $\hat{f}_h < 0$ were for $a = -1$ and $a = -2$ when $T = 100$, where the proportions of replications affected were 0.00016 and 0.091, respectively. In these cases, the estimates were removed, and the summary statistics were computed with the remainder of the replications. In view of the results of Wang et al. (2011), we also compute an estimate of a based on the Euler approximation given by

$$(x_{th} - x_{th-h}) = ahx_{th-h} + u_{th},$$

where u_{th} is a serially uncorrelated random disturbance with variance $\sigma^2 h$. We denote the estimate of a obtained using this approximation by $\hat{a}_E$.

The results appear in Table 14.1 in which, for each value of a, the mean values and standard errors (across the replications) of $\hat{a}_{ML}$ and $\hat{a}_E$ are reported, as well as the means and standard errors of $\hat{f}_h$. In the latter case, we also report the actual values of f_h. It can be seen clearly from Table 14.1 that the estimates of a using $\hat{a}_{ML}$, although slightly biased as expected, are all stable across the range of values of h, and although $\hat{a}_E$ has smaller bias than $\hat{a}_{ML}$ when $a = -0.1$, its performance in terms of bias deteriorates as a becomes more negative. This is in accordance with the results of Wang et al. (2011). It can also be seen that $\hat{a}_E$ has a smaller standard error than $\hat{a}_{ML}$ in all cases. The estimates of the discrete time autoregressive parameter f_h using $\hat{f}_h$ can be seen to depend clearly on the value of h. Although $\hat{f}_h$ is a reasonably good estimator of f_h, the implications for the dependence properties of the variable x depend very much on the sampling interval chosen; the same is not true when the temporal aggregation is taken into account.

Table 14.1 Simulation results: means and standard errors of estimators

h	$\hat{a}_{ML}$	$\hat{a}_E$	$\hat{f}_h$	f_h	$\hat{a}_{ML}$	$\hat{a}_E$	$\hat{f}_h$	f_h
$a = -0.1$					$a = -0.5$			
1	−0.1213	−0.1127	0.8873	0.9048	−0.5292	−0.4050	0.5950	0.6065
	(0.0591)	(0.0504)	(0.0504)		(0.1452)	(0.0814)	(0.0814)	
1/2	−0.1206	−0.1163	0.9418	0.9512	−0.5235	−0.4578	0.7711	0.7788
	(0.0567)	(0.0524)	(0.0262)		(0.1217)	(0.0909)	(0.0460)	
1/3	−0.1203	−0.1174	0.9609	0.9672	−0.5219	−0.4772	0.8409	0.8465
	(0.0559)	(0.0530)	(0.0177)		(0.1153)	(0.0957)	(0.0319)	
1/4	−0.1202	−0.1181	0.9705	0.9753	−0.5214	−0.4875	0.8781	0.8825
	(0.0555)	(0.0534)	(0.0133)		(0.1125)	(0.0979)	(0.0245)	
1/6	−0.1201	−0.1186	0.9802	0.9835	−0.5207	−0.4979	0.9170	0.9200
	(0.0551)	(0.0537)	(0.0090)		(0.1097)	(0.1000)	(0.0167)	
1/12	−0.1200	−0.1193	0.9901	0.9917	−0.5203	−0.5088	0.9576	0.9592
	(0.0548)	(0.0541)	(0.0045)		(0.1072)	(0.1024)	(0.0085)	
$a = -1.0$					$a = -2.0$			
1	−1.0581	−0.6388	0.3612	0.3679	−2.1249	−0.8667	0.1333	0.1353
	(0.3030)	(0.0934)	(0.0934)		(0.8435)	(0.0988)	(0.0988)	
1/2	−1.0291	−0.7989	0.6006	0.6065	−2.0554	−1.2716	0.3642	0.3679
	(0.1951)	(0.1140)	(0.0570)		(0.3853)	(0.1318)	(0.0659)	
1/3	−1.0247	−0.8644	0.7119	0.7165	−2.0343	−1.4698	0.5101	0.5134
	(0.1746)	(0.1225)	(0.0408)		(0.2986)	(0.1488)	(0.0496)	
1/4	−1.0234	−0.9003	0.7749	0.7788	−2.0291	−1.5861	0.6035	0.6065
	(0.1662)	(0.1275)	(0.0319)		(0.2682)	(0.1597)	(0.0399)	
1/6	−1.0217	−0.9377	0.8437	0.8465	−2.0244	−1.7148	0.7142	0.7165
	(0.1583)	(0.1327)	(0.0221)		(0.2420)	(0.1716)	(0.0286)	
1/12	−1.0208	−0.9777	0.9185	0.9200	−2.0219	−1.8590	0.8451	0.8465
	(0.1513)	(0.1386)	(0.0115)		(0.2205)	(0.1858)	(0.0155)	

14.4 Empirical Applications

There have been many applications of the methods of this chapter, most notably in the area of macroeconomic modelling to which, because it drove much of the early work, we devote Sect. 14.4.3 below. Representative papers include, in the areas of asset allocation, Campbell et al. (2004); consumers' demand, Bergstrom and Chambers (1990) and Chambers (1992); uncovered interest parity, Diez de los Rios and Sentana (2011); exchange rates, Renault et al. (1998); short-term interest rate models, Nowman (1997), Yu and Phillips (2001) and Phillips and Yu (2011); and empirical finance in general, Thornton and Chambers (2016). It is quite clear that, perhaps now more than ever, economic activity occurs continuously around the clock and yet, such is the undertaking required to measure this activity, published statistics cannot hope to provide real-time information. Here we introduce two applications within macroeconomics which are illustrative of the use of an exact discrete representation to resolve this tension. One involves a univariate time series, namely, consumer price inflation in the UK, while the other explores the important relationship between gross domestic product (GDP) in the USA and crude oil prices. As mentioned in Sect. 14.2, the impact of time aggregation is to induce serial correlation in the disturbances, η_t, and so the ability of a continuous time specification to explain the observed serial correlation adequately is an important test of its suitability. In order to address this issue, Bergstrom (1990, chapter 7) proposed a portmanteau-type test statistic based on the $n \times 1$ vectors of standard normal variates ξ_{th} $(t = 1, \ldots, T)$. Bergstrom's statistic is defined by

$$S_l = \frac{1}{n(T-l)} \sum_{r=1}^{l} \left(\sum_{t=l+1}^{T} \xi_{th}' \xi_{th-rh} \right)^2,$$

which has an approximate chi-squared distribution with l degrees of freedom (the number of lags used) under the null hypothesis that the model is correctly specified. For robustness, we also report the Schwarz Bayesian model selection criterion (SBC) for each model. Each of these models is deliberately narrow in their focus. Modern economies are, of course, large, complex and interconnected systems and so we finish with an overview of some of the large-scale macroeconomic modelling carried out in continuous time.

14.4.1 Consumer Prices and Inflation

In a continuous time setting, price inflation can be defined as the instantaneous rate of change of the price level, i.e. $\pi(t) = D \log p(t)$. Consider the continuous time ARMA(2,1) model for $\log p(t)$ given by

$$D^2 \log p(t) = \gamma_0 + A_1 D \log p(t) + A_0 \log p(t) + u(t) + \theta Du(t), \ t > 0, \qquad (14.34)$$

where γ_0, A_1, A_0 and θ are scalars and $u(t)$ is a mean zero uncorrelated process with variance σ_u^2. Under the condition that $A_0 = 0$, i.e. that $\log p(t)$ has a zero root in continuous time (and a unit root in discrete time), the implied law of motion for inflation becomes

$$D\pi(t) = \gamma_0 + A_1 \pi(t) + u(t) + \theta Du(t), \ t > 0. \qquad (14.35)$$

Hence $\pi(t)$ satisfies a continuous time ARMA(1,1) process which corresponds to a continuous time ARIMA(2,1,1) process for $\log p(t)$.

Estimates of (14.34) with $A_0 = 0$ were obtained using monthly data for the UK consumer price index over the period January 1996 to March 2014, a total of 219 observations. The results are given in Table 14.2. The estimates of the parameters in the CARMA(2,0) are well determined, and there is no evidence of misspecification,

Table 14.2 Estimates for inflation

	CARMA(2,0)	CARMA(2,1)
γ_0	0.0261	0.0013
	(0.0038)	(0.0006)
A_1	−14.8432	−0.7362
	(0.0046)	(0.3425)
θ	0.0000	−1.9916
		(0.9772)
σ_u	0.0562	0.0020
	(0.0027)	(0.0010)
$\log L$	909.8639	913.0956
SBC	−1803.5607	−1804.6349
S_4	0.4427	0.9942

Standard errors in parentheses; entries for S_4 are p-values

at least as measured by Bergstrom's S_4 statistic. However, addition of the MA(1) component results in a statistically significant increase in the value of the maximised log-likelihood function—the likelihood ratio statistic for testing the null hypothesis that $\theta = 0$ is equal to 6.4634 with a marginal probability of 0.0110. The p-value of the S_4 statistic is far from significant, suggesting that the inclusion of the statistically significant MA(1) component yields an improved fit; this is also the inference drawn from a comparison of the SBC values for the two models.

14.4.2 Oil Prices and the Macroeconomy

Next, we explore the relationship between US output, as measured by real GDP in tens of billions of chained 2009 dollars, and the oil price, as measured by the price of West Texas Intermediate in dollars per barrel. The data are quarterly ranging from 1986 to 2013 quarter 3 from the Federal Reserve Bank of St. Louis. In common with most authors who have examined these series, for example, Hamilton (1996), we find that both processes show strong evidence of unit root behaviour, with augmented Dickey-Fuller test statistics of −0.609 and −0.318 for GDP and the oil price, respectively, but find no evidence of a cointegrating relationship. Nonstationary but non-cointegrated data are consistent with the specification in Eq. (14.15) with $A_0 = 0$. We define the 2×1 vector $x(t) = [\mathrm{GDP}(t),\ \mathrm{Oil}(t)]'$.

We consider two candidate models nested within a continuous time ARMA(2,1) model, the continuous time ARIMA(1,1,0) which has $A_0 = 0$, $p = 2$ and $q = 0$ and the continuous time CARIMA(1,1,1), which has $q = 1$. The exact discrete representation of both models is an ARIMA(1,1,1),

$$\Delta_h x_{th} = f_0 + F_1 \Delta x_{th-h} + \eta_{th}, \quad t = 3, \ldots, T, \tag{14.36}$$

where $\Delta_h x_{th} = x_{th} - x_{th-h}$ as in (14.24), with the CARIMA(1,1,1) offering more flexibility in modelling the autocovariance structure of the discrete time disturbance η_{th}.

Results for the two models are presented in Table 14.3. The CARIMA(1,1,1) is preferred by the SBC, and the likelihood ratio test fails to reject the CARIMA(1,1,1) in favour of the CARIMA(1,1,0), with a test statistic of 24.566. The moving average coefficients in the first column of Θ are significant, reflecting the impact of lagged shocks to GDP on both GDP and oil prices. Both models have values for the Bergstrom S_1 and S_4 statistic in the acceptable region.

The literature has focussed on the question of whether changes in the oil price lead to changes in GDP, reflected in the top right element of the matrix A_1. It is noticeable that in the CARIMA(1,1,0), neither of the coefficients on the rate of change of GDP or on the oil price is significant in the equation determining the other variable. When a moving average error is introduced to capture more complicated dynamics, however, the t-ratio on the top right element of A_1 is -2.36, indicating that growth in oil prices slows GDP, while that on the bottom left is 1.975, suggesting that growth in GDP accelerates oil price growth.

As a by-product of the estimation, to aid comparison with other models, the intercept vector, f_0, and autoregressive matrix, F_1 of the exact discrete time representation are also reported. These reinforce the point that the CARIMA(1,1,1) model predicts a stronger reaction from one series to lagged changes in the other.

14.4.3 *Macroeconometric Modelling*

While many of the applications of continuous time models and methods occur today in the area of empirical finance, much of the literature's early development was driven through the desire to make advances in the area of large-scale macroeconometric modelling. This was in no small part due to Rex Bergstrom who, in collaboration with a student, Clifford Wymer, produced in Bergstrom and Wymer (1976) the first continuous time macroeconometric model, the formulation and estimation of which represents one of the landmarks in the development of modern econometrics. The economy-wide model, which comprised 13 equations (10 structural equations and 3 identities) in 35 key parameters, served as a prototype for later developments in macroeconomic modelling and the modelling of financial and commodity markets. It had innovative features beyond simply being formulated in continuous time: it was formulated as a dynamic disequilibrium model[23] involving a system of partial adjustment equations in the form of continuous time error correction equations, where each causally dependent variable continually adjusts in

[23]See Hillinger (1996) for a discussion of the history and conceptual foundations of such models in macroeconometric modelling and Wymer (1996) for a similar discussion that focuses on continuous time.

Table 14.3 Estimates of CARIMA models for GDP and oil prices

	CARIMA(1,1,0)			CARIMA(1,1,1)				
Continuous time model parameters								
	a_0	A_1		a_0	A_1		Θ	
GDP	3.9720	−0.4966	−0.3474	4.1819	−0.5107	−0.4729	−0.3648	−0.9745
	(4.3642)	(0.2368)	(0.4838)	(4.3192)	(0.1890)	(0.2004)	(0.1829)	(1.2448)
Oil price	2.8840	−0.2382	−1.2451	−5.5287	1.0499	−2.4061	0.3253	−0.7239
	(1.7099)	(0.6098)	(0.6241)	(1.7247)	(0.5317)	(0.6380)	(0.1594)	(0.7642)
log L	−737.8751			−725.5962				
SBC	1518.1359			1512.4163				
S_1	0.4328			0.6619				
S_4	0.1875			0.1945				
Exact discrete time model parameters								
	f_0	F_1		f_0	F_1			
GDP	3.1898	0.6033	−0.2960	2.8670	0.6797	−0.5649		
Oil price	1.1864	−0.0784	0.3289	−0.7302	0.2376	−0.1179		

Standard errors in parentheses; entries for S_1 and S_4 are p-values response to the deviation from its partial equilibrium level; it embodied the intensive use of economic theory and other a priori information to support a parsimonious representation in the model parameters; and its design facilitated an analysis of its steady-state and stability properties using methods developed earlier by Bergstrom (1967).[24]

An earlier comprehensive survey of continuous time macroeconomic modelling can be found in Bergstrom (1996), which includes the various stages of the Italian continuous time model of Gandolfo and Padoan (1984, 1990),[25] the economy-wide models contained in the volume edited by Gandolfo (1993) and, not least, the model by Bergstrom et al. (1992) which signified the next stage of development. This model was the first to incorporate the exact methods that are the focus of this chapter; it used more realistic, second-order partial adjustment equations using the method for higher-order systems pioneered by Bergstrom (1983) and described in Sect. 14.2.1 above; and, unlike the Bergstrom and Wymer (1976) model, it incorporated exogenous variables. Estimating this model, which comprised 14 equations with 63 parameters and 11 exogenous variables, required around a day's computing time on a CRAY X-MP/48 supercomputer, which at the time represented the cutting edge of computer technology. The Italian model was further developed into a system including non-linear equations by Gandolfo et al. (1996), although estimation was facilitated through a linear approximation about sample means; see Wymer (1993) for details of the underlying estimation method. Wymer (1993, 1997, 2012) has developed a direct, full-information maximum likelihood approach to the estimation of such non-linear systems although, given the development of this literature, the properties of this estimator must currently be inferred from those of the estimator based on a linear approximation about sample means. Starting values for the procedure are readily obtained from applying the method of maximum likelihood to this linear approximation.

The theoretical basis for what could be seen as a third-stage continuous time model was provided by Bergstrom (1997), where unobservable stochastic trends are incorporated within the system of stochastic differential equations to take advantage of insights gained from the development of unit root econometrics that occurred in the discrete time literature. The project was finally brought to fruition in a book by Bergstrom and Nowman (2007) that was published after Bergstrom's death in 2005. The model comprised a system of 18 mixed first- and second-order non-linear differential equations with 63 structural parameters, 33 long-run parameters, 27 speed of adjustment parameters and 3 drift parameters. Its linearisation about sample means results in precisely the model considered by Bergstrom (1997). The parameter estimates and speed of adjustment parameters were all plausible, and the model was seen, through an examination of its steady-state and stability properties, to generate plausible long-run behaviour. Its post-sample forecasting

[24] See Gandolfo (1981) for a textbook treatment.

[25] At the time of writing (July 2017), Pier Carlo Padoan is Italy's Minister of Economy and Finance, a position he has held since February 2014.

performance also compared favourably with a second-order VAR model with exogenous variables. The book provided a retrospect of what Rex Bergstrom achieved over a lifetime of research in the area of continuous time econometrics; a brief survey of this contribution with an emphasis on macroeconomic modelling is provided by Nowman (2009).

14.5 Concluding Comments

This chapter has aimed to provide a survey of methods of continuous time modelling based on an exact discrete time representation. Such an approach is synonymous with the name of Rex Bergstrom whose pioneering contributions were instrumental in attracting the current authors to the field. Our survey has attempted to highlight the techniques involved with the derivation of an exact discrete time representation of an underlying continuous time model, providing specific details for a second-order linear system of stochastic differential equations. Issues of parameter identification, causality, nonstationarity and mixed frequency data have also been addressed, all of which are important to consider in applications in economics and other disciplines. Although our focus has been on Gaussian estimation of the exact discrete time model, we have also discussed alternative time domain (state space) and frequency domain approaches. Computational issues have also been explored, where here the focus is on the exploitation of sparse matrices and the computation of the matrix exponential. Two new empirical applications have been included along with a discussion of applications in the field of macroeconometric modelling. While our focus is, of necessity, oriented towards economics and econometrics, we hope that the material contained in this chapter will be of interest in the social and behavioural sciences more widely.

Acknowledgements We thank an editor and two anonymous referees for helpful comments that have led to improvements in this paper and the Scottish Institute for Research in Economics for arranging facilities in the School of Economics, University of Edinburgh, for the authors to meet to work on this chapter. The first author also thanks the Economic and Social Research Council for financial support under grant number ES/M01147X/1.

Appendix

The Gauss code below was used in the simulation exercise. Note that n is used in the code as the data span and t is the sample size, whereas, in the text in Sect. 14.3, it is T and N, respectively, that are used for these quantities.

```
/* Simulation of continuous time AR(1) process
                                    at different frequencies */
new;
a=-1.0;                      /* Continuous time AR parameter    */
hv=1|1/2|1/3|1/4|1/6|1/12;/* Discrete time sampling intervals*/
n=100;                       /* Data span                       */
x0=0;                        /* Initial value                   */
nreps=100000;                /* Number of replications          */
s2=1;                      /*Continuous time innovation variance*/
rndseed 6665;                /* seed for random numbers         */

rhv=rows(hv);
tv=n./hv;
maxt=maxc(tv);
hmin=minc(hv);
hrel=hv/hmin;
eahm=exp(a*hmin);
e2ahm=exp(2*a*hmin);
eahv=exp(a*hv);
s2m=s2*(e2ahm-1)/(2*a);
sm=sqrt(s2m);
cta=zeros(nreps,rhv);        /* nreps times number of h values */
dta=cta; eta=cta; nogood=0;

for i (1,nreps,1);
    u=sm*rndn(maxt,1);
    xm=datagen(u);           /* maxt times 1                   */
    for hi (1,rhv,1);
        h=hv[hi,1];
        t=tv[hi,1];
        xh=reshape(xm,tv[hi,1],hrel[hi,1]);
        x=xh[.,hrel[hi,1]];
        bhat=x[2:t,1]/x[1:t-1,1];
        if bhat le 0; ahat=0; nogood=nogood+1;
        else; ahat=ln(bhat)/h;
        endif;
        ehat=(x[2:t,1]-x[1:t-1,1])/(h*x[1:t-1,1]);
        cta[i,hi]=ahat;
        dta[i,hi]=bhat;
        eta[i,hi]=ehat;
    endfor;
endfor;

stop;

proc datagen(e);
    local x;
    x = recserar(e, x0, eahm);
    retp( x );
endp;
```

References

Aït-Sahalia, Y. (2007). Estimating continuous time models using discretely sampled data. In R. Blundell, T. Persson, & W. K. Newey (Eds.), *Advances in economics and econometrics, theory and applications, ninth world congress of the econometric society*. Cambridge: Cambridge University Press.

Aït-Sahalia, Y., & Jacod, J. (2014). *High-frequency financial econometrics*. Princeton: Princeton University Press.

Bergstrom, A. R. (1966). Non-recursive models as discrete approximations to systems of stochastic differential equations. *Econometrica, 34*, 173–182. https://doi.org/10.2307/1909861

Bergstrom, A. R. (1967). *The construction and use of economic models*. London: English Universities Press.

Bergstrom, A. R. (1983). Gaussian estimation of structural parameters in higher order continuous time dynamic models. *Econometrica, 51*, 117–152. https://doi.org/10.2307/1912251

Bergstrom, A. R. (1984). Continuous time stochastic models and issues of aggregation over time. In Z. Griliches & M. D. Intriligator (Eds.), *Handbook of econometrics* (Vol. 2, pp. 1145–1212). Amsterdam: North-Holland. https://doi.org/10.1016/S1573-4412(84)02012-2

Bergstrom, A. R. (1985). The estimation of parameters in nonstationary higher-order continuous-time dynamic models. *Econometric Theory, 1*, 369–385. https://doi.org/10.1017/S0266466600011269

Bergstrom, A. R. (1986). The estimation of open higher-order continuous time dynamic models with mixed stock and flow data. *Econometric Theory, 2*, 350–373. https://doi.org/10.1017/S026646660001166X

Bergstrom, A. R. (1990). *Continuous time econometric modelling*. Oxford: Oxford University Press.

Bergstrom, A. R. (1996). Survey of continuous time econometrics. In W. A. Barnett, G. Gandolfo, & C. Hillinger (Eds.), *Dynamic disequilibrium modeling* (pp. 3–25). Cambridge: Cambridge University Press.

Bergstrom, A. R. (1997). Gaussian estimation of mixed-order continuous-time dynamic models with unobservable stochastic trends from mixed stock and flow data. *Econometric Theory, 13*, 467–505. https://doi.org/10.1017/S0266466600005971

Bergstrom, A. R., & Chambers, M. J. (1990). Gaussian estimation of a continuous time model of demand for consumer durable goods with applications to demand in the United Kingdom. In A. R. Bergstrom (Ed.), *Continuous time econometric modelling* (pp. 279–319). Oxford: Oxford University Press.

Bergstrom, A. R., & Nowman, K. B. (2007). *A continuous time econometric model of the United Kingdom with stochastic trends*. Cambridge: Cambridge University Press. https://doi.org/10.1017/CBO9780511664687

Bergstrom, A. R., Nowman, K. B., & Wymer, C. R. (1992). Gaussian estimation of a second order continuous time macroeconometric model of the United Kingdom. *Economic Modelling, 9*, 313–351. https://doi.org/10.1016/0264-9993(92)90017-V

Bergstrom, A. R., & Wymer, C. R. (1976). A model of disequilibrium neoclassical growth and its application to the United Kingdom. In A. R. Bergstrom (Ed.), *Statistical inference in continuous time economic models* (pp. 267–327). Amsterdam: North-Holland.

Blevins, J. R. (2017). Identifying restrictions for finite parameter continuous time models with discrete time data. *Econometric Theory, 33*, 739–754. https://doi.org/10.1017/S0266466615000353

Breitung, J., & Swanson, N. R. (2002). Temporal aggregation and spurious instantaneous causality in multiple time series models. *Journal of Time Series Analysis, 23*, 651–665. https://doi.org/10.1111/1467-9892.00284

Brockwell, P. J. (2014). Recent results in the theory and applications of CARMA processes. *Annals of the Institute of Statistical Mathematics, 66*, 647–685. https://doi.org/10.1007/s10463-014-0468-7

Campbell, J. Y., Chacko, G., Rodriguez, J., & Viceira, L. M. (2004). Strategic asset allocation in a continuous-time VAR model. *Journal of Economic Dynamics and Control, 28*, 2195–2214. https://doi.org/10.1016/j.jedc.2003.09.005

Chambers, M. J. (1990). *Durability and consumers' demand: Gaussian estimation of some continuous time models*. Unpublished doctoral dissertation, University of Essex, Colchester.

Chambers, M. J. (1992). Estimation of a continuous time dynamic demand system. *Journal of Applied Econometrics, 7*, 53–64. https://doi.org/10.1002/jae.3950070106

Chambers, M. J. (1999). Discrete time representation of stationary and non-stationary continuous time systems. *Journal of Economic Dynamics and Control, 23*, 619–639. https://doi.org/10.1016/S0165-1889(98)00032-3

Chambers, M. J. (2001). Temporal aggregation and the finite sample performance of spectral regression estimators in cointegrated systems: A simulation study. *Econometric Theory, 17*, 591–607. https://doi.org/10.1017/S0266466601173044

Chambers, M. J. (2009). Discrete time representations of cointegrated continuous time models with mixed sample data. *Econometric Theory, 25*, 1030–1049. https://doi.org/10.1017/S0266466608090397

Chambers, M. J. (2011). Cointegration and sampling frequency. *Econometrics Journal, 14*, 156–185. https://doi.org/10.1111/j.1368-423X.2010.00329.x

Chambers, M. J. (2013). Jackknife estimation of stationary autoregressive models. *Journal of Econometrics, 172*, 142–157. https://doi.org/10.1016/j.jeconom.2012.09.003

Chambers, M. J. (2016). The estimation of continuous time models with mixed frequency data. *Journal of Econometrics, 193*, 390–404. https://doi.org/10.1016/j.jeconom.2016.04.013

Chambers, M. J., & Kyriacou, M. (2013). Jackknife estimation with a unit root. *Statistics and Probability Letters, 83*, 1677–1682. https://doi.org/10.1016/j.spl.2013.03.016

Chambers, M. J., & McCrorie, J. R. (2006). Identification and estimation of exchange rate models with unobservable fundamentals. *International Economic Review, 47*, 573–582. https://doi.org/10.1111/j.1468-2354.2006.00389.x

Chambers, M. J., & McCrorie, J. R. (2007). Frequency domain estimation of temporally aggregated Gaussian cointegrated systems. *Journal of Econometrics, 136*, 1–29. https://doi.org/10.1016/j.jeconom.2006.03.005

Chambers, M. J., & Thornton, M. A. (2012). Discrete time representations of continuous time ARMA processes. *Econometric Theory, 28*, 219–238. https://doi.org/10.1017/S0266466611000181

Christiano, L. J., & Eichenbaum, M. S. (1987). Temporal aggregation and structural inference in macroeconomics. In *Carnegie-Rochester conference series on public policy* (Vol. 26, pp. 63–130).

Comte, F. (1999). Discrete and continuous time cointegration. *Journal of Econometrics, 88*, 207–226. https://doi.org/10.1016/S0304-4076(98)00025-6

Comte, F., & Renault, E. (1996). Noncausality in continuous time models. *Econometric Theory, 12*, 215–256. https://doi.org/10.1017/S0266466600006575

Corradi, V. (1997). Comovements between diffusion processes: Characterization, estimation and testing. *Econometric Theory, 13*, 646–666. https://doi.org/10.1017/S0266466600006113

Culver, W. (1966). On the existence and uniqueness of the real logarithm of a matrix. *Proceedings of the American Mathematical Society, 16*, 1146–1151. https://doi.org/10.1090/S0002-9939-1966-0202740-6

de los Rios, A. D., & Sentana, E. (2011). Testing uncovered interest parity: A continuous-time approach. *International Economic Review, 52*, 1215–1251. https://doi.org/10.1111/j.1468-2354.2011.00665.x

Engle, R. F., & Granger, C. W. (1987). Cointegration and error correction: Representation, estimation and testing. *Econometrica, 55*, 251–276. https://doi.org/10.2307/1913236

Fergusson, K., & Platen, E. (2015). Application of maximum likelihood estimation to stochastic short rate models. *Annals of Financial Economics, 10*, 1–26. https://doi.org/10.1142/S2010495215500098

Florens, J. P., & Fougère, D. (1996). Noncausality in continuous time. *Econometrica, 64*, 1195–1212. https://doi.org/10.2307/2171962

Gandolfo, G. (1981). *Quantitative analysis and econometric estimation of continuous time models*. Amsterdam: North-Holland.

Gandolfo, G. (1993). *Continuous time econometrics: Theory and applications*. London: Chapman and Hall. https://doi.org/10.1007/978-94-011-1542-1

Gandolfo, G., & Padoan, P. C. (1984). *A disequilibrium model of the real and financial accumulation in an open economy*. Berlin: Springer. https://doi.org/10.1007/978-3-642-95459-7

Gandolfo, G., & Padoan, P. C. (1990). The Italian continuous time model, theory and empirical results. *Economic Modelling, 7*, 91–132. https://doi.org/10.1016/0264-9993(90)90015-V

Gandolfo, G., Padoan, P. C., De Arcangelis, G., & Wymer, C. R. (1996). Nonlinear estimation of a nonlinear continuous time model. In W. A. Barnett, G. Gandolfo, & C. Hillinger (Eds.), *Statistical inference in continuous time economic models* (pp. 127–150). Cambridge: Cambridge University Press.

Granger, C. W. J. (1966). The typical spectral shape of an economic variable. *Econometrica, 34*, 150–161. https://doi.org/10.2307/1909859

Hamerle, A., Nagl, W., & Singer, H. (1991). Problems with the estimation of stochastic differential equations using structural equations models. *Journal of Mathematical Sociology, 16*, 201–220. https://doi.org/10.1080/0022250X.1991.9990088

Hamerle, A., Singer, H., & Nagl, W. (1993). Identification and estimation of continuous time dynamic systems with exogenous variables using panel data. *Econometric Theory, 9*, 283–295. https://doi.org/10.1017/S0266466600007544

Hamilton, J. D. (1996). This is what happened to the oil price-macroeconomy relationship. *Journal of Monetary Economics, 38*, 215–220. https://doi.org/10.1016/S0304-3932(96)01282-2

Hansen, L. P., & Sargent, T. J. (1983). The dimensionality of the aliasing problem in models with rational spectral densities. *Econometrica, 51*, 377–387. https://doi.org/10.2307/1911996

Hansen, L. P., & Sargent, T. J. (1991). Identification of continuous time rational expectations from discrete data. In L. P. Hansen & T. J. Sargent (Eds.), *Rational expectations econometrics* (pp. 219–235). Boulder, Colorado: Westview Press.

Harvey, A. C. (1989). *Forecasting structual time series models and the Kalman filter*. Cambridge: Cambridge University Press.

Harvey, A. C., & Stock, J. H. (1985). The estimation of higher-order continuous time autoregressive models. *Econometric Theory, 1*, 97–117. https://doi.org/10.1017/S0266466600011026

Harvey, A. C., & Stock, J. H. (1988). Continuous time autoregressive models with common stochastic trends. *Journal of Economic Dynamics and Control, 12*, 365–384. https://doi.org/10.1016/0165-1889(88)90046-2

Harvey, A. C., & Stock, J. H. (1989). Estimating integrated higher-order continuous time autoregressions with an application to money-income causality. *Journal of Econometrics, 42*, 313–336. https://doi.org/10.1016/0304-4076(89)90056-0

Higham, N. J. (2009). The scaling and squaring method for the matrix exponential revisited. *SIAM Review, 51*, 747–764. https://doi.org/10.1137/090768539

Higham, N. J., & Al-Mohy, A. H. (2010). Computing matrix functions. *Acta Numerica, 19*, 159–208. https://doi.org/10.1017/S0962492910000036

Hillinger, C. (1996). Dynamic disequilibrium economics: History, conceptual foundations, possible futures. In G. G. W. A. Barnett & C. Hillinger (Eds.), *Dynamic disequilibrium modeling* (pp. 27–65). Cambridge: Cambridge University Press.

Jewitt, G., & McCrorie, J. R. (2005). Computing estimates of continuous time macroeconometric models on the basis of discrete data. *Computational Statistics and Data Analysis, 49*, 397–416. https://doi.org/10.1016/j.csda.2004.05.021

Jones, R. H. (1981). Fitting a continuous time autoregression to discrete data. In D. F. Findlay (Ed.), *Applied time series analysis II* (pp. 651–682). New York: Academic.

Marcellino, M. (1999). Some consequences of temporal aggregation in empirical analysis. *Journal of Business and Economic Statistics, 17*, 129–136. https://doi.org/10.1080/07350015.1999.10524802

McCrorie, J. R. (1996). *Some topics in the estimation of continuous time econometric models*. Unpublished doctoral dissertation, University of Essex, Colchester.

McCrorie, J. R. (2000). Deriving the exact discrete analog of a continuous time system. *Econometric Theory, 16*, 998–1015. https://doi.org/10.1017/S0266466600166071

McCrorie, J. R. (2001). Interpolating exogenous variables in continuous time dynamic models. *Journal of Economic Dynamics and Control, 25*, 1399–1427. https://doi.org/10.1016/S0165-1889(99)00061-5

McCrorie, J. R. (2003). The problem of aliasing in identifying finite parameter continuous time models. *Acta Applicandae Mathematicae, 79*, 9–16. https://doi.org/10.1023/A:1025858121378

McCrorie, J. R. (2009). Estimating continuous-time models on the basis of discrete data via an exact discrete analog. *Econometric Theory, 25*, 1120–1137. https://doi.org/10.1017/S0266466608090452

McCrorie, J. R., & Chambers, M. J. (2006). Granger causality and the sampling of economic processes. *Journal of Econometrics, 132*, 311–336. https://doi.org/10.1016/j.jeconom.2005.02.002

Moler, C. B., & Van Loan, C. F. (1978). Nineteen dubious ways to compute the exponential of a matrix. *SIAM Review, 20*, 801–836. https://doi.org/10.1137/1020098

Moler, C. B., & Van Loan, C. F. (2003). Nineteen dubious ways to compute the exponential of a matrix, twenty-five years later. *SIAM Review, 45*, 3–49. https://doi.org/10.1137/S00361445024180

Nowman, K. B. (1997). Gaussian estimation of single-factor continuous time models of the term structure of interest rates. *Journal of Finance, 52*, 1695–1703. https://doi.org/10.1111/j.1540-6261.1997.tb01127.x

Nowman, K. B. (2009). Albert Rex Bergstrom's contributions to continuous time macroeconometric modeling. *Econometric Theory, 25*, 1087–1098. https://doi.org/10.1017/S0266466608090427

Oud, J. H. L., & Singer, H. (2008). Continuous time modelling of panel data: SEM versus filter techniques. *Statistica Neerlandica, 62*, 4–28. https://doi.org/10.1111/j.1467-9574.2007.00376.x

Phillips, P. C. B. (1972). The structural estimation of a stochastic differential equation system. *Econometrica, 40*, 1021–1041. https://doi.org/10.2307/1913853

Phillips, P. C. B. (1973). The problem of identification in finite parameter continuous time models. *Journal of Econometrics, 1*, 351–362. https://doi.org/10.1016/0304-4076(73)90021-3

Phillips, P. C. B. (1974). *The treatment of flow data in the estimation of continuous time systems*. Paper presented at the European Meeting of the Econometric Society, Grenoble (Available as Discussion Paper No. 60, University of Essex, Department of Economics, 1975. An abridged version was published as Chapter 15 in A. R. Bergstrom, A. J. L. Catt, M. H. Peston, & B. D. J. Silverstone (Eds.), Stability and Inflation: Essays in Memory of A. W. H. Phillips, pp. 257–274. John Wiley and Sons, New York, 1978)

Phillips, P. C. B. (1987). Time series regression with a unit root. *Econometrica, 55*, 277–301. https://doi.org/10.2307/1913237

Phillips, P. C. B. (1991). Error correction and long-run equilibrium in continuous time. *Econometrica, 59*, 967–980. https://doi.org/10.2307/2938169

Phillips, P. C. B., & Yu, J. (2005). Jackknifing bond prices. *Review of Financial Studies, 18*, 707–742. https://doi.org/10.1093/rfs/hhi018

Phillips, P. C. B., & Yu, J. (2009). Maximum likelihood and Gaussian estimation of continuous time models in finance. In T. G. Anderson, R. A. Davis, J. P. Kreiss, & T. Mikosch (Eds.), *Handbook of financial time series* (pp. 707–742). Berlin: Springer. https://doi.org/10.1007/978-3-540-71297-822

Phillips, P. C. B., & Yu, J. (2011). Corrigendum to "A Gaussian approach for continuous time models of the short term interest rate". *Econometrics Journal, 14*, 126–129. https://doi.org/10.1111/j.1368-423X.2010.00326.x

Renault, E., Sekkat, K., & Szafarz, A. (1998). Testing for spurious causality in exchange rates. *Journal of Empirical Finance, 5*, 47–66. https://doi.org/10.1016/S0927-5398(96)00017-5

Robinson, P. M. (1976). Fourier estimation of continuous time models. In A. R. Bergstrom (Ed.), *Statistical inference in continuous time economic models* (pp. 215–266). Amsterdam: North-Holland.
Robinson, P. M. (1992). Review of A. R. Bergstrom's "Continuous time econometric modelling". *Econometric Theory, 8*, 571–579. https://doi.org/10.1017/S0266466600013220
Robinson, P. M. (1993). Continuous-time models in econometrics: Closed and open systems, stocks and flows. In P. C. B. Phillips (Ed.), *Models, methods, and applications of econometrics: Essays in honor of A. R. Bergstrom* (pp. 71–90). Oxford: Blackwell.
Rozanov, Y. A. (1967). *Stationary random processes*. San Francisco: Holden-Day.
Singer, H. (1995). Analytical score function for irregularly sampled continuous time stochastic processes with control variables and missing values. *Econometric Theory, 11*, 721–735. https://doi.org/10.1017/S0266466600009701
Singer, H. (2007). Stochastic differential equation models with sampled data. In K. van Montfort, J. H. L. Oud, & A. Satorra (Eds.), *Longitudinal models in the behavioural and related sciences* (pp. 73–106). New Jersey: Lawrence Erlbaum Associates.
Stock, J. H. (1987). Temporal aggregation and structural inference in macroeconomics. In *Carnegie-Rochester conference series on public policy* (Vol. 26, pp. 131–140). https://doi.org/10.1016/0167-2231(87)90023-6
Thornton, M. A. (2009). *Information and aggregation: The econometrics of dynamic models of consumption under cross-sectional and temporal aggregation*. Unpublished doctoral dissertation, University of Essex, Colchester.
Thornton, M. A., & Chambers, M. J. (2013). Temporal aggregation in macroeconomics. In N. Hashimzade & M. A. Thornton (Eds.), *Handbook of research methods and applications in empirical macroeconomics* (pp. 289–309). Cheltenham: Edward Elgar. https://doi.org/0.4337/9780857931023.00019
Thornton, M. A., & Chambers, M. J. (2016). The exact discretisation of CARMA models with applications in finance. *Journal of Empirical Finance, 38*, 739–761. https://doi.org/10.1016/j.jempfin.2016.03.006
Thornton, M. A., & Chambers, M. J. (2017). Continuous time ARMA processes: Discrete time representation and likelihood evaluation. *Journal of Economic Dynamics and Control, 79*, 48–65. https://doi.org/10.1016/j.jedc.2017.03.012
Van Loan, C. F. (1978). Computing integrals involving the matrix exponential. *IEEE Transactions on Automatic Control, 23*, 395–404. https://doi.org/10.1109/TAC.1978.1101743
Wang, X., Phillips, P. C. B., & Yu, J. (2011). Bias in estimating multivariate and univariate diffusions. *Journal of Econometrics, 161*, 228–245. https://doi.org/10.1016/j.jeconom.2010.12.006
Wymer, C. R. (1972). Econometric estimation of stochastic differential equations. *Econometrica, 40*, 565–577. https://doi.org/10.2307/1913185
Wymer, C. R. (1976). *Continuous time models in macroeconomics: Specification and estimation*. Paper prepared for the SSRC Ford Foundation Conference on "Macroeconomic policy and adjustment in open economies". Ware: Fanhams Hall. (Published as Section I of Chapter 3 in G. Gandolfo (Ed.), Continuous time econometrics: Theory and applications, pp. 35–59, Chapman and Hall, London, 1993.)
Wymer, C. R. (1993). Estimation of nonlinear continuous-time models from discrete data. In P. C. B. Phillips (Ed.), *Models, methods, and applications of econometrics: Essays in honor of A. R. Bergstrom* (pp. 91–114). Oxford: Blackwell.
Wymer, C. R. (1996). The role of continuous time disequilibrium models in macroeconomics. In W. A. Barnett, G. Gandolfo, & C. Hillinger (Eds.), *Dynamic disequilibrium modeling* (pp. 67–123). Cambridge: Cambridge University Press.
Wymer, C. R. (1997). Structural non-linear continuous-time models in econometrics. *Macroeconomic Dynamics, 1*, 518–548. https://doi.org/10.1017/S1365100597003106
Wymer, C. R. (2012). Continuous time econometrics of structural models. *Studies in Nonlinear Dynamics and Econometrics, 16*, 1–28. https://doi.org/10.1515/1558-3708.1936

Yu, J. (2012). Bias in the estimation of the mean reversion parameter in continuous time models. *Journal of Econometrics, 169*, 114–122. https://doi.org/10.1016/j.jeconom.2012.01.004

Yu, J. (2014). Econometric analysis of continuous time models: A survey of Peter Phillips's work and some new results. *Econometric Theory, 30*, 737–774. https://doi.org/10.1017/S0266466613000467

Yu, J., & Phillips, P. C. B. (2001). A Gaussian approach for continuous time models of the short-term interest rate. *Econometrics Journal, 4*, 221–225. https://doi.org/10.1111/1368-423X.00063

Zadrozny, P. (1988). Gaussian likelihood of continuous-time ARMAX models when data are stocks and flows at different frequencies. *Econometric Theory, 4*, 108–124. https://doi.org/10.1017/S0266466600011890

Chapter 15
Implementation of Multivariate Continuous-Time ARMA Models

Helgi Tómasson

15.1 Introduction

Many natural phenomena are processes that evolve over time. Treating observations of such processes as observations of a random sample is therefore usually misleading. The concept of covariance between two variables is more complex in a time-series context because the two variables are actually a set of variables defined at each time point. The covariance between two time series is therefore much more than just a number; it is a function. The aim of time-series analysis is to analyse the dynamics of the underlying data generating process. Multivariate analysis is a formal tool for capturing dependency between variables. Multivariate time-series analysis is the simultaneous analysis of dynamics and the relations between variables.

There is an abundance of time-series textbooks for the discrete equispaced case (e.g. Box and Jenkins 1976; Brockwell and Davis 1991). For the multivariate case, there are also available textbooks (e.g. Lütkepohl 1991). The modern financial literature for continuous-time treats the mathematics of continuous-time stochastic dynamics (e.g. Björk 1998). Øksendal (1998) gives a rigorous treatment of the mathematics of stochastic differential equations. In recent years, treatment of stationary continuous-time ARMA models has become popular (e.g. Brockwell and Marquardt 2005; Garcia et al. 2011). Many of those papers have a high mathematical focus. For a summary see, e.g., Schlemm and Stelzer (2012), Fasen and Fuchs (2013), Brockwell (2014) and Kawai (2015). For a more computational approach, see, e.g., Singer (1992, 1995), Iacus and Mercuri (2015) and Tómasson (2015).

H. Tómasson (✉)
Faculty of Economics, University of Iceland, Reykjavík, Iceland
e-mail: helgito@hi.is

K. van Montfort et al. (eds.), *Continuous Time Modeling in the Behavioral and Related Sciences*, https://doi.org/10.1007/978-3-319-77219-6_15

The CARMA models are a natural parameterisation of a continuous-time version of the discrete-time ARMA models which are the main subject of many textbooks on time series. The mathematical dynamics in discrete time is based on difference equations and in continuous time the corresponding mathematics is based on differential equations. The discrete-time ARMA and the continuous-time CARMA models are linear difference/differential equations for describing stochastic dynamics. Linear differential equations of higher order can be written in state-space form as multivariate differential equations of order one.

Modelling in continuous time is in a sense closer to reality if the underlying variables evolve continuously in time. The parameterisation in continuous time is not a function of the sampling frequency as is the case in discrete-time time-series analysis. Missing observations are not an issue, and irregularly spaced observations are treated in an objective way. At each observation time point, the model gives an expected value and a prediction variance. The observed value can be compared with its expected value and a scaled deviation calculated. Oud and Voelkle (2014) give some illustrations.

This chapter summarises briefly the underlying mathematics for the setup of a multivariate time-series model. The approach is based on the idea of dynamic system given by a linear differential equation. In this chapter, only stationary models are considered. The contribution of the chapter is to derive, by use of non-rigorous intuitive mathematics, the main features of multivariate CARMA models from basic calculus, linear algebra and Fourier transforms, shown in, e.g., Kreyszig (1999), as well as elementary regression theory as shown, e.g., in Greene (2012). The approach is easily implemented into statistical software.

The data is supposed to be points on solution paths of stochastic differential equations. A basic knowledge of ordinary differential equations, linear algebra, rules on mean and variance of multivariate vectors and a brief encounter with Fourier theory is sufficient for following the derivations. Expressing an ordinary linear differential equation in state-space form is a useful tool which is briefly reviewed. For understanding a multivariate CARMA model, some understanding of the univariate CARMA is necessary. Therefore a brief review of the univariate case is given. Some aspects of interpretation of CARMA and MCARMA models are discussed. The computations are implemented in R packages, *ctarmaRcpp* for the univariate case and *mctarmaRcpp* for the multivariate case. The estimation is based on numerically maximising the log-likelihood. The log-likelihood function is calculated by use of the Kalman-filter recursions (see, e.g., Harvey 1989). As an illustration, a simple dependency structure in two dimensions based on simulated data is given. A real data set of non-synchronous measurements of temperature and CO_2 for the past 800,000 years is briefly analysed. Spectral theory concepts such as coherence and phase offer an interpretation of lag structure and correlation.

15.2 Preliminaries

The conditional expected value and the corresponding conditional variances are useful concepts in prediction. If the future is denoted by the vector $\boldsymbol{U}$ and the past by the vector $\boldsymbol{V}$, and the respective expected values and covariance matrices, by $E(\boldsymbol{U}) = \boldsymbol{\mu}_U$, $E(\boldsymbol{V}) = \boldsymbol{\mu}_V$, $V(\boldsymbol{U}) = \Sigma_U$, $V(\boldsymbol{V}) = \Sigma_V$ and $Cov(\boldsymbol{U}, \boldsymbol{V}) = \Sigma_{UV}$. Expressions for the conditional expectation and variance are given in textbooks (e.g. Greene 2012, formulas B-102a and B-102b, p. 1082):

$$E(\boldsymbol{U}|\boldsymbol{V}) = \boldsymbol{\mu}_U + \Sigma_{UV}\Sigma_V^{-1}(\boldsymbol{V} - \boldsymbol{\mu}_V) \tag{15.1}$$

$$V(\boldsymbol{U}|\boldsymbol{V}) = \Sigma_U - \Sigma_{UV}\Sigma_V^{-1}\Sigma_{UV}'. \tag{15.2}$$

The expression $\Sigma_{UV}\Sigma_V^{-1}$ is the regression coefficient of the future, $\boldsymbol{U}$, onto past, $\boldsymbol{V}$. The conditional expected value is the optimal predictor, and the variance in expression (15.2) is the mean square error of prediction.

15.2.1 State-Space Representations of a Differential Equation

The state-space idea is often a convenient way of expressing a higher-order linear dynamic system. The high-order dynamic system is transformed into a high-dimensional first-order linear differential equation. For example, a second-order differential equation

$$y'' + a_1 y' + a_2 y = 0$$

can be written as a first-order two-dimensional system

$$z' = Az, \quad z = \begin{bmatrix} y \\ y' \end{bmatrix}, \quad A = \begin{bmatrix} 0 & 1 \\ -a_2 & -a_1 \end{bmatrix},$$

which is perhaps the best known state-space representation of a differential equation. The literature of time-series analysis in discrete time contains many different representations of a given system. There are many possible state-space representations of a differential equation. An example of an equivalent representation of the system above is

$$z' = Az, \quad z = \begin{bmatrix} y \\ a_1 y + y' \end{bmatrix}, \quad A = \begin{bmatrix} -a_1 & 1 \\ -a_2 & 0 \end{bmatrix}.$$

The idea is easily generalised to higher dimensions. Given a two-dimensional differential equation of the form:

$$y_1'' + a_{11}y_1' + a_{12}y_1 = 0$$
$$y_2'' + a_{21}y_2' + a_{22}y_2 = 0$$

Define:

$$z = \begin{bmatrix} y_1 \\ y_2 \\ a_{11}y_1 + y_1' \\ a_{21}y_2 + y_2' \end{bmatrix}, \quad A = \begin{bmatrix} -a_{11} & 0 & 1 & 0 \\ 0 & -a_{21} & 0 & 1 \\ -a_{12} & 0 & 0 & 0 \\ 0 & -a_{22} & 0 & 0 \end{bmatrix} \tag{15.3}$$

A possible state-space representation is given by

$$z' - Az = 0.$$

Given an initial vector, $z(t_0)$, the solution of a multivariate, nonhomogeneous, first-order linear differential equation:

$$z' = Az + \boldsymbol{g}(t),$$

is given by

$$z(t) = \boldsymbol{e}^{A(t-t_0)}z(t_0) + \int_{t_0}^{t} \boldsymbol{e}^{A(t-u)}\boldsymbol{g}(u)du, \tag{15.4}$$

where $\boldsymbol{e}$ is the matrix-exponential function. This approach is the key idea for the computational setup of the computational framework of this chapter.

15.2.2 *The Univariate* CARMA(p, q)

The underlying concept of the CARMA model is that the dynamics are based on a linear differential equation. A common notation is as follows, starting with a linear p-th-order differential equation:

$$y^p(t) + \alpha_1 y^{p-1} + \cdots + \alpha_{p-1}y'(t) + \alpha_p y(t) = 0, \tag{15.5}$$

or in polynomial form,

$$\boldsymbol{\alpha}(D)y(t) = 0$$

with $Dy(t) = y'(t)$, $D^{(p)}y(t) = y^{(p)}(t)$ denotes the p-th derivative, and

$$\boldsymbol{\alpha}(z) = z^p + \alpha_1 z^{p-1} + \cdots + \alpha_{p-1}z + \alpha_p.$$

A stochastic version of (15.5) is implemented by replacing the zero on the right-hand side of (15.5) with a stochastic concept. A standard approach is to write:

$$\begin{aligned} \boldsymbol{\alpha}(D)y(t) &= \boldsymbol{\beta}(D)DW(t), \\ \boldsymbol{\beta}(z) &= 1 + \beta_1 z + \cdots + \beta_q z^q. \end{aligned} \tag{15.6}$$

Here $W(t)$ denotes the Wiener process, and the symbol $DW(t)$ is named white noise $(d/dtW(t))$. The model defined by (15.6) is the Gaussian CARMA(p, q) model. The discrete-time ARMA model is:

$$Y_t = \phi_1 Y_{t-1} + \cdots + \phi_p Y_{t-p} + \varepsilon_t - \theta_1 \varepsilon_{t-1} - \cdots - \theta_q \varepsilon_{t-q}, \tag{15.7}$$

or in polynomial form:

$$\begin{aligned} \boldsymbol{\Phi}(B)Y_t &= \boldsymbol{\Theta}(B)\varepsilon_t, \quad BY_t = Y_{t-1}, \\ \boldsymbol{\Phi}(z) &= 1 - \phi_1 z - \cdots - \phi_p z^p, \\ \boldsymbol{\Theta}(z) &= 1 - \theta_1 z \cdots - \theta_q z^q, \end{aligned}$$

where Y_t is the observed process and ε_t is the unobserved innovation process. Among typical assumptions about the distribution is that the ε_t's are independent, identically distributed, zero-mean normal or a white noise with finite variance.

Equations (15.6) and (15.7) look similar, and that similarity is reflected in the spectral density functions:

$$f_c(\omega) = \frac{\sigma^2}{2\pi} \frac{\boldsymbol{\beta}(i\omega)\boldsymbol{\beta}(-i\omega)}{\boldsymbol{\alpha}(i\omega)\boldsymbol{\alpha}(-i\omega)},$$

in the continuous-time case, and in the discrete-time case:

$$f_d(\omega) = \frac{\sigma^2}{2\pi} \frac{\boldsymbol{\Theta}(\exp(i\omega))\boldsymbol{\Theta}(\exp(-i\omega))}{\boldsymbol{\Phi}(\exp(i\omega))\boldsymbol{\Phi}(\exp(-i\omega))}.$$

The overall variance in the stationary CARMA(p, q) process is the integrated spectrum,

$$V(Y(t)) = \int_{-\infty}^{\infty} f_c(\omega) \mathrm{d}\omega,$$

and a similar formula holds for the stationary ARMA(p, q). Graphical inspection of the spectral functions gives an idea of the range of frequencies which are the most important contributors to the variance of $Y(t)$. A virtue of fitting ARMA models to time-series data is that the spectral function is a rational function and therefore knowledge of the parameters gives an easy way to calculate the spectrum.

The introduction of the polynomials $\boldsymbol{\alpha}$ and $\boldsymbol{\beta}$, in the continuous-time case, and the polynomials $\boldsymbol{\Phi}$ and $\boldsymbol{\Theta}$, in the discrete-time case, serves the purpose of defining the dynamic nature of the process. Just as in the case of deterministic differential equation, the roots of the polynomials describe characteristic features. If the roots of $\boldsymbol{\alpha}$ are complex, then that represents a particular cyclical structure. A real-data example of the application of CARMA models for describing the cyclical nature of an electricity market is given by Garcia et al. (2011).

15.3 The Multivariate CARMA Model

In order to understand a complicated model, it is helpful to analyse a simple special case. The continuous-time version of the AR(1) model is the CAR(1). The simplest differential equation is the first-order linear case. The stochastic counterpart is the Ornstein-Uhlenbeck:

$$dY(t) = -\alpha Y(t)dt + \sigma dW(t), \quad \alpha > 0, \quad \sigma > 0.$$

Conditional on a starting value, $y(t_0)$, the solution is:

$$Y(t) = \exp(-\alpha(t - t_0))y(t_0) + \sigma \int_{t_0}^{t} \exp(-\alpha(t - s))dW(s),$$

$$E(Y(t)|Y(t_0) = y(t_0)) = \exp(-\alpha(t - t_0))y(t_0),$$

$$V(Y(t)|Y(t_0) = y(t_0)) = \frac{-\sigma^2}{2\alpha}(\exp(-2\alpha(t - t_0)) - 1).$$

Here $W(t)$ is assumed to be a standard Wiener process, but the only feature of the Wiener process used in this intuitive example is the property of independent increments, and $E(W(0)) = 0$, and $V(W(t)) = t$. That is, normality, and continuity, of sample paths is not used. Stationarity implies that $\alpha > 0$, and α is interpreted as the speed of reversion to the unconditional mean, 0. If the process is far from equilibrium, i.e. $|Y(t_0)|$ is large, then a large step towards equilibrium is expected. If the value of α is small, it will take a long time for the process to approach its overall mean.

When considering two processes, $Y_1(t)$ and $Y_2(t)$, it is possible to suspect that:

$$dY_1(t) = \alpha_{11}Y_1(t)dt + \alpha_{12}Y_2(t)dt + \sigma dW(t), \quad \alpha_{11} < 0,$$

i.e. the value of $Y_2(t)$ affects the speed of $Y_1(t)$'s reversion to its mean (zero). In this case, one could say that $Y_2(t)$ has a causal impact on $Y_1(t)$. A causal formulation of this type can be in both directions. A two-dimensional Ornstein-Uhlenbeck can be written as:

$$\begin{aligned} dY_1(t) &= \alpha_{11}Y_1(t)dt + \alpha_{12}Y_2t + \sigma_1 dW_1(t), \\ dY_2(t) &= \alpha_{21}Y_1(t)dt + \alpha_{22}Y_2t + \sigma_2 dW_2(t), \\ E(dW_1(t)dW_2(t)) &= \rho\sigma_1\sigma_2 dt. \end{aligned} \tag{15.8}$$

Equation system (15.8) is more conveniently written in matrix form:

$$dY(t) = AY(t)dt + dW(t), \quad E(dW(t)dW(t)') = \Sigma dt,$$
$$\Sigma = \begin{bmatrix} \sigma_1^2 & \rho\sigma_1\sigma_2 \\ \rho\sigma_1\sigma_2 & \sigma_2^2 \end{bmatrix}, \quad A = \begin{bmatrix} \alpha_{11} & \alpha_{12} \\ \alpha_{21} & \alpha_{22} \end{bmatrix}.$$

The dependency structure, the autocorrelation, in a two-dimensional CAR(1) depends on the causal parameters, α_{12} and α_{21}, and the correlation, ρ, between the innovations. A generalisation of the univariate CARMA model to a multivariate CARMA can be written formally as:

$$\begin{aligned} \mathbf{y}^{(p)}(t) &= A_1\mathbf{y}^{(p-1)}(t) + \cdots + A_p\mathbf{y}(t) + d\mathbf{W}(t) + B_1 d\mathbf{W}^{(2)}(t) + \cdots \\ &\quad + B_q d\mathbf{W}^{(q+1)}(t), \; V(d\mathbf{W}(t)d\mathbf{W}(t)') = \Sigma dt, \end{aligned}$$

with:

$$A_i = \begin{bmatrix} \alpha_{11,i} & \alpha_{12,i} & \cdots & \alpha_{1d,i} \\ \alpha_{21,i} & \alpha_{22,i} & \cdots & \alpha_{2d,i} \\ \vdots & \vdots & \ddots & \vdots \\ \alpha_{d1,i} & \cdots & \cdots & \alpha_{dd,i} \end{bmatrix}, \quad i = 1, \ldots, p,$$

$$B_j = \begin{bmatrix} \beta_{11,j} & \cdots & \cdots & \beta_{1d,j} \\ \beta_{21,j} & \cdots & \ddots & \vdots \\ \vdots & \vdots & \ddots & \vdots \\ \beta_{d1,j} & \cdots & \cdots & \beta_{dd,j} \end{bmatrix}, \quad j = 1, \ldots, q.$$

The parameters are $p + q$ $d \times d$ matrices, the A_i's and B_j's, and $(d(d+1)/2)$ parameters in the covariance matrix of the innovations Σ. The number of parameters that describe the covariance structure in the system is thus $pd^2 + qd^2 + d(d+1)/2$. The use of polynomials in the differential operator is less tractable in the multivariate case. The derivation of second moment properties such as spectrum and autocorrelation function is easier using a state-space representation. State-space representations are also useful for getting some intuition of interpretation of the parameters, the AR parameters, A_i; the moving-average (MA) parameters, B_i; and $\mathbf{\Sigma}$ the covariance matrix of the innovations. For the simple Ornstein-Uhlenbeck, CARMA(1,0), the interpretation is easy, the parameter α describes the force towards equilibrium (the mean) and σ is the standard deviation of the innovation process. In higher-order CARMA(p, q), the AR parameters play the same role as the coefficient in a differential equation. Direct interpretation is non-intuitive, but the roots of the characteristic polynomial will describe cyclical properties of the solution pattern. The MA parameters can be interpreted as the impact of the innovation on the derivatives of the solution pattern. Again values of the parameters are not easily interpreted. Each CARMA(p, q) is nested (has exactly the same auto-covariance and spectrum) in a CARMA$(p+1, q+1)$. The latter might have very different parameter values. The multidimensional case is illustrated in Eq. (15.8). From (15.8), it is seen that an off-diagonal element of A_1 will capture the impact of a level in one coordinate on change in another coordinate. Similarly the off-diagonal element in the MA parameters will describe the impact of an innovation in one coordinate on a derivative of another. Therefore the solution pattern of a true CARMA(p, q) will be less smooth than the solution pattern of a pure AR model. The off-diagonal elements of $\mathbf{\Sigma}$ are the covariances of the innovations. For $\mathbf{\Sigma}$ two special cases are of interest, that is, off-diagonal elements zero and off-diagonal elements representing a correlation coefficient 1 between elements of the innovation vector. If the correlation of some of the coordinates in the innovation vector is 1, the matrix $\mathbf{\Sigma}$ is not of full rank, i.e. the observation vector is of higher dimension than the innovation vector. Such a case could be termed a dynamic factor model.

15.3.1 Scaling Issues

In discrete time, the scaling of the time is given. The time points are equispaced and the time variable is a sequence of integers. The sampling frequency will affect the values of the ARMA parameters in discrete time. In continuous time, the sampling frequency does not affect the parameter values of the ARMA representation, but the definition of the time will. That is, when time is transformed from, say, seconds to minutes, the parameter values of the ARMA representation will be affected. The impact of transformations of the time scale is best understood by studying the spectral density. The spectral density of a particular CARMA process is:

$$f(\omega) = \frac{\sigma^2}{2\pi} \frac{\boldsymbol{\beta}(i\omega)\boldsymbol{\beta}(-i\omega)}{\boldsymbol{\alpha}(i\omega)\boldsymbol{\alpha}(-i\omega)} d\omega. \tag{15.9}$$

Table 15.1 Impact of scaling of time on CARMA parameters

	α_1	α_2	β_1	σ
Original time-scale	2	40	0.15	8
Time multiplied by 10	0.2	0.4	1.5	0.253
Time multiplied by 0.1	20	4000	0.015	252.98

The units of ω are radians per time unit. If the time scale is multiplied by a constant c, i.e. $\omega_* = c\,\omega$, then the spectral density of the time-transformed process will be

$$f(\omega_*) = \frac{{\sigma_*}^2}{2\pi} \frac{\boldsymbol{\beta}_*(i\omega_*)\boldsymbol{\beta}_*(-i\omega_*)}{\boldsymbol{\alpha}_*(i\omega_*)\boldsymbol{\alpha}_*(-i\omega_*)} d\omega_*. \tag{15.10}$$

The vectors $\boldsymbol{\alpha}_*$ and $\boldsymbol{\beta}_*$ in Eq. (15.10) are derived by solving for the corresponding powers of ω in (15.9). Solving for the $(\beta_j)_*$'s is straightforward, $(\beta_j)_* = c^j \beta_j$. The $(\alpha_j)_*$'s have to be scaled such that the coefficient of the highest power of the polynomial in the denominator is one, i.e. $(\alpha_j)_* = c^{-j}\alpha_j$. Then $\sigma* = c^{-(p-1/2)}\sigma$. The term $-1/2$ in the scaling transform of σ is due to the Jacobian of the transform. An example of the impact of scaling of a simple CARMA(2,1) model is shown in Table 15.1. In numerical work, a proper scaling of the time axis can be helpful.

In the univariate model, scaling of the observed variable will affect the numerical value of σ. Scaling of the observations can influence the numerical manageability of σ. The multivariate model contains the covariance matrix of the innovations. Numerical improvement can be obtained by scaling the observed variables such that the innovation standard deviations of each dimension are of similar magnitude.

15.4 Linear Stochastic Differential Equations

Suppose that the dynamics of a $pd \times 1$ state vector, $\boldsymbol{X}(t)$, are given by:

$$\begin{aligned} d\boldsymbol{X}(t) &= \boldsymbol{A}\boldsymbol{X}(t)dt + \boldsymbol{R}d\boldsymbol{W}(t), \\ E(d\boldsymbol{W}(t)d\boldsymbol{W}'(t)) &= \boldsymbol{\Sigma}dt. \end{aligned} \tag{15.11}$$

The first part of this equation corresponds to the deterministic differential equation referred to in (15.4). The second part, $\boldsymbol{R}d\boldsymbol{W}(t)$, is a residual stochastic component. In this chapter, $\boldsymbol{W}(t)$ is supposed to be a multivariate Wiener process, but the arguments apply to all uncorrelated increment processes with $V(d\boldsymbol{W}(t)) = \boldsymbol{\Sigma}dt$.

Conditional on the value at time s, $\boldsymbol{X}(s)$, the solution (compare with (15.4)) of the stochastic differential equation (15.11) is:

$$\begin{aligned} \boldsymbol{X}(t) &= \boldsymbol{e}^{\boldsymbol{A}(t-s)}\boldsymbol{X}(s) + \int_s^t \boldsymbol{e}^{\boldsymbol{A}(t-u)}\boldsymbol{R}d\boldsymbol{W}(u), \quad \boldsymbol{e} \text{ matrix exponent,} \\ E(\boldsymbol{X}(t)|\boldsymbol{X}(s)) &= \boldsymbol{e}^{\boldsymbol{A}(t-s)}\boldsymbol{X}(s), \quad t > s. \end{aligned} \tag{15.12}$$

Equation (15.12) is a regression. The regression coefficient (see (15.1)) is equal to:

$$Cov(\boldsymbol{X}(t), \boldsymbol{X}(s))(V(\boldsymbol{X}(s))^{-1} = \boldsymbol{\Gamma}_X(t-s)\boldsymbol{\Gamma}_X(0)^{-1}.$$

It is therefore clear that the auto-covariance function of $\boldsymbol{X}(t)$ is $\boldsymbol{\Gamma}_X(\tau) = \boldsymbol{e}^{\boldsymbol{A}\tau}\boldsymbol{\Gamma}_X(0)$, where $\tau = t - s$ and $\boldsymbol{\Gamma}_X(0)$ is the stationary covariance matrix of $\boldsymbol{X}(t)$. Basic rules on variance give:

$$V(\boldsymbol{X}(t)) = \boldsymbol{e}^{\boldsymbol{A}(t-s)}V(\boldsymbol{X}(s))\boldsymbol{e}^{\boldsymbol{A}(t-s)'} + \underbrace{V\left(\int_s^t \boldsymbol{e}^{\boldsymbol{A}(t-u)}\boldsymbol{R}d\boldsymbol{W}(u)\right)}_{V(\boldsymbol{X}(t)|\boldsymbol{X}(s))}$$

i.e. as is also clear from (15.2),

$$V(\boldsymbol{X}(t)|\boldsymbol{X}(s)) = \boldsymbol{\Gamma}_X(0) - \boldsymbol{e}^{\boldsymbol{A}(t-s)}\boldsymbol{\Gamma}_X(0)\boldsymbol{e}^{\boldsymbol{A}(t-s)'}.$$

The unconditional stationary variance, $\boldsymbol{\Gamma}_X(0)$, solves the equation system:

$$\boldsymbol{A}\boldsymbol{\Gamma}_X(0) + \boldsymbol{\Gamma}_X(0)\boldsymbol{A}' = -\boldsymbol{R}\boldsymbol{\Sigma}\boldsymbol{R},$$

which, by using standard results on Kronecker products, can be transformed into:

$$((I \otimes \boldsymbol{A}) + (\boldsymbol{A} \otimes I))\text{vec}(\boldsymbol{\Gamma}_X(0)) = -\text{vec}(\boldsymbol{R}\boldsymbol{\Sigma}\boldsymbol{R}'). \tag{15.13}$$

15.5 Derivation of Spectrum

For convergence reasons, it is convenient to define the Fourier transform of a stochastic process $\boldsymbol{X}(t)$ over a finite interval as:

$$\mathscr{F}_X(\omega) = \frac{1}{\sqrt{2\pi}}\frac{1}{\sqrt{2T}}\int_{-T}^{T}\boldsymbol{X}(t)\exp(-i\omega t)\mathrm{d}t$$

and the corresponding spectral density as:

$$f_X(\omega) = \lim_{T\to\infty} E(\mathscr{F}_X(\omega)\overline{\mathscr{F}_X(\omega)}),$$

$$f_X(\omega) = \frac{1}{2\pi}\int_{-\infty}^{\infty}\exp(-i\omega\tau)\Gamma_X(\tau)\mathrm{d}\tau.$$

Equation (15.11) can be rewritten as:

$$d\boldsymbol{X}(t) - \boldsymbol{A}\boldsymbol{X}(t)dt = \boldsymbol{R}d\boldsymbol{W}(t),$$

$$(I_{pd\times pd}D - \boldsymbol{A})\boldsymbol{X}(t) = \boldsymbol{R}d\boldsymbol{W}(t), \tag{15.14}$$

where D is the differential operator. Using a basic rule from calculus (Kreyszig 1999), $\mathscr{F}_{\mathrm{d}X}(\omega) = i\omega\mathscr{F}_X(\omega)$, and taking Fourier transform of (15.14) gives:

$$(i\omega I_{pd\times pd} - \boldsymbol{A})\mathscr{F}_X(\omega) = \boldsymbol{R}\mathscr{F}_{d\mathbf{W}}(\omega).$$

The spectrum can be interpreted as the variance of the Fourier transform. So, using standard results on Fourier transform of the Dirac-delta function (the auto-covariance of white noise) shows the fact that:

$$\lim_{T\to\infty} E(\mathscr{F}_{d\mathbf{W}}(\omega)\overline{\mathscr{F}_{d\mathbf{W}}(\omega)}) = \frac{\boldsymbol{\Sigma}}{2\pi},$$

and the spectrum of $\boldsymbol{X}(t)$ is therefore:

$$f_X(\omega) = (i\omega I_{pd\times pd} - \boldsymbol{A})^{-1}\frac{\boldsymbol{R}\boldsymbol{\Sigma}\boldsymbol{R}'}{2\pi}(-i\omega I_{pd\times pd} - \boldsymbol{A}')^{-1}.$$

Another way of deriving the spectrum is to use the fact that $E(\boldsymbol{X}(t+h)\boldsymbol{X}(t)') = \boldsymbol{e}^{Ah}\boldsymbol{\Gamma}_X(0)$, and $E(\boldsymbol{X}(t-h)\boldsymbol{X}(t)') = \boldsymbol{\Gamma}_X(0)\boldsymbol{e}^{Ah'}$, for $h > 0$. The Laplace transform of $\boldsymbol{e}^{At}$ is $\int_0^\infty e^{-st}\boldsymbol{e}^{At}dt = (sI_{pd\times pd} - \boldsymbol{A})^{-1}$. Using that, it is straightforward to show:

$$f_X(\omega) = \frac{1}{2\pi}\left((-i\omega I_{pd\times pd} - \boldsymbol{A})^{-1}\boldsymbol{\Gamma}_X(0) + \boldsymbol{\Gamma}_X(0)(i\omega I_{pd\times pd} - \boldsymbol{A}')^{-1}\right).$$

The diagonal of the matrix $f_X(\omega)$ contains the spectral functions of each series. The off-diagonal element, the cross-spectrum, is a measure of covariance in the frequency domain. If $f_{ij}(\omega)$ is the cross-spectrum between coordinates i and j, the coherence is given by:

$$coh_{ij}(\omega) = \frac{|f_{ij}(\omega)|^2}{f_{ii}(\omega)f_{jj}(\omega)},$$

and the phase is given by:

$$phase_{ij}(\omega) = \mathrm{atan}(-\mathrm{Im}(f_{ij}(\omega)/\mathrm{Re}(f_{ij}(\omega)))$$

where Im and Re give the imaginary and real part of a complex number, respectively. Elementary Fourier theory shows that if $X_2(t)$ is a lagged version of $X_1(t)$, i.e. $X_2(t) = X_1(t-l)$, then

$$f_{12}(\omega) = \exp(-il\omega)f_{11}(\omega).$$

It is therefore clear that if one coordinate is the lagged version of the other, the coherence is one, and the phase is $l\omega$.

15.5.1 Interpretation of the Spectrum

Interpretation of covariances and autocorrelations is in general hard. In addition, the estimators of the various AR and MA parameters will typically be very correlated, so the direct interpretation of the estimated CARMA parameters is also hard. The visual interpretation of the spectrum is much easier. In one dimension, a peak in the spectrum at a certain frequency ω_0 (radians per unit of time) is interpreted as a large proportion of variation due to cycle of length $2\pi/\omega_0$ units of time. In the multivariate case, the coherence can be interpreted as a squared correlation of frequencies, i.e. a high coherence at a particular frequency suggests that cycles of the corresponding frequency tend to stick together. The slope of the phase function suggests lagging of particular frequencies. For example, if the phase function has constant slope, one variable is largely a lagged version of another, i.e. a lag at all frequencies. The value of the slope is the lag.

15.5.2 Choice of State-Space Representation

There are many possible state-space representations of a multivariate CARMA model (see, e.g., Schlemm and Stelzer 2012). Here a state-space representation based on (15.3), described in Zadrozny (1988), is used. This representation is chosen because the untransformed MA parameters enter the dynamic state equation (15.16). The representation in Zadrozny (1988) is:

$$\boldsymbol{y}(t) = \boldsymbol{C}\boldsymbol{X}(t), \quad \boldsymbol{C} \quad pd \times d \tag{15.15}$$

$$d\boldsymbol{X}(t) = \boldsymbol{A}\boldsymbol{X}(t)dt + \boldsymbol{R}d\boldsymbol{W}, \quad \boldsymbol{C} = \begin{bmatrix} I_{d\times d} \\ 0_{d\times d} \\ \vdots \\ 0_{d\times d} \end{bmatrix} \tag{15.16}$$

$$\boldsymbol{A} = \begin{bmatrix} A_1 & I_{d\times d} & 0_{d\times d} & \cdots & \cdots & 0_{d\times d} \\ A_2 & 0_{d\times d} & I_{d\times d} & 0_{d\times d} & \cdots & 0_{d\times d} \\ \vdots & 0_{d\times d} & 0_{d\times d} & \ddots & \vdots & \vdots \\ A_{p-1} & 0_{d\times d} & 0_{d\times d} & \cdots & \cdots & I_{d\times d} \\ A_p & 0_{d\times d} & 0_{d\times d} & \cdots & \cdots & 0_{d\times d} \end{bmatrix}, \quad \boldsymbol{R} = \begin{bmatrix} B_q \\ \vdots \\ B_1 \\ I_{d\times d} \end{bmatrix}.$$

A useful property of this state-space representation is that restrictions on the MA parameters can be applied at the state-space level. An example of restrictions on the parameter space is, e.g., putting some parts of the A_i's and B_j's equal to zero.

Looking at the state-space representation of the CARMA model offers an intuitive interpretation. The AR parameters give a direct causal interpretation similar to the parameters of a deterministic differential equation. A plausible interpretation of the MA parameters is that the innovation shock affects the slope/derivative of possibly all coordinates of the system. The covariance matrix of the innovation can be interpreted as the size and connection between the various coordinates of the innovation process. In the case of estimation from data, the parameter estimates are likely to be highly correlated so it is hard to give a clear interpretation based on estimates.

The conditional expected values, and corresponding covariances, at discrete non-synchronous time points can be calculated by the Kalman-filter algorithm and repeated application of Eqs. (15.1) and (15.2). The spectrum of $\boldsymbol{y}_t$ is readily calculated as a function of the spectrum for $\boldsymbol{X}_t$:

$$f_{\boldsymbol{y}}(\omega) = \boldsymbol{C} f_{\boldsymbol{X}}(\omega)\boldsymbol{C}'.$$

The non-synchronous case is solved by letting the $\boldsymbol{C}$ matrix depend on the sampling. That is, if coordinate i is observed, the $\boldsymbol{C}$ matrix consists of nothing but zeros except for coordinate ii which is set to one. The measurement equation (15.15) with that particular $\boldsymbol{C}$ reflects that coordinate i is measured at that particular time point.

With this setup, the implementation of the Kalman filter is straightforward. Solving Eq. (15.13), for $\boldsymbol{\Gamma}_X(0)$, can be numerically problematic. In some cases, it is possible to derive explicit algorithms to calculate $\boldsymbol{\Gamma}_X(0)$ (see, e.g., Tsai and Chan 2000). In practice, it is worthwhile to observe that $\boldsymbol{A}$ and I are both sparse matrices, so sparse-matrix algorithms can be used to invert $(I \times \boldsymbol{A}) + (\boldsymbol{A} \times I)$. Implementation of the Kalman filter is sketched in Appendix 1.

15.5.3 Computer Packages for Data Analysis

In the univariate case, a data set is of the form $(\boldsymbol{y}(t_1), \ldots, \boldsymbol{y}(t_n))$. In statistics, all data analysis is seen through a particular kind of model or family of models. The univariate CARMA parameters are a set of AR and MA parameters defining the dynamic nature of the process and a parameter describing the scale of the innovations. It is practical to keep the data, the values of the observations, the timing of the observations and the model parameters together in a bundle. In programming languages such as R, this is easily accomplished by defining this bundle as an object. The calculations used in this chapter have been bundled into an R package *ctarmaRcpp* for the univariate case and *mctarmaRcpp* for the multivariate case. The multivariate version is analogous to the univariate one and designed to be logically compatible. The initial multivariate object is a set of univariate objects of type `ctarma` which are merged into a multivariate object of the type `mctarma`, where the initial parameter values in the multivariate object are such that the coordinates of the observed series are assumed uncorrelated. The interface of the R packages is briefly shown in Appendix 2.

15.6 Illustration of Multivariate CARMA Estimation

The parameter space of a stationary multivariate ARMA model is complicated. Estimation of parameters is based on numerical maximisation of the likelihood function. First each coordinate is estimated with a univariate CARMA model. Then the estimates of the univariate model are used as a starting value of the multivariate model.

15.6.1 An Example of Coherence and Phase

Consider an Ornstein-Uhlenbeck (CARMA(1,0)), $X_1(t)$:

$$dX_1 = -\alpha X_1 dt + \sigma dW,$$

and a l ($l > 0$) time-period lagged version of $X_1(t)$, $X_2(t)$, is defined as:

$$X_2(t) = X_1(t - l).$$

The auto-covariance function of X_1 and X_2 is:

$$\gamma(\tau) = \sigma^2 \frac{\exp(-2\alpha\tau)}{2\alpha},$$

and the cross-covariance:

$$\gamma_{X_1,X_2}(\tau) = \sigma^2 \frac{\exp(-2\alpha(\tau + l))}{2\alpha}, \qquad \tau > -l,$$

$$\gamma_{X_1,X_2}(\tau) = \sigma^2 \frac{\exp(-2\alpha(-\tau - l))}{2\alpha}, \qquad \tau < -l.$$

Calculation of the Fourier transform gives the spectrum:

$$f_{X_1}(\omega) = f_{X_2}(\omega) = \frac{\sigma^2}{2\pi} \frac{1}{\alpha^2 + \omega^2},$$

and the cross-spectrum:

$$f_{X_1,X_2}(\omega) = \frac{\sigma^2}{2\pi} \frac{\exp(-il\omega)}{\alpha^2 + \omega^2}.$$

Because $|f_{X_1,X_2}(\omega)| = f_{X_1}(\omega)$, it is clear that $coh(\omega) = 1$ and the phase:

$$phase(\omega) = \text{atan}\left(-\frac{\text{Im}(f_{X_1,X_2}(\omega))}{\text{Re}(f_{X_1,X_2}(\omega))}\right) = l\omega.$$

It is of interest to see how well this lag structure can be captured with a realisation of a multivariate ARMA model.

15.6.2 An OU Process and a Lagged Copy

An OU process X_1 was simulated over a period of 100 units of time. The coefficient α was chosen such that $\exp(-\alpha) = 1/2$ ($\alpha \simeq 0.69$), i.e., that half-life is one time period. The intervals between observations are i.i.d. exponential with mean 1/10, i.e. on average 10 observations per time period. The $X_2(t)$ process was then defined as the same numbers as $X_1(t-l)$, i.e. the observation time points of X_2 are one period later ($l = 1$) than X_1. The first time point where X_1 observed is $t = 0.070$ and $X_1(0.070) = -0.75$, and the first time point where X_2 is observed is $t = 1.070$ and $X_2(1.070) = -0.75$.

First a few ctarma objects containing X_1 and X_2 are created using the *ctarmaRcpp* package.

```
set.seed(12345689)
a=log(2)
sigma=1
nn=1000
tt1=cumsum(rexp(nn,1))/10
b=1

y1=carma.sim.timedomain(tt1,a,b,sigma)

y2=y1
tt2=tt1+1

my1=ctarma(ctarmalist(y1,tt1,a,b,sigma))
my2=ctarma(ctarmalist(y2,tt2,a,b,sigma))

my1e=ctarma.maxlik(my1)
my2e=ctarma.maxlik(my2)
```

Now the objects `my1e` and `my2e` contain the maximum-likelihood estimates of the CARMA(1,0) model. The log-likelihood is calculated by:

```
> ctarma.loglik(my1e)
[1] 52.61891
```

There are many ways such that the CARMA(1,0) can be embedded in a higher-order CARMA. The function `ctarma.new` creates an equivalent object with one more AR term and one more MA term.

```
> ctarma.loglik(ctarma.new(my1e))
[1] 52.61891
```

Using this feature repeatedly,

```
my2=ctarma.new(my1e)
my2e=ctarma.maxlik(my2)
my3=ctarma.new(my2e)
my3e=ctarma.maxlik(my3)
my4=ctarma.new(my3e)
my4e=ctarma.maxlik(my4)
```

higher-order CARMA models can be fitted. The estimated ARMA parameters and log-likelihood for CARMA(4,3) are:

```
> my4e$ahat
[1] 3.237786 5.369379 6.158908 1.837239
> my4e$bhat
[1] 1.0000000 1.1202398 0.7557877 0.2705627
> my4e$sigma
[1] 3.673917
> ctarma.loglik(my4e)
[1] 53.04228
```

The nature of the overfitting is illustrated in Fig. 15.1. When the true order, one AR parameter, is estimated, ($p = 1, q = 0$), the shape of the spectral curve is the expected smoothly decaying function. When a more complex model is estimated, the extra effort of the more complicated model results in a more irregular shape of the estimated spectrum at low frequencies. It is clear that the uncertainty is located at the low frequencies, i.e. we can inherently only observe few low-frequency cycles.

In the multivariate package *mctarmaRcpp*, a multivariate object is created by merging two univariate objects. For example, a multivariate CARMA(4,3) was created using two estimated univariate CARMA(4,3) objects. The estimated univariate models are nested in an initial multivariate object. That multivariate object is then used as a starting point for estimating the multivariate model.

The two univariate CARMA(4,3) `my14e` and `my24e` objects are merged into a bivariate non-synchronous CARMA(4,3) object by:

```
cc=cbind(as.list(my14e),as.list(my24e))
library(mctarmaRcpp)
mct4=mctobjnonsync(cc)
```

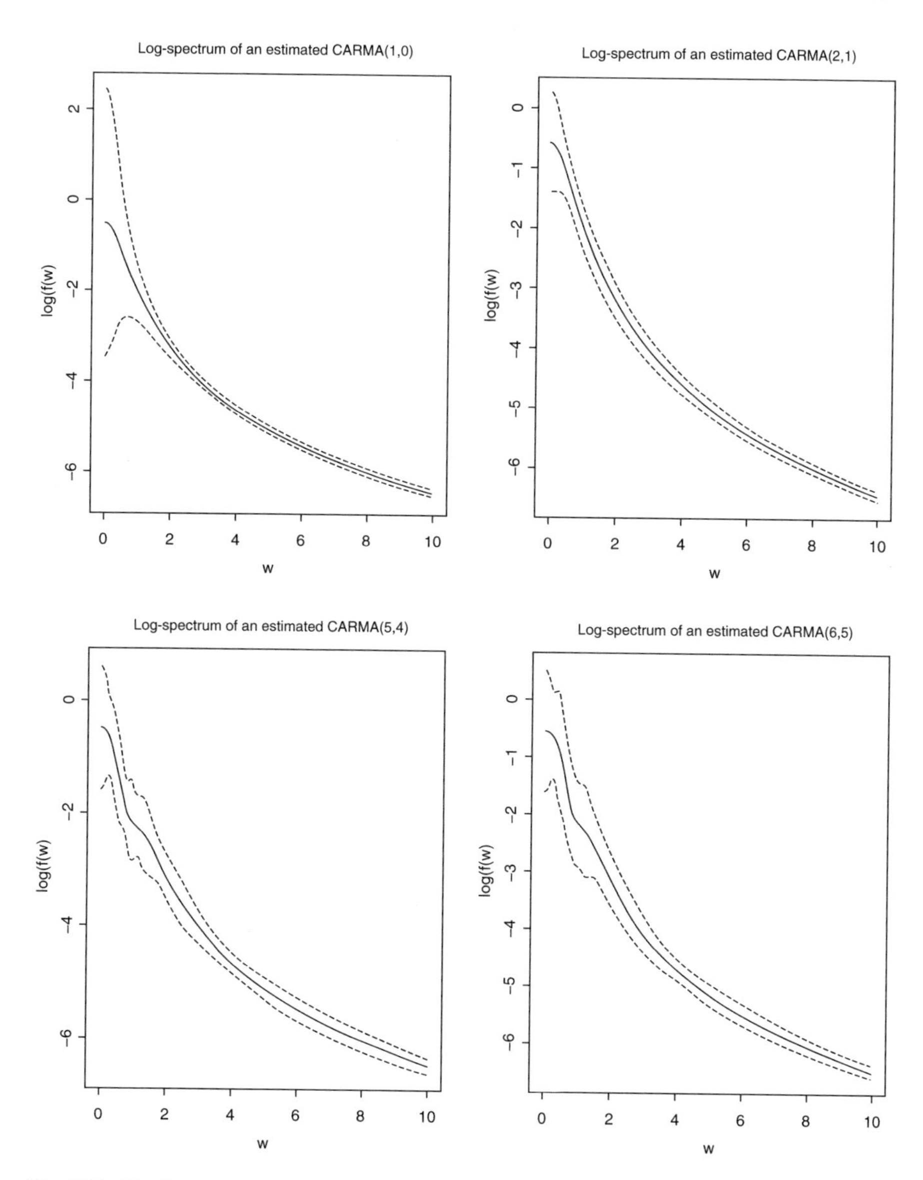

Fig. 15.1 The log-spectrum and approximate confidence intervals based on estimates of several CARMA models for a simulated Ornstein-Uhlenbeck process

The only difference between the objects, `my14e` and `my24e`, is that the time has been translated one time unit, as X_2 is just a lagged copy of X_1. This bivariate CARMA(4,3) object, `mct4`, is composed of two estimated univariate CARMA(4,3) objects. The parameterisation of this bivariate CARMA(4,3) object reflects the assumption of independence of the two coordinates (X_1, X_2). The log-likelihood value of the bivariate object is therefore the sum of the log-likelihood of univariate objects:

```
> mctarmanonsync.loglik(mct4)/2
[1] 53.04228
```

The multivariate (non-synchronous) object contains the observed series, time point of observation and which coordinate of the vector is observed:

```
> cbind(mct4e$y,mct4e$tt,mct4e$z)[13:19,]
           [,1]     [,2] [,3]
[1,] -0.9942879 1.391931    2
[2,] -1.1332209 1.536528    2
[3,]  0.6258719 1.564668    1
[4,] -0.9815641 1.564864    2
[5,] -0.6691930 1.608199    2
[6,]  0.5880817 1.610724    1
[7,]  0.6950109 1.739455    1
```

That is, the 13th row is observed at time 1.391931, has the value −0.9942879 and is an observation of X_2. The 15th row is from series X_1. The initial bivariate CARMA(4,3) object contains the matrices of the parameters: A, R and Σ. These are based on the univariate estimates. In this case, the parameters describing the dynamics of X_1 are exactly the same as those describing the dynamics of X_2, because X_2 is just a lagged version of X_1. Non-trivial parts of these matrices in the initial (assuming two uncorrelated series) value of the multivariate object are in this case:

$$A = \begin{bmatrix} -3.24 & 0 \\ 0 & -3.24 \\ -5.37 & 0 \\ 0 & -5.37 \\ -6.16 & 0 \\ 0 & -6.16 \\ -1.84 & 0 \\ 0 & -1.84 \end{bmatrix}, \quad R = \begin{bmatrix} 0.27 & 0 \\ 0 & 0.27 \\ 0.76 & 0 \\ 0 & 0.76 \\ 1.12 & 0 \\ 0 & 1.12 \\ 1 & 0 \\ 0 & 1 \end{bmatrix}, \quad \Sigma = \begin{bmatrix} 13.50 & 0.00 \\ 0.00 & 13.50 \end{bmatrix}$$

The X_1 and X_2 contain exactly the same values; the time increments are exactly the same, so that the univariate estimates are exactly the same. The covariance structure between X_1 and X_2 is then estimated by maximising the log-likelihood.

```
mct4e=mctarmanonsync.maxlik(mct4)
```

Now this new mctarma object, `mct4e`, contains the maximum-likelihood estimate of the bivariate CARMA(4,3) model. The log-likelihood values of the models assuming independence and the bivariate model are

```
> mctarmanonsync.loglik(mct4)
[1] 106.0846
> mctarmanonsync.loglik(mct4e)
[1] 633.7593
```

The number of parameters in the two univariate models is 16, and in the bivariate model, it is 31. A substantial increase in the log-likelihood by the addition of 15 parameters reflects the fact that X_1 and X_2 are highly related. The coherence function based on the estimated CARMA(4,3) is shown in Fig. 15.2. The estimated coherence is close to true value of 1 in large parts of the low-frequency range. The scale is in radian per time period, i.e. $2\pi \simeq 6.3$ is about one cycle per time unit. Figure 15.3 shows the estimated phase of the estimated CARMA(4,3). In the low frequencies, the slope of the phase curve is close to the theoretical value 1, reflecting (the fact) that $X_2(t)$ is lagged version of $X_1(t)$.

In this example, all frequencies were lagged equally, i.e. the whole $X_2(t)$ was a delayed version of $X_1(t)$. The bivariate CARMA approach is able to capture this delay (in this case) to a high degree. This is clear from studying the phase and coherence graphs. This two-dimensional system is driven by a one-dimensional innovation. One coordinate is just a lagged version of the other. This is captured by $\hat{\Sigma}$, the estimated covariance matrix of innovations. The correlation in this matrix is in this case very close to 1 (0.999999). In this case, it will not add much to estimate a model of a higher degree. If in practice such a simple relation, as a one period lag, was suspected, perhaps, a more natural way would be to estimate the lag directly.

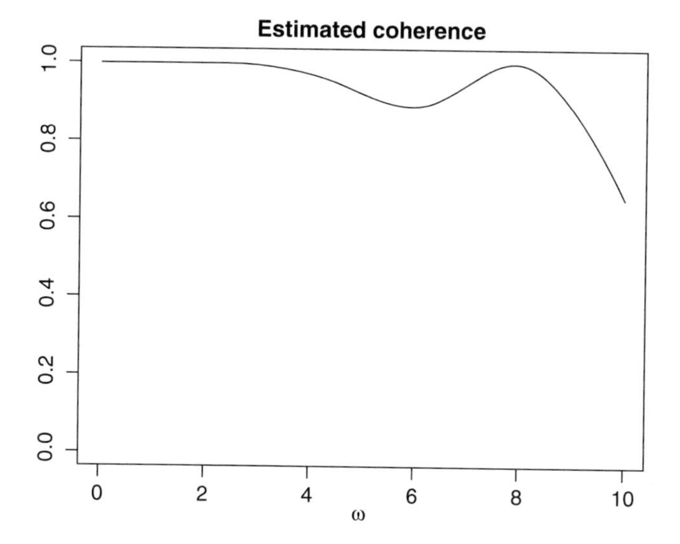

Fig. 15.2 Estimated coherence of a CARMA(1,0) and its lagged version. The estimates are based on estimates of a CARMA(4,3)

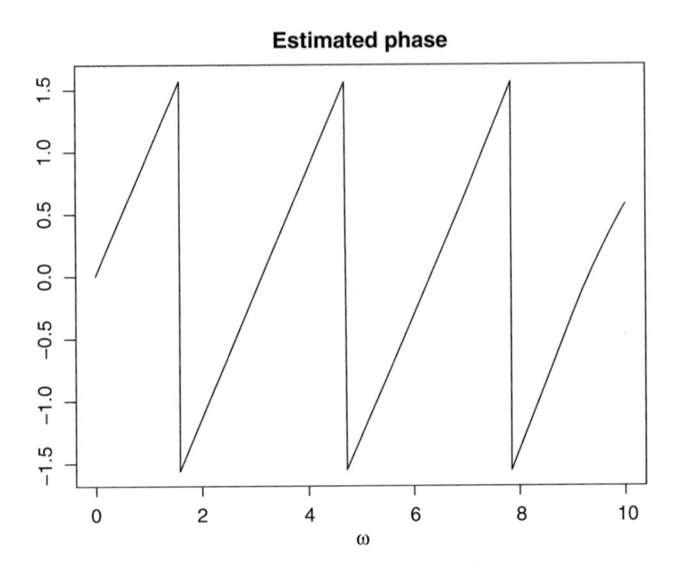

Fig. 15.3 The estimated phase of CARMA(1,0) and its lagged version. The estimates are based on estimates of a CARMA(4,3)

15.6.3 A Real-Data Example

Data for the illustration are the 800,000-year historical series of temperature and CO_2. The series are non-synchronous and unevenly spaced in time. The data are given in Jouzel et al. (2007) and Luthi et al. (2008). The temperature series consists of 5788 observations. The average time between measurements is 138 years, and the standard deviation is 173 years. For the CO_2, we have 1095 observations. The average time between observations is 730 years, and the standard deviation is 672 years. Both series were detrended with a regression line. The temperature series has virtually no trend, but there is a slight trend in the CO_2 series. The time series are shown in Fig. 15.4.

A feature of the R package *ctarmaRcpp* is that it is possible to create a *ctarma* object that contains the data, series and time of observation and the set of parameters in a univariate CARMA model. This chapter only deals with stationary models. That is, the real part of the eigenvalues of the $\boldsymbol{A}$ matrix is negative. For numerical optimisation, a valid starting value of the parameters is necessary. The complexity of the parameter space therefore requires caution in choice of starting values.

Therefore a stepwise-forward procedure is implemented in the *ctarmaRcpp* package. The stepwise-forward approach is based on the fact that the CARMA(p, q) model is nested in the CARMA$(p + 1, q + 1)$ model. So each estimated CARMA(p, q) model can be (at least) equally well fitted by a CARMA$(p+1, q+1)$ model. By using starting values in a CARMA$(p + 1, q + 1)$ model based on an existing CARMA(p, q), it is ensured that a more complicated model will always give a better fit. A stepwise-backward approach based on Belcher et al. (1994) is

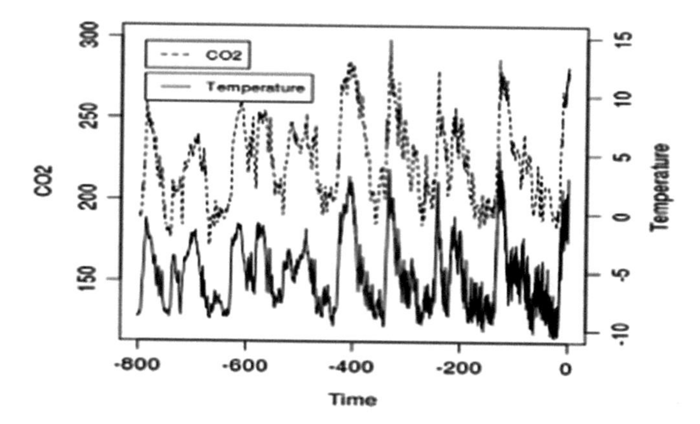

Fig. 15.4 Evolution of temperature and CO_2 in the last 800,000 years

also feasible. That approach was also implemented in an earlier version, the *ctarma* package used for calculations in Tómasson (2015).

First, the data series and observation time point are combined in a *ctarma* object, for example, in a CARMA(1,0) (Ornstein-Uhlenbeck) object with $\alpha = 1$ and $\sigma = 1$. That is, the start can be

```
library(ctarmaRcpp)
m1=ctarma(ctarmalist(y,tt,1,1,1))
```

And then the CAR(1) model is estimated:

```
m1e=ctarma.maxlik(m1)
```

Now the object `m1e` contains the original series and the maximum-likelihood estimate of the CARMA(1,0) model. Then an equivalent CARMA(2,1) model is created:

```
m2=ctarma.new(m1e)
> ctarma.loglik(m1e)
[1] -8580.44
> ctarma.loglik(m2)
[1] -8580.44
```

The object `m2` represents a CARMA(2,1) with exactly the same auto-covariance as the estimated CARMA(1,0). The `m2` object gives sensible starting values for estimation of a CARMA(2,1). The log-likelihood value based on the maximum-likelihood estimate of CARMA(1,0) is exactly the same as of this particular CARMA(2,1). Using the ctarma.maxlik function, one can calculate the maximum-likelihood estimate of CARMA(2,1).

```
m2e=ctarma.maxlik(m2)
> ctarma.loglik(m2e)
[1] -5696.489
```

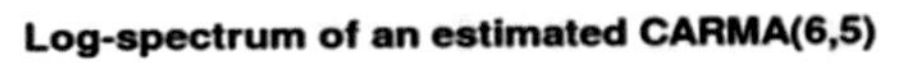

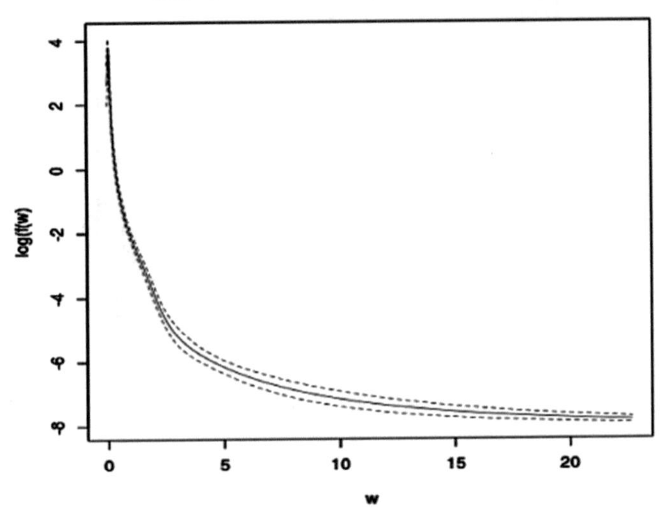

Fig. 15.5 The log spectrum of an estimated CARMA(6,5) for temperature

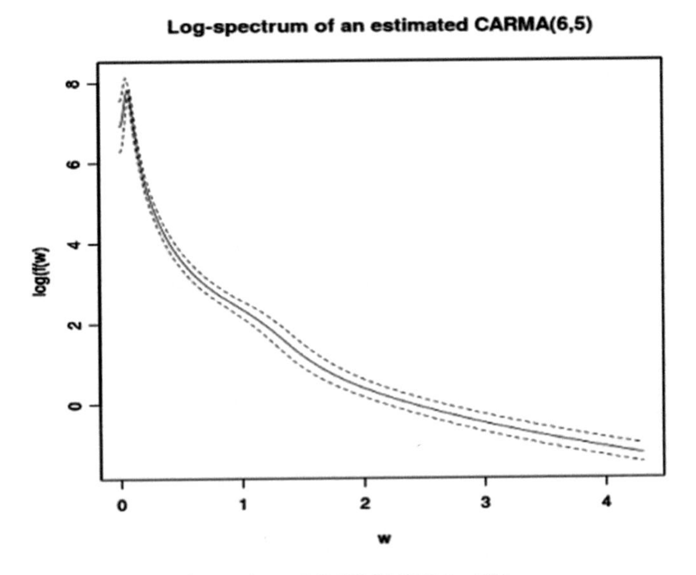

Fig. 15.6 The log spectrum of an estimated CARMA(6,5) for CO_2

Now the object `m2e` contains the maximum-likelihood estimate of the CARMA(2,1). This procedure is repeated just as in the illustration example of CARMA(1,0). The logs of the spectral density based on estimates of CARMA(6,5) models are shown in Figs. 15.5 and 15.6.

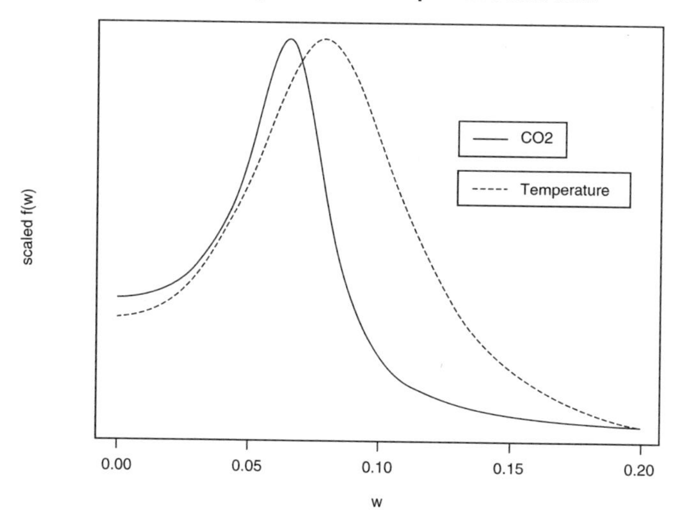

Fig. 15.7 A low-frequency view of the scaled spectrum of temperature and CO_2

The main feature of both series is the high spectrum at low frequency. Taking a closer look at the scaled estimated spectrum at low frequencies, Fig. 15.7 suggests that the variation of the CO_2 is more concentrated at lower frequencies than of the temperature. The peak of the CO_2 suggests a cycle of about 100,000 years. Similarly the figure suggests that the cycle of the temperature is about 80,000 years. Not only is the peak of the spectrum for CO_2 at a lower frequency, but a relatively larger share of the variance of the CO_2 seems due to frequencies close to zero. That might be a result of improper modelling of the trend in the CO_2. The estimated spectrum at low frequencies is sensitive to model choice, indicating a high level of uncertainty at low frequencies.

The most prominent features of the estimated multivariate model (Figs. 15.8 and 15.9) reflect those of the two univariate series. The high values of the spectrum are concentrated at the low frequencies. The coherence suggests some correlation at low frequencies, but the phase does not give a clear impression on which series is leading. Taking the slope of estimated phase at the frequency range, $0.5 < \omega < 2.5$ (cycle of 2500–12,500 years), suggests temperature lags 200 years (the slope is 0.2) based on the CARMA(4,3) and that CO_2 lags 2 (virtually zero) years based on the CARMA(6,5). At this frequency band, the coherence is low, so that the possible impact of one series on the other at those frequencies is little. The high coherence is concentrated at the very low frequencies. The estimated correlation of the innovations is 0.6 based on the CARMA(4,3) and 0.1 based on the CARMA(6,5).

The physical interpretation is that these data reveal very little information on an eventual causality in the relation between CO_2 and temperature.

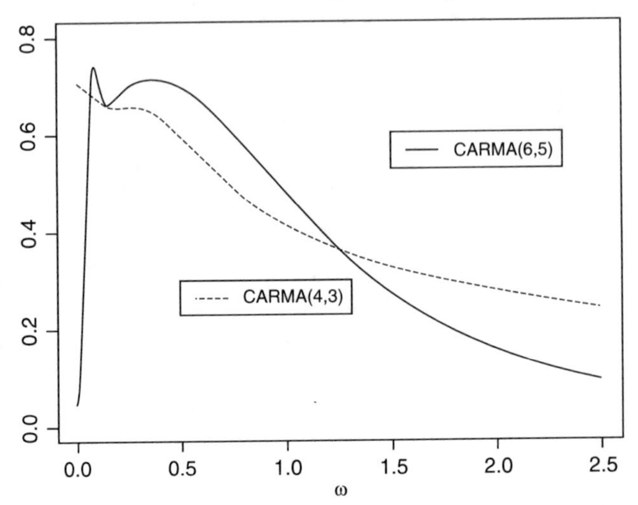

Fig. 15.8 Coherence between temperature and CO_2 estimated with CARMA(4,3) and CARMA(6,5) models

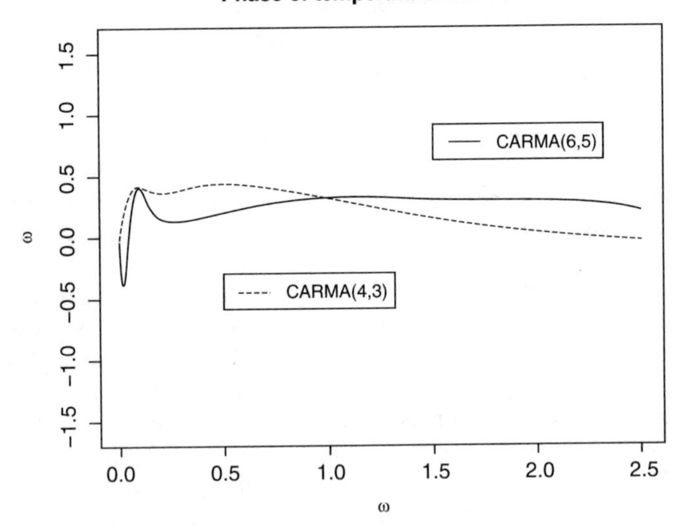

Fig. 15.9 Phase between temperature and CO_2 estimated with CARMA(4,3) and CARMA(6,5) models

15.7 Discussion

In this chapter, the relationship between the statistical background of ARMA time-series models and the general linear model was reviewed. By use of state-space representation, a generalisation to higher dimensions is straightforward. The dynamics captured by linear differential equations is extended by introducing a stochastic component. The solution of such differential equations yields a continuous-time stochastic process with prescribed properties. These properties are defined by three matrices, the autoregressive (AR) parameters, $\boldsymbol{A}$; the moving-average (MA) parameters, $\boldsymbol{R}$; and $\boldsymbol{\Sigma}$ the covariance matrix of the innovation process. This covariance matrix can be interpreted precisely as any other covariance matrix, i.e. square roots of diagonal are standard deviations, and correlations of the innovations can be calculated off-diagonally. It is conceivable that $\boldsymbol{\Sigma}$ is not of full rank, that is, we are dealing with the situation that the innovation process has a lower dimension than the observed process. In such case, we can talk about a dynamic stochastic factor model.

As in the univariate case, direct interpretation of AR and MA parameters is non-trivial. One can say that the AR parameters have a meaning similar to that in a differential equation; e.g. an off-diagonal element in A_i suggests some causal impact. Likewise the MA parameters suggest a pathway from the innovations to derivatives of the solution of the stochastic differential equation. However, the numerical values of the ARMA parameters are hard to interpret. Therefore the spectral approach is important. The spectral concept was briefly reviewed, introducing interpretable concepts such as coherence and phase.

Data analysis based on estimates of a statistical model requires computational procedures. Calculating the maximum-likelihood estimates requires numerical optimisation. Such procedures are sensitive to starting values, and therefore a sequence of nested models has been designed and implemented in R programs. The design is based on the creation of R objects which can be sequentially used. The multivariate ARMA is much more complicated. The creation of the multivariate R object is based on importing univariate R objects. That multivariate R object gives a valid starting point for maximisation of the log-likelihood of the multivariate model. The final multivariate R object will contain maximum-likelihood estimates of the multivariate model, and an objective statement on the covariance in a multivariate model is possible.

The concept Nyquist frequency is used in discrete-time time-series analysis. It describes how informative data can be on certain frequencies in the spectrum. Variations at frequencies higher than the Nyquist frequency will be mapped to a lower frequency just because of the sampling frequency. This is called aliasing in time-series textbooks. When time points are sampled randomly, the continuous-time approach is alias-free. However, as is clear from the examples, the informative value of data will depend on the density and range of the sampled time points. If a precise estimate of the nature of a particular cycle is wanted, it is necessary to observe several cycles. The very low frequency ($\omega \simeq 0$) reflects a very long cycle, i.e. a kind of trend, so that data will typically only represent a fraction of that cycle. Similarly, if precise information of variations at very high frequencies is wanted, high-frequency data is needed. That is, some sequences of observation which are measured densely in time are needed.

The CARMA approach described in this chapter is basically a technical approach. The theoretical covariances, i.e. coherence/phase functions, are approximated by the CARMA parameterisation. As with all parametric models, the interpretation makes most sense within some limits of data. The spectral interpretation offers a straightforward interpretation. If some of the coordinates are overfitted, i.e. the $p + q$ value is large, there is a risk for non-smooth spectral estimates at some frequencies.

The real-data example in this chapter introduces an objective way of giving a statement on the correlation of temperature and CO_2. It is clear that a more precise statement could be given, if more data were available. The observation period, 800,000 years, may seem to be a long time. However, in the context of time series, we are measuring cyclical characteristics of a process. If precise information of cycles is wanted, we shall need to observe many cycles. If information on high-frequency properties is wanted, high density sampling is needed. In the climate data of the last 800,000 years, we see essentially 10–12 large cycles. Comparing the real-data example with the simulation analysis, it seems clear that the relationship between temperature and CO_2 is weak with no signs of a simple lag structure. The multivariate CARMA approach described in this chapter assumes stationarity. For implementing a particular trend structure, further studies are required.

Appendix 1: Implementation of Kalman Filter

The likelihood function is calculated by use of the Kalman-filter recursions. The dynamics are defined by Eqs. (15.15) and (15.16). The sample selection matrix $\boldsymbol{C}$ contains 1 at coordinate ii when coordinate i is observed at time point t_i and is zero otherwise. Given information at time t_{i-1},the optimal estimate of the state vector is $\boldsymbol{X}(t_{i-1}|t_{i-1})$ and the corresponding variance matrix $P_X(t_{i-1}|t_{i-1})$. The prediction step is:

$$\begin{aligned}
\boldsymbol{X}(t_i|t_{i-1}) &= \boldsymbol{e}^{A(t_i-t_{i-1})}\boldsymbol{X}(t_{i-1}|t_{i-1}),\\
P(t_i|t_{i-1}) &= \boldsymbol{e}^{A(t_i-t_{i-1})}P(t_{i-1}|t_{i-1})\boldsymbol{e}^{A'(t_i-t_{i-1})}\\
&\quad +\Gamma_X(0) - \boldsymbol{e}^{A(t_i-t_{i-1})}\Gamma_X(0)\boldsymbol{e}^{A'(t_i-t_{i-1})},\\
\boldsymbol{F}(t_i|t_{i-1}) &= \boldsymbol{C}(t_i)P(t_i|t_{i-1})\boldsymbol{C}'(t_i),\\
\hat{\boldsymbol{y}}(t_i|t_{i-1}) &= \boldsymbol{C}(t_i)\boldsymbol{X}(t_i|t_{i-1}).
\end{aligned}$$

The corresponding updating equations:

$$X(t_i|t_i) = X(t_i|t_{i-1}) + P(t_i|t_{i-1})C(t_i)F^*(t_i|t_{i-1})(y(t_i) - \hat{y}(t_i|t_{i-1})),$$
$$P(t_i|t_i) = P(t_i|t_{i-1}) - P(t_i|t_{i-1})C(t_i)F^*(t_i|t_{i-1})C'(t_i)P(t_i|t_{i-1}).$$

Here $F^*(t_i|t_{i-1})$ denotes a generalised inverse of $F(t_i|t_{i-1})$. The normal likelihood is easily calculated in the non-synchronous case, because at each time point, observed data are essentially univariate. The observed value is compared to its predicted value. The corresponding variance is the non-zero element of element of $F(t_i|t_{i-1})$).

Appendix 2: A Brief Description of the R Packages

The R packages are based on objects. In the *ctarmaRcpp* package, a *ctarma* object contains the observed series, y=$y(t_1), \ldots, y(t_n)$; the time points of observations, tt=$t_1, \ldots, t_n$; the AR parameters, a=$(\alpha_1, \ldots, \alpha_p)$; the MA parameters b=$(\beta_1, \ldots, \beta_q)$; and the standard deviation of the innovations, sigma=σ.

```
 names(ctarma1)
[1] "y"      "tt"      "a"      "b"      "sigma"
>ctarma1$y[1:5]
[1] -1.330619311  0.215559470  0.218091305 -0.004512067
-0.124261787
>ctarma1$tt[1:5]
[1] 0.08360339 0.51537932 0.65898476 0.77687963 0.86278019
>> ctarma1$a
[1]  2 40
> ctarma1$b
[1] 1.00 0.15
>ctarma1$sigma
[1] 8
```

The package *ctarmaRcpp* contains functions on *ctarma* objects for, e.g., calculating and maximising the log-likelihood. Spectral densities can also be calculated and plotted.

In the multivariate *mctarmaRcpp*, a second compatible (same number of AR and MA coefficients) *ctarma* object ctarma2 can be merged with ctarma1 into a bivariate *mctarma* object.

```
> cc=cbind(as.list(ctarma0),as.list(ctarma2))
> mctobj=mctobjnonsync(cc)
> mctobj$A        [,1] [,2] [,3] [,4]
[1,]   -2    0    1    0
[2,]    0   -2    0    1
[3,]  -40    0    0    0
[4,]    0  -40    0    0
```

```
> mctobj$R
     [,1] [,2]
[1,] 0.15 0.00
[2,] 0.00 0.15
[3,] 1.00 0.00
[4,] 0.00 1.00
> mctobj$sigma
     [,1] [,2]
[1,]   64    0
[2,]    0   64
```

That is, the multivariate object `mctobj` now contains the matrix of AR parameters A, the matrix of MA parameters R and the variance matrix of the innovations Σ. The object also contains a vector measurements `y`, an ordered vector of time points of measurements `tt` and a vector `z` denoting which coordinate of the $\boldsymbol{y}$ vector was observed.

```
cbind(mctobj$y,mctobj$tt,mctobj$z)[1:5,]
           [,1]       [,2] [,3]
[1,] -0.6597594 0.03740642    2
[2,] -1.3306193 0.08360339    1
[3,]  0.9946577 0.27623398    2
[4,]  1.1272836 0.31624703    2
[5,]  1.2388762 0.32568241    2
```

The observations are non-synchronous, i.e. only one coordinate of the $\boldsymbol{y}$ vector is observed at each time point. Observations 1 and 3–5 are of coordinate 2; the second is of coordinate 1.

References

Belcher, J., Hampton, J., & Tunnicliffe Wilson, G. (1994). Parameterization of continuous autoregressive models for irregularly sampled time series data. *Journal of the Royal Statistical Association, Series B, 56*(1), 141–155.

Björk, T. (1998). *Arbitrage theory in continuous time*. Oxford: Oxford University Press. https://doi.org/10.1093/0198775180.001.0001

Box, G. E. P., & Jenkins, G. M. (1976). *Time series analysis: Forecasting and control*. San Fransisco: Holden Day.

Brockwell, P. J. (2014). Recent results in the theory and applications of carma processes. *Annals of the Institute of Statistical Mathematics, 66*(4), 647–685. https://doi.org/10.1007/s10463-014-0468-7

Brockwell, P. J., & Davis, R. A. (1991). *Time series: Theory and methods*. New York: Springer. https://doi.org/10.1007/978-1-4419-0320-4

Brockwell, P. J., & Marquardt, T. (2005). Levy-driven and fractionally integrated ARMA processes with continuous time parameter. *Statistica Sinica, 15*, 477–494.

Fasen, V., & Fuchs, F. (2013). Spectral estimates for high-frequency sampled continuous-time autoregressive moving average processes. *Journal of Time Series Analysis, 34*(5), 532–551. https://doi.org/10.1111/jtsa.12029

Garcia, I., Klüppelberg, C., & Müller, G. (2011). Estimation of stable CARMA models with an application to electricity spot prices. *Statistical Modelling, 11*(5), 447–470. https://doi.org/10.1177/1471082X1001100504

Greene, W. H. (2012). *Econometric analysis* (7th ed.). Upper Saddle River, NJ: Pearson.

Harvey, A. C. (1989). *Forecasting, structural time series models and the Kalman filter*. Cambridge: Cambridge University Press. https://doi.org/10.1017/cbo9781107049994

Iacus, S. M., & Mercuri, L. (2015). Implementation of Lévy Carma model in Yuima package. *Computational Statistics, 30*(4), 1111–1141. https://doi.org/10.1007/s00180-015-0569-7

Jouzel, J., et al. (2007). EPICA Dome C ice core 800kyr deuterium data and temperature estimates. IGBP PAGES/World Data Center for Paleoclimatology Data Contribution Series # 2007-091. NOAA/NCDC Paleoclimatology Program, Boulder CO, USA.

Kawai, R. (2015). Sample path generation of Lévy-driven continuous-time autoregressive moving average processes. *Methodology and Computing in Applied Probability, 19*(1), 175–211. https://doi.org/10.1007/s11009-015-9472-5

Kreyszig, E. (1999). *Advanced engineering mathematics* (8th ed.). London: Wiley (Linear fractional transformation, p. 692).

Luthi, D., Floch, M. L., Bereiter, B., Blunier, T., Barnola, J.-M., Siegenthaler, U., et al. (2008). High-resolution carbon dioxide concentration record 650,000-800,000 years before present. *Nature, 453*, 379–382. https://doi.org/10.1038/nature06949

Lütkepohl, H. (1991). *Introduction to multiple time series analysis*. Berlin: Springer. https://doi.org/10.1007/978-3-662-02691-5

Øksendal, B. (1998). *Stochastic differential equations: An introduction with applications* (5th ed.). Berlin: Springer. https://doi.org/10.1007/978-3-662-03620-4

Oud, J. H. L., & Voelkle, M. C. (2014). Do missing values exist? incomplete data handling in cross-national longitudinal studies by means of continuous time modeling. *Quality & Quantity, 48*(6), 3271–3288. https://doi.org/10.1007/s11135-013-9955-9

Schlemm, E., & Stelzer, R. (2012). Multivariate CARMA processes, continuous-time state space models and complete regularity of the innovations of the sampled processes. *Bernoulli, 18*(1), 46–64. https://doi.org/10.3150/10-BEJ329

Singer, H. (1992). Continuous-time dynamical systems with sampled data, errors of measurment and unobserved components. *Journal of Time Series, 14*(5), 527–544. https://doi.org/10.1111/j.1467-9892.1993.tb00162.x

Singer, H. (1995). Analytical score function for irregularly sampled continuous time stochastic processes with control variables and missing values. *Econometric Theory, 11*(4), 721–735. https://doi.org/10.1017/S0266466600009701

Tómasson, H. (2015). Some computational aspects of Gaussian CARMA modelling. *Statistics and Computing, 25*(2), 375–387. https://doi.org/10.1007/s11222-013-9438-9

Tsai, H., & Chan, K. (2000). A note on the covariance structure of a continuous-time ARMA process. *Statistica Sinica, 10*, 989–998.

Zadrozny, P. (1988). Gaussian likelihood of continuous-time ARMAX models when data are stocks and flows at different frequencies. *Econometric Theory, 4*(1), 108–124. https://doi.org/10.1017/S0266466600011890

Chapter 16
Langevin and Kalman Importance Sampling for Nonlinear Continuous-Discrete State-Space Models

Hermann Singer

16.1 Introduction

Nonlinear models occur, when interactions between variables (including self-interactions) are considered. For example, the relation between depression scores and cognitive variables may be influenced by life events (Singer 1986). In the simplest case, one obtains product terms, which may be viewed as first-order approximations of a general nonlinear interaction function, expanded by a multidimensional Taylor series. Generally, nonlinear interactions result from linear specifications, when the parameters of the linear model depend on the variables.

The well-known chaotic Lorenz model, an approximate three-variable model of turbulence (Lorenz 1963; Schuster and Just 2006), contains simple product terms and yet generates a complicated motion, where the influence of the initial condition is very strong. Thus, even in a deterministic system, prediction is only possible for short time periods, since the initial condition must be represented as a localized random distribution, determined by the measurement error. Additionally, random equation errors must be added, since the dynamical equation is usually only an approximation to the true motion. Moreover, data frequently cannot be collected continuously.

There are many examples in the literature. For example, Thomas and Martin (1976) consider a continuous-time nonlinear model of mother-child interaction, in which the dynamics of the dyad are both self-regulatory and interactive. The

H. Singer (✉)
Department of Economics, FernUniversität in Hagen, Hagen, Germany
e-mail: Hermann.Singer@FernUni-Hagen.de

K. van Montfort et al. (eds.), *Continuous Time Modeling in the Behavioral and Related Sciences*, https://doi.org/10.1007/978-3-319-77219-6_16

difference in behavioral levels[1] serves as state-dependent interaction parameter. This in turn yields multiplicative interaction terms. However, the empirical analysis relies on discrete-time models, where derivatives are replaced by differences. This approximate approach, which is widespread, only works for small sampling intervals and may be strongly biased. Moreover, only deterministic models are specified, and the error term is added in an ad hoc manner to the discrete model. However, in the course of an exact discretization, the error term contains the dynamics of the original specification.

A nonlinear model for oscillatory finger motions is presented in Molenaar and Newell (2003). Similar to the Ginzburg-Landau model (see Sect. 16.5.2), the dynamics for the variable "relative phase" (a so-called order parameter) are given by the gradient of a potential function, where the number of stable states depends on the ratio of free parameters, leading to a bifurcation when the ratio crosses a critical value. The dynamic model is given in continuous time, including a random equation error (white noise), but the fitting of the unknown parameters to empirical data is done by a discrete-time state-space model. Again, the transition between the continuous-time model and the discrete-time data is not a seamless process, since the discrete-time model is a complicated, explicitly unknown function of the original dynamics. Even in linear systems, the derived "exact discrete model" (EDM) is a linear model with complicated nonlinear parameter restrictions. Starting with Åström (1970) and Bergstrom (1976a), there is a rich literature on the estimation of sampled continuous linear systems. Even more so in nonlinear systems, one cannot, in most cases, compute exact discrete models or the transition density between measurement times explicitly. A multiplicative EDM is discussed in Sect. 16.5.1.

Therefore, maximum likelihood parameter estimation, and filtering and smoothing (state estimation), of sampled linear and nonlinear stochastic differential equations is of high practical importance, since most data sets are only collected at discrete, often large sampling times, especially in economics and the social sciences (Bergstrom 1976b; Doreian and Hummon 1976; Newell and Molenaar 2014).

Continuous-time models have the advantage in that no artificial time discretization must be introduced. For example, in conventional time series and panel analysis, the dynamical specification is determined by the sampling interval of the measurement scheme, which is arbitrarily given by the research design. However, the dynamics should be formulated independent of the measurements (Bartlett 1946).

Discrete-time measurements can be treated by joining a measurement model to the continuous process model (continuous-discrete state-space model, cf. Jazwinski 1970). Thus, the state, dependent on continuous time, is only observed at certain, possibly unequally spaced, measurement times. Moreover, only parts of the state vector may be observed and are superposed by measurement noise. Nevertheless,

[1] Mother behaviors—look at infant, smile, vocalize, touch, or hold infant. Infant behaviors—look at mother, smile, vocalize, touch or hold mother, fuss/cry.

exact ML estimation is possible, if one can compute the transition density between the sampling intervals.

In the linear case,[2] the approach was pioneered by Bergstrom (1983, 1988), who introduced the "exact discrete model," an exact autoregressive scheme valid at the measurement times, but with nonlinear parameter restrictions due to discrete sampling. From this difference equation, the Gaussian likelihood function can be computed.

In applications, a measurement model is convenient, which includes observation noise and a linear transformation of the latent states to the measurements, in analogy to factor analysis. In this case, the likelihood function can be computed recursively either with Kalman filter methods (Hamerle et al. 1991; Harvey and Stock 1985; Jones 1984; Jones and Tryon 1987; Singer 1990, 1993, 1995, 1998), or by representing the dynamic state-space model as a structural equation model (SEM) (Oud and Jansen 2000; Oud and Singer 2008; Singer 2012). In both approaches, the parameter matrices of the discrete-time model are nonlinear matrix functions of the original parameters of the continuous-time model. One can also formulate the EDM on a latent time grid which is finer than the sampling interval. Then, the nonlinear parameter functions can be linearized, and their values over the whole sampling interval are implicitly generated either by the filter (cf. Singer 1995) or by the SEM equations. With this so-called oversampling device, also linear[3] SEM software can be used (for a comparison of linear and nonlinear oversampling, see Singer 2012). Clearly, the introduction of latent states between the measurement times enlarges the structural matrices which may lead to a slowing down of computations or large approximation errors in the likelihood function (Singer 2010).

In the nonlinear case, the transition density cannot be computed analytically in general and is explicitly known only for some special cases, e.g., the square root model (Feller 1951). Therefore, in general one must use numerical methods to obtain approximate transition densities or moments thereof, for the sampling times determined by the measurements.

There are a variety of approaches to obtain approximations, e.g., Kalman filtering (Arasaratnam et al. 2010; Ballreich 2017; Chow et al. 2007; Särkkä et al. 2013; Singer 2002, 2011), analytical approximations (Aït-Sahalia 2002, 2008; Chang and Chen 2011; Li 2013), Monte Carlo methods (Beskos et al. 2006; Chow et al. 2016; Elerian et al. 2001; Fuchs 2013; Girolami and Calderhead 2011; Pedersen 1995; Särkkä et al. 2013; Singer 2002; Stramer et al. 2010), and numerical solutions of the Fokker-Planck equation (Risken 1989; Wei et al. 1997). In this chapter, only a few comments are in order.

The algorithms based on Kalman filtering use a recursive computation of the a priori density (time update) and the a posteriori density (measurement update).

---

[2]In order to avoid misunderstandings, one must distinguish between (non)linearity in the continuous-time dynamical specification (differential equation) w.r.t. the state variables, and in the derived "exact discrete model" w.r.t. the parameters.

[3]W.r.t. the parameters.

The likelihood function is obtained as normalizing constant of the posterior in the Bayes formula, which incorporates the new measurements. The main problem in continuous time is the computation of the transition density over the (large) measurement interval. One can insert intermediate time slices in the time update and use a short-time approximation (Euler density). The resulting high-dimensional integral can be simulated. The posterior distribution is obtained from the a priori samples by using an importance resampling procedure. This is a simple example of particle filtering methods. Unfortunately, the resampling introduces noise leading to nondifferentiable likelihood surfaces (cf. Pitt 2002). One can use a modified resampling based on the cumulative distribution function approximated as a polygonal line, which yields a much smoother likelihood surface (Malik and Pitt 2011).

Alternatively, the Fokker-Planck equation, which is a partial differential equation for the transition density, can be solved numerically over the sampling interval (see Sect. 16.4.2). The formal solution, which involves the operator exponential function, can also be treated by using computer algebra programs. This leads to analytical expressions, but one must assume state-independent diffusion functions or the possibility of an appropriate transform. Instead of treating the densities directly, one can consider moment equations. These may be solved using Taylor expansions or numerical integration. These approaches are widespread in the engineering literature. Their drawback is frequently the restriction to the first two moments (but see Singer 2008a).

Another approach is the direct Bayesian sampling of the parameters from the posterior distribution. Clearly, one needs the distribution of the measurements given the parameters (likelihood). Again, one can insert latent states and use analytical approximations for the transition density. A well-known problem is the decline of acceptance rates for volatility parameters when inserting too many latent variables. In this case, one can use transformations or use better approximations for the transition density. Alternatively, the likelihood can be approximated by particle filtering and inserted into a Metropolis-Hastings algorithm (Flury and Shephard 2011).

In this paper, the likelihood function is computed nonrecursively by integrating out the latent variables of the state-space model, such that only the marginal distribution of the measurements remains. This task is performed by using a Langevin sampler (Apte et al. 2007, 2008; Hairer et al. 2007, 2009, 2005; Langevin 1908; Roberts and Stramer 2001, 2002) combined with importance sampling. Alternatively, Gaussian integration is used, based on the output of a Kalman smoother. This results in a smooth dependence on the parameters, facilitating quasi-Newton algorithms for finding the maximum likelihood estimate. The unknown importance density is estimated from the sampler data in several ways.

We derive a conditional Markov representation which is estimated using kernel density and regression methods. Alternatively, the true importance density is replaced by an estimated Gaussian density. From this, new data are generated which have the same second-order properties (mean and covariance function) as the original Langevin data. As a variant, the mean and covariance function are directly computed by a Kalman smoother, avoiding Langevin sampling. This approach will be called extended Kalman sampling (EKS).

Alternatively, the likelihood is calculated by numerical integration, using Euler transition kernels and transition kernels derived from the Fokker-Planck equation (finite differences and integral operator approach).

In Sect. 16.2, the continuous-discrete state-space model is introduced. Then, Langevin path sampling is discussed in a finite-dimensional approach. We analytically compute the negative gradient of the potential $-\log p(\eta|z)$ [log probability density of latent states given the data], which serves as the drift function of a Langevin equation. It is not assumed that the diffusion function of the state-space model is state independent. Section 16.4 discusses the maximum likelihood approach and the determination of the likelihood function. In Sect. 16.4.1, two methods for computing the importance density are discussed, namely, an estimation approach and a reference measure method. Then, in Sect. 16.4.2, the likelihood is calculated by numerical integration. In Sect. 16.5, applications such as the model of geometrical Brownian motion and the bimodal Ginzburg-Landau model are considered, and the Fokker-Planck and the importance sampling Monte Carlo approach are compared with each other. In two appendices, analytical drift functions and a continuum limit for the Langevin sampler are considered.

16.2 Continuous-Discrete State-Space Model

Continuous-time system dynamics and discrete-time measurements (at possibly irregular measurement times $t_i, i = 0, \ldots, T$) can be unified by using the *nonlinear continuous-discrete state-space model* (Jazwinski 1970, ch. 6.2)

$$dY(t) = f(Y(t), x(t), \psi)dt + G(Y(t), x(t), \psi)dW(t) \tag{16.1}$$

$$Z_i = h(Y(t_i), x(t_i), \psi) + \epsilon_i;\ i = 0, \ldots, T. \tag{16.2}$$

In (16.1), the state vector $Y(t) \in \mathbb{R}^p$ is a continuous-time random process and the nonlinear drift vector and diffusion matrix functions $f : \mathbb{R}^p \times \mathbb{R}^q \times \mathbb{R}^u \to \mathbb{R}^p$ and $G : \mathbb{R}^p \times \mathbb{R}^q \times \mathbb{R}^u \to \mathbb{R}^p \times \mathbb{R}^r$ depend on a u-dimensional parameter vector ψ. Furthermore, $x(t) \in \mathbb{R}^q$ are deterministic exogenous (control) variables. The system errors in (16.1) are increments $dW(t) = W(t + dt) - W(t)$ of the Wiener process $W(t) \in \mathbb{R}^r$. Its formal derivative is Gaussian white noise $\zeta(t) = dW(t)/dt$ with zero mean and covariance function $E[\zeta_\rho(t)\zeta_{\rho'}(s)] = \delta_{\rho\rho'}\delta(t - s);\ \rho, \rho' = 1, \ldots, r$, where $\delta(t - s)$ is the Dirac delta function (cf. Lighthill 1958), and $\delta_{\rho\rho'}$ is the Kronecker delta symbol.[4] Thus the process errors are independent for the times $t \neq s$ and the components $\rho \neq \rho'$.

The random initial condition $Y(t_0)$ is assumed to have a density $p_0(y, \psi)$ and is independent of $dW(t)$. The nonlinear state equation (16.1) is interpreted in the sense

[4] One has $\int \delta(x - x')\phi(x')dx' = \phi(x)$ and $\sum_{\rho'} \delta_{\rho\rho'}\phi_{\rho'} = \phi_\rho$.

of Itô. This means that the multiplicative noise term $G(Y(t))dW(t)$, as it stands, is only a symbol without mathematical meaning. In this paper, it is assumed that it can be interpreted as $G(Y(t))[W(t+dt) - W(t)]$ (the increments are directed to the future), but there are other possibilities. For a discussion of these subtle questions, I refer to the literature (Arnold 1974; Van Kampen 1981).

Finally, the error term $\epsilon_i \sim N(0, R(x(t_i), \psi))$ in the measurement equation (16.2) is a discrete-time white noise, independent for the times of measurement. It is assumed to be independent of the system error $dW(t)$ (cf. Jazwinski 1970, pp. 209–210). The measurement (output) function h maps the state to the observations. For example, only the strength (modulus) of Y may be observed. Thus, (16.2) may be interpreted as dynamic nonlinear factor analysis.

In contrast to linear systems, an analytical solution of (16.1) cannot be found in general. Some solvable equations are listed in Kloeden and Platen (1992, ch. 4.4). In most cases, one must resort to numerical approximations.

16.3 Langevin Path Sampling: Finite-Dimensional Approach

16.3.1 Likelihood and Langevin Equation

In order to obtain finite-dimensional probability densities of the state $Y(t)$ w.r.t. Lebesgue measure and a numerically feasible approach, the SDE (16.1) is first replaced by the Euler-Maruyama approximation

$$\begin{aligned} \eta_{j+1} &= \eta_j + f(\eta_j, x_j, \psi)\delta t + G(\eta_j, x_j, \psi)\delta W_j \\ &:= \eta_j + f_j \delta t + g_j \delta W_j \end{aligned} \tag{16.3}$$

on a temporal grid $\tau_j = t_0 + j\delta t;\ j = 0, \ldots, J = (t_T - t_0)/\delta t$ (cf. Kloeden and Platen 1992, chs. 10.2, 14.1 and Stuart et al. 2004). Thus the stochastic differential equation (16.1) is approximated by a difference equation. This kind of approximation is constructed in analogy to deterministic differential equations. One can show that the approximation η_j converges to the state $Y(\tau_j)$ for small discretization interval δt (Kloeden and Platen 1992, ch. 9). The approximation error of the Euler scheme could be displayed by a superscript $\eta_j^{\delta t}$, but this is dropped for simplicity.

The process noise in (16.3) is given as $\delta W_j = z_j \sqrt{\delta t}$, $z_j \sim N(0, I_r)$, *i.i.d.*, and the control variables are $x_j = x(\tau_j)$. The state variables η_j are measured at times $t_i = \tau_{j_i}$, $j_i = (t_i - t_0)/\delta t$ according to

$$Z_i = h(\eta_{j_i}, x_{j_i}, \psi) + \epsilon_i. \tag{16.4}$$

Since $\eta_{j+1}|\eta_j$ is conditionally Gaussian, the finite-dimensional density of the latent states $\eta = \{\eta_J, \ldots, \eta_0\}$ is given by [for clarity dropping the parameter and the exogenous variables; setting $\tilde{\eta} := \{\eta_J, \ldots, \eta_1\}$]

$$p(\eta) = \prod_{j=0}^{J-1} p(\eta_{j+1}|\eta_j)\; p(\eta_0) = p(\tilde{\eta}|\eta_0)p(\eta_0) \tag{16.5}$$

$$:= Z^{-1}e^{-S}\; p(\eta_0) \tag{16.6}$$

$$S = \tfrac{1}{2}\sum_{j=0}^{J-1}(\eta_{j+1} - \eta_j - f_j\delta t)'(\Omega_j\delta t)^{-1}(\eta_{j+1} - \eta_j - f_j\delta t) \tag{16.7}$$

$$Z = \prod_{j=0}^{J-1} |2\pi\Omega_j\delta t|^{1/2}, \tag{16.8}$$

where $\Omega_j = g_j g_j'$ is assumed to be nonsingular.[5] In order to compute the likelihood function of the measured data $z = \{z_T, \ldots, z_0\}$, one can augment the density function of states η_{j_i} at measurement times t_i with imputed variables, e.g., the latent states between measurements, leading to the integral representation

$$p(z) = \int p(z|\eta)p(\eta)d\eta. \tag{16.9}$$

This has the advantage in that the density $p(\eta)$ can be computed explicitly by the simple Euler density (16.5). A disadvantage is the introduction of many latent variables, since δt must be small. To improve the density approximation for larger δt, one could use the methods of (Aït-Sahalia 2002, 2008) and Li (2013) to obtain an analytical asymptotic expansion of p, or the so-called local linearization (LL) method, which is exact for linear systems (Shoji and Ozaki 1998a). The latter method has the advantage that the density approximation integrates to unity (cf. Stramer et al. 2010).

The resulting high-dimensional integration will be accomplished by Monte Carlo methods and, in comparison, by numerical integration. A direct approximation of (16.9) by the mean

$$p(z) \approx N^{-1}\sum_{l=1}^{N} p(z|\eta_l)$$

is extremely inefficient, since the unconditional random vectors $\eta_l \sim p(\eta)$ in the sum are mostly suppressed by the measurement density $p(z|\eta) =$

[5] Otherwise, one can use the singular normal distribution (cf. Mardia et al. 1979, ch. 2.5.4, p. 41). In this case, the generalized inverse of Ω_j is used and the determinant $|\cdot|$, which is zero, is replaced by the product of positive eigenvalues. Singular covariance matrices occur, for example, in autoregressive models of higher order, when the state vector contains derivatives of a variable.

$\prod_{i=0}^{T} \phi(z_i, h(\eta_{j_i}), R_i)$. This may be thought of defining windows in phase space where the trajectories η_l must pass through (cf. Kac 1980). Using the idea of importance sampling, one can rewrite (16.9) as

$$p(z) = \int \frac{p(z|\eta)p(\eta)}{p(\eta|z)} p(\eta|z) d\eta, \tag{16.10}$$

where $p(\eta|z)$ is the optimal importance density (Kloeden and Platen 1999, p. 519). Thus one must be able to draw conditional trajectories $\eta_l|z$, a task which can be accomplished by the Langevin approach (Parisi and Wu 1981).

Introducing a $(J+1)p$-dimensional random process $\eta(u) = \eta_{j\alpha}(u); j = 0, \ldots, J, \alpha = 1, \ldots, p$ and a potential $\Phi(\eta) := -\log p(\eta)$, one may consider a *Langevin equation* (Langevin 1908) in the fictitious (artificial) time dimension u

$$d\eta(u) = -\partial_\eta \Phi(\eta) du + \sqrt{2}\, dW(u), \tag{16.11}$$

where W is a $(J+1)p$-dimensional Wiener process and $\partial_\eta := \frac{\partial}{\partial \eta}$ is the gradient. The physical time dimension t occurring in (16.1) is enfolded in the components $\eta_{j\alpha}(u); j = 0, \ldots, J$ of the state. The concept of a *potential* stems from physics, where gravitational or electric fields are described by negative gradients of scalar functions. In the present context, one considers an abstract space defined by the variables η_j. For example, if the density p is a normal distribution $N(0, 1)$, $-\log p \propto \frac{1}{2}\eta^2$, and one considers the random motion in a quadratic potential well. Clearly, the Euler density (16.5) generates a complicated diffusion process.

Under the assumption of ergodicity,[6] the (autocorrelated) trajectory $\eta(u)$ asymptotically ($u \to \infty$) is in equilibrium and draws data from the stationary distribution

$$p_{stat}(\eta) = e^{-\Phi(\eta)} = p(\eta).$$

This may be seen by considering the stationary Fokker-Planck equation

$$\partial_u p(\eta, u) = \sum_{j\alpha} \partial_{\eta_{j\alpha}} [(\partial_{\eta_{j\alpha}} \Phi(\eta)) p(\eta, u) + \partial_{\eta_{j\alpha}}\ p(\eta, u)] = 0 \tag{16.12}$$

for the density $p(\eta, u)$ (see, e.g., Risken 1989, chs. 5.2, 6.0). Of course, one can directly draw independent vectors η from (16.3). The advantage of (16.11) is the possibility of drawing from the posterior $p(\eta|z)$, by using

$$\Phi(\eta|z) = -\log p(\eta|z) = -[\log p(z|\eta) + \log p(\eta) - \log p(z)] \tag{16.13}$$

[6]In statistical mechanics, one assumes the equivalence of time averages and ensemble averages (cross sections of identical systems).

as potential. The last term, which is the desired quantity (16.10), drops out by computing the gradient. Keeping a continuum limit $\delta t \to 0$ for Eq. (16.3) in mind (see Appendix 2), the partial derivatives in the Langevin equation are scaled by the time discretization interval δt, leading to

$$d\eta(u) = -\delta_\eta \Phi(\eta|z)du + \sqrt{2}\, dW(u)/\sqrt{\delta t}. \tag{16.14}$$

Here we set $\delta_\eta := \frac{\partial}{\partial \eta \delta t} := \frac{\delta}{\delta \eta}$ in view of the functional derivative $\frac{\delta}{\delta \eta(t)}$.

16.3.2 Drift Function

In order to obtain an efficient sampling algorithm, the drift function in (16.14)

$$\delta_\eta \log p(\eta|z) = \delta_\eta[\log p(z|\eta) + \log p(\eta)]$$

must be computed analytically. For details, see Appendix 1, in particular (16.44), (16.52). Consider the term $\log p(\eta)$ (cf. Eq. (16.5)). The Lagrangian (16.7), also called Onsager-Machlup functional (Onsager and Machlup 1953) may be rewritten as

$$S = \frac{1}{2}\sum_{j=0}^{J-1} \delta\eta'_j (\Omega_j \delta t)^{-1} \delta\eta_j \tag{16.15}$$

$$- \sum_{j=0}^{J-1} f'_j \Omega_j^{-1} \delta\eta_j + \frac{1}{2}\sum_{j=0}^{J-1} f'_j \Omega_j^{-1} f_j \delta t \tag{16.16}$$

$$:= S_0 + S_1 + S_2,$$

$\delta\eta_j := \eta_{j+1} - \eta_j$. In a system without drift ($f = 0$), only the first term S_0 remains, corresponding to a random walk.[7] Therefore, the density $p(\eta)$ can be factorized as $p(\eta) = p_0(\eta)\, \alpha(\eta)$ where

$$\alpha(\eta) = \frac{p(\eta)}{p_0(\eta)} = \exp\left\{ \sum_{j=0}^{J-1} f'_j \Omega_j^{-1} \delta\eta_j - \frac{1}{2}\sum_{j=0}^{J-1} f'_j \Omega_j^{-1} f_j \delta t \right\} \tag{16.17}$$

[7]In the case of a state-dependent diffusion matrix, $\eta_{j+1} = \eta_j + G(\eta_j, x_j, \psi)\delta W_j$ generates a more general martingale process. Expression (16.16) remains finite in a continuum limit (see Appendix 2).

is the density ratio and $p_0(\eta) = Z^{-1}\exp(-S_0)p(\eta_0)$ is the density of the driftless process including the initial condition. Thus one has the decomposition $\log p(\eta) = -\log Z - S_0 - S_1 - S_2 + \log p(\eta_0)$.

16.4 Maximum Likelihood Estimation

16.4.1 Monte Carlo Approach

In Sect. 16.3.1, conditional path sampling was motivated by computing the density function (16.9) of the data z efficiently. Considering $L(\psi) = p(z; \psi)$ as a function of the parameter vector ψ, the maximum likelihood estimator is obtained as $\hat{\psi} = \arg\max_{\psi} L(\psi)$ (cf. Basawa and Prakasa Rao 1980). In Eq. (16.10), namely,

$$p(z) = \int \frac{p(z|\eta)p(\eta)}{p(\eta|z)} p(\eta|z)d\eta := E[g(\eta, z)|z], \tag{16.18}$$

the expectation over the invariant (stationary) distribution may be estimated as fictitious-time average

$$\hat{p}(z; U) = U^{-1}\int_0^U g(\eta(u), z)du \approx N^{-1}\sum_{l=0}^{N-1} g(\eta_l, z) := \hat{p}(z; \delta u, N) \tag{16.19}$$

if $\eta(u)$ is ergodic. The sum results as an approximation of the integral with grid size δu, $U = N\delta u$, $u_l = l\delta u$, $\eta_l = \eta(u_l)$. In (16.19), the importance (smoothing) density $p(\eta|z)$ is unknown, whereas conditional trajectories $\eta_l \sim p(\eta|z)$ can be drawn from the Langevin equation (16.14) in equilibrium. One could also simulate independent replications $\eta_l(u), l = 0, \ldots, N-1$, evaluated at large u.

The Langevin equation for the simulation of $\eta(u)$ in the fictitious time dimension u is a nonlinear Itô equation with $(J+1)p$ components. It can be solved approximately using several numerical schemes, e.g., the Heun method or a fourth-order Runge-Kutta scheme (Gard 1988; Kloeden and Platen 1999; Rümelin 1982). Here we use an Ozaki scheme (Ozaki 1985) with Metropolis mechanism (Hastings 1970) in order to ensure a correct stationary distribution. This scheme is exact for linear drift functions.

In order to improve the relaxation of the Langevin equation to equilibrium, it can be scaled with a positive definite kernel matrix K. The transformed equation

$$d\eta(u) = -K\delta_\eta\Phi(\eta|z)du + \sqrt{2}K^{1/2}\,dW(u)/\sqrt{\delta t}, \tag{16.20}$$

$K = K^{1/2}(K^{1/2})'$ yields the same stationary distribution, as may be seen from the stationary Fokker-Planck equation (16.12). The kernel may be even time and state dependent, i.e., $K = K(\eta, u)$ (cf. Okano et al. 1993; Parisi and Wu 1981), but then additional terms and convolution integrals are required. In the constant case, one has (setting $\partial_{j\alpha} := \partial_{\eta_{j\alpha}}$ and summing over double indices)

$$
\begin{aligned}
0 &= \partial_{j\alpha} K_{j\alpha;l\beta}(\partial_{l\beta}\Phi)p + \partial_{j\alpha}\partial_{l\beta} K^{1/2}_{j\alpha;u\gamma} K^{1/2}_{u\gamma;l\beta} p \\
&= \partial_{j\alpha}\Big\{K_{j\alpha;l\beta}(\partial_{l\beta}\Phi)p + K_{j\alpha;l\beta}\partial_{l\beta} p\Big\} = \text{div} J
\end{aligned}
$$

Therefore, setting the probability current $J = 0$, one obtains the equilibrium distribution $p_{stat}(\eta|z) = e^{-\Phi(\eta|z)}$, independent of K (see, e.g., Risken 1989, chs. 5.2, 6.0). If the Langevin equation (16.14) were linear with drift $H\eta\delta t$ (i.e., $\Phi = \frac{1}{2}\eta'(-H)\eta\delta t^2$ with symmetric negative definite H), one can use $K = -H^{-1}$ as kernel to obtain an equation with decoupled drift components $d\eta = -(\eta\delta t)du + \sqrt{2}K^{1/2}\, dW(u)/\sqrt{\delta t}$. For example, the leading term (16.40) is linear. If the linear drift is differentiated w.r.t η, one obtains the matrix $H = \delta_\eta(H\eta\delta t)$. In the nonlinear case, the idea is to compute the Hessian matrix $H_{j\alpha;j'\alpha'} = -\delta_{\eta_{j\alpha}}\delta_{\eta_{j'\alpha'}}\Phi(\eta|z)$ at a certain point $\eta(u_0)$ and use $K = -H^{-1}$ as kernel.

16.4.1.1 Determination of the Importance Density

For computing the likelihood function (16.18), the importance density $p(\eta|z)$ must be determined. Several approaches have been used:

1. Approximation of the optimal $p(\eta|z)$ by a density estimator $\hat{p}(\eta|z)$, using kernel density or regression methods.

   One can use a density estimate

$$
\hat{p}(\eta|z) = N^{-1}\sum_{l=1}^{N}\kappa(\eta - \eta_l;\, H), \tag{16.21}
$$

   where κ is a kernel function and H is a suitably chosen smoothing parameter. In this article a multivariate Gaussian kernel $\kappa(y, H) = \phi(y; 0, h^2 S)$ is used, where $e = 1/(p+4)$, $A = (4/(p+2))^e$; $h = An^{-e}$ and S is the sample covariance matrix (Silverman 1986, p. 78 ff.).

   The density estimate (16.21) seems to be natural, but the dimension of η is very high, namely, $(J+1)p$, $J = (t_T - t_0)/\delta t$. It turns out that the estimation quality is not sufficient. Therefore the Markov structure of the state-space model must be exploited first. We use the Euler-discretized state-space model (16.3)

$$
\begin{aligned}
\eta_{j+1} &= f(\eta_j)\delta t + g(\eta_j)\delta W_j \\
z_i &= h(y_i) + \epsilon_i,
\end{aligned}
$$

(where the dependence on x_j and ψ is dropped for simplicity) in combination with the Bayes formula

$$p(\eta|z) = p(\eta_J|\eta_{J-1}, \ldots, \eta_0, z)\; p(\eta_{J-1}, \ldots, \eta_0|z).$$

Now it can be shown that η_j is a conditional Markov process

$$p(\eta_{j+1}|\eta_j, \ldots, \eta_0, z) = p(\eta_{j+1}|\eta_j, z). \tag{16.22}$$

To see this, we use the conditional independence of the *past* $z^i = (z_0, \ldots, z_i)$ and *future* $\bar{z}^i = (z_{i+1}, \ldots z_T)$ given $\eta^j = (\eta_0, \ldots, \eta_j)$. One obtains

$$p(\eta_{j+1}|\eta^j, z^i, \bar{z}^i) = p(\eta_{j+1}|\eta^j, \bar{z}^i) = p(\eta_{j+1}|\eta_j, \bar{z}^i)$$
$$p(\eta_{j+1}|\eta_j, z^i, \bar{z}^i) = p(\eta_{j+1}|\eta_j, \bar{z}^i)$$
$$j_i \leq j < j_{i+1}$$

since

(i) The transition density $p(\eta_{j+1}|\eta^j, z^i, \bar{z}^i)$ is independent of past measurements, given the past true states, and only the last state η_j must be considered (Markov process).
(ii) The transition density $p(\eta_{j+1}|\eta_j, z^i, \bar{z}^i)$ is independent of past measurements.

Thus we have proved $p(\eta_{j+1}|\eta^j, z^i, \bar{z}^i) = p(\eta_{j+1}|\eta_j, z^i, \bar{z}^i)$ □

With the representation (16.22), it is sufficient to estimate the density function $p(\eta_{j+1}|\eta_j, z)$ with a low-dimensional argument η_j instead of the full $\eta = (\eta_0, \ldots, \eta_J)$. The estimation can be accomplished by using either

a. Density estimation methods as in (16.21), or by
b. Regression methods.

In the latter case, the Euler density is modified to the form

$$p(\eta_{j+1}, \delta t|\eta_j, z) \approx \phi(\eta_{j+1}; \eta_j + (f_j + \delta f_j)\delta t, (\Omega_j + \delta\Omega_j)\delta t) \tag{16.23}$$

where the correction terms are estimated using the data $\eta_j \equiv \eta_j|z$ from the Langevin sampler. It is assumed that the conditional states fulfill the Euler-discretized SDE (cf. 16.22)

$$\delta\eta_{j+1} = [f(\eta_j) + \delta f_j(\eta_j)]\delta t + [g(\eta_j) + \delta g_j(\eta_j)]\delta W_j \tag{16.24}$$

with modified drift and diffusion functions. This functions can be estimated by parametric specifications (e.g., $\delta f(x) = \alpha + \beta x + \gamma x^2$) or nonparametrically.

The introduction of drift corrections was derived analytically in Singer (2002, 2014).

2. Another approach is the choice of a (in general suboptimal) reference density $p_0(\eta|z) = p_0(z|\eta)p_0(\eta)/p_0(z)$, where the terms on the right-hand side are known explicitly. This yields the representation

$$p(z) = p_0(z) \int \frac{p(z|\eta)}{p_0(z|\eta)} \frac{p(\eta)}{p_0(\eta)} p_0(\eta|z) d\eta. \tag{16.25}$$

In this paper we use a conditional Gaussian density

$$p_0(\eta|z) = \phi(\eta|z) = \phi(\eta; E[\eta|z], V[\eta|z]) \tag{16.26}$$

where the conditional moments are estimated from the

a. Langevin sampler data $\eta_l = \eta_l|z$. Alternatively, one can use a
b. Kalman smoother

to obtain approximations of the conditional moments. We use algorithms based on the Rauch-Tung-Striebel smoother with Taylor expansions or using integration (unscented transform or Gauss-Hermite integration; see, e.g., Jazwinski 1970; Särkkä 2013, ch. 9).

In both cases, one must (re)sample the data from $p_0 = \phi(\eta|z)$ to compute the likelihood estimate

$$\hat{p}(z) = N^{-1} \sum_l \frac{p(z|\eta_l)p(\eta_l)}{p_0(\eta_l|z)} \tag{16.27}$$
$$\eta_l \sim p_0.$$

16.4.1.2 Score Function

The score function $s(z) := \partial_\psi \log p(z)$ (dropping ψ) can be estimated by using a well-known formula of Louis (1982). One can write

$$\begin{aligned} s(z) &= \partial_\psi \log \int p(z, \eta) d\eta = p(z)^{-1} \int s(z, \eta) p(z, \eta) d\eta \\ &= E[s(z, \eta)|z], \end{aligned} \tag{16.28}$$

$s(z, \eta) := \partial_\psi \log p(z, \eta) = \partial_\psi p(z, \eta)/p(z, \eta)$, with the estimate

$$\hat{s}(z) = N^{-1} \sum_l s(z, \eta_l). \tag{16.29}$$

From this a quasi-Newton algorithm, $\psi_{k+1} = \psi_k + F_k^{-1}\hat{s}_k(z)$ can be implemented. For example, one can use BFGS secant updates for F_k.

16.4.1.3 Bayesian Estimation

A full Bayesian solution to the estimation problem can be obtained by the decomposition of the posterior

$$p(\eta, \psi|z) = p(z, \eta, \psi)/p(z) = p(z|\eta, \psi)p(\eta|\psi)p(\psi)/p(z). \quad (16.30)$$

From $\log p(\eta, \psi|z)$, one obtains a system of Langevin equations of the form (see Eq. (16.14))

$$d\eta(u) = \delta_\eta \log p(\eta, \psi|z)du + \sqrt{2}\, dW_1(u)/\sqrt{\delta t} \quad (16.31)$$

$$d\psi(u) = \partial_\psi \log p(\eta, \psi|z)du + \sqrt{2}\, dW_2(u), \quad (16.32)$$

where $W_1 : (J + 1)p \times 1$ and $W_2 : u \times 1$ are independent Wiener processes. For large "time" u, one obtains correlated random samples from $p(\eta, \psi|z)$. The drift term $\partial_\psi \log p(\eta, \psi|z)$ coincides with the score $s(z, \eta)$ except for the prior term $\partial\psi \log p(\psi)$, since $p(\eta, \psi|z) = p(z, \eta|\psi)p(\psi)/p(z)$.

16.4.2 *Numerical Integration Approach*

The Monte Carlo approach will be compared with a method based on Riemann integration combined with transition kernels, which are computed in three different ways, namely, by using (1) the Euler transition kernel and the matrix exponential of the Fokker-Planck operator, which is represented (2) as a matrix of finite differences or (3) as an integral operator.

16.4.2.1 Transition Kernel Approach

The $(J+1)p$-dimensional integral in the likelihood $p(z) = \int p(z|\eta)p(\eta)d\eta$ can be computed without Monte Carlo integration, at least for small dimensions p of η_j. One can write

$$p(z) = \int \prod_{i=0}^{T-1} \left[p(z_{i+1}|\eta_{j_{i+1}}) \prod_{j=j_i}^{j_{i+1}-1} p(\eta_{j+1}|\eta_j) \right] p(z_0|\eta_0)p(\eta_0)d\eta,$$

by using the conditional independence of $Z_i|\eta$ and the Markov property of η (see Eqs. (16.3)–(16.4)). The likelihood expression can be represented recursively by the Kalman updates (time update, measurement update, conditional likelihood)

$$p(\eta_{j_{i+1}}|Z^i) = \left[\prod_{j=j_i}^{j_{i+1}-1} \int p(\eta_{j+1}|\eta_j)d\eta_j\right] p(\eta_{j_i}|Z^i)$$

$$p(\eta_{j_{i+1}}|Z^{i+1}) = p(z_{i+1}|\eta_{j_{i+1}})p(\eta_{j_{i+1}}|Z^i)/p(z_{i+1}|Z^i)$$

$$p(z_{i+1}|Z^i) = \int p(z_{i+1}|\eta_{j_{i+1}})p(\eta_{j_{i+1}}|Z^i)d\eta_{j_{i+1}}$$

$$i = 0, \ldots, T$$

with initial condition $p(\eta_0|Z^0) = p(z_0|\eta_0)p(\eta_0)/p(z_0)$; $p(z_0) = \int p(z_0|\eta_0)p(\eta_0)$ $d\eta_0$ and $Z^i := \{z_i, \ldots, z_0\}$ (data up to time t_i). Thus one has the prediction error decomposition

$$p(z) = \prod_{i=0}^{T-1} p(z_{i+1}|Z^i)p(z_0)$$

by the Bayes formula. Actually, the Kalman representation is more general since it is also valid for densities $p(\eta_{j+1}|\eta_j, Z^i)$, $p(z_{i+1}|\eta_{j_{i+1}}, Z^i)$ depending on lagged measurements (cf. Liptser and Shiryayev 2001, vol. II, ch. 13).

The p-dimensional integrals will be approximated as Riemann sums (or using Gauss-Legendre integration)

$$\int p(\xi|\eta)p(\eta|\zeta)d\eta \approx \sum_k p(\xi|\eta_k)p(\eta_k|\zeta)\delta\eta \approx \sum_k p(\xi|\eta_k)p(\eta_k|\zeta)w_k$$

on a p-dimensional grid of supporting points η_k, i.e., $k = \{k_1, \ldots, k_p\}$, $k_\alpha = 0, \ldots, K_\alpha$ is a multi-index, and $\eta_k = \{\eta_{k_1}, \ldots, \eta_{k_p}\}$ is a p-vector inside a p-dimensional interval $[a, b] = [a_1, b_1] \times \ldots \times [a_p, b_p]$, with coordinates

$$\eta_{k_\alpha} = a_\alpha + k_\alpha \delta\eta_\alpha \tag{16.33}$$

$\eta_{K_\alpha} = b_\alpha, \alpha = 1, \ldots, p$, and volume element $\delta\eta = \prod_{\alpha=1}^{p} \delta\eta_\alpha$.

16.4.2.2 Euler Transition Kernel

The Euler transition kernel $p(\xi_l|\eta_k)$ for the difference equation (16.3) can be viewed as a matrix T_{lk}. It is given by the normal density $\phi(\xi; \eta + f(\eta)\delta t, \Omega(\eta)\delta t)$ on the grid points ξ_l, η_k. The dependence on the time index j, the exogenous variables $x_j = x(\tau_j)$, and the lagged data Z^i is omitted here.

A better approximation of the true transition kernel $p(y_{j+1}, \tau_{j+1}|y_j, \tau_j)$ of the original SDE (16.1) can be achieved through an expansion of the drift function using Itô's lemma (Shoji and Ozaki 1997, 1998b; Singer 2002)

$$f(y) \approx f(y_j) + f_y(y_j)(y - y_j) + \frac{1}{2} f_{yy}(y_j)\Omega(y_j)(t - t_j). \tag{16.34}$$

The approach is known under the label local linearization (LL). Inserting (16.34) into the moment equations , one obtains the linear differential equations $\dot{\mu} = E[f] \approx f_j + A_j(\mu - y_j) + c_j(t - t_j)$, $\dot{\Sigma} = A_j\Sigma + \Sigma A_j' + \Omega_j$ with the Jacobian terms $A_j := f_y(y_j), c_j := \frac{1}{2} f_{yy}(y_j)\Omega(y_j)$ (for details, see Singer 2002). The second-order term c_j only leads to contributions of order δt^2. Thus an improved transition kernel is given by

$$\begin{aligned} p(\eta_{j+1}|\eta_j) &= \phi(\eta_j; \mu_{j+1}, \Sigma_{j+1}) \\ \mu_{j+1} &= \eta_j + A_{1j} f_j + A_j^{-1}[-I\delta t + A_{1j}]c_j \\ \text{row } \Sigma_{j+1} &= (A_j \otimes I + I \otimes A_j)^{-1}(A_j^* \otimes A_j^* - I)\text{row } \Omega_j, \end{aligned}$$

$A_j^* := \exp(A_j\delta t)$, $A_{1j} := A_j^{-1}(A_j^* - I)$, where row is the row-wise vector operator. For small δt, one recovers the Euler terms $\mu_{j+1} = \eta_j + f_j\delta t$; $\Sigma_{j+1} = \Omega_j\delta t$. The approaches are denoted as ETK and LLTK in the sequel.

16.4.2.3 Fokker-Planck Equation

The transition kernel can also be obtained as a short-time approximation to the solution of the Fokker-Planck equation

$$\begin{aligned} \partial_t p(y, t|y_j, \tau_j) &= -\partial_\alpha[f_\alpha p(y, t|y_j, \tau_j)] + \frac{1}{2}\partial_\alpha\partial_\beta[\Omega_{\alpha\beta} p(y, t|y_j)] \\ &:= F(y)p(y, t|y_j, \tau_j) \end{aligned}$$

with initial condition $p(y, \tau_j|y_j, \tau_j) = \delta(y - y_j)$. A summation convention over doubly occurring indices α, β is supposed. Formally, one has $p(y_{j+1}, \tau_{j+1}| y_j, \tau_j) = \exp[F(y_{j+1})\delta t]\ \delta(y_{j+1} - y_j)$, if the system is autonomous, i.e., $x(t) = x$. In this case, the measurement time interval $\Delta t_i := t_{i+1} - t_i$ may be bridged in one step $p(y_{i+1}, t_{i+1}|y_i, t_i) = \exp[F(y_{j+1})\Delta t_i]\ \delta(y_{i+1} - y_i)$. For equally spaced data, only one kernel must be computed. In the nonautonomous case, a time-ordered exponential must be considered. In this paper, it is assumed that $x(t) = x_j$ is constant in the interval $[\tau_j, \tau_{j+1})$, so $p(y_{j+1}, \tau_{j+1}|y_j, \tau_j) =$

$\exp[F(y_{j+1}, x_j)\delta t]\ \delta(y_{j+1} - y_j)$. In this section, two approximations are considered:

1. First, the spatial (y) derivatives are replaced by finite differences on a multidimensional grid y_k.
2. Second, the differential operator F is rewritten as an integral operator, and the integrals are approximated as sums.

The time dimension t is kept continuous. Such mixed approaches are called lattice approximation, semi-discretization or method of lines (cf. Jetschke 1991; Schiesser 1991; Yoo 2000). In both cases, one obtains a linear matrix differential equation which can be solved by the matrix exponential function.

Finite Differences Using finite differences, one may write

$$-\partial_\alpha[f_\alpha p(y)] \approx -\delta_{k_1 l_1} \dots \nabla_{k_\alpha l_\alpha} \dots \delta_{k_p l_p}[f_\alpha(y_l)p(y_l)]$$

with multi-index $l = \{l_1, \dots, l_p\}$, $y_l = \{y_{l_1}, \dots, y_{l_p}\} = a + l\delta y$ (see 16.33) and two-sided differences $\nabla_{k_\alpha l_\alpha} = (\delta_{k_\alpha+1;l_\alpha} - \delta_{k_\alpha-1;l})/(2\delta y_\alpha)$. The diffusion term is

$$\frac{1}{2}\partial_\alpha\partial_\beta[\Omega_{\alpha\beta}p(y)] \approx \frac{1}{2}\delta_{k_1 l_1} \dots \nabla_{k_\alpha l_\alpha} \dots \nabla_{k_\beta l_\beta} \dots \delta_{k_p l_p}[\Omega_{\alpha\beta}(y_l)p(y_l)]$$

with the replacement (diagonal terms $\alpha = \beta$): $\nabla^2_{k_\alpha l_\alpha} \to \Delta_{k_\alpha l_\alpha} = (\delta_{k_\alpha+1;l_\alpha} - 2\delta_{k_\alpha l_\alpha} + \delta_{k_\alpha-1;l_\alpha})/(\delta y^2_\alpha)$ (centered second difference). Thus the Fokker-Planck operator is replaced by the matrix

$$F_{kl} = -\delta_{k_1 l_1} \dots \nabla_{k_\alpha l_\alpha} \dots \delta_{k_p l_p} f_\alpha(y_l) + \frac{1}{2}\delta_{k_1 l_1} \dots \nabla_{k_\alpha l_\alpha} \dots \nabla_{k_\beta l_\beta} \dots \delta_{k_p l_p}\Omega_{\alpha\beta}(y_l). \quad (16.35)$$

Usually the multi-indices are flattened to a $K = \prod_\alpha K_\alpha$-dimensional index. Clearly, one obtains a high-dimensional matrix problem. The transition kernel on the grid points is written as matrix (possibly depending on x_j)

$$p(\eta_{j+1;k}|\eta_{jl}) = [\exp(F(x_j)\delta t)]_{kl}/\delta\eta$$

where the matrix exponential function may be evaluated using several methods, including Taylor series and eigen methods (Moler and Van Loan 2003)

Integral Operator Alternatively, the differential operator $F(y)$ can be represented as integral operator

$$F(y)p(y) = \int F(y, y')p(y')dy'$$

with integral kernel $F(y, y') = F(y)\delta(y - y') = L(y')\delta(y - y') = L(y', y)$ (Risken 1989, p. 69). Here $L(y) = F^*(y)$ is the backward operator, and $\delta(y - y')$ is the Dirac delta function (cf. Lighthill 1958). The differential operation $q(y) =$

$F(y)p(y) = \int F(y, y')p(y')dy$ is thus transformed to an integral operation. It may be approximated on the p-dimensional grid y_k as matrix-vector product

$$q(y_k) = \sum_l F(y_k, y_l)p(y_l)\delta y$$

$$F(y_k, y_l) = F(y_k)\delta(y_k - y_l) := F_{kl}/\delta y.$$

Explicitly, one obtains the distribution (cf. 16.35)

$$F(y, y') = -\delta(y_1 - y_1') \ldots \partial_{y_\alpha}\delta(y_\alpha - y_\alpha') \ldots \delta(y_p - y_p') f_\alpha(y)$$
$$+\frac{1}{2}\delta(y_1 - y_1') \ldots \partial_{y_\alpha}\delta(y_\alpha - y_\alpha') \ldots \partial_{y_\beta}\delta(y_\beta - y_\beta') \ldots \delta(y_p - y_p')\Omega_{\alpha\beta}(y).$$

The delta function is interpreted as a function sequence $\delta_n(y)$ with the property $\lim_{n\to\infty} \int \delta_n(y - y')\phi(y')dy' = \phi(y)$ for any test function ϕ (Lighthill 1958). In numerical applications, one must use a certain term of this delta sequence with appropriate smoothness properties, leading to a free parameter n of the numerical procedure. For example, one can use the series $\delta_n(y-y') = \sum_{m=0}^n \chi_m(y)\chi_m(y')$ for a complete orthonormal system χ_m or the truncated Fourier transform $\delta_n(y - y') = \int_{-n}^n \exp(2\pi i(y - y')k)dk$.

If one writes $F(y)p(y) = F(y)\int \delta(y - y')p(y')dy$, the term $p(y, \delta_n) = \int \delta_n(y - y')p(y')dy$ may be interpreted as a functional. In this guise the procedure was called DAF (distributed approximating functional; cf. Wei et al. 1997) using Hermite functions (oscillator eigenfunctions) χ_m.

If the delta functions on the grid points $y_{k_\alpha}, y_{l_\alpha}$ are replaced by $\delta(y_{k_\alpha} - y_{l_\alpha}) \to \delta_{k_\alpha l_\alpha}/\delta y_\alpha$, $\delta'(y_{k_\alpha} - y_{l_\alpha}) \to (\delta_{k_\alpha+1,l_\alpha} - \delta_{k_\alpha-1,l_\alpha})/(2\delta y_\alpha^2)$, $\delta''(y_{k_\alpha} - y_{l_\alpha}) \to (\delta_{k_\alpha+1,l_\alpha} - 2\delta_{k_\alpha l_\alpha} + \delta_{k_\alpha-1,l_\alpha})/\delta y_\alpha^3$, one recovers the finite difference approximation (16.35). This choice corresponds to the delta sequence $\delta_{\delta y}(y) = \chi_A(y)/\delta y$, $A = [-\delta y/2, \delta y/2]$ where $\chi_A(x)$ is the indicator function of the set A. In this case the free parameter $n = \delta y$ is naturally given by the spatial discretization interval δy. The Euler transition kernel does not require the choice of free parameters (it is naturally given by the time discretization δt). The spatial discretization should be of order $\sqrt{\Omega \delta t}$.

16.5 Applications

16.5.1 Geometrical Brownian Motion (GBM)

The SDE

$$dy(t) = \mu y(t)dt + \sigma y(t)\,dW(t) \tag{16.36}$$

is a popular model for stock prices, used by Black and Scholes (1973) for modeling option prices. It contains a multiplicative noise term $y\,dW$ and is thus bilinear. The form $dy(t)/y(t) = \mu dt + \sigma\, dW(t)$ shows that the simple returns are given by a constant value μdt plus white noise. For the log returns, we set $x = \log y$, and use Itô's lemma to obtain $dx = dy/y + 1/2(-y^{-2})dy^2 = (\mu - \sigma^2/2)dt + \sigma dW$. This shows that the log returns contain the Wong-Zakai correction, i.e., $\tilde{\mu} = \mu - \sigma^2/2$. From this, we obtain the exact solution

$$y(t) = y(t_0)e^{\tilde{\mu}(t-t_0)+\sigma[W(t)-W(t_0)]}, \tag{16.37}$$

which is a *multiplicative* exact discrete model with log-normal distribution. The exact transition density is thus

$$p(y,t|y_0,t_0) = y^{-1}\phi\left(\log(y/y_0); \tilde{\mu}(t-t_0), \sigma^2(t-t_0)\right) \tag{16.38}$$

The model was simulated using $\mu = 0.07$, $\sigma = 0.2$ and $\delta t = 1/365$. Only monthly data were used (Fig. 16.1).

The Langevin sampler output is displayed in Fig. 16.2. Due to a convergence kernel (inverse Hessian of the potential Φ), all trajectories relax to equilibrium at

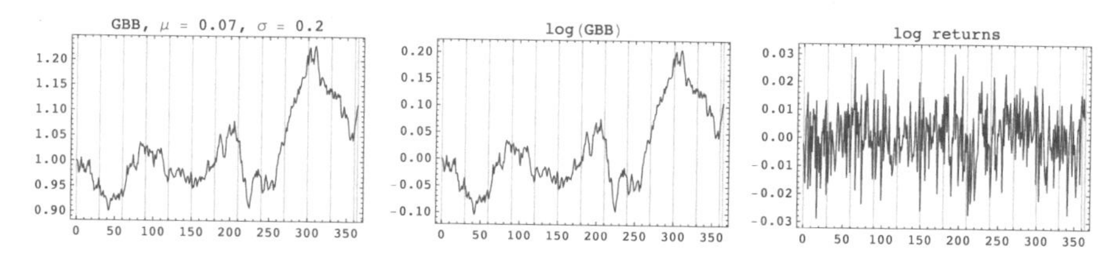

Fig. 16.1 Geometrical Brownian motion GBM: trajectory and log returns. Vertical lines: measurement times

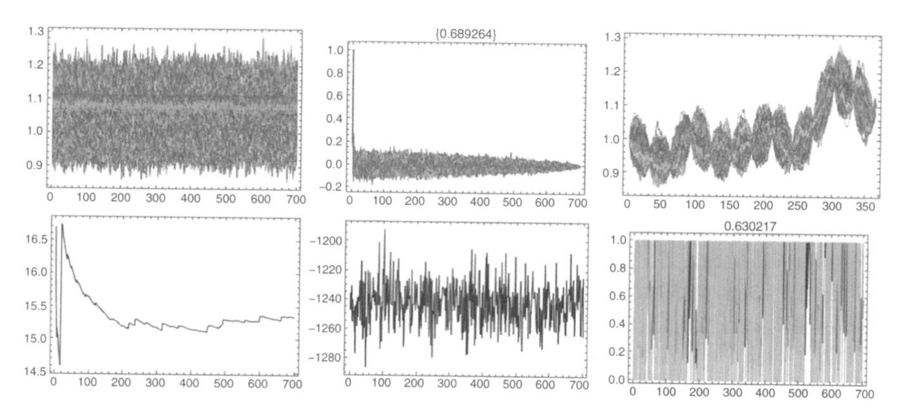

Fig. 16.2 Langevin sampler. Conditionally Gaussian importance density $\hat{p}_2 = \prod_j \phi(\eta_{j+1}, \eta_j|z)/\phi(\eta_j|z)$. From top, left: (1,1): trajectory η_{jl} over l (replications), (1,2): autocorrelation of η_{jl}, (1,3): trajectories η_{jl} over j, (2,1): convergence of estimator $\hat{p}(z)$, (2,2): potential $\log p(\eta_{jl})$ over l, (2,3): average acceptance probability and rejection indicator for Metropolis algorithm

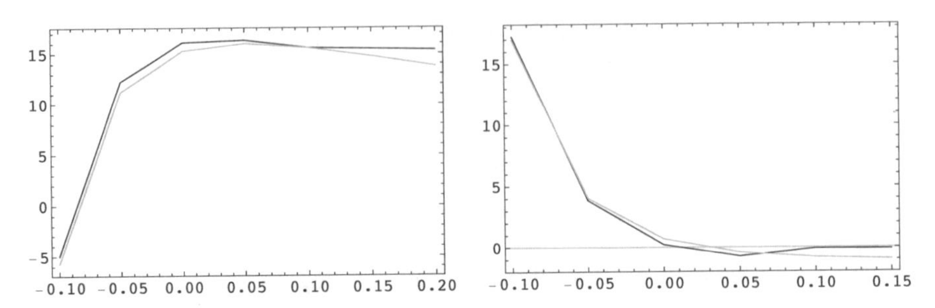

Fig. 16.3 Likelihood (left) and score (right) as a function of $\sigma - 0.2$, $\hat{p}_2$ = conditional kernel density. Thin lines: exact log-likelihood

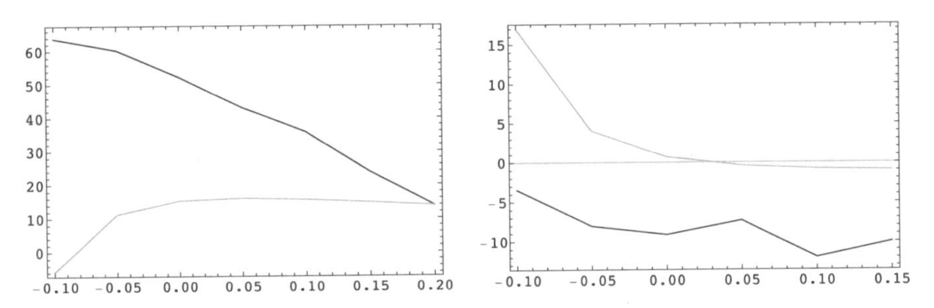

Fig. 16.4 Likelihood and score, $\hat{p}_2$ = full kernel density

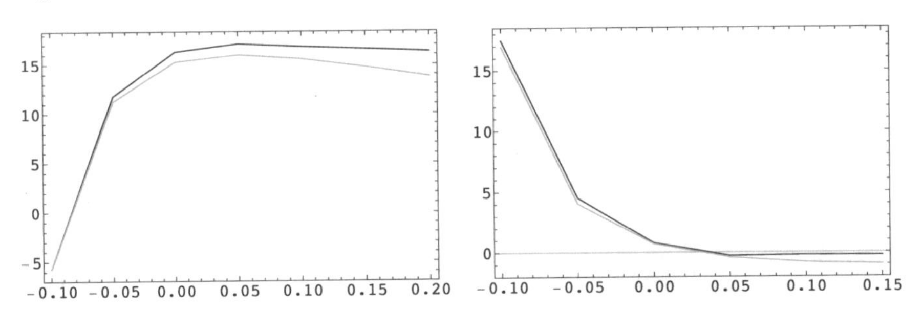

Fig. 16.5 Likelihood and score, $\hat{p}_2$ = conditionally Gaussian transition density

about the same rate, and the autocorrelation is small (first row, second columnn). We sampled $M = 2000$ trajectories, dropping 30% at the beginning to avoid non-stationarity and used only every second one (thinning). This reduces autocorrelation in the sample. Thus the effective sample size is $M' = 700$ (Figs. 16.3 and 16.4).

We obtain a smooth likelihood surface with small approximation error (Figs. 16.5, 16.6, 16.8, 16.9, 16.10). Clearly, the usage of the full kernel density (16.21) yields bad results (Fig. 16.4). In contrast, the conditional Markov representation (16.22) works well (Fig. 16.3). One can also use a conditional Gaussian density in (16.22) or a linear GLS estimation of the drift correction δf_j and diffusion correction $\delta \Omega_j$ (Fig. 16.6; see Eq. (16.23)). If the diffusion

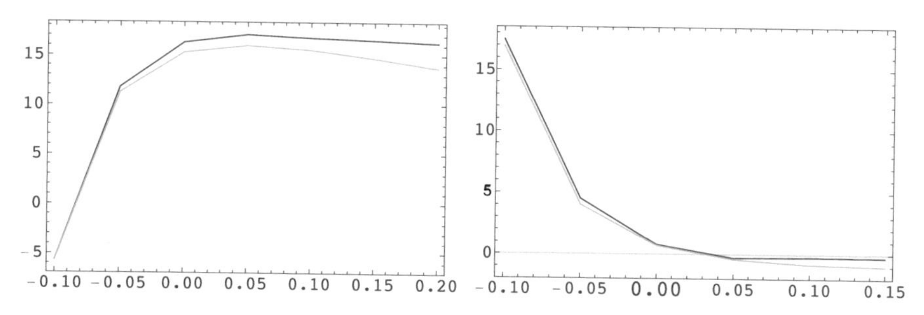

Fig. 16.6 Likelihood and score, $\hat{p}_2$ = linear GLS estimation of drift and diffusion corrections $\delta f_j, \delta\Omega_j$ (Eq. (16.23))

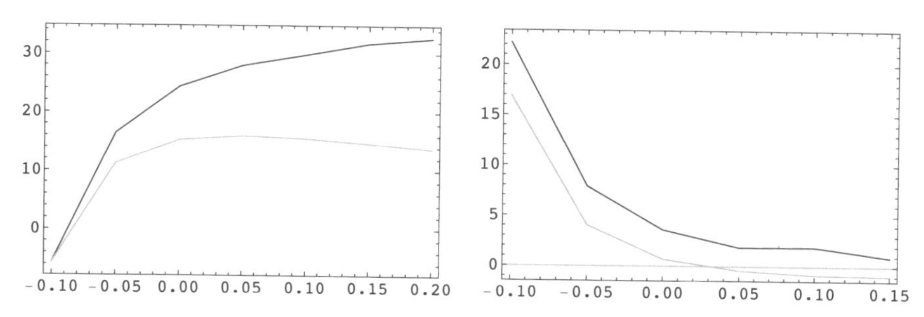

Fig. 16.7 Likelihood and score, $\hat{p}_2$ = linear GLS, diffusion matrix without correction

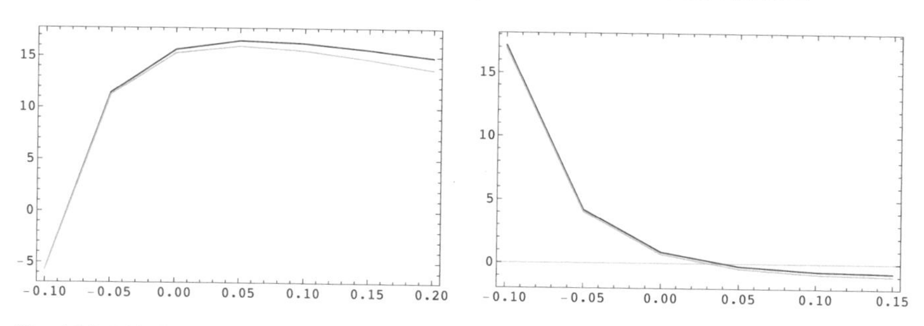

Fig. 16.8 Likelihood and score, Langevin/Gauss-resampling

matrix is not corrected ($\delta\Omega_j = 0$), the likelihood function just increases and does not attain a maximum (Fig. 16.7). Slightly better results are obtained using the Langevin sampler with subsequent Gauss-resampling (Fig. 16.8; see Eq. (16.26), (a)). Importance sampling using an extended Kalman smoother (extended Kalman sampling EKS, see (16.26)) yields very good results (Fig. 16.9). Finally, the transition kernel filter (TKF) with an Euler transition kernel is displayed, where the integration range is $\{y_0, y_1, dy\} = \{0.7, 1.5, 0.0025\}$. This leads to a 321×321 transition matrix with 103,041 elements (see Sect. 16.4.2.2). However, entries smaller than 10^{-8} were set to 0, and only 15,972 values were stored in a sparse array. Again, the results are very accurate. In this one-dimensional example, the

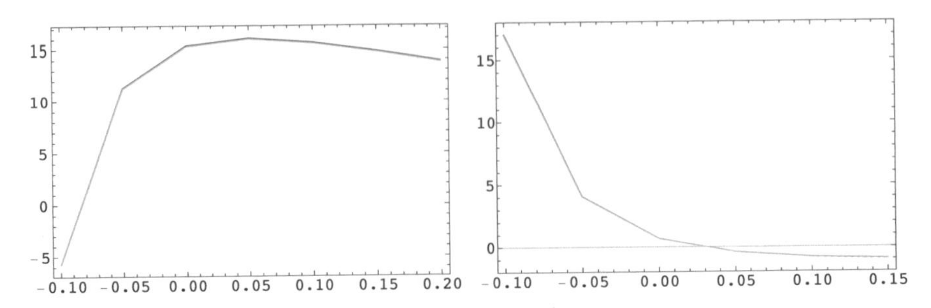

Fig. 16.9 Likelihood and score, extended Kalman sampling

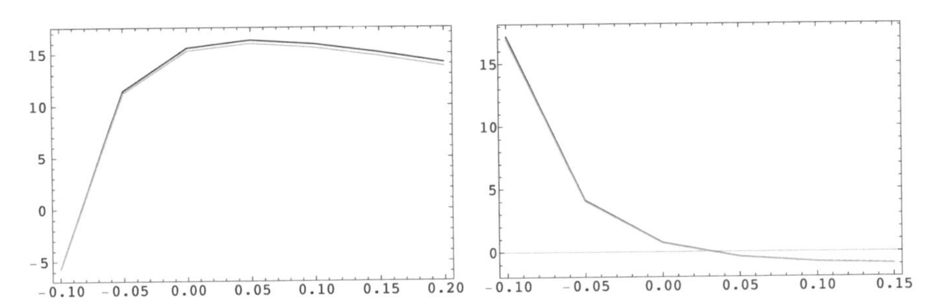

Fig. 16.10 Likelihood and score, transition kernel filter with Euler transition kernel

TKF is most efficient, while the EKS takes four times more CPU time. The methods based on the conditional Markov property take much more CPU time, especially when using a kernel density estimate for $p(\eta_{j+1}|\eta_j, z)$ (see 16.22). In a higher-dimensional state-space $\mathbb{R}^p$, however, the grid-based TKF soon gets infeasible, since we have K^p supporting points $\eta_k, k = (k_1, \ldots, k_p)$ in $\mathbb{R}^p$. The transition kernel matrix is even of dimension $K^p \times K^p$ (Fig. 16.10).

16.5.2 Ginzburg-Landau Model

The Ginzburg-Landau model

$$dY = -[\alpha Y + \beta Y^3]dt + \sigma dW(t) \tag{16.39}$$

with measurement equation $Z_i = Y_i + \epsilon_i$ at sampling times t_i is a well-known nonlinear benchmark model, since the stationary distribution

$$p_0(y) \propto \exp[-(2/\sigma^2)\Phi(y)],\ \Phi(y) := \tfrac{1}{2}\alpha y^2 + \tfrac{1}{4}\beta y^4$$

can exhibit a bimodal shape (see Fig. 16.11; Särkkä et al. 2013; Singer 2002, 2011). Originally arising from the theory of phase transitions in physics, it can also be used

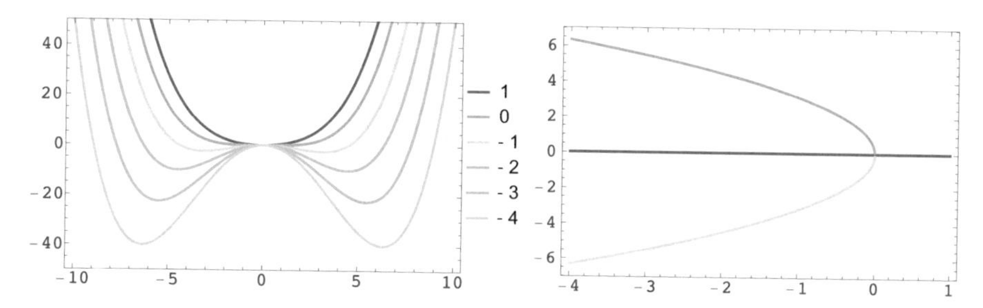

Fig. 16.11 Ginzburg-Landau model. Left: plot of potential Φ as function of order parameter y for several $\alpha = 1, 0, \ldots, -5$. For positive α values, one minimum at $y = 0$ occurs. Two minima and one maximum for negative values. Hopf bifurcation at $\alpha = 0$. Right: pitchfork diagram over α (3 extrema for $\alpha < 0$)

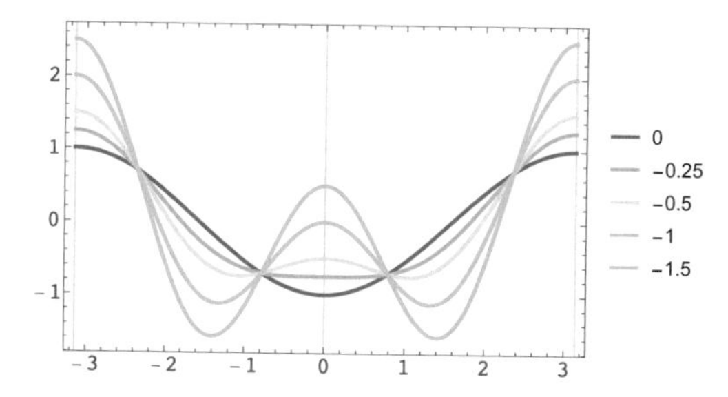

Fig. 16.12 Potential $V(y) = -a\cos(y) - b\cos(2y)$ of a model for human hand movement. The value b/a controls the number of stable minima. Below $b/a = -0.25$, only one minimum occurs

as a model in economics and the social sciences for bistable systems (Herings 1996). The minima of the "potential" Φ may be interpreted as stable states of the variable y (a so-called order parameter), describing collective properties of an interacting system (Haken 1977; Weidlich and Haag 1983). Indeed, a Taylor expansion of the potential used in Haken et al. (1985); Molenaar and Newell (2003) gives a quadratic-quartic form near $y = 0$ (see Fig. 16.12). The negative gradient $-\partial_y \Phi(y)$ of the potential (a force) is the drift function in the dynamic equation (16.39). If the parameter α varies from negative to positive values, at the critical value $\alpha = 0$ a qualitative change (phase transition) of the system occurs, namely, the two minima at $\pm\sqrt{-\alpha/\beta}$ merge into one minimum at 0 (Fig. 16.11). The true parameters were set to the values $\psi_0 = \{\alpha_0, \beta_0, \sigma_0\} = \{-1, 0.1, 2\}$, $R = \text{Var}(\epsilon_i) = 0.01$, and the trajectories were simulated with a discretization interval $\delta t = 0.1$ using an Euler-Maruyama approximation (cf. Kloeden and Platen 1999). The data were sampled at times $t_i \in \{0, 2, 4, \ldots, 20\}$.

Figures 16.13 and 16.14 show the simulated likelihood surface as a function of parameter $\sigma - \sigma_0$ using a Gaussian distribution as approximate importance density. The mean $E(\eta_j|Z)$ and covariance matrix $\text{Cov}(\eta_j|Z)$, $j = 0, \ldots, J = 200$ were computed either using an extended or an unscented Kalman smoother. For the conditional covariance matrix $\text{Cov}(\eta_j, \eta_{j'}|Z) : (J+1)p \times (J+1)p$, two variants of

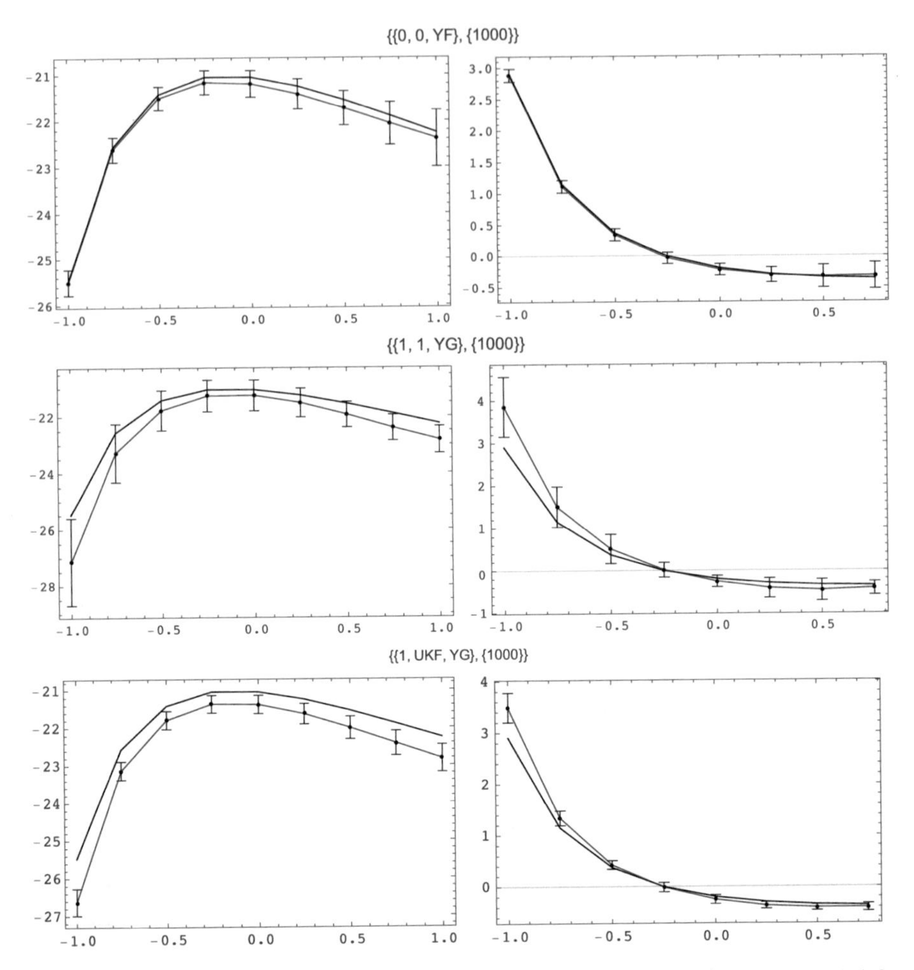

Fig. 16.13 Ginzburg-Landau model. Extended Kalman sampling, $M = 1000$. Likelihood (left) and score (right) as function of $\sigma - \sigma_0$. Reference trajectory = smoother solution. First row: Brownian bridge. Second row: extended Kalman smoother. Third row: unscented Kalman smoother. Line without error bars: transition kernel filter with Euler transition kernel

the linearization can be used: computation of the Jacobians $\partial f(y)/\partial y$ and $\partial h(y)/\partial y$ either along the filtered or along the smoothed trajectory. Furthermore, one can set $f \equiv 0$ in the smoother, which corresponds to a Brownian bridge (cf. Durham and Gallant 2002). The error bars were obtained using ten likelihood surface replications with different random seeds. Clearly, the Brownian bridge performs best, and the likelihood surface is smooth as a function of the parameter (Figs. 16.13 and 16.14, first line, right). This is in contrast to methods based on particle filters (cf. Pitt 2002; Singer 2003). In this case, a modified resampling method must be applied (Malik and Pitt 2011). It should be noted that not only the mean over the likelihood surfaces,

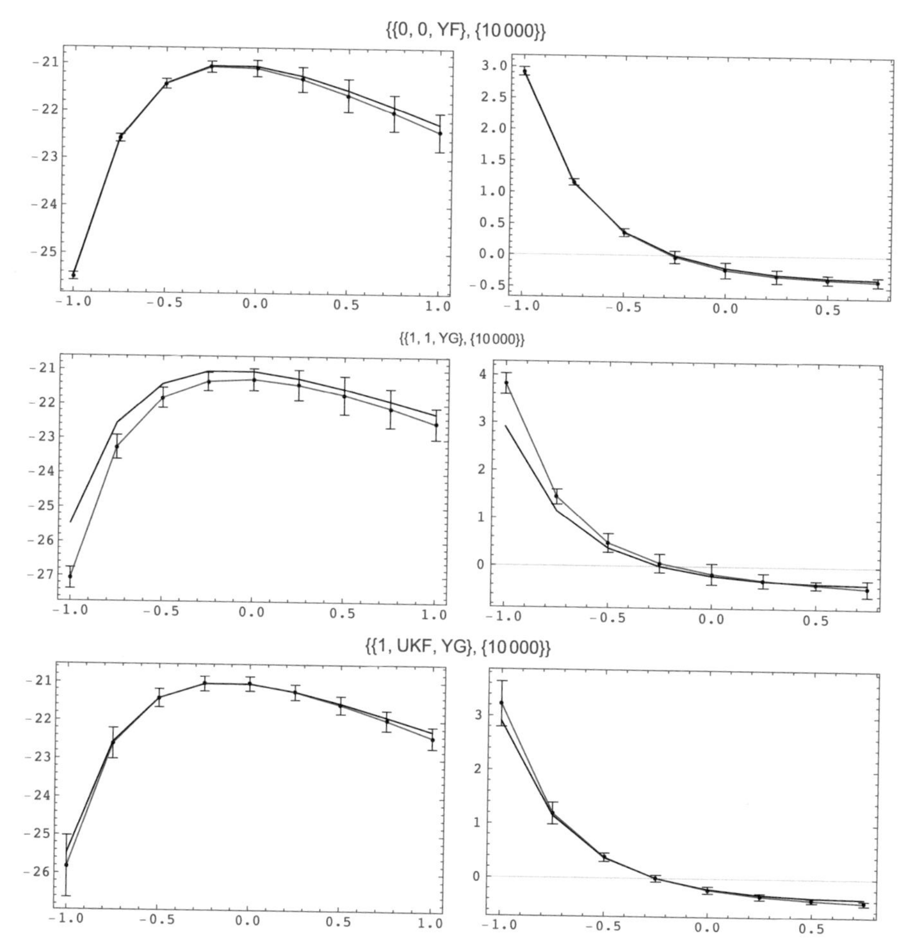

Fig. 16.14 Ginzburg-Landau model. Extended Kalman sampling, $M = 10{,}000$. Likelihood (left) and score (right) as function of $\sigma - \sigma_0$. Reference trajectory = smoother solution. First row: Brownian bridge. Second row: extended Kalman smoother. Third row: unscented Kalman smoother. Line without error bars: transition kernel filter with Euler transition kernel

but also each single one is a smooth curve, which facilitates the usage in Newton-type algorithms or in Langevin samplers for Bayesian estimation.

The superiority of the Brownian bridge sampler can be understood from Fig. 16.15. The Kalman smoother (first line, left) tends to predict values in the potential wells of $\Phi(y)$ (cf. Singer 2005, 2008b), whereas the exact smoothing solution computed from the Langevin sampler (first line, second column, $N = 2000$ replications) predicts more or less straight lines between the measurements. This behavior is exactly produced by the Brownian bridge (second line), since here the drift was set to $f = 0$. The unscented Kalman smoother also gives a conditional mean similar to the Langevin sampler (third line, left and middle). Also, it can be

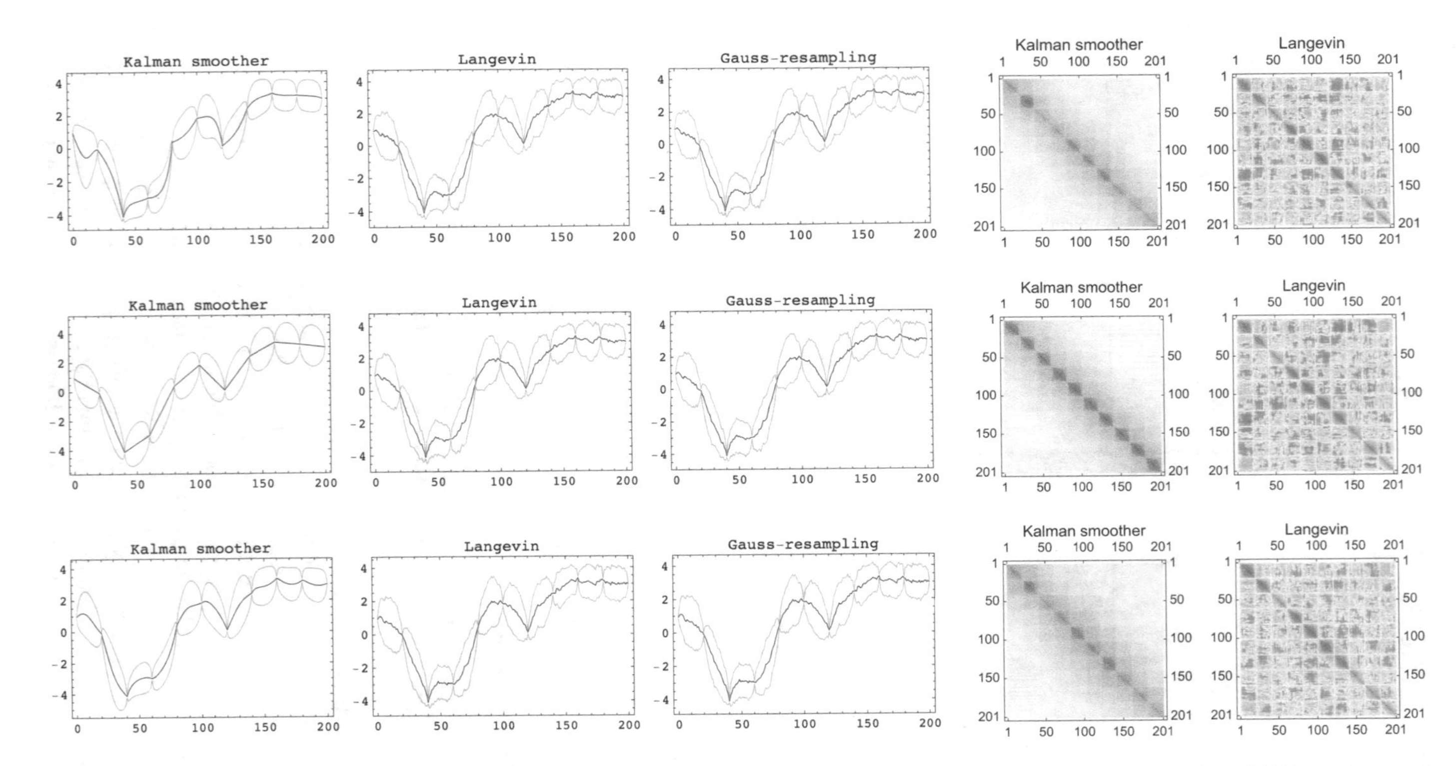

Fig. 16.15 Ginzburg-Landau model. First line, from left: Conditional mean ± standard deviation: (11) extended Kalman smoother, (12) Langevin sampler, (13) Gauss-resampling, conditional covariance matrix: (14) extended Kalman smoother, (15) Langevin sampler. Second line, Brownian bridge. Third line, unscented Kalman smoother (see text)

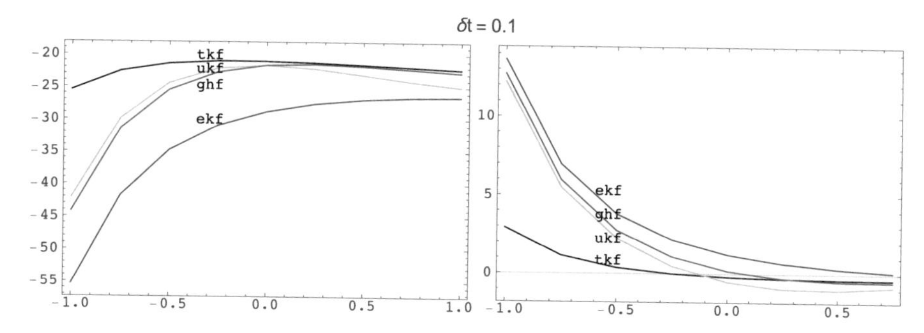

Fig. 16.16 Ginzburg-Landau model. Likelihood of TKF, EKF, UKF, and GHF as function of $\sigma - \sigma_0$

seen that the estimated conditional covariance matrix from the sampler data is quite noisy (Fig. 16.15, right column).

Thus, the conditional mean given by the Brownian bridge and the unscented Kalman smoother is (at least in this example) nearer to the exact smoothing solution (Langevin sampler) than the Kalman smoother, producing a better approximate importance density (16.26). This, in turn, improves the importance sampling results.

Since the extended Kalman sampler is based on an extended Kalman smoother, it is interesting to inspect the likelihood surfaces produced by the EKF and other integration-based filters, such as the unscented Kalman filter (UKF) and the Gauss-Hermite filter (GHF). The comparison with the exact TKF/ETK is displayed in Fig. 16.16. It is seen that the integration-based filters are superior to the Taylor-based EKF. Still, the likelihood surfaces strongly deviate from the exact TKF solution.

As noted above, the EKF moment equation $\dot{\mu} = E[f(y)] \approx f(\mu)$ gives solutions in the minima of $\Phi(y)$. This does not happen for the integration-based equation $\dot{\mu} = E[f(y)] \approx \sum \alpha_l f(\eta_l)$, where μ tends to zero (cf. Singer 2005, 2008b). Therefore, an integration-based smoother should give better results, as is the case.

Table 16.1 serves to compare the ML estimates given by the several estimation methods with the reference solution TKF/ETK. The likelihood function was maximized using a quasi-Newton algorithm with numerical score and BFGS secant updates. Convergence was assumed when the score $||s_k||_1$ and the step $||\psi_{k+1} - \psi_k||_1$ were both smaller than $\epsilon = 10^{-2}$. The standard errors were computed using the negative inverse Hessian (observed Fisher information) after convergence. The results of the Kalman sampler with $M = 10{,}000$ replications are very similar to the reference solution TKF/ETK. A smaller sample size ($M = 1000$) gives good results for the Brownian bridge sampler, the unscented Kalman sampler, and the Gauss-Hermite sampler. The extended Kalman smoother leads to bad results in this case. The integration-based filters UKF and GHF produce somewhat biased ML estimates, whereas the EKF, again, performs worst, especially for parameter α. Clearly, these findings are only preliminary. In a simulation study with $S = 100$ replications, the methods

Table 16.1 ML estimates for several estimation methods

	TKF/ETK		EKF		UKF, $\kappa = 0$		GHF, $m = 3$	
True	est	std	est	std	est	std	est	std
−1	−0.5005	0.5407	−0.1039	0.2432	−0.349	0.7421	−0.349	0.7421
0.1	0.0609	0.0568	0.0364	0.0415	0.0406	0.0642	0.0406	0.0642
2	1.6157	0.4739	1.543	0.3801	1.7353	0.7355	1.7353	0.7355

	EKS		BBS		UKS, $\kappa = 0$		GHS, $m = 3$	
	est	std	est	std	est	std	est	std
Extended Kalman sampling, $M = 1000$								
−1	−1.0676	0.3985	−0.4176	0.4457	−0.4176	0.4457	−0.4762	0.4357
0.1	0.1166	0.0439	0.0518	0.0464	0.0518	0.0464	0.0572	0.0424
2	1.73	0.3122	1.5631	0.4245	1.5631	0.4245	1.5969	0.4239
Extended Kalman sampling, $M = 10{,}000$								
−1	−0.4864	0.4659	−0.4908	0.5061	−0.4501	0.4449	−0.4498	0.4468
0.1	0.0581	0.0437	0.06	0.0533	0.0555	0.0454	0.0554	0.0454
2	1.5882	0.4169	1.6101	0.4606	1.5899	0.4352	1.5888	0.4286

The transition kernel filter with Euler transition kernel (TKF/ETK) (top, left) serves as reference value. *EKF* extended Kalman filter, *UKF* unscented Kalman filter, *GHF* Gauss-Hermite Kalman filter, *EKS* extended Kalman sampler, *BBS* Brownian bridge sampler ($f = 0$), *UKS* unscented Kalman sampler, *GHS* Gauss-Hermite sampler

1. Kalman filtering (extended Kalman filter, unscented Kalman filter),
2. Numerical integration with transition kernels filters: Euler (TKF/ETK), finite differences (TKF/FD), integral operator (TKF/INT)
3. Kalman smoother sampling (extended Kalman smoothing EKS, Brownian bridge (BBS). and unscented Kalman smoothing (UKS)

are compared (Table 16.2). The data were simulated as described above and sampled with intervals $\Delta t = 1, 2$. Thus the data sets are of length $T + 1 = 11, 21$.

The sample statistics of the approximate ML estimators are presented in Table 16.2. Generally, for the smaller sampling interval, the estimates display smaller bias and root-mean-square error, computed as $\text{RMSE} = \sqrt{\text{bias}^2 + \text{std}^2}$, where $\text{bias} = \bar{\hat{\psi}} - \psi_0$ is the difference of the sample mean of ML estimator $\hat{\psi}$ and the true value, and std^2 is the sample variance.

In terms of speed, the Kalman filter methods are fastest, followed by the transition kernel filter and the Kalman sampler. However, in higher dimensions p of the state space, the performance of the grid-based approach declines, since one has an exponential increase of sample points (see Sect. 16.4.2), whereas the sampler works with matrices of order $Jp \times Jp$. Generally, the estimates of parameters β and σ are weakly biased. However, parameter α, which controls the bimodality of the potential (see Fig. 16.11), is more difficult to estimate. For the EKF, the mean is near the true value, but the standard error is large, leading to high RMSE values. An inspection of the histogram and other statistics reveals a skewed distribution.

Table 16.2 Simulation study

meth.	par.	value	$\Delta t = 1$ mean	med	std	bias	RMSE	$\Delta t = 2$ mean	med	std	bias	RMSE
Kalman filter												
EKF	α	−1	−0.956	−0.536	1.268	0.044	1.269	−0.955	−0.369	1.322	0.045	1.322
	β	0.1	0.148	0.087	0.165	0.048	0.172	0.201	0.093	0.237	0.101	0.258
	σ	2	2.229	2.027	0.765	0.229	0.799	2.746	2.284	1.411	0.746	1.596
UKF	α	−1	−1.586	−0.865	1.965	−0.586	2.051	−1.414	−0.354	2.068	−0.414	2.109
	β	0.1	0.161	0.086	0.184	0.061	0.194	0.142	0.054	0.181	0.042	0.186
	σ	2	2.135	2.058	0.559	0.135	0.575	2.127	1.966	0.721	0.127	0.732
Transition kernel filter												
ETK	α	−1	−1.392	−1.108	1.028	−0.392	1.100	−1.460	−1.073	1.201	−0.460	1.287
	β	0.1	0.149	0.118	0.115	0.049	0.125	0.154	0.113	0.128	0.054	0.139
	σ	2	2.171	2.116	0.483	0.171	0.512	2.207	2.041	0.710	0.207	0.739
FD	α	−1	−1.300	−1.077	0.831	−0.300	0.884	−1.474	−1.066	1.268	−0.474	1.353
	β	0.1	0.139	0.113	0.091	0.039	0.099	0.155	0.112	0.131	0.055	0.142
	σ	2	2.157	2.120	0.466	0.157	0.491	2.221	2.060	0.723	0.221	0.756
INT	α	−1	−1.298	−1.076	0.829	−0.298	0.881	−1.470	−1.066	1.262	−0.470	1.347
	β	0.1	0.139	0.113	0.091	0.039	0.099	0.154	0.112	0.130	0.054	0.141
	σ	2	2.157	2.120	0.465	0.157	0.491	2.220	2.059	0.722	0.220	0.755

(continued)

Table 16.2 (continued)

meth.	par.	value	$\Delta t = 1$ mean	med	std	bias	RMSE	$\Delta t = 2$ mean	med	std	bias	RMSE
Extended Kalman sampling												
$M = 1000$												
EKS	α	−1	−0.837	−0.823	0.281	0.163	0.325	−0.646	−0.648	0.231	0.354	0.422
	β	0.1	0.088	0.086	0.027	−0.012	0.029	0.071	0.066	0.031	−0.029	0.043
	σ	2	1.931	1.893	0.344	−0.069	0.351	1.804	1.740	0.453	−0.196	0.494
BBS	α	−1	−0.893	−0.873	0.333	0.107	0.350	−0.824	−0.812	0.381	0.176	0.420
	β	0.1	0.094	0.088	0.035	−0.006	0.035	0.091	0.086	0.045	−0.009	0.046
	σ	2	1.921	1.887	0.357	−0.079	0.366	1.932	1.931	0.504	−0.068	0.508
UKS	α	−1	−0.877	−0.820	0.340	0.123	0.361	−0.701	−0.717	0.272	0.299	0.404
	β	0.1	0.095	0.090	0.036	−0.005	0.037	0.075	0.072	0.032	−0.025	0.041
	σ	2	1.941	1.911	0.364	−0.059	0.368	1.847	1.863	0.459	−0.153	0.484
$M = 2000$												
BBS	α	−1	−0.961	−0.964	0.364	0.039	0.366	−0.861	−0.862	0.393	0.139	0.417
	β	0.1	0.102	0.096	0.041	0.002	0.041	0.092	0.086	0.042	−0.008	0.043
	σ	2	1.956	1.910	0.367	−0.044	0.369	1.932	1.869	0.439	−0.068	0.445
$M = 5000$												
BBS	α	−1	−1.024	−1.040	0.430	−0.024	0.431	−1.009	−0.998	0.420	−0.009	0.420
	β	0.1	0.108	0.104	0.045	0.008	0.046	0.103	0.101	0.040	0.003	0.040
	σ	2	1.965	1.932	0.352	−0.035	0.354	2.011	1.952	0.467	0.011	0.467
$M = 10{,}000$												
BBS	α	−1	−1.018	−0.997	0.413	−0.018	0.413	−0.893	−0.950	0.392	0.107	0.407
	β	0.1	0.107	0.104	0.043	0.007	0.044	0.095	0.093	0.040	−0.005	0.041
	σ	2	1.977	1.974	0.346	−0.023	0.346	1.941	1.921	0.456	−0.059	0.460

Sampling intervals $\Delta t = 1, 2$. Comparison of Kalman filter, transition kernel filter and extended Kalman sampling. See text

Indeed, the median is far from the true value $\alpha_0 = -1$. The performance of the UKF is disappointing. Again, the distribution of $\hat{\alpha}$ is strongly left-skewed. There are some samples with very low estimates. Slightly better results are obtained for the Gauss-Hermite filter (not reported).

The transition kernel filter as well yields biased estimates for α, again with a left-skewed distribution. The finite difference and the integral operator method perform very similar. An inspection of the median shows that it is near to the true values.

The best results are obtained with the Kalman smoother sampling. From Table 16.1 and Figs. 16.13 and 16.14, a sample size of $M = 1000$ is not sufficient to give a good approximation of the likelihood surface. Therefore, for the Brownian bridge sampler, simulations with $M = 2000, 5000, 10{,}000$ were performed. Especially for the large sampling interval $\Delta t = 2$, the bias could be reduced with increasing M. One must keep in mind, however, that the data sample size is small, so the (unknown) exact ML estimator may be biased as well.

Other estimation methods for the importance density, such as Langevin sampling with kernel density or regression methods, are very time-consuming when a reasonable approximation for the likelihood surface should be achieved. These methods are presently under further study.

16.6 Conclusion

We analytically computed the drift function of a Langevin sampler for the continuous-discrete state-space model, including a state-dependent diffusion function. In the continuum limit, a stochastic partial differential equation is obtained. From this, we can draw random vectors from the conditional distribution of the latent states, given the data. This sample can be used for the estimation of the unknown importance density and in turn to the determination of a variance-reduced MC estimator of the likelihood function. Moreover, one obtains a numerical solution of the optimal smoothing problem.

The unknown importance density was estimated from the sampler data using kernel density and regression methods. Alternatively, a Gaussian reference density with suboptimal properties but known analytical form was used. Methods based on transition kernels and the Fokker-Planck equation generally gave good results but seem to be restricted to low-dimensional state spaces. In the geometrical Brownian motion model, well known from finance, all methods gave encouraging results.

However, in a strongly nonlinear system (Ginzburg-Landau model), estimation methods for the importance density based on kernel density and regression methods performed disappointingly, whereas the extended Kalman sampler EKS (using a Gaussian importance density determined by extended Kalman smoothers) gave smooth likelihood surfaces near to the true ones.

In further work, the methods will be tested with further simulation studies and extended to higher-dimensional models such as the three-dimensional Lorenz model well known from chaos theory.

More generally, one can compute variance-reduced estimates of functionals involving Itô processes such as the Feynman-Kac formula used in finance and quantum theory (Singer 2016).

Appendix 1: Langevin Sampler: Analytic Drift Function

Notation

In the following, the components of vectors and matrices are denoted by Greek letters, e.g., $f_\alpha, \alpha = 1, \ldots, p$, and partial derivatives by commas, i.e., $f_{\alpha,\beta} := \partial f_\alpha / \partial \eta_\beta = \partial_\beta f_\alpha = (f_\eta)_{\alpha\beta}$. The Jacobian matrix $\partial f/\partial \eta$ is written as f_η and its βth column as $(f_\eta)_{\bullet\beta}$. Likewise, $\Omega_{\alpha\bullet}$ denotes row α of matrix $\Omega_{\alpha\beta}$ and $\Omega_{\bullet\bullet} = \Omega$ for short.

Latin indices denote time, e.g., $f_{j\alpha} = f_\alpha(\eta_j)$. Furthermore, a sum convention is used for the Greek indices (i.e., $f_\alpha g_\alpha = \sum_\alpha f_\alpha g_\alpha$). The difference operators $\delta = B^{-1} - 1, \nabla = 1 - B$, with the backshift $B\eta_j = \eta_{j-1}$ are used frequently. One has $\delta \cdot \nabla = B^{-1} - 2 + B := \Delta$ for the central second difference.

Functional Derivatives

The functional $\Phi(y)$ may be expanded to first order by using the functional derivative $(\delta\Phi/\delta y)(h) = \int (\delta\Phi/\delta y(s))h(s)ds$. One has $\Phi(y + h) - \Phi(y) = (\delta\Phi/\delta y)(h) + O(\|h\|^2)$.

A discrete version is $\Phi(\eta) = \Phi(\eta_0, \ldots, \eta_J)$ and $\Phi(\eta + h) - \Phi(\eta) = \sum_j [\partial\Phi(\eta)/\partial(\eta_j \delta t)] h_j \delta t + O(\|h\|^2)$. As a special case, consider the functional $\Phi(\eta) = \eta_j$. Since $\eta_j + h_j - \eta_j = \sum (\delta_{jk}/\delta t) h_k \delta t$ one has the continuous analogue $y(t) + h(t) - y(t) = \int \delta(t - s)h(s)ds$, thus $\delta y(t)/\delta y(s) = \delta(t - s)$.

State-Independent Diffusion Coefficient

First we assume a state-independent diffusion coefficient $\Omega_j = \Omega$, but later we set $\Omega_j = \Omega(\eta_j, x_j)$. This is important, if the Lamperti transformation does not lead to constant coefficients in multivariate models.[8] In components, the term (16.15) reads

$$S_0 = \frac{1}{2}\sum_{j=0}^{J-1}(\eta_{j+1;\beta} - \eta_{j\beta})(\Omega_{\beta\gamma}\delta t)^{-1}(\eta_{j+1;\gamma} - \eta_{j\gamma}),$$

Note that $(\Omega_{\beta\gamma}\delta t)^{-1} \equiv [(\Omega\delta t)^{-1}]_{\beta\gamma}$ and the semicolon in $\eta_{j+1;\beta}$ serves to separate the indices; it is not a derivative. Differentiation w.r.t. the state $\eta_{j\alpha}$ yields ($j = 1, \ldots, J-1$)

$$\partial S_0/\partial(\eta_{j\alpha}\delta t) = -\Omega^{-1}_{\alpha\gamma}\delta t^{-2}(\eta_{j+1;\gamma} - 2\eta_{j\gamma} + \eta_{j-1;\gamma}) \tag{16.40}$$

In vector notation, we have $\partial S_0/\partial(\eta_j\delta t) = -\Omega^{-1}\delta t^{-2}\Delta\eta_j$. On the boundaries $j = 0, j = J$ we obtain

$$\partial S_0/\partial(\eta_{0\alpha}\delta t) = -\Omega^{-1}_{\alpha\gamma}\delta t^{-2}(\eta_{1\gamma} - \eta_{0\gamma})$$

$$\partial S_0/\partial(\eta_{0\alpha}\delta t) = \Omega^{-1}_{\alpha\gamma}\delta t^{-2}(\eta_{J\gamma} - \eta_{J-1;\gamma})$$

Next, the derivatives of $\log\alpha(\eta)$ are needed. One gets

$$\partial S_1/\partial(\eta_{j\alpha}\delta t) = -\delta t^{-1}[f_{j\beta,\alpha}\Omega^{-1}_{\beta\gamma}\delta\eta_{j\gamma} - \Omega^{-1}_{\alpha\gamma}(f_{j\gamma} - f_{j-1;\gamma})]$$

or in vector form, using difference operators

$$\partial S_1/\partial(\eta_{j\alpha}\delta t) = -\delta t^{-1}[f'_{j\bullet,\alpha}\Omega^{-1}\delta\eta_j - \Omega^{-1}\delta f_{j-1}], \tag{16.41}$$

where $f_{j\bullet,\alpha}$ is column α of the Jacobian $f_\eta(\eta_j)$. The second term yields

$$\begin{aligned}\partial S_2/\partial(\eta_{j\alpha}\delta t) &= \partial/\partial\eta_{j\alpha}\frac{1}{2}[f_{j\beta}\Omega^{-1}_{\beta\gamma}f_{j\gamma}] = f_{j\beta,\alpha}\Omega^{-1}_{\beta\gamma}f_{j\gamma}\\ &= f'_{j\bullet,\alpha}\Omega^{-1}f_j.\end{aligned} \tag{16.42}$$

Finally, one has to determine the drift component corresponding to the measurements, which is contained in the conditional density $p(z|\eta)$. Since it was assumed

[8] These are called irreducible diffusions. A transformation $z = h(y)$ leading to unit diffusion for z must fulfil the system of differential equations $h_{\alpha,\beta}g_{\beta\gamma} = \delta_{\alpha\gamma}, \alpha, \beta = 1, \ldots, p; \gamma = 1, \ldots, r$. The inverse transformation $y = v(z)$ fulfills $v_{\alpha,\gamma}(z) = g_{\alpha\gamma}(v(z))$. Thus $v_{\alpha,\gamma\delta} = g_{\alpha\gamma,\epsilon}v_{\epsilon,\delta} = v_{\alpha,\delta\gamma} = g_{\alpha\delta,\epsilon}v_{\epsilon,\gamma}$. Inserting v, one obtains the commutativity condition $g_{\alpha\gamma,\epsilon}\,g_{\epsilon\delta} = g_{\alpha\delta,\epsilon}\,g_{\epsilon\gamma}$, which is necessary and sufficient for reducibility. See Kloeden and Platen (1992, ch. 10, p. 348), Aït-Sahalia (2008).

that the error of measurement is Gaussian (see 16.2), we obtain

$$p(z|\eta) = \prod_{i=0}^{T} p(z_i|\eta_{j_i}) = \prod_{i=0}^{T} \phi(z_i; h_i, R_i),$$

where $\phi(y; \mu, \Sigma)$ is the multivariate Gaussian density, $h_i = h(\eta_{j_i}, x_{j_i})$ is the output function and $R_i = R(x_{j_i})$ is the measurement error covariance matrix. Thus the derivative reads (matrix form in the second line)

$$\begin{aligned} \partial \log p(z|\eta)/\partial(\eta_{j\alpha}\delta t) &= \sum_{i=0}^{T} h_{i\gamma,\alpha} R^{-1}_{i\beta\gamma}(z_{i\beta} - h_{i\beta})(\delta_{jj_i}/\delta t) \\ &= \sum_{i=0}^{T} h'_{i\bullet,\alpha} R_i^{-1}(z_i - h_i)(\delta_{jj_i}/\delta t) \end{aligned} \quad (16.43)$$

The Kronecker symbol δ_{jj_i} only gives contributions at the measurement times $t_i = \tau_{j_i}$. Together we obtain for the drift of the Langevin equation (16.14)

$$\begin{aligned} \delta_\eta \log p(\eta|z) &= \delta_\eta[\log p(z|\eta) + \log p(\eta)] \\ &= (16.43) - (16.40 + 16.41 + 16.42) + \delta_\eta \log p(\eta_0). \end{aligned} \quad (16.44)$$

Here, $p(\eta_0)$ is an arbitrary density for the initial latent state.

State-Dependent Diffusion Coefficient

In the case of $\Omega_j = \Omega(\eta_j, x_j)$ the expressions get more complicated. The derivative of S_0 now reads

$$\begin{aligned} \partial S_0/\partial(\eta_{j\alpha}\delta t) = \delta t^{-2}[&\Omega^{-1}_{j-1;\alpha\beta}\delta\eta_{j-1;\beta} - \Omega^{-1}_{j\alpha\beta}\delta\eta_{j\beta} \\ &+\frac{1}{2}\delta\eta_{j\beta}\Omega^{-1}_{j\beta\gamma,\alpha}\delta\eta_{j\gamma}], \end{aligned} \quad (16.45)$$

$\Omega^{-1}_{j\beta\gamma,\alpha} \equiv (\Omega^{-1})_{j\beta\gamma,\alpha}$. A closer relation to expression (16.40) may be obtained by the Taylor expansion

$$\Omega^{-1}_{j-1;\alpha\beta} = \Omega^{-1}_{j\alpha\beta} + \Omega^{-1}_{j\alpha\beta,\gamma}(\eta_{j-1;\gamma} - \eta_{j\gamma}) + O(\|\delta\eta_{j-1}\|^2) \quad (16.46)$$

leading to

$$\begin{aligned} \partial S_0/\partial(\eta_{j\alpha}\delta t) &= -\Omega^{-1}_{j\alpha\beta}\delta t^{-2}(\eta_{j+1;\beta} - 2\eta_{j\beta} + \eta_{j-1;\beta}) \\ &\quad -\Omega^{-1}_{j\alpha\beta,\gamma}\delta t^{-2}\delta\eta_{j-1;\beta}\delta\eta_{j-1;\gamma} + O(\delta t^{-2}\|\delta\eta_{j-1}\|^3) \\ &\quad +\frac{1}{2}\Omega^{-1}_{j\beta\gamma,\alpha}\delta t^{-2}\delta\eta_{j\beta}\delta\eta_{j\gamma}. \end{aligned} \tag{16.47}$$

In the state-dependent case, also the derivative of the Jacobian term $\log Z^{-1} = -\frac{1}{2}\sum_j \log|2\pi\Omega_j\delta t|$ is needed. Since the derivative of a log determinant is

$$\partial \log|\Omega|/\partial\Omega_{\alpha\beta} = \Omega^{-1}_{\beta\alpha},$$

one obtains

$$\partial \log Z^{-1}/\partial(\eta_{j\alpha}\delta t) = -\frac{1}{2}\delta t^{-1}\Omega^{-1}_{j\beta\gamma}\Omega_{j\beta\gamma,\alpha} = -\frac{1}{2}\delta t^{-1}\mathrm{tr}[\Omega_j^{-1}\Omega_{j,\alpha}],$$

$\Omega_{j,\alpha} = \Omega_{j\bullet\bullet,\alpha}$ for short. Using the formula $\Omega_j\Omega_j^{-1} = I$; $\Omega_{j,\alpha} = -\Omega_j\Omega^{-1}_{j,\alpha}\Omega_j$, we find

$$\partial \log Z^{-1}/\partial(\eta_{j\alpha}\delta t) = \frac{1}{2}\delta t^{-1}\mathrm{tr}[\Omega^{-1}_{j,\alpha}\Omega_j]. \tag{16.48}$$

The contributions of S_1 and S_2 are now (see 16.16)

$$\partial S_1/\partial(\eta_{j\alpha}\delta t) = \tag{16.49}$$
$$-\delta t^{-1}[f_{j\beta,\alpha}\Omega^{-1}_{j\beta\gamma}\delta\eta_{j\gamma} - (\Omega^{-1}_{j\alpha\gamma}f_{j\gamma} - \Omega^{-1}_{j-1;\alpha\gamma}f_{j-1;\gamma}) + f_{j\beta}\Omega^{-1}_{j\beta\gamma,\alpha}\delta\eta_{j\gamma}]$$

$$\partial S_2/\partial(\eta_{j\alpha}\delta t) = f_{j\beta,\alpha}\Omega^{-1}_{j\beta\gamma}f_{j\gamma} + \frac{1}{2}f_{j\beta}\Omega^{-1}_{j\beta\gamma,\alpha}f_{j\gamma}. \tag{16.50}$$

It is interesting to compare the terms in (16.45, 16.49, 16.50) depending on the derivative $\Omega^{-1}_{j\beta\gamma,\alpha}$, which read in vector form

$$\frac{1}{2}\delta t^{-2}\mathrm{tr}[\Omega^{-1}_{j,\alpha}\delta\eta_j\delta\eta_j'] - \delta t^{-1}\mathrm{tr}[\Omega^{-1}_{j,\alpha}\delta\eta_j f_j'] + \frac{1}{2}\mathrm{tr}[\Omega^{-1}_{j,\alpha}f_j f_j'],$$

and the Jacobian derivative (16.48). The terms can be collected to yield

$$\frac{1}{2}\delta t^{-2}\mathrm{tr}\{\Omega^{-1}_{j,\alpha}[\Omega_j\delta t - (\delta\eta_j - f_j\delta t)(\delta\eta_j - f_j\delta t)']\}, \tag{16.51}$$

as may be directly seen from the Lagrangian (16.7).

In summary, the Langevin drift component $(j\alpha)$, $j = 0, \ldots J; \alpha = 1, \ldots, p$ is in vector-matrix form

$$
\begin{aligned}
\delta_{\eta_{j\alpha}} \log p(\eta|z) &= \delta_{\eta_{j\alpha}}[\log p(z|\eta) + \log p(\eta)] \qquad (16.52)\\
&= \sum_{i=0}^{T} h'_{i\bullet,\alpha} R_i^{-1}(z_i - h_i)(\delta_{jj_i}/\delta t)\\
&\quad + \delta t^{-2}[\Omega^{-1}_{j\alpha\bullet}\delta\eta_j - \Omega^{-1}_{j-1;\alpha\bullet}\delta\eta_{j-1}]\\
&\quad + \delta t^{-1}[f'_{j\bullet,\alpha}\Omega_j^{-1}\delta\eta_j - (\Omega^{-1}_{j\alpha\bullet}f_j - \Omega^{-1}_{j-1;\alpha\bullet}f_{j-1})]\\
&\quad - f'_{j\bullet,\alpha}\Omega_j^{-1} f_j\\
&\quad + \frac{1}{2}\delta t^{-2}\mathrm{tr}\{\Omega^{-1}_{j,\alpha}[\Omega_j\delta t - (\delta\eta_j - f_j\delta t)(\delta\eta_j - f_j\delta t)']\}\\
&\quad + \delta_{\eta_{j\alpha}} \log p(\eta_0).
\end{aligned}
$$

Here, $h_{i\bullet,\alpha}$ is column α of Jacobian $h_\eta(\eta_{j_i})$, $\Omega^{-1}_{j\alpha\bullet}$ is row α of $\Omega(\eta_j)^{-1}$, $\Omega^{-1}_{j,\alpha} := \Omega^{-1}_{j\bullet\bullet,\alpha}$, and $f_{j\bullet,\alpha}$ denotes column α of Jacobian $f_\eta(\eta_j)$.

Appendix 2: Continuum Limit

The expressions in the main text were obtained by using an Euler discretization of the SDE (16.1), so in the limit $\delta t \to 0$, one expects a convergence of η_j to the true state $y(\tau_j)$ (see Kloeden and Platen 1999, ch. 9). Likewise, the $(J+1)p$-dimensional Langevin equation (16.14) for $\eta_{j\alpha}(u)$ will be an approximation of the stochastic partial differential equation (SPDE) for the random field $Y_\alpha(u, t)$ on the temporal grid $\tau_j = t_0 + j\delta t$.

A rigorous theory (assuming constant diffusion matrices) is presented in the work of Reznikoff and Vanden-Eijnden (2005); Hairer et al. (2005, 2007); Apte et al. (2007); Hairer et al. (2011). In this section it is attempted to gain the terms, obtained in this literature by functional derivatives, directly from the discretization, especially in the case of state-dependent diffusions. Clearly, the finite-dimensional densities w.r.t. Lebesgue measure lose their meaning in the continuum limit, but the idea is to use large but finite J, so that the Euler densities $p(\eta_0, \ldots, \eta_J)$ are good approximations of the unknown finite-dimensional densities $p(y_0, \tau_0; \ldots; y_J, \tau_J)$ of the process $Y(t)$ (cf. Stratonovich 1971, 1989, Bagchi 2001 and the references cited therein).

Constant Diffusion Matrix

First we consider constant and (nonsingular) diffusion matrices Ω. The Lagrangian (16.15) attains the formal limit (Onsager-Machlup functional)

$$S = \frac{1}{2}\int dy(t)'(\Omega dt)^{-1}dy(t) \tag{16.53}$$

$$-\int f(y)'\Omega^{-1}dy(t) + \frac{1}{2}\int f(y)'\Omega^{-1}f(y)dt. \tag{16.54}$$

If $y(t)$ is a sample function of the diffusion process $Y(t)$ in (16.1), the first term (16.53) does not exist, since the quadratic variation $dy(t)dy(t)' = \Omega dt$ is of order dt. Thus we have $dy(t)'(\Omega dt)^{-1}dy(t) = \text{tr}[(\Omega dt)^{-1}\,dy(t)dy(t)'] = \text{tr}[I_p] = p$. Usually, (16.53) is written as the formal expression $\frac{1}{2}\int \dot{y}(t)'\,\Omega^{-1}\dot{y}(t)dt$, which contains the (nonexisting) derivatives $\dot{y}(t)$. Moreover, partial integration yields

$$-\frac{1}{2}\int y(t)'\Omega^{-1}\ddot{y}(t)dt \tag{16.55}$$

so that $C^{-1}(t,s) = \Omega^{-1}(-\partial^2/\partial t^2)\delta(t-s)$ is the kernel of the inverse covariance (precision) operator of $Y(t)$ (for drift $f = 0$; i.e., a Wiener process). Indeed, since

$$\partial^2/\partial t^2 \min(t,s) = -\delta(t-s), \tag{16.56}$$

the covariance operator kernel $C(t,s)$ is

$$C(t,s) = \Omega(-\partial^2/\partial t^2)^{-1}\delta(t-s) = \Omega \min(t,s).$$

Thus, $p(y) \propto \exp[-\frac{1}{2}\int y(t)'\Omega^{-1}\ddot{y}(t)dt]$ is the formal density of a Gaussian process $Y(t) \sim N(0,C)$.

In contrast, the terms in (16.54) are well defined and yield the Radon-Nikodym derivative (cf. 16.17)

$$\alpha(y) = \exp\left\{\int f(y)'\Omega^{-1}dy(t) - \frac{1}{2}\int f(y)'\Omega^{-1}f(y)dt\right\}. \tag{16.57}$$

This expression can be obtained as the ratio of the finite-dimensional density functions $p(y_J,\tau_J,\ldots,y_1,\tau_1|y_0,\tau_0)$ for drifts f and $f = 0$, respectively, in the limit $\delta t \to 0$ (cf. Wong and Hajek 1985, ch. 6, p. 215 ff.). In this limit, the (unkown) exact densities can be replaced by the Euler densities (16.5). Now, the terms of the Langevin equation (16.14) will be given. We start with the measurement term (16.43), $\alpha = 1,\ldots,p$

$$\delta \log p(z|y)/\delta y_\alpha(t) = \sum_{i=0}^{T} h'_{i\bullet,\alpha}R_i^{-1}(z_i - h_i)\delta(t-t_i) \tag{16.58}$$

where the scaled Kronecker delta $(\delta_{jj_i}/\delta t)$ was replaced by the delta function (see Appendix 1). Clearly, in numerical implementations, a certain term of the delta sequence $\delta_n(t)$ must be used (cf. Lighthill 1958). Next, the term stemming from the driftless part (16.40) is

$$-\delta S_0/\delta y_\alpha(t) = \Omega^{-1}_{\alpha\bullet}\ddot{y}(t) = \Omega^{-1}_{\alpha\bullet} y_{tt}(t),$$

or $\Omega^{-1} y_{tt}(t)$ in matrix form, which corresponds to (16.55). The contributions of S_1 are (cf. 16.41)

$$-\delta S_1/\delta y_\alpha(t) = f(y)_{\beta,\alpha}\Omega^{-1}_{\beta\gamma} dy_\gamma(t)/dt - \Omega^{-1}_{\alpha\gamma} df_\gamma(y)/dt.$$

The first term is of Itô form. Transformation to Stratonovich calculus (Apte et al. 2007, sects. 4, 9) yields

$$h_{\alpha\beta} dy_\beta = h_{\alpha\beta} \circ dy_\beta - \frac{1}{2} h_{\alpha\beta,\gamma}\Omega_{\beta\gamma} dt \tag{16.59}$$

$$df_\alpha = f_{\alpha,\beta} dy_\beta + \frac{1}{2} f_{\alpha,\beta\gamma}\Omega_{\beta\gamma} dt = f_{\alpha,\beta} \circ dy_\beta \tag{16.60}$$

Thus, we obtain

$$\begin{aligned}
-\delta S_1/\delta y_\alpha(t) &= f(y)_{\beta,\alpha}\Omega^{-1}_{\beta\gamma} \circ dy_\gamma(t)/dt - \frac{1}{2} f(y)_{\beta,\alpha\beta} \\
&\quad -\Omega^{-1}_{\alpha\gamma} f(y)_{\gamma,\delta} \circ dy_\gamma(t)/dt \\
&= (f_y'\Omega^{-1} - \Omega^{-1} f_y) \circ y_t(t) - \frac{1}{2}\partial_y[\partial_y \cdot f(y)]
\end{aligned}$$

where $\partial_y \cdot f(y) = f_{\beta,\beta} = \mathrm{div}(f)$. Finally we have (cf. 16.42)

$$-\delta S_2/\delta y(t) = -f_y'\Omega^{-1} f$$

and $\delta_{y(t)} \log p(y(t_0)) = \partial_{y_0} \log p(y_0)\delta(t - t_0)$. Putting all together, one finds the Langevin drift functional (in matrix form)

$$\begin{aligned}
-\frac{\delta\Phi(y|z)}{\delta y(t)} &:= F(y|z) \\
&= \sum_{i=0}^{T} h_{iy}'(y) R_i^{-1}(z_i - h_i(y))\delta(t - t_i) \\
&\quad +\Omega^{-1} y_{tt} + (f_y'\Omega^{-1} - \Omega^{-1} f_y) \circ y_t
\end{aligned}$$

$$-\frac{1}{2}\partial_y[\partial_y \cdot f(y)] - f_y' \Omega^{-1} f$$
$$+\partial_{y_0} \log p(y_0)\delta(t - t_0)$$

and the SPDE (cf. Hairer et al. 2007)

$$dY(u,t) = F(Y(u,t)|z))du + \sqrt{2}\, dW_t(u,t), \tag{16.61}$$

where $W_t(u,t) = \partial_t W(u,t)$ is a cylindrical Wiener process with $E[W_t(u,t)] = 0$, $E[W_t(u,t)\ W_s(v,s)'] = I_p \min(u,v)\delta(t-s)$, and $W(u,t)$ is a Wiener field (Brownian sheet). See, e.g., Jetschke (1986); Da Prato and Zabczyk (1992, ch. 4.3.3). The cylindrical Wiener process may be viewed as continuum limit of $W_j(u)/\sqrt{\delta t}$, $E[W_j(u)/\sqrt{\delta t}\ W_k'(v)/\sqrt{\delta t}] = I_p \min(u,v)\delta t^{-1}\delta_{jk}$.

State-Dependent Diffusion Matrix

In this case, new terms appear. Starting with the first term in (16.47), one gets

$$-\Omega^{-1}_{j\alpha\beta}\delta t^{-2}(\eta_{j+1;\beta} - 2\eta_{j\beta} + \eta_{j-1;\beta}) \to -\Omega(y(t))^{-1} \circ \ddot{y}(t).$$

The second term in (16.47) contains terms of the form $h_j\ (\eta_j - \eta_{j-1})$ which appear in a backward Itô integral. Here we attempt to write them in symmetrized (Stratonovich) form. It turns out that the Taylor expansion (16.46) must be carried to higher orders. Writing (for simplicity in scalar form)

$$\Omega^{-1}_{j-1}\delta\eta_{j-1} - \Omega^{-1}_j \delta\eta_j := h_{j-1}\delta\eta_{j-1} - h_j\delta\eta_j$$

and expanding around η_j

$$h_{j-1} = h_j + \sum_{k=1}^{\infty} \frac{1}{k!} h_{j,k}(\eta_{j-1} - \eta_j)^k$$

one obtains

$$h_{j-1}\delta\eta_{j-1} - h_j\delta\eta_j = h_j(\delta\eta_{j-1} - \delta\eta_j) + \sum_{k=1}^{\infty} \frac{(-1)^k}{k!} h_{j,k}\delta\eta_{j-1}^{k+1}. \tag{16.62}$$

To obtain a symmetric expression, $h_{j,k}$ is expanded around $\eta_{j-1/2} := \frac{1}{2}(\eta_{j-1} + \eta_j)$. Noting that $\eta_j - \eta_{j-1/2} = \frac{1}{2}\delta\eta_{j-1}$, we have

$$h_{j,k} = \sum_{l=0}^{\infty} \frac{(\frac{1}{2})^l}{l!} h_{j-1/2,k+l}\delta\eta_{j-1}^l \tag{16.63}$$

and together

$$h_j(\delta\eta_{j-1} - \delta\eta_j) + \sum_{k=1,l=0}^{\infty} \frac{(-1)^k(\frac{1}{2})^l}{k!\,l!} h_{j-1/2,k+l}\delta\eta_{j-1}^{k+l+1}. \tag{16.64}$$

Multiplying with δt^{-2} and collecting terms to order $O(\delta t^2)$, one gets the continuum limit

$$-\Omega^{-1} \circ \ddot{y} - \Omega_\eta^{-1} \circ \dot{y}^2 - \tfrac{1}{24}\Omega_{\eta\eta\eta}^{-1}\Omega^2. \tag{16.65}$$

The last term in (16.47) is absorbed in the expression (16.51).

The continuum limit of the first two terms in the derivative of S_1 (see (16.49) is

$$-f(y)_{\beta,\alpha}\Omega(y)_{\beta\gamma}^{-1} dy_\gamma(t)/dt + d[\Omega(y)_{\alpha\gamma}^{-1} f_\gamma(y)]/dt.$$

Transforming to Stratonovich calculus (16.59–16.60) yields

$$\begin{aligned} &-\{f(y)_{\beta,\alpha}\Omega(y)_{\beta\gamma}^{-1} - [\Omega(y)_{\alpha\beta}^{-1} f_\beta(y)]_{,\gamma}\} \circ dy_\gamma(t)/dt \\ &\qquad + \frac{1}{2}[f(y)_{\beta,\alpha}\Omega(y)_{\beta\gamma}^{-1}]_{,\delta}\Omega_{\gamma\delta}. \end{aligned} \tag{16.66}$$

Equation (16.50) yields

$$\delta S_2/\delta y_\alpha(t) = f(y)_{\beta,\alpha}\Omega(y)_{\beta\gamma}^{-1} f(y)_\gamma + \frac{1}{2} f(y)_\beta \Omega(y)_{\beta\gamma,\alpha}^{-1} f(y)_\gamma. \tag{16.67}$$

The last term to be discussed is (16.51). Formally,

$$\begin{aligned} &\frac{1}{2}\delta t^{-2}\mathrm{tr}\ \{\Omega_{,\alpha}^{-1}[\Omega dt - (dy - f dt)(dy - f dt)']\} \\ &\quad = \frac{1}{2}\mathrm{tr}\{\Omega_{,\alpha}^{-1}[\Omega\delta t^{-1} - (\dot{y} - f)(\dot{y} - f)']\}. \end{aligned} \tag{16.68}$$

From the quadratic variation formula $(dy - f dt)(dy - f dt)' = \Omega dt$, it seems that it can be dropped. But setting $\delta\eta_j - f_j\delta t = g_j z_j\sqrt{\delta t}$ (from the Euler scheme, see (16.3)), one gets

$$X := \frac{1}{2}\delta t^{-1}\mathrm{tr}\{\Omega_{j,\alpha}^{-1}\Omega_j\ (I - z_j z_j')\}$$

In scalar form, one has $X := \frac{1}{2}\delta t^{-1}\Omega_{j,\alpha}^{-1}\Omega_j\ (I - z_j^2)$ which is χ_1^2-distributed, conditionally on η_j. One has $E[1 - z^2] = 0$; $\mathrm{Var}(1 - z^2) = 1 - 2 + 3 = 2$, thus $E[X] = 0$ and $\mathrm{Var}[X] = \frac{1}{2}\delta t^{-2} E[\Omega_{j,\alpha}^{-2}\Omega_j^2]$.

Therefore, the drift functional in the state-dependent case is

$$
\begin{aligned}
-\frac{\delta\Phi(y|z)}{\delta y(t)} &:= F(y|z) \\
&= (16.58) - (16.65) - (16.66) - (16.67) + (16.68) \\
&\quad + \partial_{y_0} \log p(y_0)\delta(t - t_0)
\end{aligned}
$$

Discussion

The second-order time derivative (diffusion term w.r.t. t) $\Omega^{-1}y_{tt}$ in the SPDE (16.61) resulted from the first term (16.53) in the Lagrangian corresponding to the driftless process (random walk process). Usually this (in the continuum limit), infinite term is not considered and removed by computing a density ratio (16.17) which leads to a well-defined Radon-Nikodym density (16.54). On the other hand, the term is necessary to obtain the correct SPDE. Starting from the Radon-Nikodym density (16.57) for the process $dY(t) = fdt + GdW(t)$ at the outset, it is not quite clear how to construct the appropriate SPDE. Setting for simplicity $f = 0$ and dropping the initial condition and the measurement part, Eq. (16.61) reads

$$
dY(u,t) = \Omega^{-1}\, Y_{tt}(u,t)du + \sqrt{2}\, dW_t(u,t).
$$

This linear equation (Ornstein-Uhlenbeck process) can be solved using a stochastic convolution as ($A := \Omega^{-1}\partial_t^2$)

$$
Y(u,t) = \exp(Au)Y(0,t) + \int_0^u \exp(A(u-s))\sqrt{2}\, dW_t(s,t).
$$

(cf. Da Prato 2004, ch. 2). It is a Gaussian process with mean $\mu(u) = \exp(Au)E[Y(0)]$ and variance $Q(u) = \exp(Au)\mathrm{Var}(Y(0))\exp(A^*u) + \int_0^u \exp(As)2\exp(A^*s)ds$ where A^* is the adjoint of A. Thus the stationary distribution ($u \to \infty$) is the Gaussian measure $N(0, Q(\infty))$ with $Q(\infty) = -A^{-1} = -\Omega\cdot[\partial_t^2]^{-1}$, since $A = A^*$. But this coincides with $C(t,s) = \Omega \min(t,s)$, the covariance function of the scaled Wiener process $G \cdot W(t)$ (see (16.56); $\Omega = GG'$). Thus, for large u, $Y(u,t)$ generates trajectories of $GW(t)$. More generally ($f \neq 0$), one obtains solutions of SDE (16.1). A related problem occurs in the state-dependent case $\Omega(y)$. Again, the term $\int dy'(\Omega dt)^{-1}dy$ yields a second-order derivative in the SPDE, but after transforming to symmetrized Stratonovich form, also higher-order terms appear (16.64), (16.65).

Moreover, the differential of Ω^{-1} in the Lagrangian (16.53)–(16.54) imports a problematic term similar to (16.53) into the SPDE, namely, $\frac{1}{2}(\dot{y}-f)'(\Omega^{-1})_y(\dot{y}-f)$, which can be combined with the derivative of the Jacobian (cf. 16.68). Formally, it is squared white noise where the differentials are in Itô form. A procedure similar to (16.63), i.e.,

$$h_{j,k} = \sum_{l=0}^{\infty} \frac{(-\frac{1}{2})^l}{l!} h_{j+1/2,k+l} \delta\eta_j^l \tag{16.69}$$

can be applied to obtain Stratonovich-type expressions. Because of the dubious nature of these expressions, only the quasi-continuous approach based on approximate finite-dimensional densities and Langevin equations is used in this paper.

References

Aït-Sahalia, Y. (2002). Maximum likelihood estimation of discretely sampled diffusions: A closed-form approximation approach. *Econometrica, 70*(1), 223–262. https://doi.org/10.1111/1468-0262.00274

Aït-Sahalia, Y. (2008). Closed-form likelihood expansions for multivariate diffusions. *Annals of Statistics, 36*(2), 906–937. https://doi.org/10.1214/009053607000000622

Apte, A., Hairer, M., Stuart, A. M., & Voss, J. (2007). Sampling the posterior: An approach to non-Gaussian data assimilation. *Physica D: Nonlinear Phenomena, 230*(1–2), 50–64. https://doi.org/10.1016/j.physd.2006.06.009

Apte, A., Jones, C. K. R. T., Stuart, A. M., & Voss, J. (2008). Data assimilation: Mathematical and statistical perspectives. *International Journal for Numerical Methods in Fluids, 56*(8), 1033–1046. https://doi.org/10.1002/fld.1698

Arasaratnam, I., Haykin, S., & Hurd, T. (2010). Cubature Kalman filtering for continuous-discrete systems: Theory and simulations. *IEEE Transactions on Signal Processing, 58*, 4977–4993. https://doi.org/10.1109/TSP.2010.2056923

Arnold, L. (1974). *Stochastic differential equations*. New York: Wiley.

Åström, K. J. (1970). *Introduction to stochastic control theory*. Mineola, NY: Courier Corporation.

Bagchi, A. (2001). Onsager-Machlup function. In *Encyclopedia of mathematics*. Berlin: Springer. https://www.encyclopediaofmath.org/index.php/Onsager-Machlup_function

Ballreich, D. (2017). *Stable and efficient cubature-based filtering in dynamical systems*. Berlin: Springer International Publishing.

Bartlett, M. S. (1946). On the theoretical specification and sampling properties of autocorrelated time-series. *Journal of the Royal Statistical Society (Supplement), 7*, 27–41. https://doi.org/10.2307/2983611

Basawa, I. V., & Prakasa Rao, B. L. S. (1980). *Statistical inference for stochastic processes*. London: Academic Press.

Bergstrom, A. R. (1976a). Non-recursive models as discrete approximations to systems of stochastic differential equations. In A. R. Bergstrom (Ed.), *Statistical inference in continuous time models* (pp. 15–26). Amsterdam: North Holland.

Bergstrom, A. R. (Ed.). (1976b). *Statistical inference in continuous time economic models*. Amsterdam: North Holland.

Bergstrom, A. R. (1983). Gaussian estimation of structural parameters in higher order continuous time dynamic models. *Econometrica: Journal of the Econometric Society, 51*(1), 117–152. https://doi.org/10.2307/1912251

Bergstrom, A. R. (1988). The history of continuous-time econometric models. *Econometric Theory, 4*, 365–383. https://doi.org/10.1017/S0266466600013359

Beskos, A., Papaspiliopoulos, O., Roberts, G. O., & Fearnhead, P. (2006). Exact and efficient likelihood-based inference for discretely observed diffusion processes (with discussion). *Journal of the Royal Statistical Society Series B, 68*, 333–382. https://doi.org/10.1111/j.1467-9868.2006.00552.x

Black, F., & Scholes, M. (1973). The pricing of options and corporate liabilities. *Journal of Political Economy, 81*, 637–654. https://doi.org/10.1086/260062

Chang, J., & Chen, S. X. (2011). On the approximate maximum likelihood estimation for diffusion processes. *The Annals of Statistics, 39*(6), 2820–2851. https://doi.org/10.1214/11-AOS922

Chow, S.-M., Ferrer, E., & Nesselroade, J. R. (2007). An unscented Kalman filter approach to the estimation of nonlinear dynamical systems models. *Multivariate Behavioral Research, 42*(2), 283–321. https://doi.org/10.1080/00273170701360423

Chow, S.-M., Lu, Z., Sherwood, A., & Zhu, H. (2016). Fitting nonlinear ordinary differential equation models with random effects and unknown initial conditions using the stochastic approximation expectation–maximization (SAEM) algorithm. *Psychometrika, 81*(1), 102–134. https://doi.org/10.1007/s11336-014-9431-z

Da Prato, G. (2004). *Kolmogorov equations for stochastic PDEs*. Basel: Birkhäuser. https://doi.org/10.1007/978-3-0348-7909-5

Da Prato, G., & Zabczyk, J. (1992). *Stochastic equations in infinite dimensions*. New York: Cambridge University Press. https://doi.org/10.1017/CBO9780511666223

Doreian, P., & Hummon, N. P. (1976). *Modelling social processes*. New York, Oxford, Amsterdam: Elsevier.

Durham, G. B., & Gallant, A. R. (2002). Numerical techniques for simulated maximum likelihood estimation of stochastic differential equations. *Journal of Business and Economic Statistics, 20*, 297–316. https://doi.org/10.1198/073500102288618397

Elerian, O., Chib, S., & Shephard, N. (2001). Likelihood inference for discretely observed nonlinear diffusions. *Econometrica, 69*(4), 959–993. https://doi.org/10.1111/1468-0262.00226

Feller, W. (1951). Two singular diffusion problems. *Annals of Mathematics, 54*, 173–182. https://doi.org/10.2307/1969318

Flury, T., & Shephard, N. (2011). Bayesian inference based only on simulated likelihood: particle filter analysis of dynamic economic models. *Econometric Theory, 27*(5), 933–956. https://doi.org/10.1017/S0266466610000599

Fuchs, C. (2013). *Inference for diffusion processes: With applications in life sciences*. Berlin: Springer. https://doi.org/10.1007/978-3-642-25969-2

Gard, T. C. (1988). *Introduction to stochastic differential equations*. New York: Dekker.

Girolami, M., & Calderhead, B. (2011). Riemann manifold Langevin and Hamiltonian Monte Carlo methods. *Journal of the Royal Statistical Society: Series B (Statistical Methodology), 73*(2), 123–214. https://doi.org/10.1111/j.1467-9868.2010.00765.x

Hairer, M., Stuart, A. M., & Voss, J. (2007). Analysis of SPDEs arising in path sampling, part II: The nonlinear case. *Annals of Applied Probability, 17*(5), 1657–1706. https://doi.org/10.1214/07-AAP441

Hairer, M., Stuart, A. M., & Voss, J. (2009). Sampling conditioned diffusions. In J. Blath, P. Morters, & M. Scheutzow (Eds.), *Trends in stochastic analysis* (pp. 159–186). Cambridge: Cambridge University Press. https://doi.org/10.1017/CBO9781139107020.009

Hairer, M., Stuart, A. M., & Voss, J. (2011). Signal processing problems on function space: Bayesian formulation, stochastic PDEs and effective MCMC methods. In D. Crisan & B. Rozovsky (Eds.), *The Oxford handbook of nonlinear filtering* (pp. 833–873). Oxford: Oxford University Press.

Hairer, M., Stuart, A. M., Voss, J., & Wiberg, P. (2005). Analysis of SPDEs arising in path sampling, part I: The Gaussian case. *Communications in Mathematical Sciences, 3*(4), 587–603. https://doi.org/10.4310/CMS.2005.v3.n4.a8

Haken, H. (1977). *Synergetics*. Berlin: Springer.

Haken, H., Kelso, J. A. S., & Bunz, H. (1985). A theoretical model of phase transitions in human hand movements. *Biological Cybernetics, 51*(5), 347–356. https://doi.org/10.1007/BF00336922

Hamerle, A., Nagl, W., & Singer, H. (1991). Problems with the estimation of stochastic differential equations using structural equations models. *Journal of Mathematical Sociology, 16*(3), 201–220. https://doi.org/10.1080/0022250X.1991.9990088

Harvey, A. C., & Stock, J. (1985). The estimation of higher order continuous time autoregressive models. *Econometric Theory, 1*, 97–112. https://doi.org/10.1017/S0266466600011026

Hastings, W. K. (1970). Monte Carlo sampling methods using Markov chains and their applications. *Biometrika, 57*(1), 97–109. https://doi.org/10.1093/biomet/57.1.97

Herings, J. P. (1996). *Static and dynamic aspects of general disequilibrium theory*. Boston: Springer. https://doi.org/10.1007/978-1-4615-6251-1

Jazwinski, A. H. (1970). *Stochastic processes and filtering theory*. New York: Academic Press.

Jetschke, G. (1986). On the equivalence of different approaches to stochastic partial differential equations. *Mathematische Nachrichten, 128*, 315–329. https://doi.org/10.1002/mana.19861280127

Jetschke, G. (1991). Lattice approximation of a nonlinear stochastic partial differential equation with white noise. In *International series of numerical mathematics* (Vol. 102, pp. 107–126). Basel: Birkhäuser. https://doi.org/10.1007/978-3-0348-6413-8_8

Jones, R. H. (1984). Fitting multivariate models to unequally spaced data. In E. Parzen (Ed.), *Time series analysis of irregularly observed data* (pp. 158–188). New York: Springer. https://doi.org/10.1007/978-1-4684-9403-7_8

Jones, R. H., & Tryon, P. V. (1987). Continuous time series models for unequally spaced data applied to modeling atomic clocks. *SIAM Journal of Scientific and Statistical Computing, 8*, 71–81. https://doi.org/10.1137/0908007

Kac, M. (1980). *Integration in function spaces and some of its applications*. Pisa: Scuola normale superiore.

Kloeden, P. E., & Platen, E. (1992). *Numerical solution of stochastic differential equations*. Berlin: Springer. https://doi.org/10.1007/978-3-662-12616-5

Kloeden, P. E., & Platen, E. (1999). *Numerical solution of stochastic differential equations*. Berlin: Springer. (corrected third printing)

Langevin, P. (1908). Sur la théorie du mouvement brownien [On the theory of Brownian motion]. *Comptes Rendus de l'Academie des Sciences (Paris), 146*, 530–533.

Li, C. (2013). Maximum-likelihood estimation for diffusion processes via closed-form density expansions. *The Annals of Statistics, 41*(3), 1350–1380. https://doi.org/10.1214/13-AOS1118

Lighthill, M. J. (1958). *Introduction to Fourier analysis and generalised functions*. Cambridge: Cambridge University Press. https://doi.org/10.1017/CBO9781139171427

Liptser, R. S., & Shiryayev, A. N. (2001). *Statistics of random processes* (Vols. I and II, 2nd ed.). New York: Springer

Lorenz, E. (1963). Deterministic nonperiodic flow. *Journal of the Atmospheric Sciences, 20*, 130. https://doi.org/10.1175/1520-0469(1963)020$⟨$0130:DNF$⟩$2.0.CO;2

Louis, T. A. (1982). Finding the observed information matrix when using the EM algorithm. *Journal of the Royal Statistical Association B, 44*(2), 226–233.

Malik, S., & Pitt, M. K. (2011). Particle filters for continuous likelihood evaluation and maximisation. *Journal of Econometrics, 165*(2), 190–209. https://doi.org/10.1016/j.jeconom.2011.07.006

Mardia, K. V., Kent, J. T., & Bibby, J. M. (1979). *Multivariate analysis*. London: Academic Press.

Molenaar, P., & Newell, K. M. (2003). Direct fit of a theoretical model of phase transition in oscillatory finger motions. *British Journal of Mathematical and Statistical Psychology, 56*(2), 199–214. https://doi.org/10.1348/000711003770480002

Moler, C., & Van Loan, C. (2003). Nineteen dubious ways to compute the exponential of a matrix, twenty-five years later. *SIAM Review, 45*(1), 1–46. https://doi.org/10.1137/S00361445024180

Newell, K. M., & Molenaar, P. C. M. (2014). *Applications of nonlinear dynamics to developmental process modeling*. Hove: Psychology Press.

Okano, K., Schülke, L., & Zheng, B. (1993). Complex Langevin simulation. *Progress of Theoretical Physics Supplement, 111*, 313–346. https://doi.org/10.1143/PTPS.111.313

Onsager, L., & Machlup, S. (1953). Fluctuations and irreversible processes. *Physical Review, 91*(6), 1505–1515. https://doi.org/10.1103/PhysRev.91.1505

Oud, J. H. L., & Jansen, R. A. R. G. (2000). Continuous time state space modeling of panel data by means of SEM. *Psychometrika, 65*, 199–215. https://doi.org/10.1007/BF02294374

Oud, J. H. L., & Singer, H. (2008). Continuous time modeling of panel data: SEM versus filter techniques. *Statistica Neerlandica, 62*(1), 4–28.

Ozaki, T. (1985). Nonlinear time series and dynamical systems. In E. Hannan (Ed.), *Handbook of statistics* (pp. 25–83). Amsterdam: North Holland.

Parisi, G., & Wu, Y.-S. (1981). Perturbation theory without gauge fixing. *Scientia Sinica, 24*, 483.

Pedersen, A. R. (1995). A new approach to maximum likelihood estimation for stochastic differential equations based on discrete observations. *Scandinavian Journal of Statistics, 22*, 55–71.

Pitt, M. K. (2002). *Smooth particle filters for likelihood evaluation and maximisation* (Warwick economic research papers No. 651). University of Warwick. http://wrap.warwick.ac.uk/1536/

Reznikoff, M. G., & Vanden-Eijnden, E. (2005). Invariant measures of stochastic partial differential equations and conditioned diffusions. *Comptes Rendus Mathematiques, 340*, 305–308. https://doi.org/10.1016/j.crma.2004.12.025

Risken, H. (1989). *The Fokker-Planck equation* (2nd ed.). Berlin: Springer. https://doi.org/10.1007/978-3-642-61544-3

Roberts, G. O., & Stramer, O. (2001). On inference for partially observed nonlinear diffusion models using the Metropolis–Hastings algorithm. *Biometrika, 88*(3), 603–621. https://doi.org/10.1093/biomet/88.3.603

Roberts, G. O., & Stramer, O. (2002). Langevin diffusions and Metropolis-Hastings algorithms. *Methodology and Computing in Applied Probability, 4*(4), 337–357. https://doi.org/10.1023/A:1023562417138

Rümelin, W. (1982). Numerical treatment of stochastic differential equations. *SIAM Journal of Numerical Analysis, 19*(3), 604–613. https://doi.org/10.1137/0719041

Särkkä, S. (2013). *Bayesian filtering and smoothing* (Vol. 3). Cambridge: Cambridge University Press. https://doi.org/10.1017/CBO9781139344203

Särkkä, S., Hartikainen, J., Mbalawata, I. S., & Haario, H. (2013). Posterior inference on parameters of stochastic differential equations via non-linear Gaussian filtering and adaptive MCMC. *Statistics and Computing, 25*(2), 427–437. https://doi.org/10.1007/s11222-013-9441-1

Schiesser, W. E. (1991). *The numerical method of lines.* San Diego: Academic Press.

Schuster, H. G., & Just, W. (2006). *Deterministic chaos: An introduction.* New York: Wiley.

Shoji, I., & Ozaki, T. (1997). Comparative study of estimation methods for continuous time stochastic processes. *Journal of Time Series Analysis, 18*(5), 485–506. https://doi.org/10.1111/1467-9892.00064

Shoji, I., & Ozaki, T. (1998a). Estimation for nonlinear stochastic differential equations by a local linearization method 1. *Stochastic Analysis and Applications, 16*(4), 733–752. https://doi.org/10.1080/07362999808809559

Shoji, I., & Ozaki, T. (1998b). A statistical method of estimation and simulation for systems of stochastic differential equations. *Biometrika, 85*(1), 240–243. https://doi.org/10.1093/biomet/85.1.240

Silverman, B. W. (1986). *Density estimation for statistics and data analysis.* London: Chapman and Hall. https://doi.org/10.1007/978-1-4899-3324-9

Singer, H. (1986). *Depressivität und gelernte Hilflosigkeit als Stochastischer Prozeß [Depression and learned helplessness as stochastic process]* (Unpublished master's thesis). Universität Konstanz, Konstanz, Germany. http://www.fernuni-hagen.de/imperia/md/content/ls_statistik/diplomarbeitsinger.pdf

Singer, H. (1990). *Parameterschätzung in zeitkontinuierlichen dynamischen Systemen [Parameter estimation in continuous time dynamical systems].* Konstanz: Hartung-Gorre-Verlag.

Singer, H. (1993). Continuous-time dynamical systems with sampled data, errors of measurement and unobserved components. *Journal of Time Series Analysis, 14*(5), 527–545. https://doi.org/10.1111/j.1467-9892.1993.tb00162.x

Singer, H. (1995). Analytical score function for irregularly sampled continuous time stochastic processes with control variables and missing values. *Econometric Theory, 11*, 721–735. https://doi.org/10.1017/S0266466600009701

Singer, H. (1998). Continuous panel models with time dependent parameters. *Journal of Mathematical Sociology, 23*, 77–98. https://doi.org/10.1080/0022250X.1998.9990214

Singer, H. (2002). Parameter estimation of nonlinear stochastic differential equations: Simulated maximum likelihood vs. extended Kalman filter and Itô-Taylor expansion. *Journal of Computational and Graphical Statistics, 11*(4), 972–995. https://doi.org/10.1198/106186002808

Singer, H. (2003). Simulated maximum likelihood in nonlinear continuous-discrete state space models: Importance sampling by approximate smoothing. *Computational Statistics, 18*(1), 79–106. https://doi.org/10.1007/s001800300133

Singer, H. (2005). *Continuous-discrete unscented Kalman filtering* (Diskussionsbeiträge Fachbereich Wirtschaftswissenschaft No. 384). FernUniversität in Hagen. http://www.fernunihagen.de/lsstatistik/publikationen/ukf2005.shtml

Singer, H. (2008a). Generalized Gauss-Hermite filtering. *Advances in Statistical Analysis, 92*(2), 179–195. https://doi.org/10.1007/s10182-008-0068-z

Singer, H. (2008b). Nonlinear continuous time modeling approaches in panel research. *Statistica Neerlandica, 62*(1), 29–57.

Singer, H. (2010). SEM modeling with singular moment matrices. Part I: ML-estimation of time series. *Journal of Mathematical Sociology, 34*(4), 301–320. https://doi.org/10.1080/0022250X.2010.509524

Singer, H. (2011). Continuous-discrete state-space modeling of panel data with nonlinear filter algorithms. *Advances in Statistical Analysis, 95*, 375–413. https://doi.org/10.1007/s10182-011-0172-3

Singer, H. (2012). SEM modeling with singular moment matrices. Part II: ML-estimation of sampled stochastic differential equations. *Journal of Mathematical Sociology, 36*(1), 22–43. https://doi.org/10.1080/0022250X.2010.532259

Singer, H. (2014). Importance sampling for Kolmogorov backward equations. *Advances in Statistical Analysis, 98*(4), 345–369. https://doi.org/10.1007/s10182-013-0223-z

Singer, H. (2016). *Simulated maximum likelihood for continuous-discrete state space models using Langevin importance sampling* (Diskussionsbeiträge Fakultät Wirtschaftswissenschaft No. 497). Paper presented at the 9th International Conference on Social Science Methodology (RC33), 11–16 September 2016, Leicester, UK. http://www.fernuni-hagen.de/wirtschaftswissenschaft/forschung/beitraege.shtml

Stramer, O., Bognar, M., & Schneider, P. (2010). Bayesian inference for discretely sampled Markov processes with closed-form likelihood expansions. *Journal of Financial Econometrics, 8*(4), 450–480. https://doi.org/10.1093/jjfinec/nbp027

Stratonovich, R. L. (1971). On the probability functional of diffusion processes. In *Selected translations in mathematical statistics and probability, Vol. 10* (pp. 273–286).

Stratonovich, R. L. (1989). Some Markov methods in the theory of stochastic processes in nonlinear dynamic systems. In F. Moss & P. McClintock (Eds.), *Noise in nonlinear dynamic systems* (pp. 16–71). Cambridge: Cambridge University Press. https://doi.org/10.1017/CBO9780511897818.004

Stuart, A. M., Voss, J., & Wiberg, P. (2004). Conditional path sampling of SDEs and the Langevin MCMC method. *Communications in Mathematical Sciences, 2*(4), 685–697. https://doi.org/10.4310/CMS.2004.v2.n4.a7

Thomas, E. A. C., & Martin, J. A. (1976). Analyses of parent-infant-interaction. *Psychological Review, 83*(2), 141–156. https://doi.org/10.1037/0033-295X.83.2.141

Van Kampen, N. G. (1981). Itô vs. Stratonovich. *Journal of Statistical Physics, 24*, 175–187. https://doi.org/10.1007/BF01007642

Wei, G. W., Zhang, D. S., Kouric, D. J., & Hoffman, D. K. (1997). Distributed approximating functional approach to the Fokker-Planck equation: Time propagation. *Journal of Chemical Physics, 107*(8), 3239–3246. https://doi.org/10.1063/1.474674

Weidlich, W., & Haag, G. (1983). *Quantitative sociology*. Berlin: Springer.

Wong, E., & Hajek, B. (1985). *Stochastic processes in engineering systems*. New York: Springer. https://doi.org/10.1007/978-1-4612-5060-9

Yoo, H. (2000). Semi-discretization of stochastic partial differential equations on R1 by a finite-difference method. *Mathematics of Computation, 69*(230), 653–666. https://doi.org/10.1090/S0025-5718-99-01150-3

Index

K. van Montfort et al. (eds.), *Continuous Time Modeling in the Behavioral and Related Sciences*, https://doi.org/10.1007/978-3-319-77219-6

Printed in the United States
By Bookmasters